SYSTEMS ENGINEERING

WITH ECONOMICS, PROBABILITY, AND STATISTICS

Second Edition

C. JOTIN KHISTY
JAMSHID MOHAMMADI
ADJO A. AMEKUDZI

ISBN-13: 978-1-60427-055-6

Printed and bound in the U.S.A. Printed on acid-free paper.

10 9 8 7 6 5 4 3 2 1

Library of Congress Cataloging-in-Publication Data

Khisty, C. Jotin, 1928-
Systems engineering with economics, probability, and statistics / by C. Jotin Khisty, Jamshid Mohammadi, and Adjo A. Amekudzi.—2nd ed.
p. cm.
Rev. ed. of: Fundamentals of systems engineering / C. Jotin Khisty, Jamshid Mohammadi.
Includes bibliographical references and index.
ISBN 978-1-60427-055-6 (hardcover : alk. paper)
1. Systems engineering. I. Mohammadi, Jamshid. II. Amekudzi, Adjo A. III. Khisty, C. Jotin, 1928- Fundamentals of systems engineering. IV. Title.
TA168.K46 2012
620.001'1—dc23

2011043688

Direct all inquiries to J. Ross Publishing, Inc., 5765 N. Andrews Way, Fort Lauderdale, FL 33309.

Phone: (954) 727-9333
Fax: (561) 892-0700
Web: www.jrosspub.com

Contents

Systems Engineering with Economics, Probability, and Statistics, Second Edition

C. Jotin Khisty
Illinois Institute of Technology, Chicago

Jamshid Mohammadi
Illinois Institute of Technology, Chicago

Adjo A. Amekudzi
Georgia Institute of Technology, Atlanta

Dedication

To
Lena, Mohan, Suman, and Vijaya
CJK

Shar, Negar, and Cameron
JM

My parents
AAA

Preface to Second Edition

It appears that our lives are filled with a constant barrage of demands to make decisions, large and small. Obviously, since there is a consequence attached to every decision we make, there is a corresponding burden to justify our choices. This question of making correct decisions is all the more critical for engineers, planners, and managers who are called upon to make decisions involving millions of dollars, not only for public projects but also for clients in the private sector. Now, more than ever, such decision making is beset with complexity, unpredictability, and uncertainty. In addition, our knowledge-base about the dynamics of systems may not be as clear and complete as many imagine, and our value systems may also be in dispute. The first edition of this book was published to address some of these problems and to facilitate good decision making.

It has been about ten years since the first edition of this book was published, and the purpose of this edition, like the first, remains generally the same. However, we have made some major changes—three new chapters have been added that include a new chapter consisting of case studies on Systems Thinking, plus two new chapters, one on Sustainability and the other on case studies on Sustainability. These three chapters, we are sure, provide readers with a great deal of new knowledge and perspectives in these areas of systems that have become essential in dealing with today's problems. We have also made minor changes to nearly all of the other chapters. Armed with these tools, methods, and procedures, engineers, planners, and managers will be better prepared not only to make good decisions, but to take the leadership to extend their influence into spheres of policy making and improving the quality of life.

At this time, the authors would like to thank many of their undergraduate and graduate students who have contributed so much to our understanding of systems thinking and its ramifications, methods of decision making under uncertainties, and critical thinking in design and many other areas related to systems engineering. In a way, they have all been our best instructors. Also the authors' appreciation goes to many of their colleagues and fellow faculty members at Illinois Institute of Technology and Georgia Institute of Technology. Dr. P. S. Sriraj, University of Illinois-Chicago, deserves special thanks for helping the authors with the four case studies included in Chapter 13. We also extend thanks to Mr. Tim Pletscher, Senior Acquisitions Editor at J. Ross Publishing, who has helped us all along the process of putting this book through the hoops.

C. Jotin Khisty
Jamshid Mohammadi
Adjo A. Amekudzi

Preface to First Edition

Society depends, to a large extent, on sociopolitical and managerial decision making relying, directly or indirectly, on the advice of engineers. The planning, designing, constructing, operating, and maintaining of engineering facilities, in the private and public sectors, represents the expenditure of billions of dollars every year. Yet, many engineers who are responsible for this massive expenditure have little or no formal training or education in the fundamentals of economics or in the basic concepts of systems analysis, so essential for dealing with these kinds of decisions. However, this picture is rapidly changing with the revision of curricula for professional schools of engineering all over North America and abroad.

The main objective of this book is to present the fundamental principles of economics, probability, statistics, and systems analysis to engineering sophomore, junior, and senior students. While there are scores of excellent textbooks dealing with all four components individually, there are comparatively few textbooks covering these areas in a single volume. Our experience has been that engineering students need to be exposed to the fundamentals of these topics early on in their undergraduate training so that they can apply the knowledge gained in these areas when they take advanced courses toward their bachelor's degree, or in some cases toward their master's degree. This textbook attempts to integrate the power of quantitative analysis with the conceptual richness of capital budgeting and microeconomics into systems engineering. The contents of this textbook provide students the basic concepts and tools that have proven most useful in engineering and planning contexts.

Some of the special features of the book need to be highlighted. First, a wide range of topics is covered, all drawn from the *systems approach* standpoint. Second, the emphasis is on presenting the fundamental concepts and their practical engineering applications, unencumbered by complicated mathematics. Third, realizing that our *best* practice is far from perfect and our theories incomplete, we have included many open-ended, value-laden, real-world exercises at the end of every chapter. Fourth, although the conventional practice is for engineers to normally deal with natural and physical systems, we have included a chapter on soft systems thinking and analysis. This inclusion was in response to the emergence of a range of formal methodologies that aim not just to produce *optimal* solutions, but to facilitate an enriched decision-making process most suitable for application in an uncertain world. To our knowledge, this is the first attempt at including such material in an undergraduate textbook.

Chapters 1 and 2 are introductory in nature, focusing on the natural, physical, and human systems. They describe the nature of problems likely to be encountered in engineering practice. Chapters 3 and 4 cover the basic topics on engineering economics and the fundamental tools of microeconomics, respectively. The principles of probability are taken up next in Chapters 5, 6, and 7. Each of these chapters covers one of several basic areas of the theory of probability, while Chapters 8 and 9 deal with the principles of applied statistics. The application of probability and statistics in engineering is in such areas as decision making, design under uncertainty, data analysis and interpretation, and system safety analysis. In this respect, we assume that engineering problem-solving practice, to a large extent, depends on engineering knowledge and the ability to interpret data, and, as such, knowledge of probability and statistics plays an essential role in the decision-making process. Chapters 8 and 9 focus on applied statistics and discuss such topics as the analysis of engineering data, empirical estimation of statistical parameters, hypothesis testing, and correlation analysis. Even though courses in probability and statistics (usually offered by the department of mathematics/statistics) are included in engineering curricula, our experience has been that students taking such courses invariably end up having difficulty in applying their knowledge in these areas to engineering problems. Our objective in including these chapters is to emphasize the practical applications and relevance of the basic concepts to engineering design.

Chapters 10 and 11 serve as introductory chapters on hard systems engineering, while Chapter 12 deals with Soft Systems Thinking. As is well known, systems analysis developed out of wartime military operations planning, particularly during World War II, and has since been the dominant traditional approach underlying systems engineering, systems analysis, management science, and operations research since at least the last fifty years. In essence, hard systems engineering is concerned with the application of quantitative analysis to engineering and managerial decision making for obtaining an optimal or best feasible solution. Soft systems thinking, on the other hand, is a more recent development, and in contrast to hard systems analysis, admits that there are multiple perceptions of reality and of solving wicked, messy, complex, and ill-structured problems characteristic of contemporary engineering and planning problems.

These are particularly encountered where socioeconomic and political ramifications are predominant.

We have presented the material through ample illustrative engineering and managerial worked examples. This will surely motivate practicing engineers and students to grasp the essential concepts for analysis and design. In view of the major worldwide thrust on distance learning and self-study by individuals who need a self-contained textbook dealing with the fundamentals of systems engineering, with economics, probability and statistics, we are confident that this textbook will be ideal.

There are a number of audiences for this book. It is quite possible that students could complete the major topics included in this book in two 3-credit semester courses. Instructors in architecture, construction management, chemical, electrical, mechanical, or industrial engineering, as well as in urban planning, in addition to civil engineering, could formulate basic one- or two-semester course(s) to meet their own specific requirements. This textbook could be effectively used in graduate courses too, particularly in construction management and transportation engineering programs. A solutions manual is available.

Many people warrant acknowledgment, individually or jointly, from the authors, and for personal or work-related reasons. We wish to thank our colleagues and students in the Department of Civil and Architectural Engineering, Illinois Institute of Technology (IIT), Chicago, for their help and advice. For the first author, this book began as a set of notes as early as 1978 when he was a faculty member at Washington State University (WSU), Pullman, Washington. Many batches of graduate and undergraduate students at WSU and IIT have contributed in one form or another in the development of this book. In more recent years, five of my former students, Dr. P. S. Sriraj, Cemal Ayvalik, Turan Arslan, Raymond Tellis, and Sagar Sonar, helped this author put the book together. My sincere thanks go to them. The second author expresses his gratitude to his colleagues, Dr. Anatol Longinow of Wiss, Janney, Elstner Associates of Northbrook, Illinois, and Joseph F. Braun of Systems & Electronics of Elk Grove, Illinois, for providing an opportunity and reposing their trust in him to apply many of the statistical methods in Chapters 8 and 9 in real-world engineering data analysis problems. Finally, this book would not have taken shape without the admirable help, support, and advice provided by Laura Curless, Scott Disanno, and their staff at Prentice-Hall.

As with any textbook containing a vast amount of numerical work together with scores of examples and exercises, we would appreciate it if errors and inconsistencies are brought to our notice.

C. Jotin Khisty
Jamshid Mohammadi

This book has free material available for download from the Web Added Value™ resource center at *www.jrosspub.com*

At J. Ross Publishing we are committed to providing today's professional with practical, hands-on tools that enhance the learning experience and give readers an opportunity to apply what they have learned. That is why we offer free ancillary materials available for download on this book and all participating Web Added Value™ publications. These online resources may include interactive versions of material that appears in the book or supplemental templates, worksheets, models, plans, case studies, proposals, spreadsheets and assessment tools, among other things. Whenever you see the WAV™ symbol in any of our publications, it means bonus materials accompany the book and are available from the Web Added Value Download Resource Center at www.jrosspub.com.

Downloads for *Systems Engineering with Economics, Probability and Statistics, Second Edition* consist of a solutions manual available to instructors covering those exercise problems that require numerical solutions.

Mapping the Terrain of the Systems Approach

1

1.1 INTRODUCTION

In essence, this book is about using the systems approach to make decisions. It answers the basic question: How can we choose the best course of action, taking into account the goals we are trying to achieve and the constraints that limit our action, by such factors as time, labor, money, and the policies set by the government or by a private organization? Our main purpose is to give the widest possible overview of systems engineering to a beginning engineering student and to explain how a combination of the principles of probability and statistics, economics, and systems analysis can be used for solving engineering problems related to planning, design, and management.

This chapter maps out the terrain of what is covered in succeeding chapters and also describes some preliminary definitions connected with science and systems engineering. How would an engineer minimize the capital and maintenance cost of a long-span bridge? How could an engineer advise his client how to maximize (or optimize) the total income from a high-rise building? What would be the best way to maximize the safety of the railroad system running through your city? Should the government subsidize persons buying an electric car to boost the *economy of scale* of electric car manufacturers? Should the city extend the light-rail system in San Diego and what would be the implications and consequences attached to this decision? These are the kinds of questions you, as an engineer, planner, or manager, will have to face when you take up a responsible position with a public or private undertaking. To tackle such questions you will need a basic knowledge of economics (both capital budgeting and microeconomics), the principles of probability and statistics, and a working knowledge of systems engineering. All of these topics are covered in this book.

A professional engineer must understand and apply the basic laws of mathematics, physics, chemistry, and economics for planning, designing, managing, and operationalizing engineering works. With hundreds of different recognized engineering specialties, a simple yet comprehensive definition of engineering is:

> Engineering is the profession in which knowledge of the mathematical and physical sciences gained by study, experience, and practice is applied with judgment to develop ways to utilize economically, the materials and forces of nature for the progressive well being of society (Crandall and Seabloom, 1970).

It is a concern related to economics that distinguishes engineering from pure science. While economic considerations may be of little or no concern to the pure scientist, the function of the engineer is to utilize the principles of economics to achieve a more efficient and economical end product, such as highways, buildings, water-supply systems, and so on. And it is the evolution of this end product from its conception to its final production, using the creative processes, that is known as engineering design. Design is both an art and a science in that it is a creative problem solving process in which the engineer works within the bounds of a limited monetary budget, a prescribed time line, and specific laws and regulations to convert data, information, technical know-how, combined with his/her ideas, into an accepted product. When an engineering design is finally approved by those authorized to do so, the finished design can then be implemented (Crandall and Seabloom, 1970).

1.2 THE NATURE OF SCIENCE

All engineers invariably take several courses in mathematics and science because these courses form the backbone of engineering science. In a broad sense, science is a way of acquiring testable knowledge about the world. It is now recognized that the knowledge we gain from the scientific approach is provisional and probabilistic, because it is possible that additional experiments carried out by scientists may alter what we already know. Naturally all theories and laws that we currently know are really approximations of the truth within a certain domain of validity.

Some important characteristics of the scientific approach are:

Hypothesis setting and testing: Scientists make propositions or suppositions for reasoning, investigation, or experimentation, for a limited number of variables. Experiments are then conducted to test the hypothesis, holding all other variables constant. If the hypothesis turns out to be correct, it adds to our current knowledge base. If not, the results are rejected.

Replicability: Scientific knowledge must be as objective as possible, which means that a number of observers performing the same experiment, independent of one another but under the same conditions and assumption, should be able to replicate results and verify the observations. This is the scientists' way of verifying (or validating) or falsifying (or rejecting) a proposed hypothesis.

Refutability: While it is impossible for scientists to conduct all possible experiments on a particular topic, due to time constraints, it is important to perform valid experiments using appropriate scientific techniques to decide between competing hypotheses. Although many scientists tend to have their theories corroborated by effective scientific techniques, it is quite possible that these theories could be refuted through a series of additional experiments.

Reductionism. The real world under study is so complex and messy that scientists can only perform simple experiments to capture and comprehend it. As a result, scientists experiment with small units or entities of the real world that can explain cause and effect in a linear way. This style of thinking and experimentation, called reductionism, isolates the phenomenon under investigation from its environment, which eventually produces a *mechanistic* view of the world (Checkland, 1999; Flood and Carson, 1993).

According to the scientific method, all genuine inquiry and knowledge should be based on hard facts, experimentation, and explanation. It goes further in believing that the methods of science are applicable to all enquiry, especially that of the human and social sciences. This traditional scientific approach has been debated and attacked by many scientists and philosophers, and we take up this debate while considering soft systems thinking in Chapter 12.

1.3 ENGINEERING PLANNING, DESIGN, AND MANAGEMENT

The planning and designing of a product are basic tasks undertaken by engineers to produce an end product. Planning is the arrangement of specific steps for the attainment of an objective. It is a future-oriented and prescriptive process because it assumes our ability to control our own destiny, at least within certain limits. In the context of engineering, planning generally involves the arrangement of spatial patterns over time. However, it must be remembered that it is not the spatial patterns that are planning: they are just the objects of a process. Management, on the other hand, is the skillful use of means (e.g., technology) to accomplish certain ends (e.g., objectives). Engineering design, defined by the Accreditation Board for Engineering and Technology (ABET), is:

> The process of devising a system, component, or process to meet desired needs. It is a decision-making process (often iterative), in which the basic sciences, mathematics, and the engineering sciences are applied to convert resources optimally to meet these stated needs (ABET, 2009).

1.4 THE SYSTEMS APPROACH

With the rapid technological advances made in every sphere of inquiry, engineers, planners, managers, decision makers, and even pure scientists realized that the complexity of real-world problems could not be handled by simply applying the traditional scientific method, which had its limitations. This is particularly true when dealing with social systems or engineering problems that have a social or human component. Indeed, if you look around for an engineering problem without the human factor, you would be hard pressed to find one. So then, where do we begin? Or better still, where *should* we begin? We will begin with a simple basic definition of the systems approach:

> The systems approach represents a broad-based, systematic approach for problem solving and is particularly geared toward solving complex problems that involve systems. A system is a set of interrelated parts—components—that perform a number of functions to achieve common goals. Systems analysis is the application of the scientific method modified to capture the *holistic* nature of the real world in order to solve complex problems. In fact, the systems approach ought to be called the *systemic approach*; systemic in the sense that it offers systemic (holistic rather than piecemeal) as well as systematic (step-by-step rather than intuitive) guidelines for engineers to follow (Flood and Carson, 1993; Jackson, 2000).

Goals are desired end states, and operational statements of goals are called objectives that should be measurable (where possible) and attainable. Feedback and control are essential for the effective performance of a system. The development of objectives may in itself involve an iterative process. Objectives will generally suggest their own appropriate measures of effectiveness (MOEs). A MOE is a measurement of the degree to which each alternative action satisfies the objective. Measures of the benefits foregone or the opportunities lost for each of the alternatives are called measures of costs (MOC). MOCs are the consequences of decisions. A criterion relates the MOE to the MOC by stating a decision rule for selecting among several alternative actions whose costs and effectiveness have been determined. One particular type of criterion, a standard, is a fixed objective: the lowest (or highest) level of performance acceptable. In other words, a standard represents a cutoff point beyond which performance is rejected (Khisty and Lall, 2003). The following example will help you to understand the basic concepts.

Example 1.1 A medium-sized city with a population of 250,000 plans to investigate the implementation of a public transport system. This is a first-cut preliminary look to be accomplished in, say, a couple of days. Your task is to provide a sample set of goals, objectives, alternatives, MOCs, and MOEs to demonstrate to citizens in your neighborhood how one could begin thinking about these issues.

Solution A sample set of goals, objectives, alternatives, and MOEs could be framed with the help of citizen groups:

Goal: To provide an accessible, safe, reliable, and cost-effective transport system

Objectives:

1. To carry people at a minimum operating speed of 25 mph during peak hours
2. To have fares that would compete with the cost of operating a private car

Continues

Example 1.1: *Continued*

Alternatives A. Regular bus system (RB)
B. Street car system (SC)
C. Light-rail system (LR)

MOEs
1. Cost to ride (in cents/ride): should be less than half of the cost of using an automobile
2. Punctuality: delay should be less than ±5 min
3. Speed: should be the same speed as a private car (25 mph)
4. Area-wide coverage: service should be approximately a 10 min walking distance from residence

MOCs Capital and maintenance cost of providing the service in relative dollar terms is:

$$RB = 20x;\ SC = 40x;\ LR = 60x, \text{ where } x \text{ is in millions}$$

Assessment Set up a matrix and assign scores between 1 and 5, representing poor to excellent values, respectively:

Alternatives	MOEs	1	2	3	4	Total effectiveness
A		4	3	3	5	= 15
B		3	3	2	3	= 11
C		4	5	5	2	= 16

The matrix indicates that the light-rail system is the best of the three systems considered, but, when we compare the scores along with the costs, we find that the cost/unit of effectiveness is:

$$\text{For } A = 20x/15 = 1.33x;\ B = 40x/11 = 3.64x;$$

$$\text{and for } C = 60x/16 = 3.75x$$

This indicates that the bus system is the best.

Conclusion

The best alternative based on the cost per unit effectiveness is alternative A, which is the lowest of the three.

Discussion

This assessment has been completed merely to demonstrate how a first-cut assessment could be performed. Additional data gathering and calculations would be needed to make an accurate judgment regarding which alternative would be best for the city. Input from citizens who would use the system would also be important.

1.5 STEPS IN SYSTEMS ANALYSIS

Some of the basic steps recommended for performing an analysis are:

- Recognizing community problems and values
- Establishing goals and defining the objectives
- Establishing criteria
- Designing alternative actions to achieve Step 2
- Evaluating the alternative actions in terms of effectiveness and costs
- Selecting an alternative action in keeping with the goals and objectives, criteria, standards, and value sets established, through iteration, until a satisfactory solution is reached

A simplified analysis process is illustrated in Figure 1.1, and the hierarchical interrelationships among values, goals, objectives, and criteria are shown in Figure 1.2.

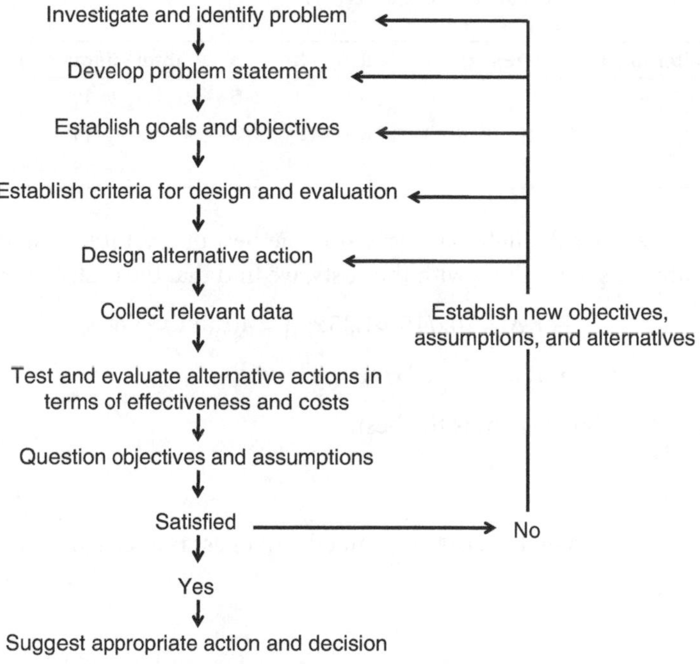

Figure 1.1 Flow chart of system analysis process

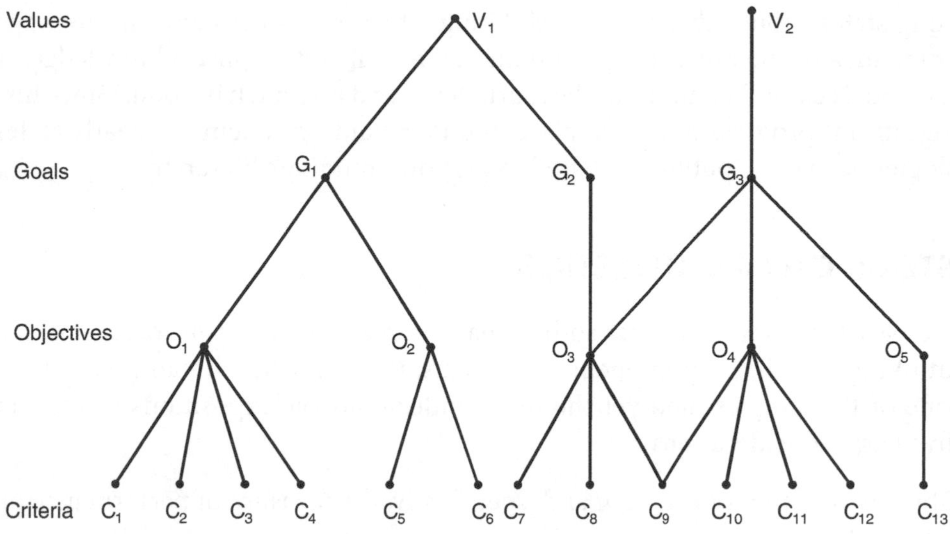

Figure 1.2 Hierarchical interrelations among values, goals, objectives, and criteria

1.6 CLASSIFICATION OF SYSTEMS

There is no standard classification scheme for systems. Boulding's (1956) hierarchy, ranked in an increasing order of complexity, was one of the first attempts at classification. Each level is said to include, in some way, the lower levels, but also to have its own emergent properties, although there is no *scale of complexity* attached to this hierarchy.

The most important distinction is between living and nonliving systems and the different types of relationship in the organizational connections involved within each hierarchy, including the lower hierarchies. Table 1.1 displays Boulding's hierarchy.

Table 1.1 Boulding's hierarchy of systems

Level	Description	Characteristics	Examples
1	Structures	Static, spatial framework	Atom, crystal, bridge
2	Clockworks	Predetermined motion	Solar system, clocks, machines
3	Control	Closed-loop control mechanisms	Thermostats
4	Open systems	Structurally self maintaining	Cells
5	Genetic systems	Society of cells	Plants
6	Animals	Nervous systems, self awareness	Birds and beasts
7	Humans	Self consciousness, knowledge, language	Human beings
8	Sociocultural systems	Roles, values, communication	Family, community, society
9	Transcendental systems	Beyond our knowledge	Religion

Boulding states, "One advantage of exhibiting a hierarchy of systems in this way is that it gives us some idea of the present gaps in both theoretical and empirical knowledge. Adequate theoretical models extend up to about the fourth level, and not much beyond." Since his remarks in 1956, significant progress has taken place in understanding systems, at nearly all levels, and new typologies have been published, based in part on Boulding's hierarchy.

1.7 SYSTEMS CHARACTERISTICS

During the past forty years, several authors have suggested basic considerations concerning systems thinking. Of all the proponents of systems, C. West Churchman (1968) has given us perhaps one of the simplest and yet the most understandable expositions of the subject. He outlines five basic considerations:

1. Objectives of the total system and, specifically, the measure of performance of the system itself
2. System environment
3. Resources of the system
4. Components of the system
5. Management of the system

These five considerations are not meant to be all-inclusive, but they capture most of the properties that engineers generally encounter in their practice. Let's discuss each of the points.

1. *Objectives* are those goals or ends that the system is working toward. Hence goal-seeking (or teleology) is a characteristic of systems. While the determination of objectives for mechanical systems is comparatively easy, those for human activity systems can be considerably more difficult. One must be cautious to distinguish between stated objectives and the real objectives of the system. For example, a student may say that her objective is to gain knowledge in order to master the subject being studied. Her real objective, in fact, simply may be to obtain good grades. In real life, objectives need to be operationalized, and to do this properly they need to be quantifiable, in some way or another, in order to measure the performance of the system. Quantification helps us to measure to what degree the system's objectives are being realized.

2. *Environment* constitutes all that is *outside* of the system. First, the environment includes all that lies outside of the system's control. Second, the environment also includes all that determines, in part at least, the manner of the system's performance.

3. *Resources* are all the means available to the system for the execution of the activities necessary for goal realization. Resources include all the things that the system can change and use to its own advantage. In human systems, one can also include the opportunities that are available to the system, besides human labor, money, and materials.

4. *Components* consist of the elements that accomplish the missions, jobs, or activities that the system has to perform to realize its objectives.

5. *Management* includes two basic functions. First, it is planning the system which involves all aspects of systems previously encountered, that is, its goals, environment, utilization of resources, and components and activities. Second, it is controlling the system which involves both the examination of the execution of plans as well as planning for change. Plans must be subject to periodic review and reevaluation because no plan can remain static throughout the life of a project. Associated with the planning and control function is the notion of information flow and feedback, often characteristic of cybernetic systems.

1.8 SYSTEMS ANALYSIS AND DECISION MAKING

We have described Churchman's systems approach for dealing with problems. You may notice that in contrast to the methods used in the pure sciences, the objectives of the systems approach is to recommend a course of action to the decision maker in addition to merely understanding the problem. Systems analysis is truly a continuous cycle of defining objectives, then designing alternative systems to achieve those objectives, evaluating the alternatives in terms of their effectiveness and cost, questioning the objectives and other assumptions underlying the analysis, seeking new alternatives and establishing new objectives, and the iteration goes on indefinitely. This cyclical process bears out the idea of the rational methodology to problems of choice.

As all of us know, we are asked to make decisions on a daily basis, not only for ourselves but for others. Professionals in all walks of life have to make difficult decisions and even more difficult predictions, and everybody expects them to make the very best decisions. But are we able to make the best decisions under such constraints as time limitations, funding restrictions, lack of resources, and political pressures? In situations of certainty, the decision maker is supposed to have complete knowledge of everything needed to make that decision. In other words, he has complete knowledge of the value of the outcomes and the occurrence of the states of nature. In situations of risk he knows the value of the outcomes and the relative probabilities of the states of nature. Under uncertainty, the values of the outcomes may possibly be known, but no information is available on the probability of the events.

1.9 MODELS AND MODEL-BUILDING

In its simplest sense a model is a representation of reality. The model is arrived at through the process of abstracting from reality those aspects with which one is concerned. For example, a map of Chicago, showing the street system, the locations of schools, hospitals, and major places of interest, including the suburbs, all drawn to scale, is a model of the city of Chicago. This map can be put to use to solve any number of problems an engineer (or a lay person) may have. A model is useful in a practical sense when it accurately duplicates the behavior of the real world system. One thing to remember is that models are neither true nor false; their value is judged by the contribution they make to our understanding of the systems they represent. Also, a model cannot represent every aspect of reality because the model is at best an approximation of the real object or situation.

Models may be classified by their correspondence to the system being modeled. For example, physical models retain some of the characteristics of the system they represent. A model of a house could, for instance, be made of plywood, scaled down, which might look almost like the house you may want to build. Photographs and blueprints are further examples of this category and are referred to as iconic models. Models constructed from a set of physical objects not found in the real system are called physical analogues. For example, an electrical system may be constructed to behave like a water distribution system for a city. Schematic or analogue models, in the shape of flow charts or organizational charts, are frequently employed by engineers, using lines and symbols that are mere abstractions of the physical world. Mathematical models of systems consist of sets of equations whose solutions explain or predict changes in the state of the system. Our primary interest is in mathematical models.

We have an opportunity to examine various types of models in succeeding chapters, but it may be helpful to keep in mind that models are capable of portraying infinitely complex problems arising in engineering. These are presented in the form of charts, graphs, equations, and symbolic representation, representing a system of inputs, processing and transforming system, and outputs as indicated in Figure 1.3. Here, a system can be observed as one containing a *black box*, connecting an input to an output. If the output is simply related to the input and not affected by any kind of feedback to the input, it is said to be a feed-forward system. If, on the other hand, a part of the output is fed back to the input, it is referred to as a feedback system, as shown in Figure 1.3.

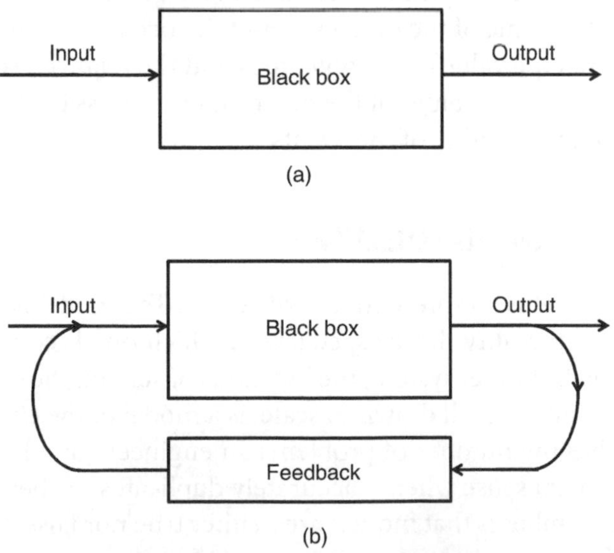

Figure 1.3 Black box or transfer function model: (a) without feedback and (b) with feedback

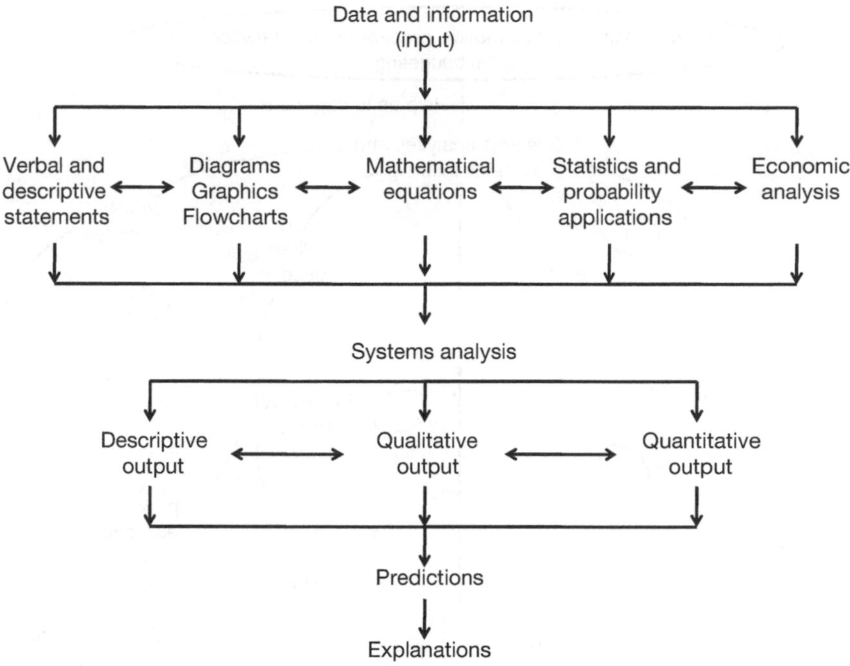

Figure 1.4 The modeling picture

Further insight into the modeling process is given in Figure 1.4. Starting with data and information regarding a system under investigation, an engineer can use these inputs to harness a variety of tools, such as graphs and flowcharts and, through mathematical equations to economic analysis, to perform an analysis.

Proper interpretation of this analysis can result in predictions and explanations of the system under consideration. In the chapters that follow we revisit these tools.

SUMMARY

The engineer has professional obligations to society and, therefore, has to exercise judgment, make decisions, and accept responsibility for his actions. Society has also become aware of the consequential impacts of engineering work. This requires that engineers realize the nature of social goals and objectives and appreciate the value of the systems approach.

This chapter has introduced the student to the general nature of science and the differences between engineering and science. The systems approach was explained and the various steps in systems analysis were outlined. Because systems occurring in engineering are so varied, the nature of iconic, analogue, and mathematical models were described. A short description of

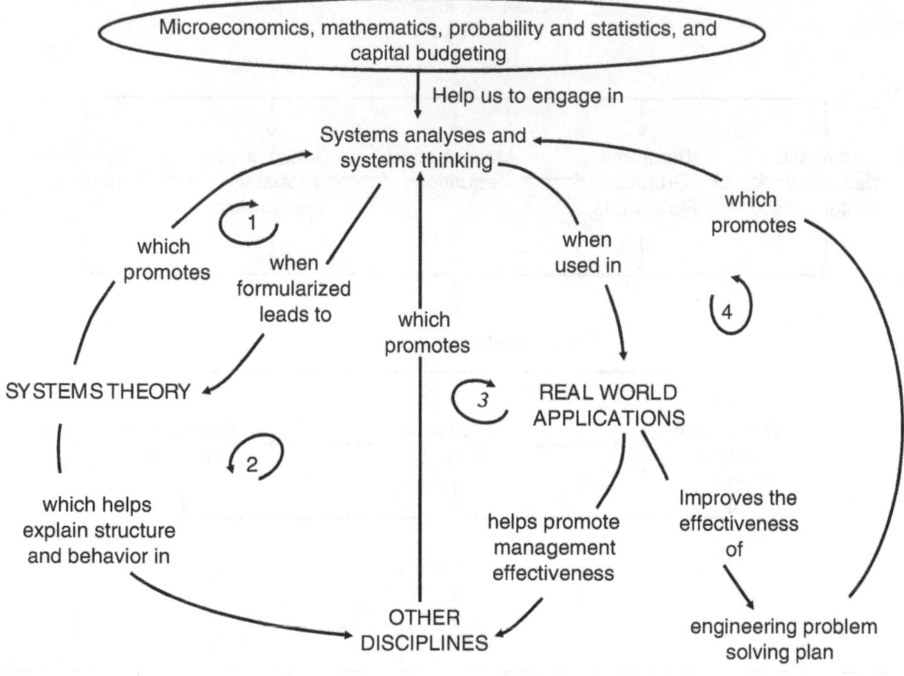

Figure 1.5 Systems thinking's development (adapted from Flood and Carson, 1993, p. 4)

models and model-building was also introduced—just enough to set the ball rolling to deal with the next chapter on problem solving.

In summary, systems thinking is a useful framework to keep in mind when dealing with complex problems in a holistic way. Systems thinking promotes systems theory, enables us to practice good management, provides us with the insight to plan for the future, and helps us to understand the structure and behavior of systems. The interlinked cycles of systems thinking, systems theory, and systems applications are shown in Figure 1.5. Further elaboration on many of the concepts described in this chapter can be found in Jackson (2000) and Midgley (2000).

REFERENCES

ABET (2009). *Criteria for Accrediting Engineering Programs*, Accreditation Board for Engineering and Technology, Baltimore, MD: ABET.

Boulding, K. E. (1956). "General Systems Theory—The Skeleton of Science," *Management Science* 2, 197-208.

Checkland, P. (1999). *Systems Thinking, Systems Practice*, Chichester, UK: John Wiley & Sons.

Churchman, C. W. (1968). *The Systems Approach*, New York, NY: Dell Press.

Crandall, K. C. and R. W. Seabloom (1970). *Engineering Fundamentals*, New York, NY: McGraw-Hill.
Flood, R. L. and E. R. Carson (1993). *Dealing with Complexity*, 2nd ed., New York, NY: Plenum Press.
Jackson, M. C. (2000). *Systems Approaches to Management*, New York, NY: Kluwer Academic.
Khisty, C. J. and B. K. Lall (2003). *Transportation Engineering: An Introduction*, Upper Saddle River, NJ: Prentice-Hall.
Midgley, G. (2000). *Systemic Intervention*, New York, NY: Kluwer Academic.

EXERCISES

1. Suppose you and a scientist were discussing the flooding of a part of your neighborhood. How would you convince him/her that the systems approach was more practical to use as opposed to the scientific method?

2. A major road underpass near your neighborhood suffers from occasional floods all year round, blocking traffic. Set up sample goals, objectives, alternatives, measures of effectiveness, and measures of cost, and then draw up conclusions as to what should be done to alleviate the problem.

3. The intersection of Pine and Oak Streets is currently controlled by stop signs. Unfortunately, it has a high accident rate (two fatalities/yr and five minor accidents/mo). Due to the high flow of traffic, there is considerable delay during the morning and evening peak hours. How would you go about examining this intersection from a systems point of view?

4. By now you have probably taken several courses in mathematics, science, and engineering at your university. Pick a course of your choice and set it into the five components suggested by Churchman. Ask questions such as: What were the goals and objectives of the course? What was the major activity in the class? What did you accomplish? What was the overall outcome? and so on.

5. Suppose you have taken a good summer job in a large-sized city and have rented an apartment 10 mi from downtown where your office is located. You have the following options: to drive your own car; car pool with a friend; take the local bus; ride the subway; or take the light-rail system each day. How would you rationalize your choice so as to minimize your expenses and maximize your utility, using the systems approach?

6. Consider the following systems: a hospital; a truck; and a domestic washer and dryer. In each case indicate the inputs and outputs associated with these systems.

7. A brand new hybrid *electric* car, capable of running on gasoline and electric batteries, is being exhibited in a showroom. At any one time several hundreds of spectators are looking critically at this car, including: prospective car owners, salesmen, and mechanics, gasoline station owners; environmentalists; transport engineers; and city planners. How will each of these people view the car from their point of view with regards to complexity, sales, saving the planet from pollution, and profit-making.

8. Draw an organizational chart of the university you are currently attending, indicating the hierarchy and control of the various components such as the president, deans, professors, and office staff. Do you think this is a feed-forward or feedback system, or is it a combination of both?
9. Which of the following systems appear to you to be feed-forward or feedback systems? Some may be combinations of both. Sketch diagrams of these systems showing the inputs, outputs, and feedback: a bank account; stock market prices; your decision to buy a bicycle; and living on a budget fixed by your parents.
10. In what ways do you expect systems thinking will help you to manage your own life? Draw a sketch of this expectation making use of ideas taken from Figure 1.5.

Problem Solving and Designing in Engineering and Planning

2

2.1 INTRODUCTION

Chapter 1 sketched a broad-brush picture of the general nature of systems engineering. In this chapter we continue to elaborate and expand on some of those topics. First, we want to find out the nature of the problems encountered in engineering and the various styles that engineers use to tackle them. Next, we critically examine the nature of measurement and data that ultimately affects the models we put to use. This is followed by looking at the modeling procedures widely employed by engineers in designing their products. As with any discipline there are many basic definitions that have evolved during the course of the development of systems engineering, and these are explained as the chapter progresses. Keep in mind that we will be coming back to this chapter as we work through the rest of this book.

2.2 PROBLEM SOLVING AND DESIGNING

We confront problems nearly every day of our lives, some trivial, others a little more complicated, and a few that are truly formidable. The task is to *define the problem* and, once this is done, the next natural step is to solve the problem. *Problem solving* is often described as a search through a vast maze that describes the environment of possibilities. Successful problem solving involves searching the maze selectively and reducing it to manageable proportions (Simon, 1981). For example, if we are given a comparatively simple puzzle and asked to find out the numerical values of the letters, to satisfy the addition of *cross* and *roads*, given just one single clue that S = 3, how would we go about solving this alphametic?

$$
\begin{array}{rcccccc}
 & & C & R & O & S & S \\
+ & & R & O & A & D & S \\
\hline
= & D & A & N & G & E & R
\end{array}
$$

By trial and error we find that the puzzle works out to be:

$$
\begin{array}{rcccccc}
 & & 9 & 6 & 2 & 3 & 3 \\
+ & & 6 & 2 & 5 & 1 & 3 \\
\hline
= & 1 & 5 & 8 & 7 & 4 & 6
\end{array}
$$

where A = 5, C = 9, D = 1, E = 4, G = 7, N = 8, O = 2, and R = 6.

What we did in solving this alphametic problem was to conduct a *search* to explore a space of potential and partial solutions. Crossword puzzles are similar. Exploring a maze, for example, is also a process of problem solving; there is a starting point and a destination (or goal) in mind, while the intersections of passages are solution options or states. Some rational process is needed to get to the goal.

How do we solve such problems? How much time does it take? What are the alternative ways of representing the environment and conducting the search? Does the solution represent the final solution or could there be other possibilities? There are no definite answers to these questions because the knowledge base and experience of each individual problem solver is so different, and our rationalities and capabilities to solve problems are so limited. Attention has been drawn to our short-term memories by George Miller's (1956) celebrated paper on "The Magical Number Seven, Plus or Minus Two." The facts that emerge from a series of experiments conducted by psychologists are that our short-term memory only permits us to remember about seven bits of information at a time. For example, if we are not interrupted in any way, most of us can possibly remember a string of seven random numbers (at the most nine) from a directory as we ring up a telephone!

Problem solving implies a concerted effort at searching for a solution. *Designing* on the other hand is concerned with *how things ought to be*—how they ought to be in order to attain goals and to function efficiently. In other words, designing is considered a problem solving process of searching through a state space, where the states represent design solutions (Simon, 1981; Banathy, 1996). This state space may be highly complex and huge; complex in that there may be many intricate relationships between the states, and huge in the sense that there may be a plethora of states far beyond our comprehension. Problem solving is an *analytical* activity connected with science; designing is a *synthetic* skill directly connected with engineering and the artificial (or man-made) world. Remember that natural science is concerned solely with how things are, and is, therefore, descriptive, while engineering by and large is normative and connected with goals and *what ought to be*.

Designing is a fundamental, purposeful human activity, and designers who work on them are the *change agents* within society whose overall goal is to improve the human condition through physical, economic, and social change. More specifically, the design process can be described as a goal-directed activity that involves the making of decisions by satisfying a set of performance requirements and constraints. For example, if an engineer is asked to design a crane that has the capacity to lift 20 tons of load, and with a budget not to exceed $50,000, she would have to utilize her knowledge of mathematics, engineering, economics, and systems to design this lifting device (Coyne et al., 1990).

Numerous models of the design process have evolved in the past 50 years, but the one that has had general acceptance consists of three phases: (1) *analysis*, (2) *synthesis*, and (3) *evaluation*, as shown in Figure 2.1. The first phase is one of understanding the problem and setting goals and objectives. More specifically, the tasks in this phase would be to diagnose, define, and

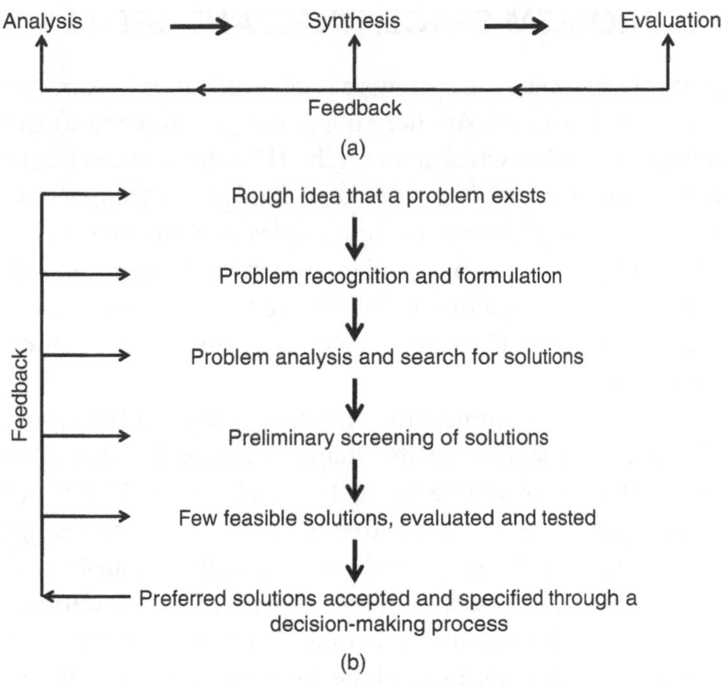

Figure 2.1 Staged process of problem solving and design

prepare all the preliminaries to truly understand the problem and match them with goals. The second phase consists of finding a variety of possible solutions that are acceptable in light of the goals set forth in Phase 1. The third phase consists of evaluating the validity of each of the solutions proposed in Phase 2 and selecting an alternative solution (or a combination of solutions if that is feasible) that satisfies the goals proposed in Phase 1. This procedure is reiterated as many times as necessary through a feedback process to revise and improve the analysis and synthesis phases. It represents the basis of a framework for understanding, formulating, analyzing, searching, designing, synthesizing, comparing, selecting, evaluating, and, finally, deciding on a course of action that is subject to budget restraints, political acceptance, social conditions, or other conditions set forth by the decision maker who may or may not be the design engineer or the chief executive officer.

The systems approach, as we have seen previously, is a decision-making process for designing systems. Because we have many alternatives from which to choose, decision making is the act of deliberately selecting a course of action for making that choice. Generally, the act of decision making is an iterative process that is characteristic in problem solving, and this is obvious if we reexamine Figure 1.1 in Chapter 1.

2.3 HIERARCHY: PROBLEM-SPACE, TREES, AND SEMI-LATTICES

As we have said repeatedly, engineering systems are generally highly complex, and it is customary for them to be broken down into smaller subsystems to understand them. The nesting of systems and their subsystems is known as a hierarchy (Hutchins, 1996; Daellenbach and Flood, 2002). But, a hierarchy is much more than a concept. It implies a framework that permits complex systems to be built from simpler ones or for complex systems to be broken down into their component parts. Hierarchy helps us to organize, understand, communicate, and learn about the system and its subsystems. Given a hierarchy of systems, it is possible to sort out and arrange their corresponding goals and objectives from which a system of priorities of high- or low-level objectives could be established.

For example, if we wanted to examine the choices we have in traveling from our home to downtown, we could create a diagram in the shape of a tree, indicating the time, the mode of travel, and the routes that are available for us to travel (Rowe, 1987). In fact, this diagram, shown in Figure 2.2, is a type of decision tree that we could use in choosing the way we would like to travel. This tree diagram indicates a problem-space. It is an abstract domain containing elements that represent knowledge states to the problem under consideration. It is represented by nodes for decision points and branches for courses of action. In the case of our decision tree, we have decided to go to work during the peak period driving our car along Route 2, although we have many other options. This choice is indicated by the double lines in Figure 2.2.

However, our traditional emphasis of depicting our ideas and choices in the form of trees is not always realistic. When there are overlaps in choices, the use of semi-lattices becomes necessary. The reality of the social structure of cities, for instance, is replete with overlaps. Take, for example, the conflict between vehicle and pedestrian movement in any busy downtown. You have cars, buses, trucks, and delivery vans competing with pedestrians and bicyclists for

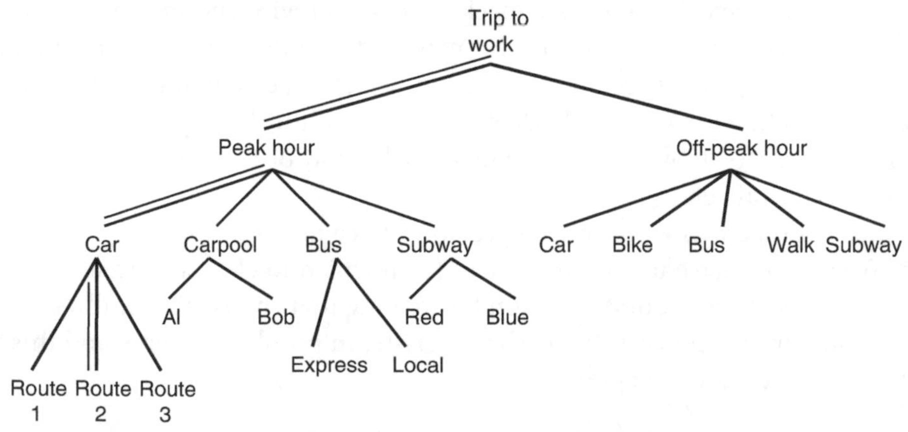

Figure 2.2 A decision tree of travel choices

movement and parking. Notice the interaction between taxis and pedestrians and between parking and motor vehicles shown in Figure 2.3.

Even a system of friends and acquaintances, and their relationships, can be represented by a semi-lattice as illustrated in Figure 2.4. Consider six individuals, Al, Bob, Chris, Dave, Ed, and Fred, represented by the letters A, B, C, D, E, and F. Their relationships are as follows: C and D are regular chess partners; D and E play tennis together on weekends; A, B, and C ski together during the winter months; B, C, and D regularly jog every morning during the summers; A, B, C, D, and E sing in a choir during the Christmas season; and all six belong to the same country club. In the corresponding Venn diagram, each set chosen to be a unit has a line drawn around it. One can see right away how much more complex a semi-lattice can be compared to a tree. Alexander (1966) says that a tree based on 20 elements can contain, at most, 19 further subsets of the 20, while a semi-lattice based on the same 20 elements can contain more than a million different subsets!

Hierarchical systems and their organizations reveal an important phenomenon that can be summarized in a nutshell: *the whole is greater than the sum of its parts*. What this means is

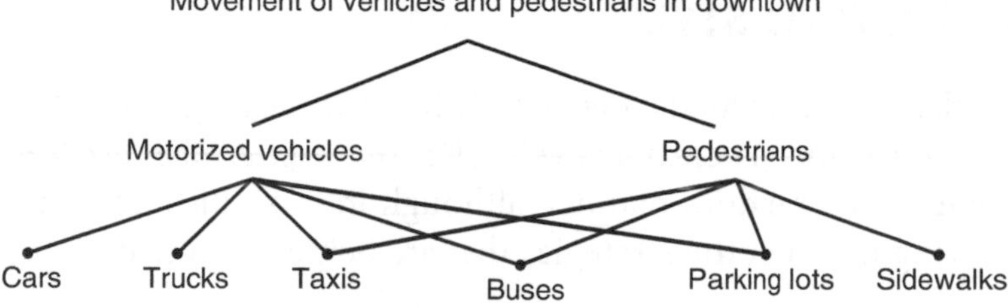

Figure 2.3 Pedestrian/vehicle movement overlap

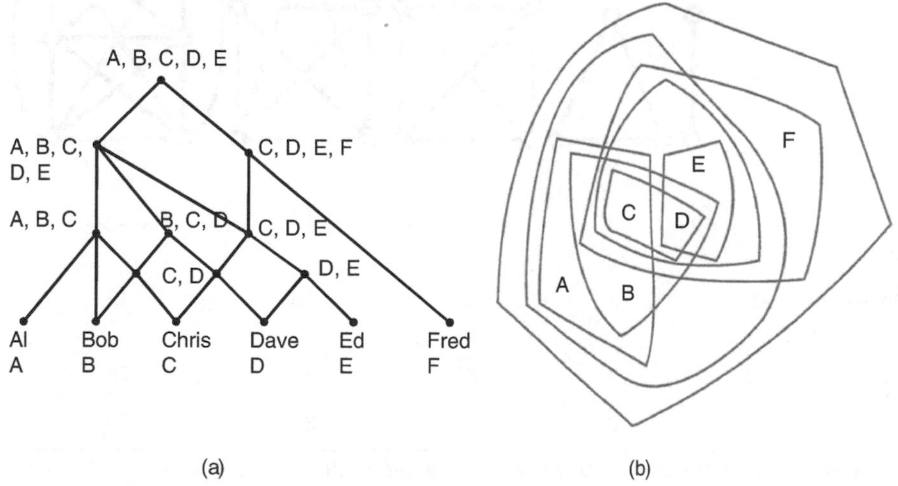

Figure 2.4 (a) Semi-lattice structure of six friends and (b) Venn diagram

that systems have *emergent* properties. For example, a human being is not the sum of his bodily parts; neither is a society a sum of its groups' members. Engineers with their mathematical training tend to believe in mathematical rigor and are reluctant to readily believe in this property, but we all know that the chemical combination of hydrogen and oxygen (which are gases) produces water that is a liquid having no correspondence with the elements that go to form it. Emergent characteristics of organizations have provided us with the term *synergy* to describe the emergence of unexpected benefits to management through group work. These benefits are not always assured: sometimes there might be disbenefits attached to well-intentioned emergence.

This is a good place to look at the meaning of system complexity. One of the simplest ways of understanding complexity is to study the number of elements and the number of relationships between the elements. When the number of parts and their possible relationships grow, the effect can be devastating as shown in the Figure 2.5. Here, *e* represents the number of elements, *r* their relationships, and *s* the states (Flood and Carson, 1993). Using the formula 2^e, *s* grows rapidly as *e* increases.

2.4 PROBLEM-SOLVING STYLES

Problem-solving styles can be divided conveniently into three categories: (1) trial-and-error; (2) generate-and-test, and (3) means-ends analysis. The *trial-and-error procedure* involves finding a solution to a problem in a random manner, although in most instances there is some sort of bounding or narrowing down of the strategies that are being examined by the problem-solver.

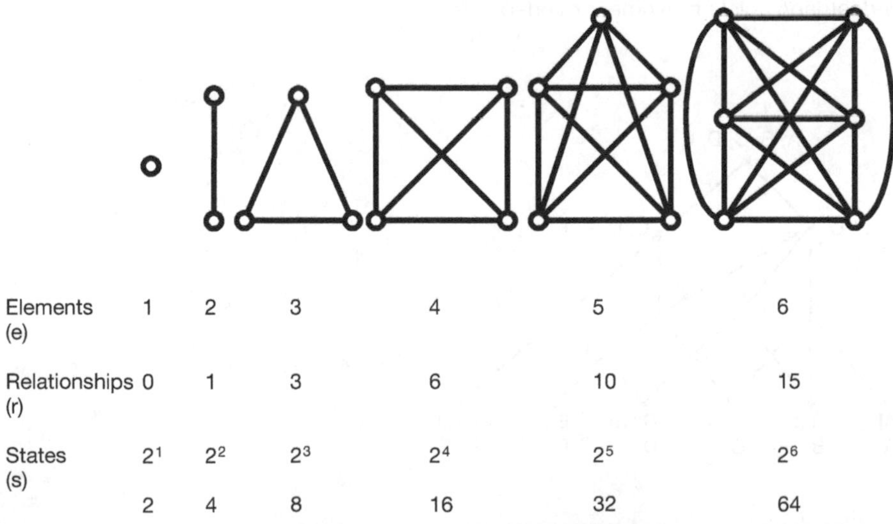

Elements (e) 1 2 3 4 5 6
Relationships (r) 0 1 3 6 10 15
States (s) 2^1 2^2 2^3 2^4 2^5 2^6
2 4 8 16 32 64

Figure 2.5 Elements, relationships, and states as a measure of complexity

Naturally, some tacit methods (involving intuition) may be used to evaluate solutions; but successive trials are made more or less independent of the results of intermediate tests. Notice that feedback is at best minimal. An example of the trial-and-error style of problem solving is playing with a jigsaw puzzle without knowing what the true picture looks like. Another example is the exercise of arranging two dozen pieces of furniture in a house with a floor area of 2000 sq. ft. to satisfy some general criteria. In both these cases, one could generate hundreds of candidate solutions by arranging and rearranging the pieces of the puzzle in a fashion that approximates the criteria prescribed until some acceptable arrangement emerges. This approach is often referred to as the black box approach referred to in Chapter 1 (Rowe, 1987).

The *generate-and-test* approach is not enormously elaborate compared to the trial-and-error procedure. However, its distinguishing feature is that it makes use of information regarding prior trials as the basis for directing further searches for an acceptable solution. This exercise of incrementally moving from worse to better solutions could be conceived as *hill climbing*, where the top of the hill is considered to be the best solution (Rowe, 1987). Figure 2.6 shows how the feedback mechanism operates in the case of the generate-and-test procedure. For a designer to determine if an error exists between what she is doing and what is needed to meet the goal, she must monitor her own activities by feeding back a portion of her designed output for comparison with her input. If the feedback tends to reduce error rather than aggravate it, the feedback is called *negative feedback*—negative because it tends to oppose what the designer is doing.

Figure 2.6 shows two methods: (1) an open-loop control and (2) a closed-loop control, where the output $O(t)$ is a function of the input $I(t)$, and the value of the multiplier K depends on the system's characteristics. Thus, $K = O(t)/I(t)$. In an open-loop system, with a positive multiplier K, the output increases in relation to the input, but there is no possibility for the system to correct itself. Naturally, such systems have to be carefully designed from the onset.

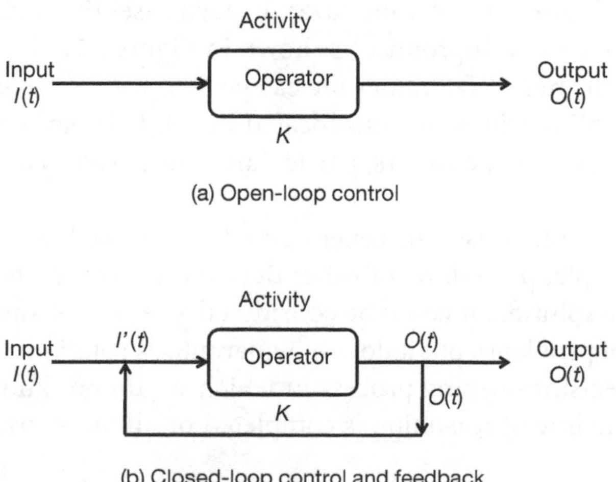

Figure 2.6 (a) Open-loop and (b) closed-loop control

With a closed-loop control, one can attain much greater reliability because the feedback serves as a self-correcting mechanism. Here $I'(t) = I(t) - O(t)$. In this case, $O(t)/I(t) = K$ and $K' = K/(1 + K)$. When $K' < 1$, we define this condition as negative feedback, which makes $O(t) < I(t)$. This means the output decreases with an increase in $I(t)$, thus providing the necessary self-correction. With positive feedback, $K' > 1$, which makes $O(t) > I(t)$. This causes $I(t)$ to increase progressively with each iteration, leading eventually to instability. In actual practice, the feedback is achieved by using only a small fraction of the output, Δt, or using the difference between the standard desired output, $S(t)$, and the output, $O(t)$, as the feedback, $\Delta(t) = S(t) - O(t)$.

Means-ends analysis is, in a way, an extension of the generate-and-test procedure, particularly for examining alternative decision rules for different problem solving situations. Remember that the means of achieving an end is really the technology or mechanism of attaining a goal. The question, then, is how do we utilize the means to achieve the goal? There are three essential components associated with this analysis: (1) a prescribed set of actions (means or technology), (2) a prescribed set of goals (or ends), and (3) a set of decision rules (Rowe, 1987). In the case of a reasonably well-defined problem where the goals and the technology are known and proven, means-ends analysis is most effective, particularly when an organizing principle is selected and applied under conditions of constancy. For instance, if an engineer wants to replace a stop sign at a busy intersection with a signaling device, all he has to do is determine the traffic volumes on the approaches together with the nature and frequency of the accidents at this intersection. If these records indicate that they exceed those prescribed by law or warrant, he can, without little additional investigation, make the necessary changes. Notice that he has a goal and the necessary technology, together with the prescribed law, to implement the change. This is a typical case of a simple problem because there is hardly any ambiguity attached to the problem or its solution. However, few, if any, engineering problems are as tidy and simple as this one. Indeed, most engineering and planning problems would fall into categories where the means, as well as the ends, are nebulous and fuzzy. In such cases the engineer will have to resort to judgment, negotiation, and compromise as shown in Figure 2.7. The problems encountered in cells B, C, and D of this matrix fall under the category of *wicked problems* that are described later in this chapter. Dealing with situations located in cell D is particularly problematic, and they are examined in subsequent chapters, particularly under soft systems analysis in Chapter 12 (Khisty, 1993).

Heuristic reasoning is often used in generate-and-test as well as in means-ends analysis. Basically, it is any principle, procedure, or other device that contributes to a reduction in the search for a satisfactory solution. It could be considered as a rule-of-thumb that often helps in solving a certain class of problems, but it does not guarantee a solution. Consequently, heuristic reasoning involves a decision-making process in which we do not know whether we actually have a solution until the line of reasoning is completed or all the steps are carried out (Rowe, 1987).

Means \ Ends	Goals and objectives: Certain	Goals and objectives: Uncertain
Certain	A Computation	C Compromise
Uncertain	B Judgment	D Chaos or *inspiration*

Figure 2.7 Means/ends configuration

Ackoff (1999) suggests that there are at least four ways of treating problems that we encounter on a day-to-day basis:

1. We can *absolve* a problem by simply ignoring that it exists or by imagining (and probably believing) that it will eventually disappear. People suffering from minor health problems usually believe that eventually such problems will disappear.
2. We can also *resolve* a problem by taking appropriate action, based on our experience and common sense, combined with our quantitative and qualitative expertise. Such a resolution results in the removal or suppression of the problem.
3. We can *optimize* the result, which is another way of tackling a problem. That is, we work out an outcome that will optimize the result through experimentation or analytical means.
4. We can *dissolve* the problem. One way of doing this is by redesigning the system containing the problem.

Consider the following situation as an example of tackling problems using Ackoff's thinking:

> A small taxi company rented out its vehicles to drivers on a daily basis but found that the vehicles were abused and misused, resulting in heavy losses to the company. In the early months of running the system, the managers ignored the losses, thinking that in due course the situation would improve. Unfortunately, the damage to the vehicles steadily increased. As a first step, management tried to resolve the problem by asking drivers to pay for all the damages they incurred. But this resulted in the company losing money because drivers were reluctant to rent vehicles from the company. Management tried to solve the problem by asking drivers to share the damage expenses on a 50/50 basis. While this strategy worked for some time, it eventually got out of hand because

the company overestimated the share from the drivers (because the recovery included depreciation of the vehicles). Drivers stopped renting and management lost money. Finally the company and the drivers came to an agreement that saved the situation. The company redesigned the rental system. Instead of renting cars on a daily basis, the company rented their vehicles on a monthly or quarterly basis along with the agreement that damages to the vehicles would be assessed by an independent party. Thus, the problem was eventually dissolved.

2.5 WICKED PROBLEMS

Attention was drawn to wicked problems earlier in this chapter. Most engineering and planning problems are really wicked problems as opposed to tame problems, and the ends or goals are already prescribed and apparent. A common example of a tame problem is the solution of a quadratic equation. The solution requires application of the rules of algebra to the specific structure that is given. In the case of wicked problems, both the ends and the means of solution are unknown at the outset of the problem solving exercise. Rittel and Weber (1973) have characterized wicked problems as:

- Definitive formulation does not exist and additional questions can always be asked, leading to continual reformulation.
- Stopping places are not defined and someone ends up saying, "That's good enough!"
- Solutions are neither true nor false, merely good or bad (e.g., there is no such thing as a true or false plan of a house, just good or bad.
- Solutions have no immediate or ultimate test.
- Each is a *one-shot* operation.
- They do not have an exhaustive set of potential solutions.
- Everyone is essentially unique.
- They can be considered a symptom of another wicked problem.

The problem of uncertainty is embedded in all wicked problems along with the problem of human rationality. More than thirty years ago Simon (1981) developed the theory of *bounded rationality and unbounded uncertainty* to describe the rational choice of alternatives, taking into account the cognitive limits of the designer and the decision maker. This means that, while the choices are technically rational, they are nevertheless bounded by practical circumstances. In contrast to maximizing behavior, Simon proposed the concept of satisfying behavior, because the human mind does not have the capability of choosing from hundreds of alternatives that might be placed in front of it. Problems connected with rationality and uncertainty seem to enter the designing and planning process at every turn in the means-ends configuration.

2.6 MEASUREMENT AND SCALING

2.6.1 Sources of Data

Before we can even make a preliminary examination of a problem situation, we need to have some basic data. Sources of data can generally be classified as obtrusive or unobtrusive. Obtrusive data-collection methods refer to procedures in which data are collected through some form of direct solicitation in which the researcher is directly or indirectly involved, such as face-to-face interviews, questionnaires, and many forms of observations. Unobtrusive methods of data collection are procedures that remove the investigator from the phenomenon being researched, such as through newspaper accounts, minutes of a meeting, and data banks. Huge data banks have been set up by the government and by industries devoted to the collection of information for quantifying basic social indicators, such as population, economic status of various sectors of society, and so on.

2.6.2 Measurement

Ultimately, all data must be subjected to some form of analysis. One way of doing this is through measurement. Measuring is a process that involves the assignment of numerals to objects or events according to rules, or the assignment of numerals to properties. Measurements can serve as models of events and relationships existing in the real world. The purpose of measurement is to represent the characteristics of observations by symbols that are related to each other in the same way that observed objects, events, or properties are related. However, it is important to remember that the real world is never exactly described by any mathematical measurement or model. All such descriptions are only approximations. Measurement is essentially required for control and prediction and is primarily a descriptive process. It allows one to qualify, order, and quantify certain events and, ultimately, to use the results as a basis for control and the prediction of actual performance.

While measurement in the scientific arena has progressed much further than one encounters in the behavioral sciences, there is a wide range in the degree of accuracy possible in both fields. Measurement, in any case, is subject to error, and an estimate of the magnitude of this error is necessary to determine whether or not the measurements obtained are usable in a practical situation. There are several possible sources of error in measurement by means of: the observer; the instrument used for measurement; the environment; and the object or situation being measured. Accuracy is the measure of the degree to which a given measurement may deviate from what actually exists. In our study of measurements, we strive to observe data by the means of refined techniques and to employ careful analysis so that we may be reasonably assured a degree of accuracy in our results. The principles underlying our methods are part of the study of mathematical probability and statistics described in Chapters 5 through 9.

Almost without exception we need to have solid and reliable data for our work. Fortunately, we appear to have an abundance of data, so much so that we are literally drowning in an overload of information. But data and information by themselves do not provide us with greater insight. Data really provide us with facts of the real world, which in turn must be sorted out to give us focused data for specific problems. Information is needed for problem solving. Our knowledge base that is built on our experience and technical know-how is what is ultimately needed for making rational judgments and decisions. Coupled with our knowledge base, we need to have intelligence, which is the ability to deal with novel situations arising in our day-to-day engineering practice (Khisty and Khisty, 1998).

2.6.3 Scales of Measurement

Scales of measurement are used almost daily by distinguishing among objects and responding appropriately to them. In engineering it is desirable to make distinctions of degree rather than quality. Four types of measurement scales are in general use (Patton and Sawicki, 1995):

1. *Nominal* data are classified into exhaustive, mutually exclusive but unordered categories. Individual items such as football players or car models are placed in a number of categories. For example, we can describe football players as numbers 29, 45, 63, and so forth, or car models can be sorted into Fords, Toyotas, or Buicks. We can also sort things by means of symbols; by assigning numbers it is possible to classify a given population into males and females with 1 = males and 2 = females. This scale contains the least information because it concerns only the allocation of a label or name. A special kind of nominal scale is the binary scale that can express a partial order such as yes/no. Also, if items or objects are compared as pairs, a binary scale can be used to denote whether an item is better/worse or larger/smaller than other items.

2. *Ordinal* data are classified into exhaustive, mutually exclusive, and ordered categories. Although it is qualitative, the ordinal scale contains the most information because the numbers in this scale give a rank order. For example, the ranking of students can be accomplished in several ways, but one particular example would be:

Students	A	B	C	D	E	F
Rank in math	4	3	1	6	2	5

3. *Interval* data are classified on a scale that permits them to be measured exactly using generally accepted units of measurements. This is truly a quantitative scale, but the origin is neither known nor defined although the intervals are the same. Some statistical tests, such as an F- or t-test can be applied. For example:

Room temperature (degrees F)	30	40	50	60

Properties such as height, temperature, time, income, intelligence quotient, and air pollution are examples measured with the help of interval scales. Again, thermal measurement is a good example. Here the choice of zero is arbitrary; and the equally spaced intervals do not have comparable magnitude (e.g., 30°F is not twice as hot as 15°F).

4. *Ratio* data are classified on a scale that permits them to be measured exactly using generally accepted units of measurements and include a nonarbitrary zero point. In this scale, the origin (e.g., zero) in a temperature scale is known or defined:

Room temperature (degrees C)   −40   −20   0   20   40   60

5. *Comparative* data are commonly used in evaluation forms, for example, on test scores:

Top   1%   5%   10%   25%   of the class

6. *Attitude* data are standardized techniques commonly used to develop scales or indices. Lickert scales provide a selection of five categories of response for each item: strongly approve, approve, undecided, disapprove, and strongly disapprove. These are ranked from 1 to 5, respectively, and the ranked items are then totaled to obtain an individual score for each item. Although Lickert scales are generally used for measuring attitudes, variations of the technique may be applied to behavioral measurement such as regularly, frequently, sometimes, rarely, or never. The semantic differential scaling technique is another method used in attitude surveys. It consists of two polar adjectives such as good-bad and positive-negative with ratings, depending on the breadth of response desired, but usually 0 to 5, 7, or 10. The respondent's score is calculated by summing the ratings of all evaluated items.

We can demonstrate how these different scales are used through a simple example. Let us consider three cities A, B, and C with populations of 50, 25, and 90 thousand respectively. Table 2.1 indicates how the scales are used and Table 2.2 provides some of their more important characteristics.

In summary, a nominal scale provides a set of names, categories, classes, groups, qualities, and general responses. An ordinal scale consists of objects in a particular order or rank. An interval scale is a set of numbers having a given, fixed interval size but no absolute zero point. Lastly, a ratio scale consists of a set of numbers having a given, fixed interval size and an absolute zero point.

We will be returning to the subject of measurement when we work on topics dealing with probability and statistics in Chapters 5 through 9.

TABLE 2.1 Examples of various scales and their applications

Scale			Example	Comments
Nominal	A	B	C	Cities
Ratio	50K	25K	90K	Population on 1/1/99
Binary	1	0	1	0 ≤ 40K, 1 ≥ 40K
Ordinal	2	1	3	Rank by population
Interval		A − B = 25K	A − C = 40K	Differences in population

TABLE 2.2 Scales of measurement

Scale	Comparisons	Typical example	Measure	Tests
Nominal	Identity	Male-female; Black-white	Mode	Chi-square
Ordinal space or order	Order	Social class; Graded quality	Median	Rank-order
Interval	Comparison of intervals	Temperature scale; Grade point average	Mean	t-test, ANOVA
Ratio	Comparison of absolute magnitudes	Units sold; Number of buyers	Geometric or harmonic mean	Same as interval

2.7 SYSTEM MODEL TYPES AND MODEL BUILDING

This section describes system models in general and provides some clues regarding how they are used in everyday applications. Some of the more familiar models used in engineering are discussed first and then the modeling approaches introduced in future chapters are described.

2.7.1 Model Types

There are so many varieties of models used in engineering and planning that any kind of description will always leave out some that should have been included. We start with the simplest and then proceed toward the more sophisticated ones (Flood and Carson, 1993).

Descriptive models are used to present detailed specifications of what needs to be accomplished for a project, plan, or design. It is a concise description of a problem situation that provides the framework for the problem solution.

Iconic, graphic, and diagrammatic models include toy trains, airplanes, dolls, teddy bears, and other three-dimensional representations of physical objects such as we all played with as children. Even photographs, graphs, maps, pictures, blueprints, and diagrams, which are, in some cases, two-dimensional representations of three-dimensional objects, fall under the category of iconic models. In more recent years, the use of *rich pictures* has become quite common. They help to capture a subjective interpretation of messy, complicated situations. Rich pictures represent and summarize findings and ideas, mostly in pictures and words. They have been successful in representing ideas and processes in soft system methodology. Examples of rich pictures are given in Chapter 12.

Analogue models represent a set of relationships through a different, but analogous, medium. A car's speedometer, for example, represents speed by analogous displacement of a gauge. Hydraulic engineers have used an analogue computer with great success to represent the behavior of a water distribution system for a city.

Mathematical or symbolic models are used without most of us knowing what they are. When we write down the equation of a straight line as $Y = AX + B$, we are making use of a mathematical model. Mathematical equations provide a repertory of ready-made representations. These representations are an effective means for predicting, communicating, reasoning, describing, and explaining, in concise language, our understanding of all kinds of relationships.

Decision models help us to make the best decisions in any given situation. Many of the mathematical and symbolic models described are used to achieve a particular objective or purpose. For instance, if we want to fly from New York to Istanbul, we can choose from at least fifteen options of reliable airlines that offer such service with some 30 different departure and arrival times between these two cities coupled with offers of stopovers in say, London, Frankfurt, or Paris. But one of our main considerations is to get the lowest fare with the best service. There are, of course, other considerations such as which airline offers the best frequent mileage option. How will we come to a decision about the choice of an airline? This book describes models that help us unravel the mysteries of decision making. An important role in such model building is to specify how the decision variables will affect the measures of effectiveness and the measures of cost. In my problem of choosing the best airline, I could specify my decision variables and match them with my decision objectives.

In addition to classifying models, there is another way of slicing the cake. This classification is through considering models as deterministic or probabilistic. *Deterministic models* are those in which all of the relevant data are assumed to be known with certainty. For instance, if we wanted to build a house and were provided ahead of time with the construction schedule for each of the 25 major tasks needed to complete the house, along with the manpower and budget, we would be able to tell with some degree of certainty when the house would be completed. This type of deterministic model is widely used adopting the Critical Path Method (CPM) described in Chapter 10, along with the use of other models belonging to this category, such as linear programming. Many of them are optimization models that help us to figure out how we could minimize the cost or maximize the output.

Probabilistic (or stochastic) *models*, on the other hand, assume that the decision maker will not know the values of some of the variables with any degree of certainty. For instance, if we did not know with any precision the exact time needed to complete some or all of the 25 tasks that went into constructing the house we want to build, we would have to modify the CPM model and resort to the somewhat more sophisticated method Program Evaluation and Review Technique (PERT) to assess the probable time of completion. PERT takes into consideration the probability of completing each task in a specified time by specifying limits on the time needed for each task. These limitations modify the original CPM into a more realistic model of the real world. Notice that the degree of certainty/uncertainty is crucial in working with probabilistic decision making. Models under this general category are widely used in engineering and planning and can be further classified according to:

- Decision under risk—each state of nature has a known objective probability.
- Decision under uncertainty—each action can result in two or more outcomes, but the probabilities of the states of nature are unknown.

- Decision under conflict—courses of action are taken against an opponent who is trying to maximize his/her goal. Decisions made under conflict fall under the area of inquiry known as *game theory*.

2.7.2 Models Used in Planning and Engineering

We have already noticed that there are differences between models in science and those in planning and engineering. We also noted that the models we encounter in science are closely connected with the natural world, while those we deal with in engineering are with the artificial (or man-made) world. Besides these two major distinctions there is yet a third difference. Science tends to generate models to understand the world *as it is*, whereas the goal of engineering is to construct models that will help us to deal with situations *as they ought to be*. This preoccupation of engineering with what ought to be is achieved through the process of analysis, synthesis, and evaluation.

To put in perspective what we have learned so far, let us consider the basic input/output model connected by means of a black-box. A simple depiction of this—an Inquiry System (IS)—is shown in Figure 2.8. An IS is a system of interrelated components for producing knowledge on a problem or issue (Mitroff and Linstone, 1993). The inputs are the basic entities that come into a system from the real world in the form of data and information and form the valid starting point of the inquiry. Next, the IS employs different kinds of operators (or models). These operators use the basic inputs to transform them into the final output of the system. One can use any number of operators to churn out the output, ranging from heuristic methods to highly sophisticated mathematics, to produce the output or result. But the most important component of this system is the guarantor or the component guaranteeing the operation of the entire system. It literally influences everything that goes on in the model.

Let us consider a simple problem: A clever community fundraiser collects donations on the basis that each successive contributor will contribute 1/20 of the total sum already collected to date. The fundraiser makes the first contribution of $100. How many contributors does she

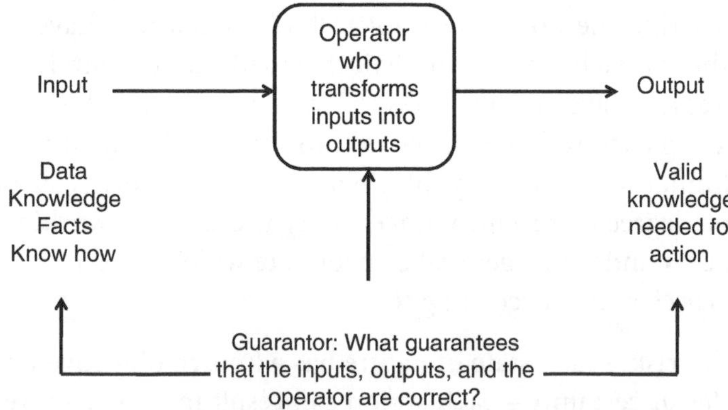

Figure 2.8 An inquiry system (adapted from Mitroff and Linstone, 1993)

need to cross the $1 million target? This is a bounded and well-structured problem without any ambiguity. We can consult a standard textbook on capital budgeting (or refer to Chapter 3) giving a formula for compound interest calculations and assign an output of $n = 189$. It would then be easy to see what the input is, what is in the black box, and also how we can be assured that the answer is right. The guarantor in this case is the formula given in the textbook on capital budgeting. We are assured that the answer is correct by working the problem backward from the output to the input via the black box. Not all problems are this simple. For example, if we were asked to define and solve the crime problem of downtown Chicago, we would have to get experts in a score of disciplines just to identify the problem, let alone solve it!

2.8 PROBLEM SOLVING THROUGH GROUP OR COMMITTEE ACTION

Engineers are often confronted with a complex problem where one can safely say that two heads, or more, are better than one. In such cases there are many advantages of using individual experts or consultants and groups of individuals who can work on a wide variety of problem situations using their collective store of information, intelligence, experience, and wisdom. As we have seen, the emergent properties of group decisions can result in spectacular outcomes. Two such methods are:

1. *Brainstorming*: It has been realized that ideas and inventions have often started from different points and yet reached similar results. While chance and accident often seem to play an important part in discovery, so too has brilliant and disciplined thinking as well as a persistence that amounts to obsession. There is a school of thought that believes that a vast accumulation of imperfect ideas and information is always lying dormant, lacking only some element to bring them to life. The advantage of collaborations of scientists, engineers, and lay people sitting together for creative thinking or for a brainstorming session was soon realized and has been used quite extensively worldwide.

Usually a group gathers together, examines a problem, and *fires off* the first ideas that come to mind. All criticism, both by word or implication, is ruled out. The wilder the ideas that emerge, the better. The objective is to get as many varied ideas as possible and to build on and improve the responses. The result of the session is a mass of ideas; most of them admittedly quite impractical, but a few worthy of closer attention. It is these few ideas that are likely to produce a *bunch of keys* for solving the problem that would not have been achieved by the exclusive use of conventional methods. There are no standard practices prescribed, but experience has indicated that great attention has to be paid to the selection and composition of the team, the presentation of the problem, an atmosphere of openness without the least hint of criticism, the recording process (which needs to be as unobtrusive as possible), and finally the assessment of ideas (Khisty, 1968).

2. *The Delphi method*: It is claimed that the Greek Oracle of Delphi was capable of coming up with forecasts of important future events based on the analysis of such things as the entrails of birds and animals. Of course, such forecasts could be interpreted in several ways, some of

them quite ridiculous and contradictory. But only the experts (who were gifted with special powers) were able to come to any kind of agreement about the interpretations.

There are many advantages to several people from various disciplines sitting together in a brainstorming session tackling a problem. However, the founders of the Delphi method at the Rand Corporation felt that a dynamic procedure for coming to a consensus regarding a problem solution through the use of questionnaires given to experts in the area of inquiry would be fruitful. The use of Delphi is not limited to consensus building, but rather on seeking relevant parameters for models of reality (what is) as well as on what ought to be or what is possible. In some ways the Delphi resembles a jury system with several characteristics worth noting. First, during a Delphi sequence there is total anonymity and responses are handled through well-worded questionnaires. Second, the committee chairperson extracts from the responses only that information relevant to the issue and then presents them to the committee. Third, a collective statistical summary is re-fed to the experts for another round of responses. Such feedback and further responses continue for as many rounds as it takes to reach some kind of consensus.

When viewed as a communication process for problem solving, there are few areas of human endeavor that do not qualify as candidates for the application of Delphi. For example, apart from the usual forecasting applications, Delphi has been widely used for evaluating budget allocation for projects, exploring various urban and regional planning options, putting together the structure of a complicated economic model, exploring social and economic goals and prioritizing them, and the list goes on (Daellenbach and Flood, 2002). We now examine the Delphi process through a simple real-world example.

Example 2.1 Four construction management experts were asked to estimate the duration for completing a complex reconstruction repair to a hydroelectric plant. Their responses (after thorough, detailed calculations) are provided. Each expert was given the same information and data. Their individual responses were confidential. The only feedback provided after each round was the mean and standard deviation:

	Rounds			
Expert	**I**	**II**	**III**	**IV**
A	12	11	9	9
B	8	8	9	9
C	7	9	10	9
D	5	7	8	9
Mean	8.00	8.75	9.00	9.00
Standard deviation	2.55	1.48	0.71	0.00

Continues

Example 2.1: *Continued*

In four rounds, the convergence of the responses from the four experts boiled down to 9 mo. It was theoretically not necessary that we wanted perfect conformity of answers from these experts. For example, had the committee chairperson specified that the convergence would be acceptable with a standard deviation of 1.00 and the mean was between 8 and 10, then it is possible that the exercise could have ended after the third round. It is the decision maker who specifies whether the collective result is acceptable and under what conditions.

SUMMARY

This chapter provides an overview of the problem solving and design process with special emphasis on their role as tools for engineers and managers. One has to be equipped to deal with *tame* problems and *wicked* problems on a day-to-day basis. For this reason, the need to classify models and understand what they are capable of doing is all important. This chapter has also dealt with data and information and ways to handle them in a preliminary way. Lastly, a brief description of problem solving using collective thinking has been included. This will prove useful in tackling complex problems requiring group opinion.

REFERENCES

Ackoff, R. (1999). *Ackoff's Best: His Classic Writings on Management*, 115-116, New York, NY: John Wiley & Sons.

Alexander, C. (1966). "A City Is Not a Tree," *Ekistics*, 139:344-348.

Banathy, B. H. (1996). *Designing Social Systems in a Changing World*, New York, NY: Plenum Press.

Coyne, R. D. et. al., (1990). *Knowledge-Based Design Systems*, New York, NY: Addison-Wesley Publishing.

Daellenbach, H. G. and R. L. Flood (2002). *The Informed Student Guide to Management Science*, London, UK: Thomson Publishers.

Flood, R. L. and E. R. Carson (1993). *Dealing with Complexity*, 2nd ed., New York, NY: Plenum Press.

Hutchins, C. L. (1996). *Systemic Thinking: Solving Complex Problems*, Aurora, CO: Professional Development Systems.

Khisty, C. J. (1968). "Brainstorming: A New Technique in solving Problems in Industry," *Engineering News* 16:2, 677-679.

———. (1993). "Citizen Participation Using a Soft Systems Perspective," in *Transportation Research Record 1400*, National Academy of Science, Washington, DC, 53-57.

Khisty, C. J. and L. L. Khisty (1998). "Using Information Systems for Mathematical Problem-solving: A New Philosophical Perspective," in *Proceedings of the Psychology of Mathematics Education*, University of Stellenbosch, Stellenbosch, South Africa, 104-111.

Miller, G. A. (1956). "The Magic Number Seven, Plus or Minus Two: Some Limits on Our Capacity for Processing Information," *Psychological Review* 63:2, 81-97.

Mitroff, I. I. and H. A. Linstone (1993). *Unbounded Mind*, New York, NY: Oxford University Press.
Patton, C. V. and D. S. Sawicki (1993). *Basic Methods of Policy Analysis and Planning,* 2nd ed., Englewood Cliffs, NJ: Prentice-Hall.
Rittel, H. and M. Weber (1973). "Dilemmas in General Theory of Planning," *Policy Science*, 4, 155-169.
Rowe, P. (1987). *Design Thinking.* Cambridge, MA: MIT Press.
Simon, H. A. (1981). *The Sciences of the Artificial*, 2nd ed., Cambridge, MA: MIT Press.

EXERCISES

1. What is your understanding of the relationship between information, data, models, and design? Illustrate this with the help of a rich picture.
2. Set up a 3 × 3 magic square and fill the nine cells with the numbers 1 through 9 (without repeating them), such that the rows, columns, and diagonals of the matrix add to 15.
3. Recall the names of your friends and set up a semi-lattice diagram of their activities—don't forget to add yourself to the number—similar to the example illustrated in the text. Draw a Venn diagram to represent the semi-lattice.
4. If you wanted to choose between three or four summer jobs you were offered, indicate by way of a tree or semi-lattice how you would make a final decision.
5. Using the feedback form of a control system, identify the appropriate parts of the following systems:
   (a) Driving an automobile along a freeway, adhering to a speed limit of 65 mph
   (b) Controlling an angry mob that is going out of control
   (c) Washing clothes in an automatic washing machine
6. Classify the following problem situations according to the means/ends matrix:
   (a) Fixing a flat tire on the freeway
   (b) Sitting in a committee meeting where one of the topics on the agenda is to debate the distribution of money for various social activities in your neighborhood
   (c) Sitting in a large committee of experts dealing with the problem of the homeless in downtown
7. How would you categorize the following situation according to the various measurement scales mentioned in the text?
   (a) Technical and economic factors of a city
   (b) Political and social factors of a city
   (c) The area of land in each square mile of a city having a slope varying from zero to 20°
   (d) The number of veh/mi traveled in Chicago in 1998

(e) Military ranks in the U.S. Army, Navy, and Air Force
(f) Ratings of cars in terms of gasoline consumption
(g) Number of people using the park by day of the week
(h) The popularity of the mayor of your city
(i) The dollar amount by which the five bids for 10 mi of a highway exceed the estimated amounts

8. Think of a truly wicked problem you have encountered or have read about in the newspapers. Briefly describe the gist of the problem and then apply Rittel and Weber's characteristics to check whether this is truly a wicked problem. (Hint: The crime problem in the inner city could be an example, but there may be better examples that you probably know about.)

9. Suppose your city's bus service has just received funding for expanding the bus lines by 25 percent. How would you set up a simple optimization model to help the planners choose the new routes and/or to modify existing routes? What minimum data would be needed upfront to set the priorities?

10. Suppose you are a student representative serving on the university parking committee with the task of finding the parking demand for students, staff, and faculty for 2015. How would you organize a brain storming session to resolve this issue or a Delphi exercise using four to five experts?

11. Think of a problem you were confronted with and analyze it to demonstrate whether you absolved, solved, resolved, or dissolved it. If you went through more than one strategy, give reasons why you did so.

12. A student in an undergraduate engineering program discovers that she is weak in mathematics. Initially she ignores the poor midterm grade she gets in math. Two weeks later she decides to attend a crash refresher course to remedy the weakness but notices that the extra time she is spending with this course is hurting her ability to do well in other courses. Finally she consults her academic adviser who suggests she reschedule her time to optimize her ability to gain good grades in all her courses. She does much better in the following semester. Analyze and comment on her strategies.

13. In the last decade hospitals have become more automated, employing modern instrumentation and computers to do most of the mechanical and tedious work. What are the advantages and disadvantages of this new trend? Are there obvious dangers of depending too much on modern technology? How would you eliminate these dangers?

This book has free material available for download from the Web Added Value™ resource center at *www.jrosspub.com*

Basic Engineering Economics and Evaluation

3

3.1 INTRODUCTION

This chapter is divided into two parts. The first part (Sections 1 through 8) introduces the basic concepts of engineering economics. This consists of a standard set of procedures for determining the relative economic value of alternative capital investments. The second part (Sections 9 through 19) extends these basic concepts and shows how engineers choose the best alternative from a set of several feasible alternatives.

Engineering economics is a branch of economics used by engineers to determine the relative economic value of a set of alternative capital investments for engineering projects. Project appraisal and cost-benefit analysis require knowledge of engineering economics. Engineers and planners are concerned with money whether its use or exchange is in the private or public sector. In the language of economics, articles produced, sold, or exchanged are known as goods. At least four sectors of production are necessary to produce a good: (1) labor, (2) land, (3) capital, and (4) enterprise. Capital includes the money, machinery, tools, and materials required to produce a good. The opportunity cost of capital is measured by the interest rate, and the interest on capital is the premium paid or received for the use of money. Here, opportunity cost represents the cost of an opportunity that is foregone because resources are used for a selected alternative and, therefore, cannot be used for other purposes. The interest rate that relates a sum of money at some date to its equivalent today is called the discount rate.

3.2 NOTATIONS

The following symbols and definitions are used unless otherwise stated:

P = Principal; a sum of money invested in the initial year, or a present sum of money
i = Interest rate per unit of time expressed as a decimal
n = Time; the number of units of time over which interest accumulates
I = Simple interest; the total sum of money paid for the use of the money at simple interest
F = Compound amount; a sum of money at the end of n units of time at interest i, made up of the principal plus the interest payable

A = Uniform series end-of-period payment or receipt that extends for n periods
S = Salvage or resale value at the end of n years

3.3 SIMPLE INTEREST

When one invests money at a *simple interest* rate of i for a period of n years, the simple interest bears the following relationship:

$$\text{Simple interest } I = P \times i \times n \tag{3.1}$$

The sum, I, will be added to the original sum at the end of the specified period, but it will remain constant for each period unless the interest changes.

Example 3.1 An amount of $2500 is deposited in a bank offering 5% simple interest per annum. What is the interest at the end of the first year and subsequent years?

Solution

$$I = (2500)(0.05)(1) = \$125$$

The interest for the second and subsequent years will also be $125. In other words, at the end of the first year, the amount will be $2625; for the second and subsequent years, it will again be $2625.

3.4 COMPOUND INTEREST

When interest is paid on the original investment as well as on the interest earned, the process is known as *compound interest*. If an initial sum, P, is invested at an interest rate, i, over a period of n years:

$$F = P(1 + i)^n \tag{3.2}$$

or

$$P = F\frac{1}{(1 + i)^n} \tag{3.3}$$

If interest i is compounded m times per period n:

$$F = P\left(1 + \frac{i}{m}\right)^{n \times m} \tag{3.4}$$

As m approaches infinity (∞), Equation 3.4 can be written as:

$$F = Pe^{i \times n} \tag{3.5}$$

or

$$P = Fe^{-i \times n} \quad (3.6)$$

Equations 3.5 and 3.6 are used when resorting to *continuous compounding*, a method frequently used. Factors $(1 + i)^n$ are $e^{i \times n}$ and called compound amount factors (CAF), and $(1 + i)^{-n}$ and $e^{-i \times n}$ are called present worth factors (PWFs) for a single payment.

Example 3.2 What is the amount of \$1000 compounded at (a) 6% per annum, (b) at 6% per every quarter, (c) at 6% per annum compounded continuously for 10 yr?

Solution

(a) Here, $i = 0.06$, $n = 10$, and $m = 1$. Equation 3.2 becomes:

$$F = 1000(1 + 0.06)^{10} = 1000(1.79085) = \$1790.85$$

(b) Here, $i = 0.06$, $n = 10$, and $m = 4$. Equation 3.4 becomes:

$$F = 1000(1 + 0.06/4)^{4\times10} = 1000(1.81402) = \$1814.02$$

(c) Here, $i = 0.06$ and $n = 10$. Equation 3.5 becomes:

$$F = 1000e^{0.06\times10} = 1000(1.82212) = \$1822.122$$

Example 3.3 What is the *effective* interest rate when a sum of money is invested at a nominal interest rate of 10% per annum, compounded annually, semiannually, quarterly, monthly, daily, and continuously?

Solution Assume the sum to be \$1 for a period of 1 yr. Then the sum at the end of 1 yr compounded:

Annually	$= 1(1 + 0.1)^{1\times1}$	= 1.10, therefore, the interest rate	= 10%
Semiannually	$= 1(1 + 0.1/2)^{1\times2}$	= 1.1025, therefore, the interest rate	= 10.25%
Quarterly	$= 1(1 + 0.1/4)^{1\times4}$	= 1.10381, therefore, the interest rate	= 10.381%
Monthly	$= 1(1 + 0.1/12)^{1\times12} = 1(1 + 0.1/2)^{1\times12}$	= 1.10471, therefore, the interest rate	= 10.471%
Daily	$= 1(1 + 0.1/365)^{1\times365}$	= 1.10516, therefore, the interest rate	= 10.516%
Continuously	$= 1 \times e^{0.1\times1}$	= 1.10517, therefore, the interest rate	= 10.517%

Discussion

The comparison illustrates the difference between nominal and effective interest rates. The effect of more frequent compounding is that the actual interest rate per year (or effective interest rate per year) is higher than the nominal interest rate. For example, continuous compounding results in an effective interest rate of 10.517% compared to a 10% nominal interest rate.

3.5 UNIFORM SERIES OF PAYMENTS

If instead of a single amount there is a uniform cash flow of costs or revenue at a constant rate, the following uniform series of payment formulas are commonly used.

3.5.1 Compound Amount Factor (CAF)

The use of the CAF helps to answer the question: What future sum (*F*) will accumulate, assuming a given annual amount of money (*A*) is invested at an interest *i* for *n* years?

$$F = A\left[\frac{(1+i)^n - 1}{i}\right] = A\left(\frac{x-1}{i}\right) \tag{3.7}$$

where $\left(\frac{x-1}{i}\right)$ is the uniform series CAF.

3.5.2 Sinking Fund Factor (SFF)

The Sinking Fund Factor (SFF) indicates how much money (*A*) should be invested at the end of each year at interest rate *i* for *n* years to accumulate a stipulated future sum of money (*F*). The SFF is the reciprocal of the CAF.

$$A = F\left[\frac{i}{(1+i)^n - 1}\right] = F\left(\frac{i}{x-1}\right) \tag{3.8}$$

where $\left(\frac{i}{x-1}\right)$ is the uniform series SSF.

3.5.3 Present Worth Factor (PWF)

The PWF tells us what amount (*P*) should be invested today at interest *i* to recover a sum of *A* at the end of each year for *n* years:

$$P = A\left[\frac{(1+i)^n - 1}{(1+i)^n i}\right] = A\left(\frac{x-1}{xi}\right) \tag{3.9}$$

where $\left(\frac{x-1}{xi}\right)$ is the uniform series PWF.

3.5.4 Capital Recovery Factor (CRF)

The Capital Recovery Factor (CRF) answers the question: If an amount of money (*P*) is invested today at an interest *i*, what sum (*A*) can be secured at the end of each year for *n* years, such that the initial investment (*P*) is just depleted?

$$A = P\left[\frac{i(1+i)^n}{(1+i)^n - 1}\right] = P\left(\frac{xi}{x-1}\right) \tag{3.10}$$

CRF is the reciprocal of PWF. Another way of considering CRF is: If a sum of *P* is borrowed today at interest rate *i*, how much *A* must be paid at the end of each period to retire the loan in *n* periods?

Example 3.4 If the interest rate is 5% per annum, what sum would accumulate after 6 yr if $1000 were invested at the end of each year for 6 yr?

Solution

$$F = A\left[\frac{(1+i)^n - 1}{i}\right] = 1000\left(\frac{(1.05)^6 - 1}{0.06}\right) = 1000 \times (6.80191) = \$6801.91$$

Example 3.5 A realtor buys a house for $200,000 and spends $1,000 per yr on maintenance for the next 8 yr. For how much should she sell the property to make a profit of $40,000? Assume $i = 12\%$ per annum.

Solution Future value of $200,000 in 8 yr:

$$F = P(1+i)^n = 200{,}000(1.12)^8 = 200{,}000(2.475963) = \$495{,}192.63$$

Future value of annual maintenance:

$$F = A\left[\frac{(1+i)^n - 1}{i}\right] = 1000\left(\frac{(1.12)^8 - 1}{0.12}\right) = 1000(12.29969) = \$12{,}299.69$$

Minimum selling price after 8 yr with a profit of $40,000:

$$= 495{,}192.63 + 12{,}299.69 + 40{,}000 = \$547{,}492.32$$

Example 3.6 The city transit system needs to set up a sinking fund for 10 buses, each costing $100,000, for timely replacements. The life of the buses is 7 yr and the interest rate is 6% per annum.

Solution

$$A = F\left[\frac{i}{(1+i)^n - 1}\right] = 100{,}000 \times 10\left(\frac{0.06}{(1.06)^7 - 1}\right) = 1{,}000{,}000(0.119135018) = \$119{,}135.02 \text{ per yr}$$

Example 3.7 A contractor wants to set up a uniform end-of-period payment to repay a debt of $1,000,000 in 3 yr, making payments every month. Interest rate = 12% per annum. What is the CRF?

Solution

$$i = \frac{0.12}{12} = 0.01 \text{ per mo and } n = 3 \times 12 = 36 \text{ mo}$$

$$A = P\left[\frac{i(1+i)^n}{(1+i)^n - 1}\right] = 1{,}000{,}000\left(\frac{0.01 \times (1.01)^{36}}{(1.01)^{36} - 1}\right) = 1{,}000{,}000(0.033214309) = \$33{,}214.31$$

3.6 UNIFORM GRADIENT SERIES

When a uniform series of a number of payments are increasing each year by a similar amount, it is possible to convert them to an equivalent uniform gradient. If the uniform increment at the end of each year is G, then all the compound amounts can be totaled to F:

$$F = \frac{G}{i}\left[\frac{(1+i)^n - 1}{i}\right] - \frac{nG}{i} \tag{3.11}$$

To convert this sum into an equivalent uniform period payment over n periods, it is necessary to substitute the sum for F in the SFF, giving:

$$A = \frac{G}{i} - \frac{nG}{i}\left[\frac{i}{(1+i)^n - 1}\right] \tag{3.12}$$

Example 3.8 The maintenance on a transit bus system amounts to \$20,000 by the end of the first year, increasing \$5000/yr for the subsequent 5 yr. What is the equivalent uniform series cost each year with interest at 5% per annum?

Solution

$$A = \frac{G}{i} - \frac{nG}{i}\left[\frac{i}{(1+i)^n - 1}\right] = \frac{5{,}000}{0.05} - \frac{5 \times 5{,}000}{0.05}\left(\frac{0.05}{(1.05)^5 - 1}\right)$$

$$= \frac{5{,}000}{0.05} - \frac{5 \times 5{,}000}{0.05}(0.1809748) = 1000{,}000 - 90{,}487.40 = \$9{,}512.60$$

Therefore, the uniform series equivalent annual cost is:

$$20{,}000 + 9{,}512.60 = \$29{,}512.60$$

for each of the 5 yr.

3.7 DISCRETE COMPOUND INTEREST FACTORS

The various factors are summarized in Table 3.1 and their relationships are shown in Figure 3.1.

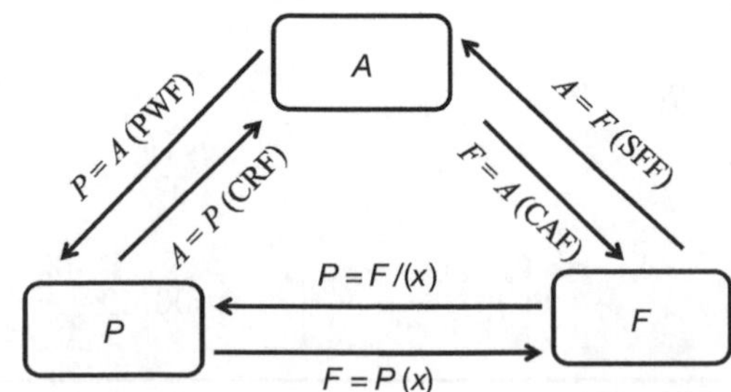

Figure 3.1 Relationships among factors

Table 3.1 Summary of factors

Symbol	Factor	Equation	Symbolic form	Find	Given
A	Single payment:				
	1. Compound amount	$F = P[x]$	$(F/P,i,n)$	F	P
	2. Present worth	$P = F[1/x]$	$(P/F,i,n)$	P	F
B	Uniform series:				
	3. Compound amount	$F = A[(x - 1)/i]$	$(F/A,i,n)$	F	A
	4. Sinking fund	$A = F[i/(x - 1)]$	$(A/F,i,n)$	A	F
	5. Present worth	$P = A[(x - 1)/ix]$	$(P/A,i,n)$	P	A
	6. Capital recovery	$A = P[ix/(x - 1)]$	$(A/P,i,n)$	A	P
C	Arithmetic gradient:				
	7. Uniform series equivalent	$A = G\{1/i - (n/i)[i/(x - 1)]\}$	$(A/G,i,n)$	A	G

Notes: The factor $(1 + i)^n$ is known as the single payment CAF and is equal to x. In a similar manner, the expression in each equation within square brackets is the factor corresponding to the description as shown in the table; for example, $[ix/(x - 1)]$ = CRF.

3.8 UNIFORM CONTINUOUS CASH FLOW AND CAPITALIZED COST

The present worth of a uniform continuous cash flow is the amount P invested at an interest rate i that will produce a cash flow of a per period for n periods. Consider a small time interval from t to $t + \delta t$ during which the flow will be $a\delta t$ and the present worth of this element is

$$(a)(\delta t)(e^{-it}).$$

The present worth of the entire flow is:

$$P = \int_0^t ae^{-it}\,dt = -\frac{a}{i}(e^{-it})\,\Big|_0^t = a\left(\frac{1 - e^{it}}{i}\right) = an\left(\frac{1 - e^{-it}}{in}\right) = anf_w \tag{3.13}$$

The term $f_w = (1 - e^{-it})$ is the PWF for uniform flow; it is useful in making comparisons of equipment and services with different service lives by reducing them to their present values.

Capitalized Cost (CC) is the present value of a series of services or equipment scheduled to be repeated every n periods to perpetuity. If P is the amount that must be paid every n period, where n is the life of the equipment, then the CC is:

$$K = P + Pe^{-in} + Pe^{-2in} + Pe^{-3in} + \ldots +$$

Summing this geometric series gives:

$$K = P\left(\frac{1}{1 - e^{-in}}\right) = P\left(\frac{e^{in}}{e^{in} - 1}\right) = P\left(\frac{1}{inf_w}\right) \tag{3.14}$$

Example 3.9 The Department of Transportation is faced with a problem. Should it rent trucks for \$480 a mo each or buy the trucks for \$33,000 each, assuming the useful life of a truck is 10 yr and $i = 14\%$?

Solution

$$\text{Present value of rented truck} = an\left(\frac{1 - e^{-in}}{in}\right)$$

where $a = 480 \times 12 = \$5{,}760$ per yr.

Therefore:

$$p = 5{,}760 \times 10\left(\frac{1 - e^{-0.14\times10}}{0.14 \times 10}\right) = 5{,}760 \times 10(0.538145) = \$30{,}997.15 < \$33{,}000$$

It is more advantageous to rent trucks.

Example 3.10 A comparison between two types of bridges has to be made using the following details:

Type	Steel	Wood
Initial cost	\$400,000	\$250,000
Paint and maintenance	\$40,000/10 yr	\$25,000/2 yr
Life	40 yr	20 yr
Interest rate	9%	9%

In the long run, which type of bridge is cheaper?

Solution

Using Equation 3.14, the CC for each bridge type is computed considering the initial cost and the periodic paint and maintenance costs. The results are summarized in the following table:

Solution	Steel bridge	Wooden bridge
First cost	$K_1 = P[1/(1 - e^{-in})]$ $= 400{,}000[1/(1 - e^{-0.09 \times 40})]$ $= 400{,}000(1.02809)$ $= 411{,}236.51$	$K_1 = 250{,}000[1/(1 - e^{-0.09 \times 20})]$ $=250{,}000(1.198031)$ $= 299{,}508.41$
Paint and maintenance	$K_2 = 40{,}000[1/(1 - e^{-0.09 \times 10})]$ $= 400{,}000(1.69512)$ $= 67{,}404.71$	$K_2 = 25{,}000[1/(1 - e^{-0.09 \times 2})]$ $= 25{,}000(6.070547)$ $= 151{,}763.68$
Total	\$478,641	\$451,272

Based on the calculations shown, the wooden bridge is cheaper.

3.9 EVALUATION

The term *evaluation* is used in planning and engineering to refer to the merits of alternative proposals. The essence of evaluation is the assessments of the comparative merits of different courses of action. One of these actions may include the *do nothing* alternative.

Engineering and planning alternatives are concerned with problems of choice between mutually exclusive plans or projects (e.g., the examination of alternative investment proposals to alleviate severe traffic congestion in the city center). Evaluation methods may be applied to the problem of choice between sets of independent plans or projects as well as mutually exclusive ones.

In general, engineering projects in the private sector are built for motives of profit, while in the public sector the purpose is, ultimately, for the raising of living standards and for social benefit, or possibly for political motives. Political motives are not amenable to economic analysis or justification. A value can, however, be put on social benefits and losses. Even if a decision determines as accurately as possible what consequences will flow from each alternative action, preferences still need to be formulated, particularly when a variety of objectives are involved. In such a case, the crux of the problem is that it is impossible to optimize in all directions at the same time. It is this ubiquitous problem, known as the *multiattribute problem*, that makes it difficult to determine preferences among outcomes.

The bottom line seems to be that whereas the basic concepts of evaluation are relatively simple, the actual process of evaluating projects or plans is complex and riddled with controversy. Although engineers and planners try to clarify difficult issues that should be considered by decision makers, it is not uncommon for decision makers to override the results of the analysis. Therefore, it should be clearly borne in mind that analysis and evaluation are primarily performed by engineers, economists, and planners, whereas the choice of an alternative is made by decision makers.

3.10 FEASIBILITY ISSUES

The solution to a problem should invariably be checked to see whether it is suitable for the situation, acceptable to the decision makers and the public, and one that can be eventually implemented, given the investment. Naturally, at some stage or another, a proposed improvement will be required to satisfy engineering, economic, financial, political, environmental, ethical, and social feasibility. For this kind of comprehensive investigation, it is necessary, in complex situations, to obtain the services of economists and sociologists to do justice to the analysis.

Engineers and planners are familiar with the large assortment of analytical tools used in evaluating alternatives. However, in most cases, decision makers demand far more information and justification on the consequences of different alternatives than what is contained in the analysis developed by technicians. Much of this additional information may be qualitative rather than quantitative.

3.11 EVALUATION ISSUES

Most evaluation methodologies utilize some form of rating system. Evaluators calculate an index or score, indicating how the welfare of society (or the quality of life) would be affected if a particular alternative was implemented. In effect, they convert all impacts into commensurate units so that they can be added and compared.

Cost-benefit analysis was one of the first evaluation methods, used extensively, that included a systematic rating procedure and has, in the last 25 years, been used widely to evaluate all types of public actions. Its rating procedure, although complex, follows the simple arithmetic of placing a monetary value on each impact. The monetary ratings are then aggregated to determine whether benefits exceed costs. The fact that the cost-benefit method has a number of weaknesses led to the development of several other evaluation techniques.

Some evaluation methodologies have established complex procedures for quantifying social welfare ratings. Most of them appear scientific and objective. It does not really matter how sophisticated these methodologies are because, ultimately, the ratings constitute value judgments. Each method has its own weakness. Indeed, each method could possibly lead to a different conclusion.

Several questions naturally stem from this broad description of evaluation methodologies:

- How should evaluation techniques be selected?
- Are economic values sufficient for weighing the costs and benefits of a contemplated public action?
- How are intangible impacts quantified?
- How can equity issues be considered in evaluation?
- How should the concept of time be treated in evaluations?
- How can discount rates be estimated?

Some of these issues are discussed in this chapter.

3.12 THE EVALUATION PROCESS

The scope of the evaluation process is outlined in Figure 3.2. The evaluation process focuses on the development and selection among several alternatives and is completed in three parts:

1. An evaluation work plan
2. A comprehensive analysis
3. A report containing the results of this analysis

The level of detail and rigor associated with each of these activities will naturally vary from project to project. Figure 3.3 shows a conceptual framework for the evaluation of improvements. It identifies the key inputs to the evaluation process, the potential impacts associated

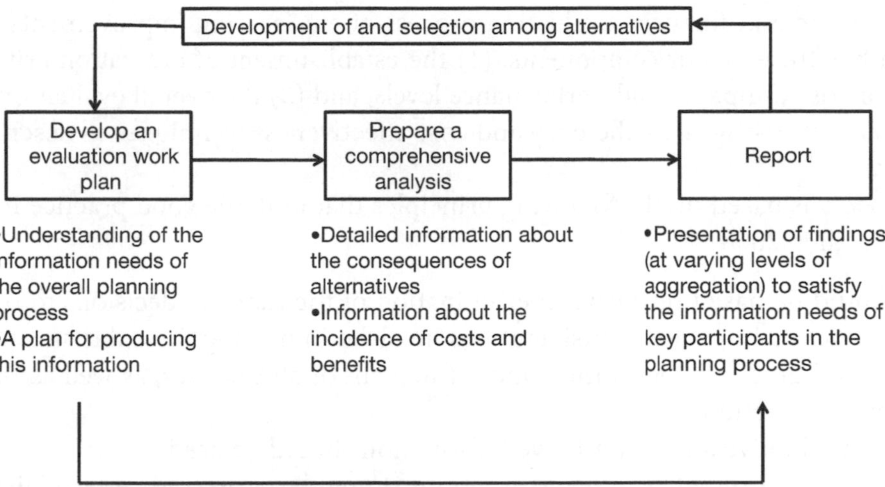

Figure 3.2 Scope of the evaluation process

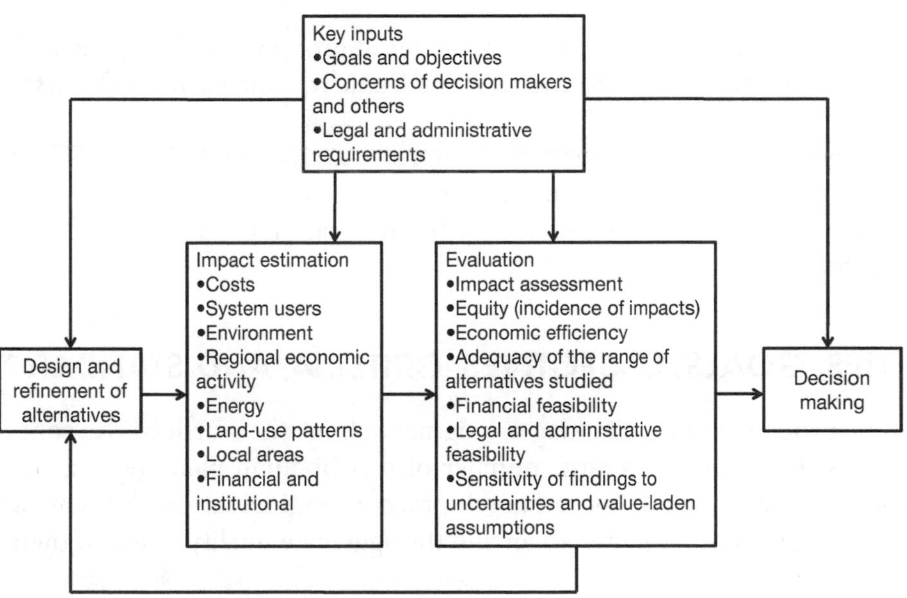

Figure 3.3 Conceptual framework for evaluation

with improvements, and the considerations in evaluating the overall merit of improvements. Figure 3.3 also illustrates the interrelationships among the activities constituting the evaluation process, including the iterative nature of evaluation associated with large-scale improvements that often require successive revision and refinement as more critical information is forthcoming about their potential impacts and feasibility.

The economic and financial evaluation process of engineering improvements reflected in this section has three major components: (1) the establishment of evaluation criteria, (2) the estimation of costs, impacts, and performance levels, and (3) the overall evaluation of alternative improvements using cost-efficiency and cost-effectiveness techniques as described later in this chapter.

This process is based on the following principles that underlie good practice in evaluation (Meyer and Miller, 2000):

- It should be based on a careful examination of the range of decisions to be made and the issues or important considerations involved in making these decisions
- It should guide the generation and refinement of alternatives, as well as the choice of a course of action
- Both qualitative and quantitative information should be used
- Uncertainties or value-laden assumptions should be addressed and explained
- All of the important consequences of choosing a course of action should be included
- Evaluation should relate the consequences of alternatives to the goals and objectives established
- The impact of each alternative on each interest group should be spelled out clearly
- Evaluation should be sensitive to the time frame in which project impacts are likely to occur
- The implementation requirements of each alternative should be carefully documented to eliminate the chance of including a *fatal fault* alternative
- Decision makers should be provided with information in a readily understandable and useful form

3.13 VALUES, GOALS, OBJECTIVES, CRITERIA, AND STANDARDS

The evaluation process involves making a judgment about the worth of the consequences of alternative plans. For example, a large number of metropolitan planning organizations have expressed the view that the planning and development of engineering facilities must be directed toward raising urban standards and enhancing the aggregate quality of life of their communities.

It may be appropriate at this stage to present a brief description of values, goals, objectives, criteria, and standards. A clear statement of their meanings follows:

A set of irreducibles forming the basic desires and drives that govern human behavior are called *values*. Because values are commonly shared by groups of people, it is possible to speak of societal or cultural values. Four of the basic values of society are the (1) desire to survive, (2) need to belong, (3) need for order, and (4) need for security.

A *goal* is an idealized end state, and although they may not be specific enough to be truly attainable, goals provide the directions in which society may like to move.

An *objective* is a specific statement that is an outgrowth of a goal. Objectives are generally attainable and are stated so that it is possible to measure the extent to which they have been attained.

Criteria result directly from the fact that the levels of attainment of objectives are measurable. In a sense, criteria are the working or operational definitions attached to objectives. They are measures, tests, or indicators of the degree to which objectives are attained. Criteria affect the quantitative characteristics of objectives and add the precision to objectives that differentiate them from goals. One particular type of criterion is a *standard*—a fixed objective [e.g., the lowest (or highest) level of performance acceptable].

The preceding chain is shown in Figure 3.4. A complex structure connects societal values to particular criteria and standards. It can be noted that there may be a high score connected with one criterion as compared to another, in which case tradeoffs between criteria may be considered. Of course, decision makers will find it comparatively easy to deal with goals and objectives with a minimum amount of internal conflicts.

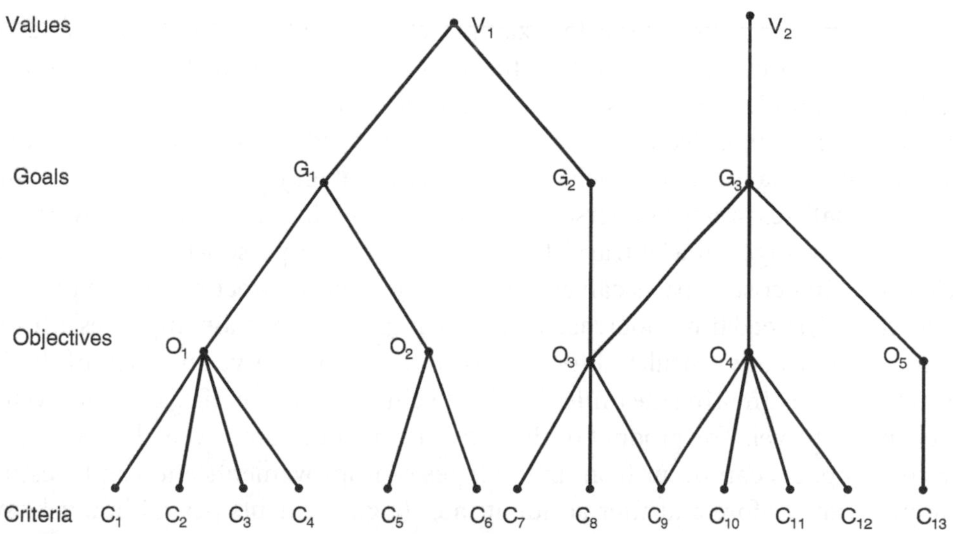

Figure 3.4 Hierarchical interrelationships among values, goals, objectives, and criteria

3.14 ESTIMATION OF COSTS, IMPACTS, AND PERFORMANCE LEVELS

3.14.1 Capital, Operating, and Maintenance Costs

Costs play a central role in economic and financial evaluation. When combined with other dollar-valued impacts and compared to other effects, they are helpful in assessing the economic efficiency (i.e., investment worthiness) of an improvement. Detailed knowledge of these costs over the life of the improvement helps to identify and assign funding sources, thus ensuring that sufficient financial resources are available to implement the improvement.

Existing procedures for estimating costs range from the application of relatively gross average cost measures (e.g., cost per route-mile for two-lane highways) to procedures that use detailed, site-specific information. Achieving a high degree of precision in estimating costs for evaluation of alternatives is important only when similar alternatives are being compared. In such cases, a more detailed cost accounting framework might be required to determine differences between alternatives.

Conversely, when alternatives differ substantially in scale or character, it is more important to establish credible upper and lower limits for the cost of each than to make precise cost estimates based on a single set of detailed assumptions for each alternative.

3.14.2 User Costs

Engineering system user costs include, for example, changes in travel-time savings, vehicle operating costs, safety, and qualitative changes in the level of service provided by the improvement.

Let's take an example from transportation. Travel-time savings will accrue from most improvements as congestion is decreased, speed limits increased, or as more convenient routes are provided. Travel-time savings for travelers who take advantage of the improvement and receive the benefit are usually expressed in person-hours of time savings and can be converted to a dollar value using an average value of travel-time (i.e., dollars per person-hour).

Vehicle operating cost impacts can accrue if shorter, more direct routes are provided, congestion is decreased, speed limits increased, or if the highway condition improves. Motor vehicle operating costs are usually calculated for the do-nothing alternative and each of the improvement alternatives by estimating the unit cost of the vehicle operation (i.e., cents per veh/mi) and the total veh/mi of travel. The product of these two is the total motor vehicle operating cost.

Safety cost impacts can occur from many types of improvements and can be estimated by applying appropriate before and after accident rates (e.g., accidents per million veh/mi) to the volume of traffic expected to use the proposed improvement. Separate estimates may be developed for fatal injury, nonfatal injury, and property-damage-only accidents if the potential impact is expected to be large or if the project's safety impact is controversial. Unit costs (e.g., dollars per accident) can then be applied to derive dollar values for accident costs.

Qualitative improvements in the level of service provided can reflect any number of effects, including better ride quality from improved pavements, improved transit reliability, or impacts that are difficult to quantify or to assign dollar costs.

3.14.3 Impacts

A wide range of possible impacts are associated with engineering improvements. These impacts include environmental as well as economic impacts and are more often described in qualitative or nonmonetary terms rather than in terms of dollars. Such impacts might include air-pollutant emissions from automobiles, noise impacts, water quality impacts, energy consumption, effects on accessibility or land-use patterns, and construction disruption (i.e., for projects that entail substantial amounts of construction activity). To the extent that these impacts represent a significant effect of the improvement, they must be identified and accounted for in the evaluation and also factored into the overall assessment.

3.14.4 Performance Levels

In many cases, formally established criteria for expected economic and financial accomplishments can serve as a valuable evaluation tool. For example, performance levels are frequently incorporated in the stated goals and objectives for a specific facility to narrow the scope of evaluation and set limits on expected results. Such criteria are also used to monitor implementation, establish budgets, and reallocate financial resources on a periodic basis.

3.15 EVALUATION OF ALTERNATIVES

The evaluation of alternatives is related to the size, complexity, and number of alternative improvements to be considered. However, answers to the following types of questions, relying on the techniques described in this chapter, should be sought for any evaluation effort:

- Are there other feasible improvements that are not considered and that might compare favorably to those alternatives that were formally evaluated?
- What alternatives were examined and eliminated in the earlier stages of the planning process?

Impact Assessment

- Have all important differences among alternatives been identified and considered?
- What are the important trade-offs in choosing among alternatives?

Equity

- What is the distribution of benefits and costs (i.e., among highway users, nearby residents, etc.)?
- Do any groups pay shares of the costs that are disproportionate to the benefits they receive?

Economic Efficiency

- Does an alternative provide sufficient benefits to justify the required expenditure? In comparison with less costly alternatives, does an alternative provide additional benefits to justify the additional funds required?
- In comparison with more costly alternatives, has sufficient weight been given to the benefits that would be forgone in the event that additional expenditure of funds is not justified?
- How do nonmonetary costs and benefits affect conclusions about economic efficiency?

Financial Feasibility

- Will sufficient funds be available to implement alternatives on schedule? From what sources?
- What is the margin of safety for financial feasibility?
- What adjustments will be necessary if this margin of safety is exceeded?

Legal and Administrative Feasibility

- Is the implementation of an alternative feasible within existing laws and administrative guidelines? If not, what adjustments might be necessary?
- What approaches will be required in later stages of the planning process? Are problems anticipated in obtaining these approvals? If these problems do materialize, can minor modifications or refinements to alternatives overcome them?

Sensitivity of Findings to Uncertainties and Value-laden Assumptions

- Are there uncertainties or value-laden assumptions that might affect the information assembled to help decision makers answer the earlier questions?
- How much would assumptions have to be changed to suggest different answers to these questions?
- How reasonable are these changes?

The answer to the many questions listed here are subjective in nature; it is quite possible that reasonable individuals might answer them differently even when given the same information.

In a comprehensive evaluation, detailed information about alternatives and their impacts should be summarized in the evaluation report. This report should document the major impacts of the performance of the alternatives analyzed and emphasize the difference in the performance of alternatives with respect to the key impact measures.

3.16 ECONOMIC AND FINANCIAL CONCEPTS

In determining the economic and financial implications of an engineering improvement, a clear understanding of several basic concepts is essential. These concepts include monetary impacts such as the cost of time, money, equipment, maintenance, and user benefits, as well as nonmonetary costs such as safety, environmental impacts, and the effects on the local economy. Each of these concepts is described briefly in what follows. When applied as part of one of the analysis techniques presented in Section 3.17, they can assist planners in determining the overall economic efficiency and effectiveness of an improvement.

Time value of money. Benefits and costs will occur at different times during the life of an improvement. However, a cost incurred today cannot be compared directly with the benefits that it will provide in some future year. The funding for an improvement should be considered as an investment that is expected to generate a return. Consequently, the costs and benefits of an improvement must be compared in equivalent terms that account for their timing. These benefits and costs can be either described in equivalent terms by calculating their present worth (for the current year or recent year for which data is available) or for an average year over the life of the improvement. Any future cost or benefit expressed in future dollars is properly converted to its corresponding present value by two factors:

1. *Discount rate.* The appropriate discount rate is an estimate of the average rate of return that is expected on private investment before taxes and after inflation. This interpretation is based on the concept that the alternative to the capital investment by the government is to leave the capital in the private sector.

2. *Inflation rate.* Inflation can be ignored in performing most aspects of the basic economic evaluation if the evaluation is prepared in present value terms, because inflation has no effect on present values. However, if the costs used are 2 or 3 years out of date, they should be adjusted to a more current year.

The only time that it might be appropriate to project future inflation in evaluating alternative improvements is when there is reasonable evidence that differential future rates of inflation might apply to important components of the evaluation. However, independent of the evaluation of alternative improvements, projections of inflation must be made for a financial feasibility analysis and development of a capital improvement program.

3.17 ANALYSIS TECHNIQUES

The objective of economic evaluation analysis techniques is to provide sufficient summary information to decision makers and interest groups to:

- Determine whether the costs of improvements are justified by the anticipated benefits (i.e., whether a proposed improvement is better than doing nothing)
- Make comparative overall assessments of different alternative improvements with each other

In complex situations involving large-scale, costly improvements, it may also be necessary to assess the distribution of benefits and costs among those affected by the improvement, such as user, operator, and societal costs.

Techniques that support the first purpose come under the category of economic efficiency analysis, or merely *efficiency analysis*, also referred to as *investment appraisal methods*. These include benefit-cost ratio, present worth, rate of return, equivalent uniform annual value, and other variations of these methods. All of them involve the translation of impacts (i.e., costs and benefits) into monetary terms. Techniques that support the second purpose come under the category of cost-effectiveness analysis. These include the use of nonmonetary effectiveness measures either to assess the relative impacts of all alternative improvements in the same terms or to hold constant the requirements that no alternatives must meet. Effectiveness measures are also used in combination with cost values in the form of ratios.

3.17.1 Economic Evaluation Methods (Efficiency Analysis)

There are four common methods of evaluating engineering projects:

1. Net present worth (NPW)
2. Equivalent uniform annual cost (EUAC)
3. Internal rate of return (IRR)
4. Benefit-cost (B/C) analysis.

The bottom line in the evaluation of individual projects is which project is the most productive. Or which project produces the highest return? Essentially, there are two indices of economic merit that can help answer these questions. The first is the B/C criterion and the second is the IRR. A brief explanation of each method follows.

By using the concept of equivalence connected with compound interest, the present worth of a single payment F, n years from now, with a discount rate of i is:

$$\text{PWSP} = \frac{F}{(1+i)^n} \tag{3.15}$$

The present worth of uniform series (PWUS) of equal annual payments is the sum of the present worth of each cost:

$$\text{PWUS} = A\left[\frac{(1+i)^n - 1}{i(1+i)^n}\right] \tag{3.16}$$

Here, A is the EUAC. The term in brackets is known as the PWUS factor. Its value has been tabulated for various combinations of i and n. Also:

$$A = \text{EUAC} = \text{PWUS}\left[\frac{i(1+i)^n}{(1+i)^n - 1}\right] \tag{3.17}$$

The term in brackets is called the CRF (see Section 3.5.4). For example, if a person borrows \$10,000 from a bank and intends to repay the amount in equal yearly installments over an 8-yr period at 5% per annum, the EUAC would be:

$$\text{EUAC} = 10{,}000\left[\frac{0.05(1+0.05)^{8}}{(1+0.05)^{8}-1}\right] = 10{,}000 \times 0.154722 = \$1{,}547.22$$

The IRR method has been proposed as an index of the desirability of projects. Naturally, the higher the rate, the better the project. By definition it is the discount rate at which the net present value of benefits equals the net present value of costs.

There are many situations in which the rate of return needs to be calculated for several alternatives. In such cases the incremental rate of return on the cash flow representing the differences between alternatives is computed. If the incremental rate of return is greater than, or equal to, the predetermined Minimum Alternative Rate of Return (MARR), choose the higher-cost alternative; otherwise, choose the lower-cost alternative.

B/C analysis is generally applied to engineering projects to ascertain the extent to which an investment will be beneficial to society. This analysis can be quite elaborate (more will be said about this method of analysis in another section). Suffice it to say here that a project that costs less than the benefits derived from the project would be eligible for consideration of being implemented ($B/C \geq 1$)

Example 3.11 Three alternatives are being considered for improving a street intersection. The annual dollar savings on account of the improvement is shown. Assume that the intersection will last for 25 yr and the interest rate is 5%. It is assumed that each of the three improvements is mutually exclusive but provides similar benefits:

Alternative	Total cost	Annual benefits
A	\$10,000	\$800
B	\$12,000	\$1,000
C	\$9,000	\$1,400

Solution *NPW method*: Use the PWF for uniform series.

NPW(A) = −10,000 + (800 × 14.094) = \$1,275.20
NPW(B) = −12,000 + (1,000 × 14.094) = \$2,094.00
NPW(C) = −19,000 + (1,400 × 14.094) = \$731.60

Therefore, select Alternative B with the highest NPW.

EUAC method: Use the CRF.

EUAC(A) = −(10,000 × 0.07095) + 800 = \$90.50
EUAC(B) = −(12,000 × 0.07095) + 1,000 = \$148.60
EUAC(C) = −(19,000 × 0.07095) + 1,400 = \$51.95

Continues

Example 3.11: *Continued*

Alternative B has the highest EUAC and should be selected.

IRR method: Compare the IRR figure with the do-nothing alternative.

$$NPW(A) = -10{,}000 + 800(P/A, i, 25\text{ yr}) = 0$$
$$(P/A, i, 25\text{ yr}) = 10{,}000/800 = 12.5 \Rightarrow i = 6.25\%$$
$$NPW(B) = -12{,}000 + 1{,}000(P/A, i, 25\text{ yr}) = 0$$
$$(P/A, i, 25\text{ yr}) = 12{,}000/1{,}000 = 12.0 \Rightarrow i = 6.70\%$$
$$NPW(C) = -19{,}000 + 1{,}400(P/A, i, 25\text{ yr}) = 0$$
$$(P/A, i, 25\text{ yr}) = 19{,}000/1{,}400 = 13.57 \Rightarrow i = 5.77\%$$

Alternative B is the best of the three because its IRR is the highest (and, of course, higher than 5%, so it is better than the do-nothing alternative).

Example 3.12 Consider two alternatives X and Y for repairs to a pumping set (in $100):

Year	X	Y
0	−20	−13.10
1	+7.76	+4.81
2	+7.76	+4.81
3	+7.76	+4.81

If the MARR is 11%, which alternative would you recommend?

Solution Take the higher initial-cost Alternative X minus the lower-cost Alternative Y:

Year	X	Y	(X − Y)
0	−20	−13.10	−6.90
1	+7.76	+4.81	+2.95
2	+7.76	+4.81	+2.95
3	+7.76	+4.81	+2.95

Solve for the incremental rate of return.

$$\text{PW of cost} = \text{PW of benefits}$$
$$6.90 = 2.95(P/A, i, 3\text{ yr})$$
$$(P/A, i, 3\text{ yr}) = 6.90/2.95 = 2.339$$

From compound interest tables or by direct calculations, IRR works out to be between 12 and 15%, which is higher than 11% (MARR). Hence, select X, which is again the one with the higher initial cost.

Continues

Example 3.12 *Continued*

While present worth analysis is the preferred technique, the annual worth (or annualized benefit, cost, and net benefit) is often used in engineering work. In these two cases the choice of the best project is made using the maximization of net benefits. In the public sector, the dominant method is to calculate the B/C ratio:

$$B/C = \frac{\text{Present worth of benefits}}{\text{Present worth of costs}} = \frac{\text{Eq. uniform annual benefits}}{\text{Eq. uniform annual costs}}$$

For a given interest rate, a $B/C \geq 1$ reflects an acceptable project. However, it must be emphasized that the maximization of the B/C ratio is *not* really the proper economic criterion; it is the maximization of net benefits that is the ultimate goal and the incremental B/C analysis is useful in achieving this goal. The following two cases are considered for incremental analysis:

1. When several alternatives are feasible either with the same benefit or cost, choose the alternative with the largest B/C ratio, subject to the ratio being greater than or equal to one. If no such alternative exists, choose the do-nothing alternative.

2. When neither the benefit nor the cost is the same for the feasible alternative, perform an incremental B/C analysis and go through the following steps:

   a. Rank-order the feasible alternatives from the lowest cost to the highest cost and number the Projects 1, 2 , . . ., *n*.

   b. For Projects 1 and 2 compute:

$$\frac{\Delta B}{\Delta C} = \frac{(PWB_2 - PWB_1)}{(PWC_2 - PWC_1)}$$

or

$$\frac{\Delta B}{\Delta C} = \frac{(EUAB_2 - EUAB_1)}{(EUAC_2 - EUAC_1)}$$

If $\frac{\Delta B}{\Delta C} \geq 1$ select Project 2 until this stage of analysis.

If $\frac{\Delta B}{\Delta C} < 1$ select Project 1 until this stage of analysis.

   c. Compute $\frac{\Delta B}{\Delta C}$ between the best project, until this stage of analysis, and the next most costly project not yet tested.

If $\frac{\Delta B}{\Delta C} \geq 1$ select the most costly project as best until this stage of analysis.

If $\frac{\Delta B}{\Delta C} < 1$ retain the less costly project as best until this stage of analysis.

   d. Repeat Step c until all projects have been tested. The surviving project is the best.

Example 3.13 Six mutually exclusive alternative projects need to be assessed for final selection. They have been ordered from lowest to highest cost. Recommend which project should be selected:

Alternatives	**1**	**2**	**3**	**4**	**5**	**6**
Benefits	120	130	195	210	250	300
Costs	75	110	140	175	240	320
B/C ratio	1.60	1.18	1.39	1.20	1.04	0.94

Compute $\frac{\Delta B}{\Delta C}$ which is essentially the slope of the line representing B versus C.

Solution

From 1 to 2: $\frac{130-120}{110-75} = \frac{10}{35} = 0.28 < 1.0$ $\quad\therefore$ retain 1

From 1 to 3: $\frac{195-120}{140-75} = \frac{75}{65} = 1.15 > 1.0$ $\quad\therefore$ select 3

From 3 to 4: $\frac{210-195}{175-140} = \frac{15}{35} = 0.24 < 1.0$ $\quad\therefore$ retain 3

From 3 to 5: $\frac{250-195}{240-140} = \frac{55}{100} = 0.55 < 1.0$ $\quad\therefore$ retain 3

Project 3 is the best with B/C = 1.39 and present worth of net benefit = $55, although Project 1 has a higher B/C ratio. Also note that Project 6 was discarded because of B/C < 1. The steps applied in this example can also be shown graphically.

3.17.2 Cost-effectiveness Analysis

Cost-effectiveness is a strategy for making decisions rather than establishing a decision rule. This approach provides a general and flexible framework for providing information to aid in the selection of alternative plans. Many of the consequences of proposed engineering plans are difficult to measure, if not intangible. Planners resorting to the benefit-cost method place dollar values on benefits and costs with the intentions of being as objective as they can be and, as a result, oversimplify the complexity of the problem.

Cost effectiveness overcomes some of these snags. Cost-effective analysis is essentially an information framework. The characteristics of each alternative are separated into two categories: (1) costs and (2) measures of effectiveness (MOE). The choice between alternatives is made on the basis of these two classes of information, eliminating the need to reduce the attributes or consequences of different alternatives to a single scalar dimension. Costs are defined in terms of all the resources necessary for the design, construction, operation, and maintenance of the alternative. Costs can be considered in dollars or in other units. Effectiveness is the degree to which an alternative achieves its objective. Decision makers can then make subjective judgments that appear best.

Cost-effective analysis arose out of the realization that it is frequently difficult to transform all major impact measures into monetary terms in a credible manner, and that important evaluation factors can often be stated in more meaningful measures than dollar costs. Basically, the method should be used when pre-established requirements exist regarding the improvement. Usually these requirements seek to establish the minimum investment (input) required for the maximum performance (output) among several alternative improvements.

One of the common general engineering performance measures used in cost-effective analysis is *level-of-service*. In practice, it is generally desirable to prepare estimates for several cost-effective measures rather than a single measure, because no single criterion satisfactorily summarizes the relative cost-effectiveness of different alternatives. For example, Table 3.2 lists cost-effectiveness indices that might be used for highway and transit evaluation. Many or all of these could be used in a particular cost-effective analysis. Several additional measures could be added, depending on local goals and objectives.

Table 3.2 List of typical cost-effective measures for evaluation

Highway
Increase in average vehicle speed per dollar of capital investment
Decrease in total vehicle delay time due to congestion per dollar of capital investment
Increase in highway network accessibility to jobs per dollar of capital investment
Decrease in accidents, injuries, and fatalities per dollar of capital investment
Change in air-pollutant emissions per dollar of capital investment
Total capital and operating costs per passenger-mile served
Transit
Increase in the proportion of the population served at a given level of service (in terms of proximity of service and frequency) per dollar of total additional cost
Increase in transit accessibility to jobs, human services, and economic centers per dollar of total additional cost
Increase in ridership per dollar of capital investment
Increase in ridership per dollar of additional operating cost
Total capital and operating cost per transit rider
Total capital and operating cost per seat-mile and per passenger-mile served
Decrease in average transit trip time (including wait time) per dollar of total additional cost

Example 3.14 Citizens (particularly children) need a transportation facility to connect their community across a six-lane freeway by means of a footbridge over the freeway or a tunnel under the freeway. The objectives of the project are (1) to provide a safe and efficient system for crossing the freeway and (2) to maintain and improve (if possible) the visual environment of

Continues

Example 3.14: *Continued*

the neighborhood. The table gives the construction and operating costs of the two alternatives. Annual costs are computed using a 25-yr service life and an 8% interest rate. It is anticipated that about 5000 pedestrians will use either of the two facilities per day:

Alternative	Total cost	Annual cost	Annual operating cost	Total annual cost
Footbridge	\$100,000	\$9,368	\$100	\$9,468
Tunnel	\$200,000	\$18,736	\$350	\$19,086

The annual difference in cost is \$9,618. The average cost per pedestrian per trip using the bridge is:

$$\frac{9{,}468}{5{,}000 \times 365} = 0.52 \text{ cents}$$

and using the tunnel:

$$\frac{19{,}086}{5{,}000 \times 365} = 1.05 \text{ cents}$$

The performance of the two alternatives may be summarized as follows:

1. The tunnel would need long, steep grades and may cause inconvenience to the elderly and handicapped. On the other hand, the bridge will be provided with stairs at either end, and this will eliminate its use by the elderly and handicapped. If ramps are to be provided, long steep grades would have to be provided at three times the cost of the bridge with stairs.
2. The tunnel is safer than the footbridge during the long winter months.
3. From the point of view of aesthetics, the tunnel is preferred to the footbridge.
4. It is possible to predict that thc tunnel would be safer than the bridge.

Another way to look at cost-effective analysis is to consider it as a way of maximizing the returns (in terms of effectiveness) of expenditures. Consider the following example:

> A city is considering eight different minibus network configurations that are mutually exclusive. The only objective is to adopt a configuration that will maximize the number of potential riders in that sector of the city. The city has established the minimum number of potential riders that the alternative will serve and also maximize the amount of money it is willing to spend. Figure 3.5 shows the results of the analysis.

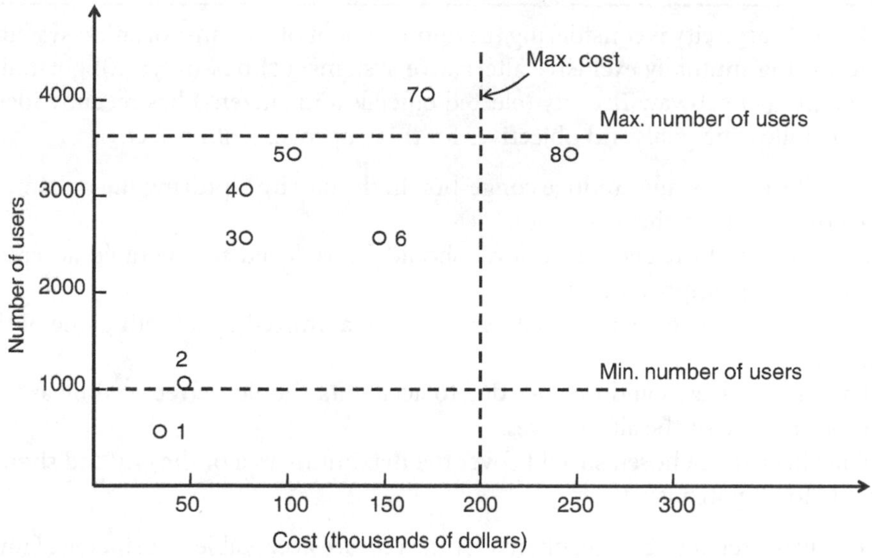

Figure 3.5 Cost-effective analysis: cost versus users

Notice that Alternatives 1, 7, and 8 do not qualify under the criteria set. Alternative 5 would be the one that serves the maximum number of users. This problem could get somewhat complicated if an additional criterion were introduced (e.g., number of hours of operation).

3.17.3 Multicriteria Evaluation Method

The examples that have been given in this chapter so far fall under the general category of single-criterion evaluation methods, because all the benefits and costs have been reduced to monetary terms. The present worth, annual cost, B/C, and IRR methods all fall into this category, because the maximization of net benefits is the single objective the analyst has in mind. However, in most engineering problems, usually one is dealing with many objectives (or criteria) that reflect the interests of the community. To incorporate several, often conflicting, objectives (or criteria) systematically, planners have developed methods to tackle this situation. One of these methods is generally called the *multicriteria evaluation method*. The steps in applying this method are best illustrated by working through an example.

Example 3.15 A large city is considering the improvement of its transportation system by adopting the following mutually exclusive alternative systems: (1) bus-only; (2)light-rail; (3) bus + heavy-rail; and (4) subway. The city (elected officials and citizens) has recently identified and adopted the following goals and objectives for this improvement project:

- The alternatives must reduce congestion in the city by capturing the maximum number of commuters in the peak hour.
- The air-pollution and noise level should be reduced to tolerable acceptance levels through this improvement.
- The total net revenue to the city should be maximized by adopting one of the alternatives.
- The rate of deaths and injuries due to accidents should be reduced by at least 50% by adopting one of the alternatives.
- The alternative chosen should cover the maximum area of the city and should be equitable to all patrons.

The alternatives prepared by consultants for the city are acceptable in principle. However, they all differ in quality, cost, alignment, coverage, and so on. City planners have formulated the following MOE for each objective:

Objective 1: percentage reduction of motor vehicles on main arterials and corridors of the city
Objective 2: percentage reduction of pollution
Objective 3: net annual revenue/annual capital cost
Objective 4: probable accident rate/million trips
Objective 5: square miles covered by the proposed network (1 mi on either side of line)

Solution

Objective and measure of effectiveness (MOE)	Alternatives			
	Bus	Light-rail	Bus + heavy-rail	Subway
1. Congestion reduction	20	(27)	25	15
2. Pollution reduction	15	30	22	(35)
3. Revenue	(25)	20	15	10
4. Accident reduction	15	20	23	(27)
5. Coverage	30	25	(40)	22

Note that each alternative has been assigned a value representing a particular MOE. Also, certain alternatives rank higher than others in a particular measure; for example, the light-rail system reduces congestion the most, by scoring the highest in the row values, 27; and the bus and heavy-rail alternative scores 40 points (the highest) in the row representing coverage. If these highest scores are matched to the corresponding relative weighting factor, one can erect

Continues

Example 3.15: *Continued*

another matrix as shown. Here the highest score for MOE 1, 27, is assigned 20 as the relative weighting factor for the light-rail system; 17 is assigned to the bus + heavy-rail alternative for MOE 5; and so on. By the same token, the other figures are calculated proportionally. The weights (out of 10) and the relative weighting factors are determined from the table. The city's elected officials (through brainstorming methods and Delphi techniques) have allocated weights on a 10-point scale to each objective as indicated:

Objective	Weight (out of 10)	Relative weighting factor (%)
1	7	20
2	5	14
3	9	26
4	8	23
5	6	17
Total	35	100

This assignment of weights is subjective (usually arrived at by committee action), but it is one way of assessing the collective opinion of a large group of people who are affected positively (or negatively) by the proposed implementation of the chosen alternative.

Next, planners, engineers, sociologists, economists, and estimators set about collecting data and costs for assigning values to each alternative, as shown in the following matrix. This step is most crucial and involves a lot of work in gathering appropriate data:

Objective and MOE	Alternatives			
	Bus	Light-rail	Bus + heavy-rail	Subway
1. Congestion reduction	15	(20)	19	11
2. Pollution reduction	6	12	9	(14)
3. Revenue	(26)	21	16	10
4. Accident reduction	13	17	20	(23)
5. Coverage	13	11	(17)	9
Total	73	81	81	67

It is obvious from the column totals that the light-rail system and the bus + heavy-rail system each score 81 points and can be declared superior in comparison to the other two alternatives. It is natural that these two alternatives will be subjected to further scrutiny and analysis before a final winner is declared.

If, for example, these two top-ranking alternatives are examined by an investigating team who perform an environmental impact study, it is quite possible that a more calculated decision can be taken. This example illustrates the difficulty in considering situations involving multicriteria application, because the outcome is sensitive to the weights assigned to the criteria.

3.17.4 Benefit-Cost Analysis

This section is an extension of economic evaluation dealt with in Section 17.1. Public expenditure decisions are generally evaluated using B/C analysis. A simple example of this type of evaluation is demonstrated here. Table 3.3 gives information on the capital cost and expected potential benefits by constructing from 5 to 30 mi of a light-rail line between downtown and the outskirts of a large city. Column 1 shows the six alternatives, and Column 2 indicates the length of each of these six alternatives. The capital cost (in units of millions of dollars) is shown in Column 3. The present value of user benefits is indicated and this has been derived assuming a money stream for 20 yr at 10 percent per annum in units of millions of dollars. Column 5 is Column 4 minus Column 3. The Column 6 indicates marginal benefits minus marginal cost (MC). The benefits and costs of the six light-rail systems are plotted in Figure 3.6. The abscissa denotes the length of the light-rail system. The marginal benefits and MC are plotted.

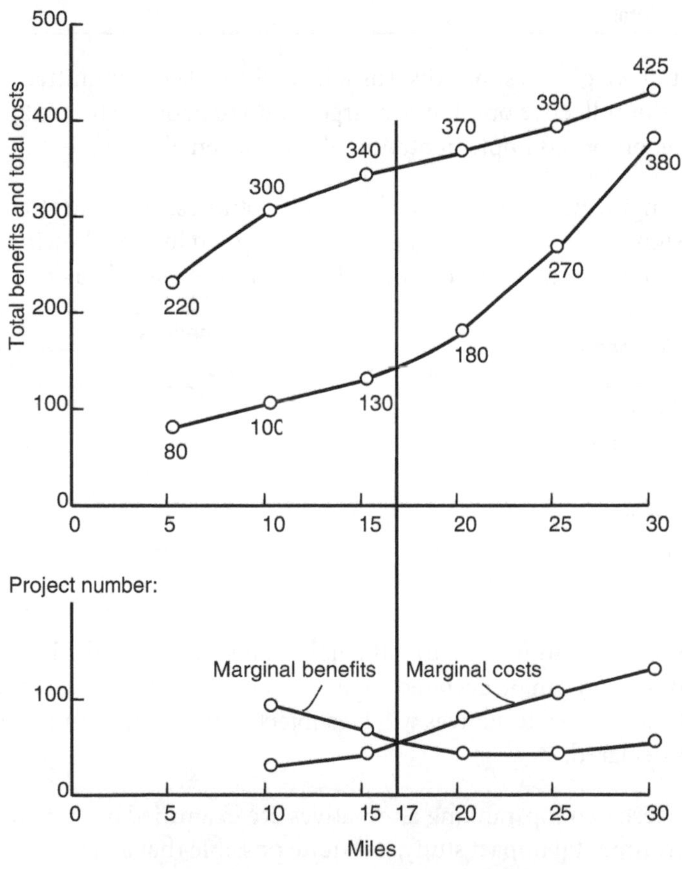

Figure 3.6 Total and marginal benefit-cost analysis

Table 3.3 Capital cost and expected benefits

(1) Alternative	(2) Miles	(3) Capital cost	(4) User savings	(5) Savings – cost	(6) Δ(saving) – Δ(cost)
1	5	80	220	140	
2	10	100	300	200	60
3	15	130	340	210	10
4	20	180	370	190	–20
5	25	270	390	120	–70
6	30	380	425	45	–75

It is obvious, by sheer inspection, that as the light-rail system is extended farther out in the suburbs of the city, the revenues (benefits) from the system become less and less attractive. The question now is: What criteria should be used in evaluating these six alternatives? Because of the nature of the data presented, the economic criterion is used to identify the best course of action.

The numbers given in Table 3.3 are for discrete lengths of the light-rail system. If the cost-benefit relations shown in Figure 3.6 were continuous, the optimum length of light-rail would be one where the marginal benefit curve cuts the MC curve. This intersection of the two curves occurs at about 17 mi from the city's center. The alternative closest to this intersection is Alternative 3 (15 mi).

Applications of B/C analysis: It is best at this stage to understand the applications of B/C analysis through practical examples. Take the case of a decision maker who is confronted with five projects whose details are provided in Table 3.4.

Case 1. Assume that the projects are mutually exclusive, but there is no funding constraint. In the absence of a budgetary constraint, it would be clearly advantageous to build Project D, which would yield a net benefit of 150.

Case 2. Now assume that the projects are no longer mutually exclusive and that a number of projects of the same category (A through E) up to the limit of the budget can be implemented.

Table 3.4 Cost-benefit and cost of project

Project	Benefits B	Cost C	Ratio B/C	Net benefit B – C	Cumulative initial cost
A	32	4	8.00	28	4
B	60	10	6.00	50	14
C	120	40	3.00	80	54
D	300	150	2.00	150	204
E	500	450	1.11	50	654

For example, A through E are alternative forms of bridges. If the budget constraint in this case were 160, then 40 bridges of Type A could be built for 160, providing 1280 as benefits and 1120 as net benefits. Similarly, 16 bridges of Type B could be built, providing (960 - 160) = 800 as net benefits. Or, four bridges of Type C could be undertaken, resulting in 320 as net benefits. Even one bridge of Type D could be considered for 150, but this would give only a 150 net benefit. Obviously, the best solution here would be to adopt 40 versions of Project A. Ratios are therefore applicable when dealing with a budget constraint.

Case 3. Assume that the budget is only 10, in which case only Projects A and B are eligible. But here again one runs into the problems of indivisibilities and, therefore, the only choice is Project B.

Case 4. Assume that a new Project C′ is added to the list as follows:

C′: Benefit = 192, Cost = 96, B/C = 2.00, B − C = 96, cumulative cost = 150

Also, assume that the budget constraint is 204 and that the projects are not mutually exclusive. Therefore, selecting projects from the top of the list and also taking into account Project C′, one should undertake A + B + C + C′ = 150. The question that comes up here is: Which is better, choosing A + B + C + C′ or choosing just D, as in Case 1, both choices aggregating 150?

Net benefit when A + B + C + C′ are considered = 28 + 50 + 80 + 96 = 254 > 150 (if D is considered). This exercise indicates how limitations of the budget change the outcome.

Example 3.16 An existing highway is 20 mi long connecting two small cities. It is proposed to improve the alignment by constructing a highway 15 mi long costing $500,000 per mi. Maintenance costs are likely to be $10,000 per mi per annum. Land acquisition costs run $75,000 per mi. It is proposed to abandon the old road and sell the land for $10,000 per mi. Money can be borrowed at 8% per annum. It has been estimated that passenger vehicles travel at 35 mph at a cost of 20¢ per mi with a car occupancy of 1.5 persons per car. What should the traffic demand for this new road be to make this project feasible if the cost of the time of the car's occupants is assessed at $10.00 per hr?

Solution

Basic costs:

Construction 15 mi at $500,000	=	$7,500,000
Land acquisition 15 mi at $75,000	=	$900,000
		$8,400,000
Deduct sale of land (old alignment)		
20 × 10,000	=	−$200,000
		$8,200,000

Continues

Example 3.16: *Continued*

It is assumed that the new road will last indefinitely. Therefore, the investment need not be repaid, only the interest.

Annual cost of initial investment:

$$8\% \text{ of } \$8{,}200{,}000 = \$656{,}000$$

Net annual income: Assume that N vehicles per annum use the road, 5 m saved at 20¢ per mi.

$$N\left(\frac{5 \times 20}{100}\right) = \$N$$

Time saved per veh/trip:

5 mi at 35 mph = 5/35 = 0.143 hr
Car occupancy = 1.5 persons/car
Value of time = $10 per hr

Therefore:

Cost saving in time at $10/hr = 0.143 × 1.5 × N × 10 = 2.14N
Maintenance savings = 5 mi × $10,000 = $50,000
Total savings = $50,000 + $$N$ + 2.14N = 50,000 + 3.14N

Benefit-cost ratio: For the project to be justified, the B/C ratio must be equal to or greater than 1. Because the project is assumed to have an infinite life:

$$B/C = \frac{\text{annual benefits}}{\text{annual costs}}$$

or

$50{,}000 + 3.14N = 656{,}000$
$N = 192{,}994$ per annum or about 528 veh/da

The project, therefore, would be justified if about 528 vehicles used the new road on a daily basis.

Discussion

The following points should be noted regarding this problem:

The social costs of land use have not been considered. For example, the land on either side of the old road would decrease in value, whereas the land adjacent to the new route would increase in value. The net result would raise the value of N.

When one works out the B/C ratio, it is generally advisable to use total benefits and total costs. In this problem, B/C was assumed to be 1 and is therefore permissible.

3.17.5 The Willingness-to-pay Concept

The willingness-to-pay concept is useful in visualizing the viewpoint of users of a system, such as a bridge or a toll road. In the case of a single-service variable, as shown in Figure 3.7a, if DD is the demand and the price (p) changes from p_1 to p_2, there will be a corresponding change in volume (v) from v_1 to v_2. The $(v_1 - v_2)$ users who abandoned the system are those who received just enough benefit from the system at price p_1 to use it.

If we consider Figure 3.7b using the same argument, reducing the price from p_0 to p_1 increases the number of users from v_0 to v_1. There are three different views on the method of measuring the amount of benefit to users.

1. Gross-benefit view:

   Total benefit to users = (E) + (A) + (D), when volume = v_0
   Total benefit to users = (E) + (A) + (D) + (B) + (C), for v_1
   Difference between the two actions = (B) + (C)

2. Consumer surplus view:

   The amount of benefit received by users beyond what they actually pay = (A) + (B)

3. User-cost view = (A) – (C)

Approximation

Gross benefit B + C = $(1/2)(p_0 + p_1)(v_1 - v_0)$
Consumer surplus A + B = $(1/2)(p_0 - p_1)(v_1 - v_0)$
User cost A – C = $p_0\, v_0 - p_1\, v_1$

An examination of these three views indicates how different the results can be.

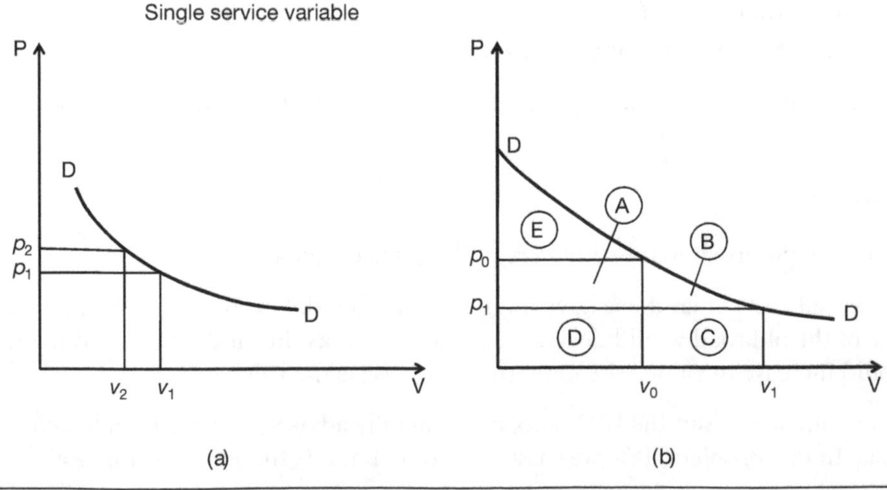

Figure 3.7 Willingness to pay

Example 3.17 A section of a busy road, 1-mi long, has a high pedestrian accident rate. It is proposed to construct a pedestrian bridge over the road for $18 million and increase the existing speed limit on the road. Details of the existing and proposed conditions are:

Details	Existing	Proposed
Vehicle operating speed	25 mph	35 mph
Peak-hour flow	3000 veh/hr	4000 veh/hr
Length of section	1 mi	1 mi
Accident rate/yr	450	25 (estimated)
Cost of driving (veh/hr)	$0.35	$0.20
Cost/accident	$3000	$3000
Peak hours affected	4 hr	4 hr

Determine the feasibility of this project if the interest rate is 6% per annum and the design life of the pedestrian bridge is 30 yr.

Solution The direct benefits derived from the construction of the pedestrian bridge can be determined by applying the concepts of consumer's surplus. The following analysis shows the gain in consumer's surplus:

Existing price = $0.35/veh/mi × 1 mi = $0.35/veh
Proposed price = $0.20/veh/mi × 1 mi = $0.20/veh

Annual number of users:

Existing: 3000 veh/hr × 4 hr/da × 365 da = 4.38 million veh/yr
Proposed: 4000 veh/hr × 4 hr/da × 365 da = 5.84 million veh/yr
Consumer's surplus as annual savings = 0.5 (0.35 – 0.20) (4.38 + 5.84) million
= $0.7665 million/yr
Annual savings from accident reduction = 3000 (450 – 25) = $1.275 million/yr
Total benefits = $2.0415 million/yr
NPW factor @ 6% for 30 yr = 13.7649
NPW = ($2.0415 million × 13.7649) – 18 = $10.1041 million

Because the NPW of this project is greater than zero, the pedestrian project is sound.

3.18 DEPRECIATION AND TAXES

In day-to-day practice, economic decisions have to consider such items as depreciation, inflation, and taxes:

Depreciation, in the accounting sense, is the loss in value of a piece of equipment over time, generally caused by wear and tear, deterioration, obsolescence, or all of the above. One can prepare a graph of the market value of the equipment with time. There are three methods of calculating depreciation:

1. Straight-line
2. Sum-of-the-years
3. Double-declining balance

Let

P = Cost of equipment
n = Depreciable life in years
S = Salvage value

and

D = Depreciation charge per period of time

Then, according to the straight-line method:

$$D = \frac{P - S}{n}$$

and by the sum-of-years method:

$$D = \frac{\text{Remaining life at the beginning of year}}{\text{Sum-of-years digits for total useful life}}(P - S)$$

where sum-of-years digits = $1 + 2 + 3 + \ldots + n = \frac{n}{2}(n + 1)$

And, by the double-declining balance:

$$D = \frac{2}{n}\,[P - (\text{Depreciation charges to date})]$$

Example 3.18 An electric motor and pump costing $10,000 has an anticipated salvage value of $2,000 at the end of its 4-yr depreciable life. Compute the depreciation schedule by the three methods.

Solution

1. Straight-line depreciation:

$$D = \frac{P - S}{n} = \frac{(10{,}000 - 2{,}000)}{4} = \$2{,}000/\text{year}$$

Continues

Example 3.18: *Continued*

2. Sum-of-years digits depreciation:

   $$\text{Sum-of-years digits} = \frac{n}{2}(n+1) = \frac{4}{2}(5) = 10$$

   $$\text{First-year depreciation} = \frac{4}{10}(8{,}000) = \$3{,}200$$

   $$\text{Second-year depreciation} = \frac{3}{10}(8{,}000) = \$2{,}400$$

   $$\text{Third-year depreciation} = \frac{2}{10}(8{,}000) = \$1{,}600$$

   $$\text{Fourth-year depreciation} = \frac{1}{10}(8{,}000) = \$800$$

3. Double-declining balance depreciation:

   $$\text{First-year depreciation} = \frac{2}{4}(10{,}000 - 0) = \$5{,}000$$

   $$\text{Second-year depreciation} = \frac{2}{4}(10{,}000 - 5{,}000) = \$2{,}500$$

   $$\text{Third-year depreciation} = \frac{2}{4}(10{,}000 - 7{,}500) = \$1{,}250$$

   $$\text{Fourth-year depreciation} = \frac{2}{4}(10{,}000 - 8{,}750) = \$625$$

Because we know that there is a salvage value of $2,000, we cannot exceed a total depreciation of more than $10,000 – 2,000 = $8,000 using double-declining balance depreciation. Therefore, the schedule is:

Year	Straight-line	Sum-of-digits	Double-declining
1	2000	3200	5000
2	2000	2400	2500
3	2000	1600	500
4	2000	800	0
Total	8000	8000	8000

Notice that using the double-declining balance depreciation method, the total depreciation in Years 1 and 2 is $7500. Thus in Year 3, only $500 depreciation is allowed (to make the total equal to $8000); no depreciation is allowed after this year.

Income tax has to be considered as part of an economic analysis. One generally starts by computing before-tax cash flow; three types of entries are considered:

1. Payment of money to purchase capital assets for which there is no direct tax imposition.
2. Periodic receipts and/or disbursements of the firm for which there is tax liability.
3. Receipts of money from the sale of capital assets for which the tax depends on the book value (e.g., cost minus depreciation taken) of the asset. If the salvage value is greater than the book value, then there is a capital gain and thus taxes have to be paid. On the other hand, if the salvage value is less than the book value, there is a capital loss. The best way to determine after-tax cash flow is to tabulate the figures as shown in the example.

Example 3.19 A company expects to receive \$30,000 each year for 20 yr from the sale of machines. There is an initial investment of \$120,000. Manufacture and sale expenses will be \$4,000 per yr. Determine the project after-tax IRR. Assume straight-line depreciation, with no salvage value, and use a 30% income tax rate.

Solution

Straight-line depreciation: $D = \dfrac{P-S}{n} = \dfrac{(120{,}000 - 0)}{20} = \$6{,}000/\text{year}:$

Year	Before-tax cash flow	Depreciation	Taxable income	Income tax 30%	After-tax cash flow
0	−120,000				−120,000
1	(30,000 – 4,000) = 26,000	6,000	20,000	6,000	14,000
2	26,000	6,000	20,000	6,000	14,000
–					
–					
30	26,000	6,000	20,000	6,000	14,000

Now, take the after-tax cash flow and compute the IRR at which the present worth of cost equals the present worth of benefits (PWBs).

$$120{,}000 = 14{,}000\,(P/A, i, 20)$$

$$(P/A, i, 20) = \frac{120{,}000}{1{,}000} = 8.5714$$

From compound interest tables, $i = 9.9\%$.

3.19 REPORTING RESULTS

A well-organized summary report can significantly influence the outcome of an economic evaluation effort. In this section, we describe the preparation of a summary report and related materials for decision makers use.

The objective of good reporting is to satisfy the information needs of decision makers and other people with a wide range of perspectives, technical understanding, and constituents. Properly conceived and executed written reports and oral presentations should enable decision makers to select, narrow, or reconfigure the alternative improvements and generally facilitate, rather than complicate, the decision-making process.

A common format for summarizing key evaluation findings has measures or impact categories listed down the left-hand side of a table and alternatives listed across the top, with entries in the table indicating the status of each alternative with respect to each evaluation measure. For those measures that are in common units (e.g., dollars), subtotals can be provided. The normal bias toward quantitative measures can be avoided by including rows summarizing the key difference among alternatives that can only be characterized by a brief phase. Table 3.5 provides an illustrative example of such a table.

Table 3.6 shows an outline of a typical report. The cover, abstract, map, and summary arguments table could be made into a small brochure of about six single-page sheets. A slightly longer version might include one-page summaries of each alternative.

Table 3.5 Illustrative summary format for presentation of evaluation results

Alternatives	0. Do nothing	1. TSM program	2. Arterial improvement	3. Bus improvement program
Mobility	Severe congestion, level of service F by 1986; transit ridership—10%	Reduced congestion, level of service C maintained; transit ridership—10%	Postponed congestion, level of service F by 1986; transit ridership—10%	Reduced congestion, level of service C maintained; transit ridership—80%
Social	Traffic will harm neighborhood	Slightly less traffic disruption	Will make pedestrian movements more difficult	Will strengthen community
Air quality	(percent reduction in pollutant concentration from 1978 levels) No sites in any alternative are expected to violate 1- or 8-hr national ambient air quality standards Total daily corridor emissions (1995)			
	CO – 60%	CO – 65%	CO – 65%	CO – 70%
	HC – 55%	HC – 55%	HC – 55%	HC – 60%
	NO_x – 40%	NO_x – 40%	NO_x – 40%	NO_x – 40%
Noise	Areas with potential for significant noise impact (80 dB or more)			
	4 areas	4 areas	6 areas	5 areas
Disruption during construction	—	Minor	2 yr of modest construction	1 yr of modest construction
Monetary evaluation	(All dollar amounts are present values in thousands of 1978 dollars based on a discount rate of 10%)			
Time	$1550	$850	$1200	$700
Vehicle operating	$400	$250	$350	$200
Accidents	$300	$150	$200	$100
User subtotal	$2250	$1250	$1750	$1000
Transit operating	$650	$750	$750	$500
Operations and maintenance (O & M) subtotal	$1450	$1650	$1600	$1050
Capital	$50	$100	$150	$200
Total transportation costs	$3750	$3000	$3500	$2250
Net benefits	—	$750	$250	$1500
Cost effectiveness	(All dollar values are present values in thousands of 1978 dollars based on a discount rate of 10%)			

Capital, O & M costs per daily person-miles of travel	$87.7	$101.7	$101.7	$71.4.0
Capital costs per hour of daily time savings compared to do nothing	—	$230	$320	$210
Additional jobs accessible within 30 min/$1,000				
Via auto	—	200	230	20
Via transit	—	250	90	460
Total both modes	—	450	320	480

Table 3.6 Typical outline of evaluation summary report

Cover: Should effectively communicate the following:
What is being studied
Location: what part of which area
What decision is being made
Who is involved: jurisdiction and/or agencies
Abstract: About two or three pages elaborating on the foregoing items, making clear choices and the major issues
Map: Showing the configuration of the alternatives evaluated
Summary of arguments pro and con for each alternative: a single table highlighting the reasons for selection or rejection of each based on the evaluation findings
Purpose of report: should identify the problem to be addressed by the proposed improvement
The alternatives:
How many were identified or arrived at?
One-page summary of each, including a more detailed description, benefits and issues
More detailed map of each alternative
Evaluation:
A summary table showing the major differences among options
More detailed tabular evaluations with text as necessary
Elaboration on the difference among options
Explanations for the lack of differences where one may be expected
Discussion and clarification of the arguments that may have been advanced by various interests or advocates
Appendix:
Summary of methodology
Annotated bibliography of available technical references
Summary explanation of relevant government regulations and programs

SUMMARY

Evaluation of engineering works improvement can have different meanings and concerns for different people, depending on the project under investigation. Twenty-five years ago, for instance, projects were evaluated by simply taking into account the economics of benefits and costs to the potential user on an aggregate basis. Today, most evaluation exercises take into consideration the environmental, social, economic, and disaggregate effects of improvements.

There are at least three basic points to consider when deciding on, and later working with, a particular analysis technique: (1) the evaluator should be fully aware of the goals and objectives of the proposed improvements and their corresponding MOE; (2) the evaluation method adopted must be clear and transparent, even to the layperson; and (3) the evaluation procedure should assist the decision maker to come to a rational determination of what is best for the community or city.

Economic evaluation methods are the most common procedures used in most engineering contexts. However, their use entails the conversion of MOE to dollar units, and this occasionally poses problems. Therefore, planners have developed several other procedures to obviate this problem.

The evaluation of improvements is considered as one of the more difficult tasks in the field of engineering economics. Not only do the principles of capital budgeting need to be understood and applied, but the evaluator needs to have a basic knowledge of microeconomics and equity. Some recent references on these and other topics are listed in the references. Readers may also benefit from reports issued by government departments and agencies such as the U.S. Department of Transportation and the Environmental Protection Agency.

REFERENCES

Blanchard, B. S. and W. J. Fabrycky (2011). *Systems Engineering and Analysis*, 5th ed., Upper Saddle River, NJ: Prentice-Hall.

Johnson, R. E, (1990). *The Economics of Building*, New York, NY: John Wiley.

Khisty, C. J. and B. K. Lall (2003). *Transportation Engineering: An Introduction*, 3rd ed., Upper Saddle River, NJ: Prentice-Hall.

Meyer, M. D. and E. J. Miller (2000). *Urban Transportation Planning*, New York, NY: McGraw-Hill.

Mishan, E. T. (1976). *Cost-Benefit Analysis*, New York, NY: Praeger.

Pilcher, R. (1992). *Principles of Construction Management*, 3rd ed., London, UK: McGraw-Hill.

Schofield, J. A. (1989). *Cost-Benefit Analysis in Urban and Regional Planning*, London, UK: Unwin Hyman.

Stokey, E. and R. Zeckhauser (1978). *A Primer for Policy Analysis*, New York, NY: W. W. Norton.

EXERCISES

1. What is the present worth of a future sum of $3500 in 10 yr with interest at 10 percent per yr?
2. A man deposits $1200, $2000, and $4000 at the end of 1, 2, and 3 yr, respectively, at 10% interest per annum. What will be the accumulation at the end of 6 yr?
3. A client wants to finance the purchase of a house costing $50,000 over a period of 10 yr. If the interest is 10 percent per annum, what will be (a) the annual payment and (b) the monthly payment?
4. A bank offers the following interest rates: (a) 6 percent compounded annually; (b) 5.9 percent compounded semiannually; (c) 5.8 percent compounded quarterly; (d) 5.5 percent compounded monthly; and (e) 5.45 percent compounded continuously. Which rate would you select to provide the highest return?
5. We want to double a large sum of inherited money. Our bank offers two interest rates: (1) 9 percent per yr compounded annually and (2) 8.70 percent per yr compounded continuously. Which one should we select and why?

6. Mary buys an automobile, promising to pay $300 per mo for 5 yr, but after 2 yr, when the 24th payment is due, she decides to make a lump-sum payment to settle the account. If the interest rate is 10% per yr, what amount will she have to pay?

7. The maintenance for a bus, whose life is 10 yr, is $1500 per yr starting the fourth year, increasing by $200 for each successive year. What is the present worth of maintenance cost? Assume $i = 10\%$ per yr.

8. A businessman asks you for a loan of money and offers to pay $20,000 at the end of 5 yr. How much should you loan him now if you expect 15 percent interest per yr on your loan?

9. Tom is thinking of buying a used car. The price is $8000 with $1000 as a down payment and the balance in 60 equal monthly installments and interest at 3 percent per mo. If payments are due at the end of each month, what is the monthly installment?

10. A wealthy businessman wants to establish a fund at his favorite university to help needy students. If he assigns that $20,000 be raised through this trust account earning 10 percent per yr interest, how much money should he set aside in his trust?

11. Two alternatives for a house improvement have the following cash flows:

	Alternatives	
Year	X	Y
0	– 40,000	– 56,000
1	+16,000	+22,000
2	+16,000	+22,000
3	+16,000	+22,000

At 6% interest rate per yr, which alternative is better? Use:

(a) Present worth analysis
(b) Future worth analysis

12. A small project having a useful life of 5 yr has five mutually exclusive alternatives. Based on an interest rate of 6 percent, which alternative should be selected?

	Alternatives				
	I	II	III	IV	V
Initial cost	800	200	400	1000	300
Uniform annual benefit	210	60	90	250	60

13. Six different alternatives have been proposed for improving the water supply for a city, but only one option can be implemented. Each option has a lifetime of 8 yr and the MARR set is 5%. All figures are in millions of dollars. Which alternative should be selected?

	P	Q	R	S	T	U
Capital cost	27	4	17	1	10	5
Annual cost	1	2.5	1.4	1.4	3	2.3
Annual benefit	9	3	8	8	5	2.5

(a) Solve by present worth analysis
(b) Solve by annual cash flow analysis
(c) Solve by incremental B/C ratio analysis
(d) Solve by incremental rate of return analysis

14. A clever community fundraiser collects funds on the basis that each successive contributor will contribute 1/20 of the total collected to date. The collector himself makes the first contribution of $100.

(a) What is expected from the 150th contributor?
(b) How many contributors does he need to cross the $1 million target?

15. What is the rate of return, if $2000 is invested to bring in a net profit of $450 for each of the six succeeding years?

16. A person invests $3000 anticipating a net profit of $200 by the end of the first year, and this profit increases by 25 percent for each of the next 6 yr. What is the rate of return?

17. A transit system is estimated to cost $2,000,000 with operating and maintenance costs of $28,000 per yr after the first year, and 10 percent incremental increases over the next 12 yr. At the end of the sixth year, it is anticipated that new equipment and buses will be needed at a total cost of $3,000,000. Calculate the present worth of costs for this system if the rate of return is 9 1/2 percent per yr.

18. Estimate the total present sum of money needed to finance and maintain a small transit system given the following details: initial cost of buses, $2,150,000; annual maintenance cost of buses, $80,000; initial cost of shelters, workshops, and so forth, $1,500,000; annual maintenance cost of shelters, $40,000; initial cost of equipment, $38,000; annual upkeep of equipment, $13,000; estimated life of buses, 7 yr; estimated life of shelters 10 yr; estimated life of equipment, 5 yr; interest rate 10 percent per yr.

19. A small section of a highway can be dealt with in three different ways: (1) it can be left as it is (do nothing); (2) an embankment/bridge combination can be used; or (3) a cutting

tunnel combination can be provided. Using whatever method is suitable, determine which alternative has the best financial advantage.

Alternative	Life	Initial cost	Maintenance
1. As is (do nothing)	Perpetual	—	$35,000
2. Embankment/bridge	Case (a) Perpetual	$170,000	$1,000/yr
	Case (b) 50 yr	$200,000	$2,000/yr
3. Cutting tunnel	Case (a) Perpetual	$75,000	$700/yr
	Case (b) 50 yr	$300,000	$2,000/yr

20. In Example 3.15, Objectives 1 through 5 have certain weights assigned. Suppose that these weights were reassigned as follows:

    Objective 1: weight 9 out of 10

    Objective 2: weight 8 out of 10

    Objective 3: weight 7 out of 10

    Objective 4: weight 6 out of 10

    Objective 5: weight 5 out of 10

    What would be the result of your analysis?

21. An old hill road (A) is used by vehicles and the current cost of vehicle and social costs amount to $3 million per yr. The maintenance cost runs $20,000 per yr. Two alternatives, Routes B and C, are now proposed and their details are given in the table. Discuss which alternative should be adopted, including the one using the old hill road.

Alternative	A	B	C
Interest on construction	—	450	900
Purchase land	—	200	180
Vehicle and social	3000	1500	900
Maintenance	20	10	15

22. Different configurations of a highway network for a new town have been proposed. The capital cost and user savings are shown in the table. Which configuration would be most economical?

Alternative	Length (mi)	Capital cost (millions of dollars)	User savings (millions of dollars)
A	32	120	130
B	40	121	170
C	45	125	185
D	49	132	195
E	54	140	198
F	59	150	210
G	66	160	212
H	70	172	220

23. A small city has four transportation alternatives for a bus depot based on three criteria: architectural worth, cost, and public opinion. The architectural features have been assessed by a set of 10 architects and urban planners. The cost has been assessed by an independent firm of construction specialists. Public opinion has been obtained through the elected officials of the city council. The serviceable life of the depot is 40 yr. The results are:

Alternative	Initial cost (millions of dollars)	Maintenance per yr (thousands of dollars)	Architects in favor (out of 10)	Elected officials in favor (out of 25)
A	2.0	9.2	9	18
B	2.8	8.3	7	14
C	2.9	7.8	6	20
D	3.4	6.3	6	17

Provide an analysis for the decision maker to come to a rational decision and discuss your results.

24. A bus company serves passenger transport between two cities having a demand function $p = 1{,}200 - 7q$, where p is the price in dollars/trip and q is the riders/da. Currently the fare is \$30/trip. The company is planning to buy new buses for a total of \$1 million and reduce the fare to \$25/trip.

   (a) What is the current and anticipated daily and annual ridership?
   (b) What would be the change in consumer's surplus?
   (c) If the design life of the new buses is 7 yr and the annual interest rate is 8 percent, what is the feasibility of this investment?

25. The local public works department has six mutually exclusive proposals for consideration. Their expected life is 30 yr at an annual interest rate of 8 percent. Details follow:

Project	1	2	3	4	5
Cost (millions of $)	980	2450	6900	750	4750
Annual maintenance cost (millions of $)	50	70	250	30	180
Annual benefit	150	50	120	95	85

(a) Rank-order the proposals based on increasing first cost, considering maintenance as a cost.

(b) Knowing that the budget of the department is limited, which project would you select?

(c) If the largest net benefit is the criterion, which project would you select?

26. Two highway routes having the following characteristics are being considered for construction:

Route	A	B
Length (mi)	32	22
Initial cost (millions of $)	15	20
Maintenance cost (millions of $)/yr.	0.16	0.18
Vehicular traffic	10,000	15,000
Speed (mph)	40	30
Value of time ($/hr)	10	10
Operating cost, veh/mi (cents)	30	40

Compare the two alternative routes and recommend which one should be implemented.

Basic Microeconomics for Engineers and Planners 4

4.1 THE SCOPE OF ECONOMICS AND MICROECONOMICS

Economics is the study of how people and society end up choosing, with or without the use of money, to employ scarce productive resources that could have alternative uses to produce various commodities and distribute them for consumption, now and in the future, among various persons and groups in society. It analyzes the costs and benefits of improving patterns of resource allocation. In essence, the discipline of economics addresses the basic problem of scarcity, because the resources available to society are limited, but our material wants are relatively unlimited.

Economists conveniently divide the broad area of economics into two main streams: micro and macro. Microeconomics is concerned with the study of economic laws that affect a firm on a small scale. It deals with the economic behavior of individual units such as consumers, firms, and resource owners. Microeconomics studies the factors that determine the relative prices of goods and inputs. This chapter describes how economists measure the response of output to changes in prices and income. On the other hand, macroeconomics is the study of the wealth of society at the national and international scale. It deals with the behavior of economic aggregates such as gross national product and the level of employment.

Planning, designing, constructing, operating, and maintaining engineering facilities represent annual commitments of hundreds of billions of dollars, yet engineers, planners, and policy analysts who are responsible for such work often have little or no formal training or education in economics.

The area covered by microeconomics is large and, therefore, the reader is urged to refer to standard books on this and applied subjects for an in-depth understanding. The topics covered in this chapter are selective and have been included to provide readers with an introduction only.

Included in this chapter are the basic concepts of demand and supply functions that are fundamental to understanding, designing, and managing engineering systems. Much of the work conducted in engineering is devoted to specifying and estimating performance function (e.g., demand and supply).

4.2 SOME BASIC ISSUES OF ECONOMICS

Before we get into the details of microeconomics, it may be useful to examine a few basic concepts of economics. There is always the fundamental set of questions: What to produce? How to produce? How much? and For whom? In fact, the price system has come about because of scarcity, and consumers must bid on what they are prepared to pay for the goods that they desire. Economic goods are scarce and limited, while free goods are available for no cost, such as free air. Also, there is a difference between consumer goods and capital goods. The former directly satisfies the demands of the consumer; the latter indirectly satisfies consumer demands (e.g., real capital).

Another important concept is opportunity cost. This concept is connected to scarcity because scarcity forces people to make choices and trade-offs. Getting more of one thing implies getting less of another when one's budget is fixed. Also, when the units of a good are interchangeable, each good should be valued by the goods' marginal value (the value of the use that would be forgone if there were one less unit of the good).

The demand curve shows the relationship between the price of a good and the quantity of the good purchased. In Figure 4.1, for example, the demand for bicycles shows that an increase in the price of bicycles from \$200 to \$250 decreases the demand from 300 per wk to 200. According to the law of demand, an increase in price decreases the quantity demanded but everything else remains the same and, as such, most demand curves are negatively sloped.

4.3 DEMAND FOR GOODS AND SERVICES

In general, a firm is an enterprise or company that produces a commodity or service for profit. If the total revenue, the product of the unit selling price times the number of items sold, is greater than the cost to produce them, the firm makes a profit; if not, it suffers a loss. Therefore, the

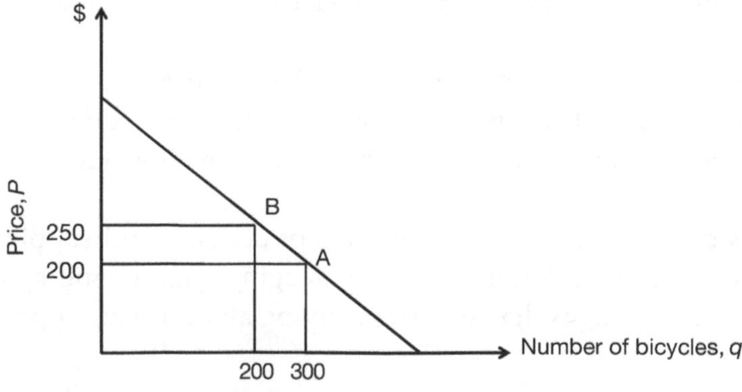

Figure 4.1 Demand for bicycles

efficiency of a firm manufacturing a commodity is an important consideration for its success. A firm's selling price of a good is equally important because it has a direct bearing on the level of production and its ability to gain a profit.

The demand for goods and services, in general, depends largely on consumers' income and the price of the particular good or service relative to other prices. For example, the demand for automobiles depends on the income of the buyers. The demand curve shows the relationship between the price of a good and the quantity of goods purchased.

A demand function for a particular product represents the willingness of consumers to purchase the product at alternative prices. An example of a linear demand function for scooters is shown in Figure 4.2a. Such a demand function is useful for predicting prices over a wide range of conditions. This demand function assumes a particular level and distribution of income, population, and socioeconomic characteristics. Note that it is an aggregate demand curve, representing the number of scooters demanded at different prices by various people. Functionally:

$$q = \alpha - \beta p \tag{4.1}$$

where q is the quantity of scooters demanded and α and β are constant demand parameters. The demand function is drawn with a negative slope expressing a familiar situation where a decrease in perceived price usually results in an increase in the quantity demanded, although this is not always true.

Figure 4.2b shows a series of shifted demand curves, representing changes in the quantity of scooters demanded due to variables other than the perceived price. Naturally, at a price p_0, one could expect different quantities q_1, q_2, and q_3 to be demanded, as the demand curve changes from D_1 to D_2 and D_3. If the curve shifts upward (from D_1 to D_3), it probably indicates an increase in income of the people buying scooters.

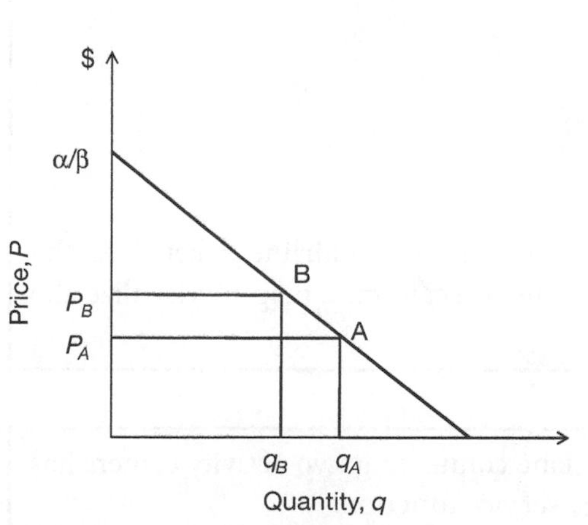

Figure 4.2a Typical demand function

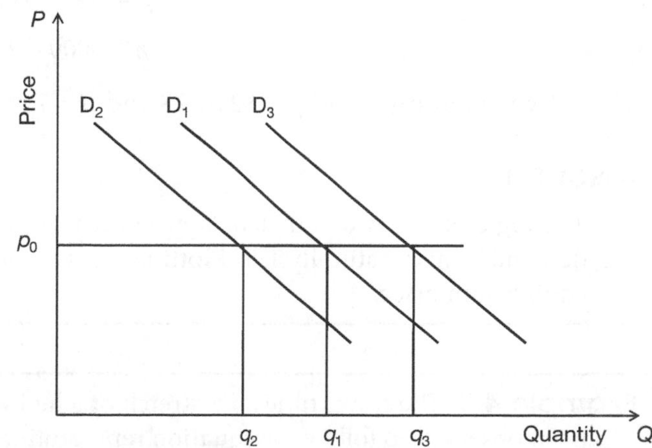

Figure 4.2b Shifted demand curves

It is important to distinguish short-run changes in quantity due to price changes represented by movement along a single demand curve, as shown in Figure 4.2a, from long-run changes due to activity or behavioral variables, represented by shifts in demand functions as shown in Figure 4.2b.

4.4 DEMAND, SUPPLY, AND EQUILIBRIUM

We have seen that the demand function is a relationship between the quantity demanded of a good and its price. In a similar manner, the supply function (or service function) represents the quantity of goods a producer is willing to offer at a given price, for example, bus seats or tons of wheat at given prices. If the demand and supply function for a particular good is known, then it is possible to deal with the concept of equilibrium. Equilibrium is said to be attained when factors that affect the quantity demanded and those that determine the quantity supplied result in being statically equal (or converging toward equilibrium). According to the law of supply, an increase in the price increases the quantity supplied while everything else remains the same. The supply curve could shift toward the right because of changes in price or production technology. Simple examples will serve to illustrate equilibrium between demand and supply.

Example 4.1 An airline company has determined the price of a seat on a particular route to be $p = 200 + 0.02n$. The demand for this route by air has been found to be $n = 5000 - 20p$, where p is the price in dollars and n are the number of seats sold per day. Determine the equilibrium price charged and the number of seats sold per day.

Solution

$$p = 200 + 0.02n$$

$$n = 5000 - 20p$$

These two equations yield $p = \$214.28$ and $n = 714$ seats.

Discussion

The logic of the two equations appears reasonable. If the price of an airline ticket rises, the demand would naturally fall. Plotting the two equations to scale may help to visualize the equilibrium price.

Example 4.2 The travel time on a stretch of a highway lane connecting two activity centers has been observed to follow the equation representing the service function:

$$t = 15 + 0.02v$$

Continues

Example 4.2: *Continued*

where t and v are measured in minutes and vehicles per hour, respectively. The demand function for travel connecting the two activity centers is $v = 4000 - 120t$

(a) Sketch these two equations and determine the equilibrium time and speed of travel.
(b) If the length of the highway lane is 20 mi, what is the average speed of vehicles traversing this length?

Solution Solving the two equations:

$$t = 15 + 0.02v$$

$$v = 4000 - 120t$$

results in the following values for v and t at equilibrium (as shown in Figure E4.2):

$$v = 647 \text{ veh/hr}$$

$$t = 27.94 \text{ min}$$

$$\text{Speed} = \frac{20 \times 60}{27.94} = 42.95 \text{ mph}$$

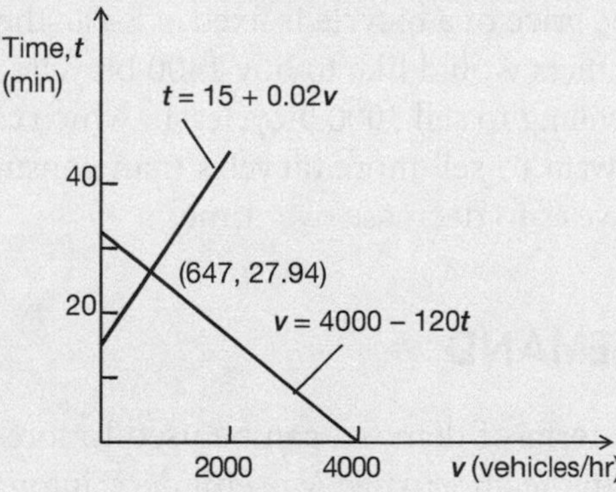

Figure E4.2 Static equilibrium of demand and supply

Discussion

It is customary to plot price, time, or price units on the vertical axis and quantity units on the horizontal axis.

If the price of a good is less than the equilibrium price, there will be a shortage of that good. Let us look at Figure 4.3 that shows the number of bicycles sold per week. At a price of $300, which is the equilibrium price, the number of bicycles supplied equals the quantity demanded so that

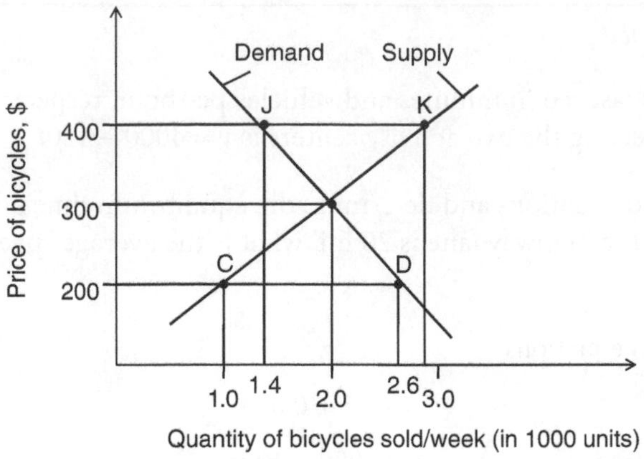

Figure 4.3 Quantity of bicycles sold per week

there is neither a shortage nor a surplus of bicycles. However, if the price is less than the equilibrium price, there will be a shortage of bicycles. For example, if the price is at $200, the market will be at point C on the demand curve. Consumers would like to buy 2600 bicycles but producers are willing to sell only 1000 bicycles (at Point D). This results in a shortage of 1600 bicycles.

By the same token, if the initial price exceeds the equilibrium price there will be a surplus of bicycles. For example, if the price of a bicycle is fixed at $400, the market will be at point J on the demand curve and consumers would like to buy 1400 bicycles. However, at point K on the supply curve, producers are willing to sell 3000 bicycles, in which case there is a surplus of 1600 bicycles. Because producers want to sell more bicycles than consumers are willing to buy, one would expect the price of bicycles to decrease over time.

4.5 SENSITIVITY OF DEMAND

Knowledge of the functional form of demand can be used to forecast changes in the quantity caused by specified changes in the short run. A useful descriptor for explaining the degree of sensitivity to a change in price (or some other factor) is the elasticity of demand (e_p).

If $q = \alpha - \beta p$ (see Equation 4.1), and e_p = the percentage change in quantity demanded that accompanies a 1 percent change in price, then:

$$e_p = \frac{\delta q/q}{\delta p/p} = \frac{\delta q}{\delta p} \times \frac{p}{q} \tag{4.2}$$

where δq is the change in quantity that accompanies δp the change in price. For a linear demand, Equation 4.2 can be used to arrive at *arc price elasticity*:

$$\text{Arc price elasticity} = \frac{\delta q}{\delta p} \times \frac{p}{q} = \frac{q_1 - q_0}{p_1 - p_0} \times \frac{(p_1 + p_0)/2}{(q_1 + q_0)/2} \tag{4.3}$$

where q_0 and q_1 represent the quantity demanded corresponding to prices p_0 and p_1, respectively. Also, for a linear demand function, we can determine the elasticity with respect to price by taking the derivative of the demand function (Equation 4.1). Thus:

$$e_p = \frac{\delta q}{\delta p} \times \frac{p}{q} = \frac{-\beta p}{q} \tag{4.4}$$

or after substitution for p in the equation:

$$e_p = 1 - \frac{\alpha}{q} \tag{4.5}$$

Example 4.3 An aggregate demand function is represented by the equation:

$$q = 200 - 10p$$

where q is the quantity of a good and p is the price per unit. Find the price elasticity of demand when:

$$q = 0, q = 50, q = 100, q = 150, \text{ and } q = 200 \text{ units}$$

corresponding to:

$$p = 20, p = 15, p = 10, p = 5, \text{ and } p = 0¢$$

Solution Using Equation 4.5:

$$e_p = 1 - \frac{\alpha}{q} \text{ where } \alpha = 200$$

$$e_5 = 1 - \frac{200}{150} = -0.333$$

$$e_{10} = 1 - \frac{200}{150} = -1$$

$$e_{15} = 1 - \frac{200}{50} = -3$$

$$e_{20} = 1 - \frac{200}{0} = -\infty$$

Figure E4.3a illustrates the demand function and the five values of p and q. When the elasticity is less than −1 (i.e., more negative than −1), the demand is described as being elastic, meaning that the resulting percentage change in quantity will be larger than the percentage change in price. In this case, demand is relatively sensitive to price change. However, when the elasticity

Continues

Example 4.3: *Continued*

is between 0 and −1, the demand is described as being inelastic or relatively insensitive. These ranges are shown in Figure E4.3b.

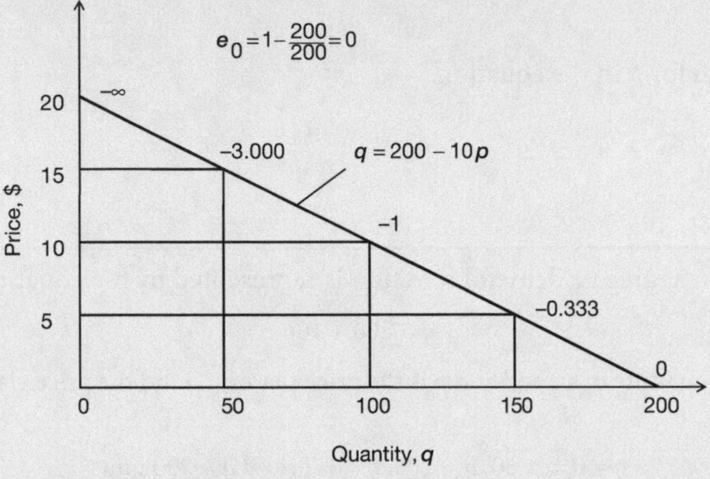

Figure E4.3a Demand function showing elasticities at various quantities

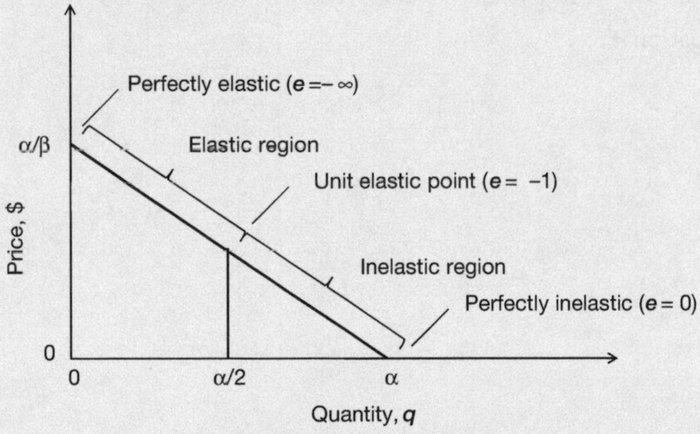

Figure E4.3b General case of a linear demand function showing elasticities

Discussion

From Figure E4.3a, it is obvious that when the price per unit is 20¢, no units are bought. Also, when nothing is charged per unit, 200 units are bought. Notice that the price elasticity for this system varies from 0 to −∞, with unit elasticity when $p = 10$.

A linear demand curve has several interesting properties. Notice, as one moves down the demand curve, the price elasticity of demand becomes smaller (i.e., more inelastic). In fact, the elasticity at a given point equals the length of the demand line segment below the point divided by the length of the line segment above it. Another point to note is that the slope of the line is constant, but the *elasticity changes* from ∞ at the top, where the demand line intersects the vertical axis, to zero, where the demand line intersects the horizontal axis. Because elasticity changes along the demand curve, it is essential to specify over what range of prices or quantity the elasticity is measured.

Example 4.4 When the admission rate to an amusement park was \$5 per visit, the average number of visits per person was 20 per yr. Since the rate has risen to \$6, the demand has fallen to 16 per yr. What is the elasticity of demand over this range of prices?

Solution

Arc price elasticity, $e_p = \frac{q_1 - q_0}{p_1 - p_0} \times \frac{(p_1 + p_0)/2}{(q_1 + q_0)/2} = -\frac{\Delta q}{\Delta p} \times \frac{(p_1 + p_0)/2}{(q_1 + q_0)/2} = -\frac{4}{1} \times \frac{(5+6)/2}{(16+20)/2} = -1.22$

Therefore, the arc price elasticity is elastic.

Discussion

Note that there are problems connected with arc price elasticity because it will differ from point elasticity, the difference increasing as Δp or Δq increase. Also note that elasticity is a unit-free measure of the percent change in quantity demanded (or supplied) for a 1% change in price.

4.6 FACTORS AFFECTING ELASTICITIES

4.6.1 Income Elasticities

Income elasticities have a special significance in engineering and are denoted by:

$$e_i = \frac{\%\text{ change in quantity of good demanded}}{\%\text{ change in income}} \tag{4.6}$$

A good is considered to be *normal* if the demand for the good goes up when a consumer's income increases ($e_i > 0$). Most goods are normal. A good is considered a *superior* good if it goes up in demand when a consumer's income increases and its share in income also goes up ($e_i > 1$). On the other hand, a good is inferior if the demand for the good goes down when a consumer's income goes up. In North America, an automobile is considered a superior good, whereas spending money on traveling by mass transit is often considered an inferior good. Gourmet food is a superior good while cheap beer is an inferior good.

4.6.2 Price Elasticities

In general, consumers buy more of a good than usual when the price goes down, and they buy less than usual when the price goes up. Some factors that affect price elasticity are:

- If a consumer spends a substantial percentage of income on, say, transportation, the more willing he will be to search diligently for a substitute, if the price of transportation goes up.
- The narrower the definition of a good, the more substitutes the good is likely to have and, thus, the more elastic its demand will be. For example, the demand for Toyotas is more elastic than the demand for automobiles and the demand for automobiles is more elastic than the demand for transportation.
- When consumers find out that availability of substitutes is easy, the more elastic the demand will be. Advertising plays an important role in making substitutes available to consumers. In the same context, the more time consumers have to find substitutes, the more elastic the demand becomes.
- Those goods that consumers consider necessities usually have inelastic demands, whereas goods considered by consumers to be luxuries usually have elastic demands. For instance, eyeglasses for a consumer are a necessary good, with few substitutes, whereas a vacation trip to Europe is a luxury good with several substitutes.

4.6.3 Elasticity and Total Revenue

It is possible to tell what the total revenue (price multiplied by output) of a firm is likely to be if the price of a unit changes:

$$e_i = \frac{\text{\% change in quantity of units demanded}}{\text{\% change in price}} \tag{4.7}$$

If $e_i > 1$, price and total revenue are *negatively* related (or demand is elastic); therefore, an increase in price will reduce total revenue, but a decrease in price will increase total revenue. If $e_i < 1$, price and total revenue are *positively* related (or demand is inelastic), in which case, an increase in price will increase total revenue and a decrease in price will decrease total revenue. If $e_i = 1$, total revenue will remain the same whether the price goes up or down.

Example 4.5 A bus company's linear demand curve is $p = 10 - 0.05q$, where p is the price of a one-way ticket and q is the number of tickets sold per hour. Determine the total revenue along the curve.

Continues

Example 4.5: *Continued*

Solution

$$p = 10 - 0.05q$$
$$R = q \times p = q(10 - 0.05q)$$

where R = total revenue

$$R = 10q - 0.05q^2$$

$$\frac{dR}{dq} = 10 - (0.05 \times 2)\,q$$

and this is equal to zero when R is maximum. Therefore, $q = 100$ when R is 500 (maximum), see Figure E4.5.

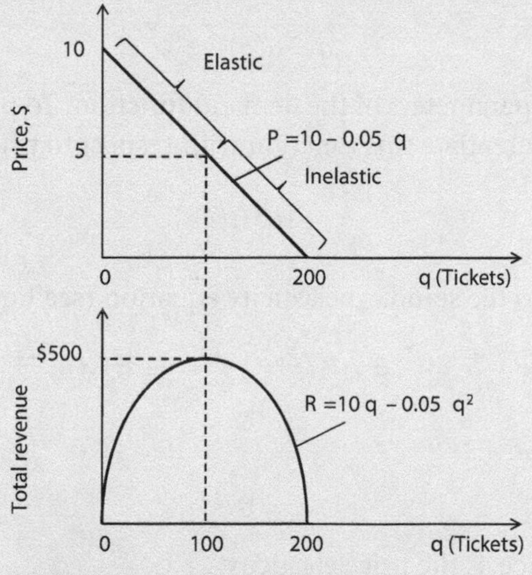

Figure E4.5 Total revenue curve

Discussion

As shown in Figure E4.5, starting from a price of $10 when hardly any tickets are sold and decreasing the price eventually to half ($5), the revenue steadily increases to a maximum of $500/hr (over the elastic portion). After that, the revenue decreases as the price further decreases and finally approaches zero when the demand approaches 200 (over the inelastic portion).

4.6.4 Price Elasticity of Supply

Just like we have calculated price elasticity of demand, we can calculate the price elasticity of supply. It is the percentage of change in the quantity of a product supplied for a 1 percent change in price. While the formulas for price elasticity of supply are identical to those for the price elasticity of demand, care has to be taken to measure quantity and price changes along its own specific curve or schedule.

4.7 KRAFT DEMAND MODEL

We occasionally come across a demand function where the elasticity of demand for a good with respect to its price is essentially constant. The demand function for such a situation corresponds to the equation:

$$q = \alpha(p)^{\beta} \tag{4.8}$$

where α and β are constant parameters of the demand function. To prove that this function has a constant elasticity, we differentiate this function with respect to price:

$$\frac{dq}{dp} = \alpha\beta(p^{\beta-1})$$

and substitute the result into the standard elasticity equation (see Equation 4.2):

$$e_p = \frac{dq}{dp} \times \frac{p}{q} = \alpha\beta(p^{\beta-1}) \times \frac{p}{q} = \alpha\beta(p)^{\beta} \times \frac{1}{q}$$

Substituting q from Equation 4.8:

$$e_p = \beta$$

Thus, β, the exponent of price, is the price elasticity.

Example 4.6 The elasticity of transit demand with respect to price has been found to be equal to −2.75, which means that a 1% increase in transit fare will result in a 2.75 decrease in the number of passengers using the system. A transit line on this system carries 12,500 passengers per day, charging 50¢ per ride. The management wants to raise the fare to 70¢ per ride. What advice would you offer to management?

Solution

$$q = \alpha(p)^{\beta}$$
$$12{,}500 = \alpha(50)^{-2.75}$$
$$\alpha = 12{,}500 \times (50)^{2.75} = 5.876 \times 10^{8}$$
$$q = 5.876 \times 10^{8} \times (70)^{-2.75} = 4{,}955$$

Continues

Example 4.6: *Continued*

Therefore, the increase in fare from 50 to 70¢ (a 40% increase) is likely to reduce the patronage on this line from 12,500 passengers per day to 4,955 (a 60.36% decrease). In terms of revenue, the results are:

$$50¢\ /\text{rider} \times 12{,}500 \text{ passengers} = \$6{,}250$$
$$70¢\ /\text{rider} \times 4{,}955 \text{ passengers} = \$3{,}468.50$$
$$\text{Loss in revenue} = \$2{,}781.50$$

Management should be advised not to increase the fare.

Discussion

In general, it has been observed that when the price is elastic (e.g., −2.75), raising the unit price will result in total loss, but lowering the price will result in total gain. The converse is also true; if the price is inelastic, raising the unit price will result in total gain, whereas lowering the unit price will result in total loss. Students may like to graph these cases to discover why.

Example 4.7 The demand function for transportation from the suburbs to downtown in a large city is:

$$Q = T^{-0.3}\, C^{-0.2}\, A^{0.1}\, I^{-0.25}$$

Where

Q = Number of transit trips
T = Travel time on transit (hours)
C = Fare on transit (dollars)
A = Cost of automobile trip (dollars)
I = Average income (dollars)

(a) There are currently 10,000 persons per hour riding the transit system at a flat fare of $1 per ride. What would be the change in ridership with a 90¢ fare? What would be the company gain or loss per hour?

(b) By auto, the trip costs $3 (including parking). If the parking charge were raised by 30¢, how would it affect the transit ridership?

(c) The average income of auto riders is $15,000 per yr. What raise in salary will riders require to cover their costs in view of the change in the parking charge noted in (b)?

Continues

Example 4.7: *Continued*

Solution

(a) This is essentially a Kraft model. The price elasticity of demand for transit trips is:

$$\frac{\delta Q/Q}{\delta C/C} = -0.2$$

This means that a 1% reduction in fare would lead to a 0.2% increase in transit patronage. Because the fare reduction is (100 – 90)/100 = 10%, one would expect an increase of 2% in patronage. Patronage would now be 10,000 + (10,000 × 0.02) = 10,200.

10,000 passengers at $1.00/ride = $10,000
10,200 passengers at $0.90/ride = $9,180

The company will lose $820 per hour.

(b) The automobile price cross elasticity of demand is 0.1, or:

$$\frac{\delta Q/Q}{\delta A/A} = 0.1$$

This means that a 1% rise in auto costs (including parking) will lead to a 0.1% rise in transit trips. A $0.30 rise is 10% of $3. Therefore, a 10% rise in auto cost would raise the transit patronage by 1%, from 10,000 to 10,100 riders.

(c) The income elasticity should be considered first:

$$\frac{\delta Q/Q}{\delta I/I} = -0.25$$

which means that a rise in income of 1% will result in a 0.25% decrease in transit patronage, or $\delta Q/Q = 1\%$ from (b). Therefore:

$$\frac{1\%}{\delta I/I} = -0.25 \text{ and } \frac{\delta I}{I} = \frac{1\%}{-0.25} = -0.04 = -4\%$$

So a 4% increase in income would cover a 30¢ increase (or 10% increase) in auto cost. If the average income was $15,000, a $600 raise in salary would change the minds of those auto drivers who were planning to ride the transit system.

4.8 DIRECT AND CROSS ELASTICITIES

The effect of change in the price of a good on the demand for the *same* good is referred to as *direct* elasticity. However, the measure of responsiveness of the demand for a good to the price of another good is referred to as *cross* elasticity.

When consumers buy more of Good A when Good B's price goes up, we say that Good A is a substitute for Good B (and Good B is a substitute for Good A). For example, when the price of gasoline goes up, travelers tend to use more transit. On the other hand, when consumers buy less of Good A when Good B's price goes up, we say that Good A is a complement to Good B. In general, complementary goods are ones that are used together. Thus, when the price of downtown parking goes up, the demand for driving a car downtown goes down; the demand for an equivalent trip by transit or by taxi to downtown goes up.

Goods are substitutes when their cross elasticities are positive, and goods are complements when their cross elasticities of demand are negative.

Example 4.8 A 15% increase in gasoline costs has resulted in a 7% increase in bus patronage and a 9% decrease in gasoline consumption in a midsized city. Calculate the implied direct and cross elasticities of demand.

Solution

Let

p_0 = Price of gas before
p_1 = Price of gas after
q_0 = Quantity of gas consumed before
q_1 = Quantity of gas consumed after

Then for *direct elasticity*:

q_0 (gas) × 0.91 = q_1 (gas)
p_0 (gas) × 1.15 = p_1 (gas)

$$e = \frac{\delta q/q}{\delta p/p} = \frac{q_0 - q_1}{p_0 - p_1} \times \frac{(p_0 + p_1)/2}{(q_0 + q_1)/2} = \frac{-0.09}{0.15} \times \frac{2.15/2}{1.91/2} = -0.675$$

Therefore, the change in gasoline consumption with respect to gasoline cost is inelastic.

b_0 = Bus patronage before
b_1 = Bus patronage after

For *cross elasticity*:

b_0 (bus) × 1.07 = b_1 (bus)
p_0 (gas) × 1.15 = p_1 (gas)

$$e = \frac{\delta b/b}{\delta p/p} = \frac{b_0 - b_1}{p_0 - p_1} \times \frac{(p_0 + p_1)/2}{(b_0 + b_1)/2} = \frac{0.07}{0.15} \times \frac{2.15/2}{2.07/2} = +0.485$$

Continues

Example 4.8: *Continued*

Discussion

In the case of direct elasticity we are calculating the percent change in gasoline consumption due to a 1% change in the price of gasoline, while in the case of cross elasticity we are calculating the percent change in bus patronage due to a 1% change in gasoline price.

4.9 CONSUMER SURPLUS

Consumer surplus is a measure of the monetary value made available to consumers by the existence of a facility. It is defined as the difference between what consumers might be willing to pay for a service and what they actually pay. A patron of a bus service pays a fare of, say, 50¢ per trip but would be willing to pay as much as 75¢ per trip, in which case, his surplus is 25¢.

The demand curve can be considered an indicator of the utility of the service in terms of price. The consumer surplus concept is shown in Figure 4.4a. The area *ABC* represents the total consumer surplus. Maximization of consumer surplus is indeed the maximization of the economic utility of the consumer. In project evaluation, the use of this concept is useful for evaluating transit systems and irrigation projects, for example.

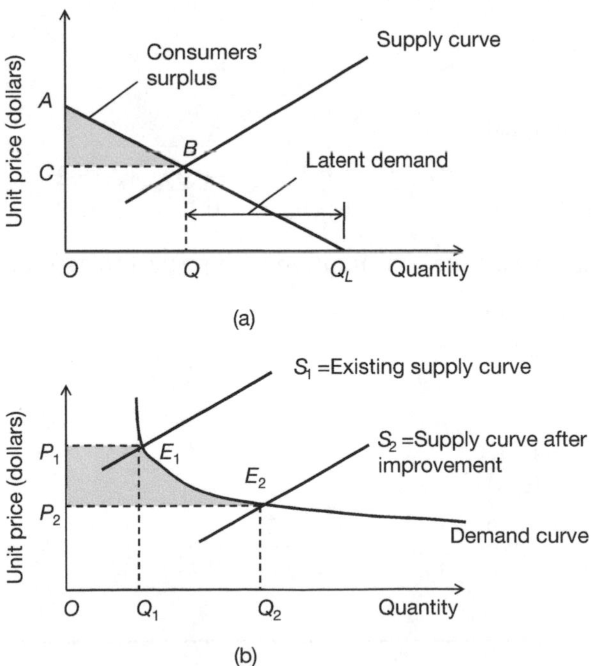

Figure 4.4 (a) Consumer surplus concept and (b) change in consumers' surplus

In general, an improvement of a facility can be measured in terms of the change in consumer surplus. Figure 4.4b indicates the case of a street having a traffic supply curve S_1, intersecting a demand curve at E_1. An additional lane has been added, shifting the supply curve to S_2 and, therefore, intersecting the demand curve at E_2. The change in consumer surplus can be quantified as the area of trapezoid $P_1P_2E_2E_1$, or $(P_1 - P_2)(Q_1 - Q_2)/2$. The consumer surplus is normally defined as the difference between the maximum amount that consumers are willing to pay for a specified quantity of a good rather than going without it. In general, referring to Figure 4.4a, *AOQB* is equal to the total community benefit, *BCOQ* is equal to the market value, and *ACB* is equal to the consumer surplus or net community benefit.

Figure 4.4a illustrates an additional concept that is useful to engineers: *latent demand*. Note that travelers between *Q* and the point of intersection of the demand function and the abscissa do not currently make trips, but they would do so if the price per trip were lower than the equilibrium price. The number of such potential travelers is popularly called latent demand. The concept can be used in several ways; for instance, a transit operator hoping to increase transit patronage by introducing a discount rate valid for nonpeak hours may like to investigate latent demand. Indeed, the quantity of trips demanded, if the price of a trip were zero (free transit), would be $Q_L - Q$ as indicated in Figure 4.4a.

Example 4.9 A bus company with an existing fleet of 100 buses that seat 40 increases its fleet size by 20% and reduces its fare of $1 to 90¢ per ride. Calculate the change in consumer surplus and the price elasticity of demand. Assume that the existing buses had a load factor of 90%, and it is anticipated that the improvement will result in a 95% load factor. Does the company lose money? Assume that all the buses in the fleet are being used during the peak hours. (Note: the vehicle load factor is a measure of seat availability and a load factor of 1.0 means that every seat is occupied.)

Solution With the existing situation:

100 buses × 40 seats × 0.90 (load factor) = 3600 persons/hr
Revenue: 3600 × 1.00 = $3600/hr

With the improved situation:

120 buses × 40 seats × 0.95 = 4560 persons/hr
Revenue: 4560 × 0.90 = $4104/hr

The company gains $4104 – $3600 = $504/hr.

Change in consumer surplus = (1.00 – 0.90)(3600 + 4560)/2 = $408/hr

Price elasticity of demand:

$$e = \frac{q_1 - q_0}{p_1 - p_0} \times \frac{(p_1 + p_0)/2}{(q_1 + q_0)/2} = \frac{3{,}600 - 4{,}560}{1.00 - 0.90} \times \frac{(1.90/2)}{8{,}160/2} = -2.235$$

Continues

Example 4.9: *Continued*

Discussion

This is an interesting situation. Even if the number of buses were not increased and the prices were lowered as indicated, resulting in an increase in the load factor, there would be an increase in total revenue of $200 because the price elasticity is elastic (−2.235). Naturally, with more buses being deployed the situation is even better. Consumer surplus is a good way to compare the two alternatives.

Example 4.10 Bob is willing to pay up to $10 to travel by bus once every month to visit his family, $8 to travel twice, and $6 to travel three times for the same purpose.

(a) If the price of a bus ticket to visit his family is $7, what is Bob's consumer surplus?

(b) If the bus company offers three tickets per month for a flat price of $19, will Bob accept the deal?

(c) What is the maximum the bus company should charge for Bob to take the three-ticket offer?

Solution

(a) Bob's individual consumer surplus is $4. For the first trip, Bob's consumer surplus is $10 − 7 = $3 and for the second trip it is $1, which adds up to $4. (He will not go for the third trip.)

(b) Yes, because for three tickets his consumer surplus is [$(10 + 8 + 6) − $19)] = $5, which is better than buying the tickets individually. (Note that $19 is the flat price for the three tickets.)

(c) At best, Bob will pay $20 for the three-ticket offer, which is the amount that would be the same consumer surplus ($4) as the option of buying the tickets separately ($24 − $20 = $4).

Discussion

Note that the bus company makes more money from Bob with the package deal. Also, note that in this problem we are calculating an individual's consumer surplus and not an aggregate (or total) consumer surplus as in Example 4.9.

4.10 COSTS

It is essential to have a knowledge of costs or the value of a product or service. To establish the true cost of a product, the analyst must determine, for example, where delivery takes place, who pays for transportation, and who pays for insurance and storage.

Before finding average costs (ACs) it is convenient to break down costs into fixed costs (FCs), variable costs (VCs), and total costs (TCs). FCs are inescapable costs and do not change with use. If a plant is producing 500 trucks per day and the plant costs $1 million to run whether one truck is produced or a hundred, the FC is $1 million. Naturally, the FC per truck produced will be reduced with increasing production, even though the FC itself remains unchanged. VCs, on the other hand, increase with output or production. If, for example, the labor cost for assembling one truck is $1,000, it is likely that this labor cost for assembling two trucks is $1,900. The TCs of production are the sum of fixed and VCs and will increase with production. For any particular level of production the AC of a single unit (one truck) can be found by dividing the TC by the number of units corresponding to the TC.

4.10.1 Laws Related to Costs

Two concepts related to costs are of importance in engineering. The first, the *law of diminishing returns* states that although an increase in input of one factor of production may cause an increase in output, eventually a point will be reached beyond which increasing units of input will cause progressively less increase in output. The second, the *law of increasing returns to scale* states that in practice the production of units is often likely to increase at a faster rate than the increase of factors of production. This phenomenon may be due to any number of factors, including technological features or the effects of specialization.

4.10.2 Average Cost

The mathematical relationship connecting the TC (C) of a product to the unit VC (c) and magnitude of the output (q) can be written as:

$$C = \alpha + \beta(q)$$

where parameter α equals the FC of production and the function $\beta(q)$ equals the VC of production and the unit VC $c = \dfrac{\beta(q)}{q}$.

The AC $\bar{c}$ of each item produced is equal to:

$$\bar{c} = \frac{C}{q} = \frac{\alpha + \beta(q)}{q} = \frac{\alpha}{q} + \frac{\beta(q)}{q} = \frac{\alpha}{q} + c \tag{4.9}$$

The relationships of the total and AC functions are shown in Figure 4.5. Notice that in this particular case, as output q increases, the AC of production decreases, and then increases at higher levels of production. When the production level reaches q', the AC is a minimum $\bar{c}$. *Economy of scale* is defined as a decrease in AC as output increases. There is an economy of scale for production levels between 0 and q'; beyond q', there is no economy of scale because the AC rises. This concept is useful to engineers in deciding whether additional capacity (or production) or growth would result in higher profits.

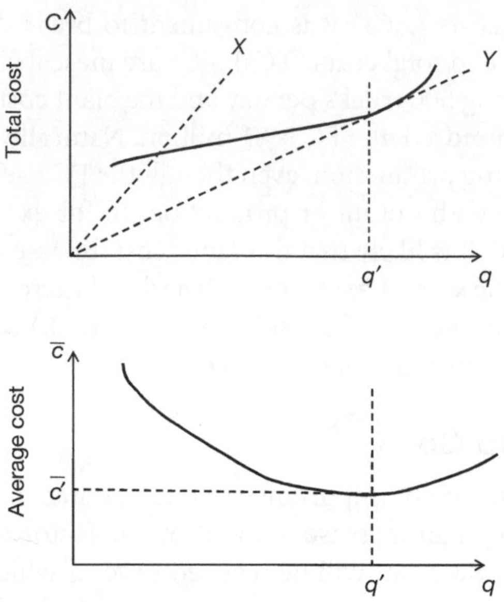

Figure 4.5 Total and average cost

4.10.3 Marginal Cost

The *marginal cost* (MC) of a product is the additional cost associated with the production of an additional unit of output. This is an important concept used several ways in engineering. An example will clarify this term. The cost of running a train system with a variable number of

Table 4.1 Costs associated with wagons per train

1 Number of wagons/train	2 Fixed cost (FC)	3 Variable cost (VC)	4 Total cost (TC)	5 Average cost (AC)	6 Marginal cost/unit (MC)
1	55	30	85	85.0	
2	55	55	110	55.0	25
3	55	75	130	43.3	20
4	55	105	160	40.0	30
5	55	155	210	42.0	50
6	55	225	280	46.7	70
7	55	315	370	52.9	90
8	55	425	480	60.0	110
9	55	555	610	67.8	130
10	55	705	760	76.0	150

wagons is given in Table 4.1. From this table of costs all other costs can be computed. Column 5 is obtained by dividing the TC given in Column 4 by the number of wagons per train shown in Column 1. The MC is calculated by subtraction of adjacent rows of TC.

The figures in Columns 1 to 4 of Table 4.1 are plotted as shown in Figure 4.6a. Similarly, the figures from Columns 5 and 6 are plotted as shown in Figure 4.6b. Note that the point of minimum cost ($40) occurs at the intersection of the AC and the MC curves. Also, the projection of this point to Figure 4.6a corresponds to the point where the gradient of the tangent drawn from the origin has a minimum slope. Note that theoretically, the AC and MC curves intersect each other at a value of x somewhere between 4 and 5. However, practically speaking, we accept a value equal to 4, where the two curves intersect each other and the minimum MC occurs.

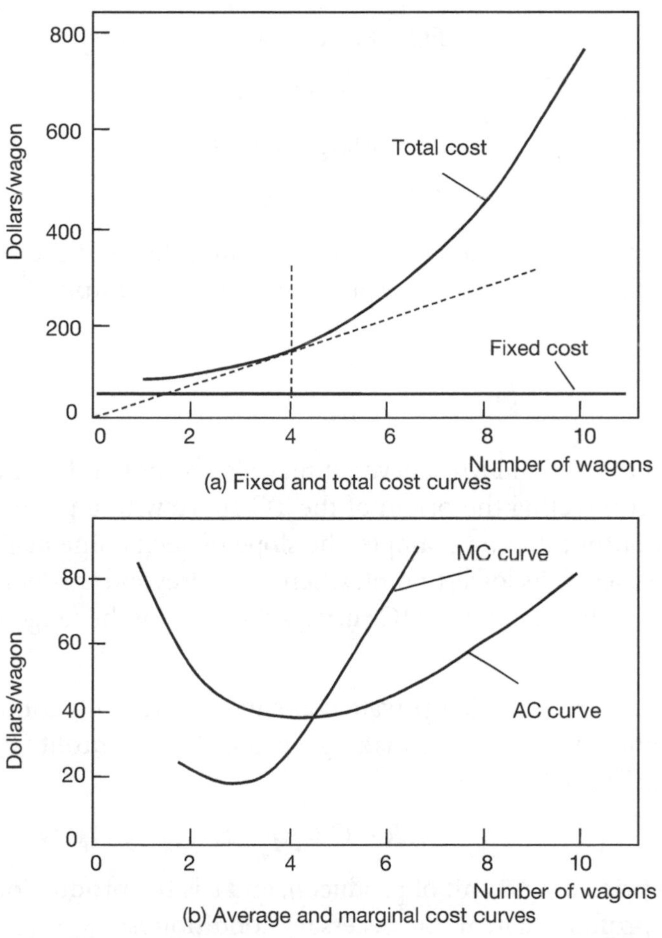

Figure 4.6 Total, average, and marginal costs

In general, we can summarize what has been demonstrated in our exercise in the train problem:

$$\text{Total cost} = TC(x) = FC + VC(x) \tag{4.10}$$

$$\text{Average cost } AC(x) = \frac{TC(x)}{x} = \frac{FC}{x} + \frac{VC(x)}{x} \tag{4.11}$$

$$\text{Marginal cost } = MC(x) = TC - TC(x-1) \tag{4.12}$$

where

TC = Total cost

FC = Fixed cost

VC = Variable cost

MC = Marginal cost

AC = Average cost

When the output is a continuous function, the differential form of the MC is used, in which the MC is the rate of change of TC with respect to a change in output. In this form, the equation is:

$$MC(x) = \frac{dTC(x)}{dx} = \frac{dVC(x)}{dx} \tag{4.13}$$

From the geometry of the AC and MC curves, it may also be noted that the AC is proportional to the slope of a line connecting the origin of the TC curve with a point on that curve corresponding to the total output. In our example, the slope of such a line begins at infinity at zero output and then decreases to its lowest point, when $x = 4$. Beyond this point of $x = 4$, the slope increases again. On the other hand, the MC curve is the slope of the tangent drawn at any point on the TC curve.

Cost and production. In general, a private company or firm will continue to produce and market a product or good as long as it is making a profit. The net profit (P) is equal to the total revenue (R) minus the TC (C).

$$P = R - C = pq - cq$$

where p is the selling price of one unit of product q, and c is the production cost of one unit.

To maximize net profits of a firm, the necessary condition is:

$$\frac{dP}{dq} = \frac{dR}{dq} - \frac{dC}{dq} = 0$$

or

$$\frac{dP}{dq} = \frac{d(pq)}{dq} - \frac{d(cq)}{dq} = 0$$

Thus:

$$\frac{d(pq)}{dq} = \frac{d(cq)}{dq}$$

Let

$$\text{MR} = \text{marginal revenue} = \frac{dR}{dq} = \frac{d(pq)}{dq}$$

$$\text{MC} = \text{marginal cost} = \frac{dC}{dq} = \frac{d(cq)}{dq}$$

Therefore

$$\text{MR} = \text{MC} \tag{4.14}$$

This equation indicates that to achieve the goal of maximizing profits, the firm should produce in such a way that MR equals MC.

Cost elasticity: The cost elasticity e_c is defined as the ratio of percentage change in cost C to the percentage change in supply q:

$$e_c = \frac{\%\text{ change in cost}}{\%\text{ change in supply}} = \frac{(\delta C/C) \times 100}{(\delta q/q) \times 100} = \frac{q}{C} \times \frac{\delta C}{\delta q}$$

In the limit when $\delta q \rightarrow 0$, $e_c = \frac{q}{C} \times \frac{dC}{dq}$. Rearranging terms:

$$e_c = \frac{dC/dq}{C/q} = \frac{\text{MC}}{\text{AC}} \tag{4.15}$$

Example 4.11 A transport company hauling goods by truck has a cost function, $C = 15q^{1.25}$ where C is the TC of supply q.

(a) Determine the AC and the MC.

(b) Prove that the cost elasticity is 1.25.

(c) Is there an economy of scale?

Continues

Example 4.11: *Continued*

Solution

(a) $\bar{c} = \dfrac{C}{q} = \dfrac{15q^{1.25}}{q} = 15q^{0.25}$ which is the AC

$$MC = \frac{dC}{dq} = (15 \times 1.25)q^{0.25} = 18.75q^{0.25}$$

(b) $e = \dfrac{MC}{AC} = \dfrac{18.75a^{0.25}}{15q^{0.25}} = 1.25$ (see Equation 4.5)

(c) Economy of scale does not exist because the AC increases with increased q.

4.11 CONSUMER CHOICE

A consumer who is operating on a fixed budget must decide how to allocate it among a large number of goods; for example, food, clothing, rent, and transportation. Her problem is to find the affordable consumption bundle that maximizes her utility. In such a situation a consumer subjectively makes use of indifference curves and a budget line. The indifference curve summarizes the consumer's subjective preferences about alternative consumer goods, while the budget line shows the limits (or constraints) imposed on the consumer.

Suppose a person gets satisfaction or utility from consuming just two goods, A and B, as shown in Figure 4.7a. To construct an indifference curve we can pick a series of points 1, 2, 3, 4, and 5 that generate the same level of satisfaction or utility for Goods A and B to this person and join these points by a smooth line. What about points 6, 7, 8, and 9, shown on this figure? All combinations above the indifference curve generate more satisfaction than those located on the curve, while those combinations below the curve generate less satisfaction. The shapes of indifference curves vary from one consumer to another.

The slope of the indifference curve is the marginal rate of substitution (MRS) between the two goods and indicates the rate at which this person is willing to substitute one good for another. Notice that the MRS changes as one goes down the curve.

An indifference map is a set of indifference curves. In general, a person's utility increases as he moves in the northeasterly direction to higher indifference curves as shown in Figure 4.7b.

A person's budget set includes all combinations of goods that he can afford given his income and the price of commodities (or goods). Considering only two goods, the budget line shows all combinations that can exhaust his budget. The slope of the budget line is the market tradeoff between two goods. The consumer's objective is to maximize utility subject to the budget constraint. In graphical terms, a person will choose a point on the highest feasible indifference curve as shown in Figure 4.8.

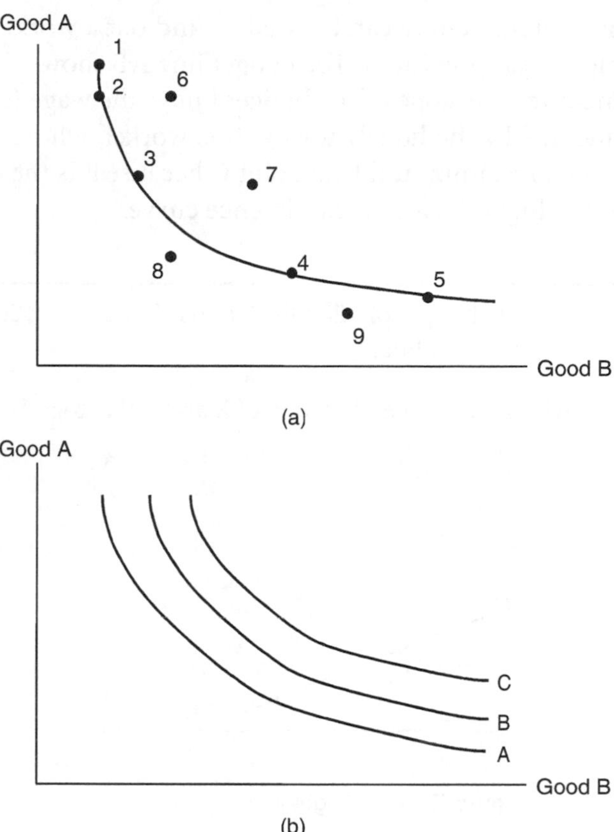

Figure 4.7 Indifference curves

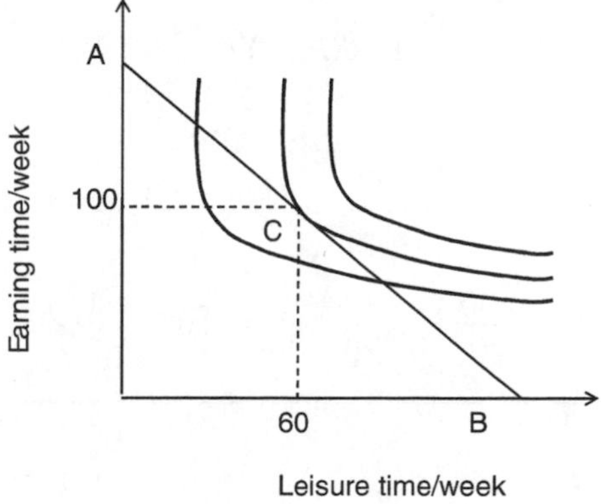

Figure 4.8 Budget line and indifference curves

In this figure the indifference curve can be used to find out a worker's choice between leisure time and labor time (or earning time). The budget line AB shows the rate at which he can trade leisure time for income. The slope of the budget line is the wage (a one-hour decrease in leisure time increases income by the hourly wage). This worker, whose budget line and indifference map are shown, will maximize utility at Point C because it is the combination of leisure and income that lies on the highest feasible indifference curve.

Example 4.12 Given a weekly budget of \$200 for Goods X and Y of \$20 and \$10 respectively, what is the equation of the budget line?

Solution The budget line, AB, includes a combination of X and Y that together are equal to \$200 (see Figure E4.12).

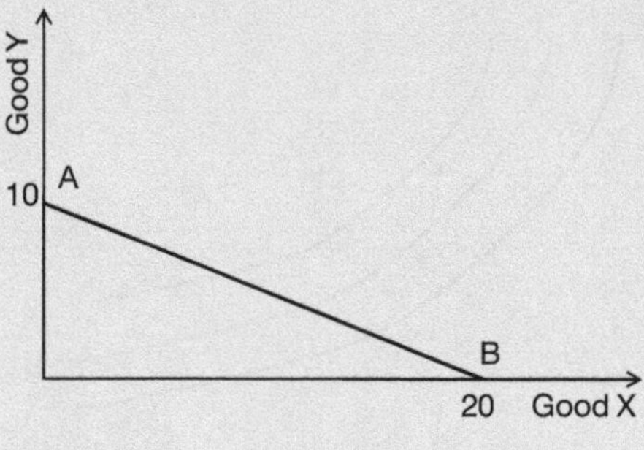

Figure E4.12 Budget line

Let I = Consumer income

Then:

$$I = XP_x + YP_y$$

or

$$Y = \frac{I - XP_x}{P_y} = \frac{I}{P_y} - \left(\frac{P_x}{P_y}\right) \cdot X$$

$$Y = \frac{200}{20} - \left(\frac{10}{20}\right) \cdot X = 10 - \left(\frac{1}{2}\right) \cdot X$$

Example 4.13 Figure E4.13 shows a budget line drawn along an indifference curve for a consumer who is primarily interested in two goods, X and Y. If the price of Good X is $50/unit:

(a) What is the consumer's income?

(b) What is the price of Good Y?

(c) What is the equation of the budget line?

(d) Find the MRS at equilibrium.

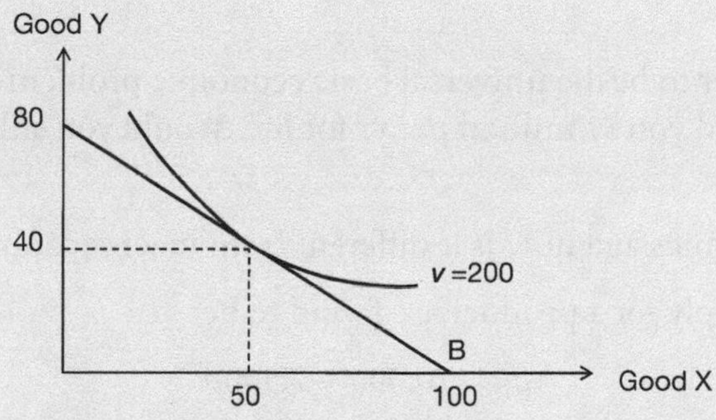

Figure E4.13 Budget and indifference curve

Solution

(a) The consumer's income (or budget) for these two goods must be 100 × 50 = $5,000 because she can afford to buy 100 units @ $50/unit of Good X if she does not choose to buy any units of Good Y.

(b) The price of Good Y per unit must be 5000/80 = $62.50

(c) The equation for the budget line:

$$P_x X + P_y Y = I, \text{ where } I = \text{Income for the two goods}$$

$$50X + 62.5Y = 5{,}000$$

(d) At equilibrium, $\text{MRS} = \frac{P_x}{P_y} = \frac{50}{62.5} = 0.8$

SUMMARY

Economics is the study of how people choose to allocate their scarce resources. Microeconomics is the branch of economics that studies the behavior of individual firms and applies it to business decision making. Terms such as demand, supply, elasticity, consumers' surplus, and latent demand are briefly described in this chapter. Readers may like to get additional information on various topics included in this chapter by referring to standard textbooks on microeconomics such as Pindyck and Rubinfield's *Microeconomics* (2008).

REFERENCE

Pindyck, R. and D. Rubinfield (2008). *Microeconomics*, 7th ed., Upper Saddle River, NJ: Prentice-Hall.

EXERCISES

1. What do you consider to be the universal basic economic problem? Suppose you won a national lottery that paid you $1 million per yr for life. Would you still be subject to economic scarcity?
2. What is microeconomics and how is it different from macroeconomics?
3. The demand and supply for a product are found to be:

$$q_D = 10{,}000 - 7{,}500p$$

and

$$q_S = 15{,}000p + 500$$

   (a) What is the equilibrium price per unit and the quantity?
   (b) If the supply changes to $q_S = 15{,}000p$, how would this affect the equilibrium price and quantity?

4. The performance function of a highway connecting a suburb with the business district can be represented by a straight line of the form $t = a + bq$ where t is the travel time in minutes, q is the traffic flow in vehicles per hour, and a and b are constants equal to 10 min and 0.01 min/veh/hr, respectively. The demand function, also represented by a straight line, is $q = c + dt$ where c and d are constants equal to 5000 veh/hr and −100 veh/hr/min, respectively.

   (a) Find the equilibrium flow (q') and the corresponding equilibrium time (t') algebraically and sketch the functions.
   (b) If the length of this highway is 22.5 mi, what is the average speed of vehicles along this highway?
   (c) It is proposed to improve this highway such that constant b is now 0.005. What would be the new values of t' and q' and the average speed on this highway?

5. The population growth and added commercial activity have affected the highway link described in Example 4.2, and improvements planned for this highway are reflected by equations representing the new conditions. The new equations are:

$$t_1 = 15 + 0.004v_1$$
$$v_1 = 4333 - 130t_1$$

   (a) Sketch the existing and proposed service and demand functions.
   (b) What are your estimates of the equilibrium time and vehicle flow on the proposed link?
   (c) If the length of this link is 20 mi, what is the average speed over the link?

6. The demand for travel over a stretch of highway is given by the function:

$$q = \frac{2000}{t + 1}$$

   where q is the travel flow, and t is the travel time (min). Plot this demand function and calculate the change in vehicle-hours of travel when travel time increases from 10 to 15 min due to road congestion.

7. The demand function for a transit system can be represented by a straight line connecting fare per person and ridership. Observations made on this system resulted in the following:

   When the fare was \$1.50 per ride, the ridership per hour was 2000; when the fare was raised to \$2 per ride, the ridership dropped to 1000. What is the equation of the demand function? What would the patronage be if the fare were:

   (a) 50¢ per ride?
   (b) Zero?

8. A bus company is charging a flat rate of 50¢ per ride to any part of the city and has a patronage of 500,000 per day. The company has decided to raise the fare to 60¢ per ride and it is estimated that 470,000 people will ride the buses. Calculate:

   (a) The arc-price elasticity
   (b) The possible total gain or loss in total revenue per day

9. Within certain limits, a bus company has a demand function connecting patronage (q) and price per ride (p) as follows:

$$q = 2125 - 1000p$$

   where q is person-trips/da, and p is the price (dollars/ride). The manager has the following options to increase the total revenue: (1) attracting additional riders by rescheduling and rerouting the bus service and thus changing the demand function to $q = 2150 - 1000p$ or

(2) encouraging more riders onto the system by reducing the fare from \$1.30 to \$1. Which option would you advise the manager to adopt and give good reasons for doing so?

10. (a) A bus company found that the price elasticity of demand for bus trips during peak hours is −0.40 for small price changes. Management would like to increase the current fare but fears that this action would lead to a reduction in patronage and that this action would also result in a loss of revenue for the company. Are these fears justified? Discuss your assessment.

    (b) If the same situation were to occur in another city where the price elasticity is −1.3, would these fears still be justified?

11. An airline company currently sells a package deal for return tickets at \$100/ticket and sells 5000 tickets per wk. Because of the high demand, the company raises its fare to \$120 per ticket hoping to raise its revenue. If the price elasticity of demand is currently −1.2, what will be the sale of tickets per week and how will this affect total revenue? What conclusions do you draw from this exercise?

12. When the supply of motor scooters falls by 10 percent, the price of scooters goes up 40 percent. What is the price elasticity of demand of scooters? What would happen if scooter distributors raise the price of scooters by 50 percent because of supply reduction?

13. Which of the following pairs of products can be considered complements and which are substitutes?

    Group A: Car batteries and automobiles

    Group B: Car tires and automobiles

    Group C: Bus travel and airplane travel

    Group D: Hot dogs and hamburgers

    Group E: Horses and carriages

    Group F: Hot dogs and buns

14. Refer to Example 4.6 in the text. If all the conditions remain the same, but the elasticity is −0.75, what advice would you offer to the management of this transit company?

15. The demand function for automobile travel along a major corridor in a medium-sized city is estimated to be:

$$Q = aA^{-2.2}B^{0.13}C^{-0.4}D^{0.75}$$

where

Q = Automobile trips per hour (peak hour)
a = Constant
A = Travel time by automobile in min
B = Travel time by bus in min
C = AC by automobile
D = AC by bus

(a) Justify the signs (+ or –) of the exponents of parameters A, B, C, and D.
(b) Because of congestion likely to occur on this corridor, the travel times of automobiles would increase by 20 percent and travel times of buses would increase by 10 percent. At the same time, auto travel costs would increase by 5 percent and bus costs decrease by 15 percent. What will be the percentage of change in auto traffic?
(c) If the AC of travel by bus increases by 10 percent, but the travel time by bus decreases by 10 percent, what would the overall percentage increase or decrease in auto travel be?

16. A transit company estimates that the cross elasticity of demand between its fast express bus and its ordinary bus is 2. Calculate the effect on the revenue received from the express service if the price of the ordinary bus service is reduced from \$75 to \$50 while the price of the express bus service remains the same.

17. Latent demand for highway travel is defined as the difference between the maximum number of trips that could be made and the number of trips that are actually made. Given a demand function $q = 1800 - 150t$ what is the time elasticity of demand when $t = 2$ min? What is the latent demand at this travel time?

18. A toll bridge is constructed but, for the time being, there is no charge for its use. If the demand for using it is $p = 120 - 2q$ where p is the price in \$, and q is the number of vehicles using the bridge per hour:

    (a) Sketch the demand curve for bridge crossings.
    (b) If the crossings are free, how many vehicles are likely to cross the bridge?
    (c) If a toll of \$20 is charged per crossing, how many vehicles will use the bridge per hour and what will be the loss in consumer's surplus?

19. A transit authority wants to improve one segment of a light-rail system by increasing the peak-hour seating by 20 percent and also by repricing the fare to achieve full utilization. The existing capacity is 2000 seats/hr at \$1/seat, and the price elasticity is –0.75. What additional consumer surplus would be generated by this action? Would the transit authority gain from this action?

20. A bus system consisting of 50 buses (55-seaters) charges $1 per ride; it has been decided to put 10 percent more buses into service. What should be the new fare per ride to achieve full utilization of capacity if the price elasticity of demand for ridership is –0.3? What will be the additional consumer surplus generated by this action?

21. The ferry service from a city to a recreational island is currently served by regular ferries and a luxury boat. Five thousand passengers per day use the regular ferries and 7000 use the luxury boat. Travel times (min) and fares ($) are:

	Travel time (min)	Fare ($)
Regular	45	1
Luxury	30	2

The linear arc-time and arc-price elasticities of demand are:

	Time	Regular fare	Time	Luxury fare
Regular	–0.03	–0.04	+0.02	+0.05
Luxury	+0.05	+0.02	–0.07	–0.20

(a) If the fare on the luxury boat is raised to $2.50, what will be the effect on ridership?
(b) If the travel time on the luxury boat is reduced to 25 min, what will be the effect on ridership?
(c) If the regular ferries increase their travel time to 50 min, what will be the effect on ridership?
(d) How will the total revenue of the service be affected by parts (a), (b), and (c)?

22. A survey of college students revealed that, in general, they value one train trip a month to a resort at $40, the second the same month at $30, the third at $20, a fourth at $15, and a fifth at $5. The survey found that students would not take more than five trips even if they were free.

(a) If the train tickets cost $25 a trip, how many trips, on average, will a typical student take?
(b) A travel club, charging monthly dues, allows students to travel free. How much, at most, would a student be willing to pay for monthly club dues?

23. A rapid transit system has estimated the following costs of operation for one of its routes, making a variety of combinations of cars per train:

Number of cars	Fixed cost ($/mi)	Variable cost ($/mi)
1	45	30
2	45	54
3	45	76
4	45	102
5	45	150
6	45	225
7	45	310

Plot the information for each combination, including the TC, the AC, and the MC of operating the system. What is the optimum number of cars per train that should be operated?

24. A concrete-mix plant needs to hire a few men at either $5/hr (if semi-skilled) or $3.50 per hr (if non-skilled). The following data are available for making a decision:

Number of workers	2	3	4	5	6	7	8
Hours of output (yd^3)	56	120	180	200	210	218	224

Additional FCs are $50/hr when semi-skilled labor is used and $60/hr when non-skilled labor is used. Determine the number of workers to be hired to minimize production costs. What is the corresponding cost per cubic yard?

25. A small bus company has an operating cost function of $C = 5 + 7q$ where q is the number of buses and C is the cost in units of $1000.
    (a) Determine the average and MC function.
    (b) Determine the elasticity function.
    (c) Does an economy of scale exist?
    (d) Would you recommend providing additional bus capacity based on (c)?

26. Refer to Exercise 25. A similar bus company in a city has a cost function of $C = 7q$. Answer parts (a) through (d) for this city.

27. An orange juice manufacturer noted that as the price of fresh oranges decreased from $1 to $0.85 per basket, the number of cans of orange-juice purchased per day increased from 70 to 100. Compute the cross elasticity of demand.

28. Draw a set of indifference curves that display the properties described below:
    (a) A person likes one fried egg served on an English muffin, and no other way
    (b) Given two brands of wine X and Y, a person likes each one of them equally well
    (c) A person likes a spoon of jam on any kind of bread

29. The number of leather shoes demanded and supplied by a company per month is:

Price ($)	Demanded (1000s)	Supplied (1000s)
60	22	14
80	20	16
100	18	18
120	16	20

(a) Graph these values and find the equilibrium price and quantity.
(b) What is the price elasticity of demand when the price is $80?
(c) What is the price elasticity of supply when the price is $80?
(d) If the shoe manufacturers association sets a price ceiling of $80, will there be a surplus or shortage of these shoes and what will be its magnitude?

30. The demand for two-bedroom apartments in a city is $q = 100 - 5p$ where q is the number of two-bedroom apartments in thousands and p is the monthly rent in 100s of dollars. The supply of these apartments is $q = 50 + 5p$

(a) What is the equilibrium rent?
(b) If the city rent control department sets a maximum monthly rent of $400, what would be the result?
(c) If it were decided to have a fixed monthly rent of $900, what would happen?

31. The price of magazines is $8/each, while the price of paperbacks is $10/each. A student with a monthly budget of $80 for books and magazines has already spent his money on four magazines which leaves him with $48 to spend on more magazines and paperbacks. Draw his budget line. If his remaining budget is spent on one magazine and four paperbacks, show his indifference curve on his budget line.

32. A woman considers movies and musical concerts as perfect substitutes.

(a) Draw a set of indifference curves showing her preference for movies and concerts.
(b) If movies cost $6 and concerts cost $12, which combination will she choose (show this on your graph) assuming her budget is $40?

Principles of Probability: Part I—Review of Probability Theory

5

5.1 INTRODUCTION

Probability plays an essential role in the evaluation and analysis of data. Almost any set of data, whether collected in the field, compiled in a laboratory, or acquired through an expert opinion survey is subject to variability. The existence of variability in a data set collected for a quantity imposes certain degrees of randomness in it. As a result, the prediction of specific limits or values for the quantity will involve uncertainty. Because of this uncertainty, the theory of probability and various methods of statistical analyses are useful in evaluating the data, identifying specific functions, and estimating parameters that can be used to properly define the quantity for which the data has been collected. Furthermore, the theory of probability can become a useful tool in treating design and decision-making problems that involve uncertainties. As an example, suppose an engineer is planning to improve an intersection for more efficient handling of the traffic. The engineer can utilize two different designs. One involves adding left-turn pockets at the intersection and utilizing left-turn traffic lights. The second method simply involves using a four-way stop sign system. It is obvious that the latter is less costly and can be implemented in a relatively short time. However, this method may involve a larger number of traffic accidents than the method utilizing the traffic lights. As a requirement for each design, the number of accidents per unit time (say, a year) for the intersection must be limited to a predetermined maximum value. Because the future accidents for both designs are uncertain, the engineer may wish to utilize a probabilistic formulation in predicting the probability that the number of accidents at the intersection will be limited to the selected maximum value. This probability is then computed for both design options. If the probability associated with the stop sign design is high, the engineer may wish to recommend the stoplight design alternative even though it involves a higher cost. The theory of probability may prove to be useful in many such problems where design and decision making are considered under an uncertain condition. This chapter reviews the basic elements of the theory of probability and is intended to provide the background needed for understanding the statistical methods that are presented in the subsequent chapters of this book. The chapter also provides several example problems as a means of introducing the role of probability in formulating an engineering problem where analysis, design, and decision making require the treatment of uncertainties and estimation of outcomes of uncertain events.

Generally, a probabilistic problem in engineering is composed of two parts: (1) the theory and model formulation and (2) the data or knowledge compilation. Theory and model formulation requires an understanding of the theory of probability and events. Data and knowledge compilation requires an understanding of statistical methods that can be used in analyzing and treating the collected data for use in the theory and model formulation. We begin the discussion with the review of the theory of probability in this chapter. The subsequent chapters continue with the theory of probability and then extend the discussion into topics in statistics and data analyses. Throughout the review of the probability theory, we present examples that may refer to model formation and/or data compilation.

5.2 EVENTS

To present a simplified description of an event and probability, we refer to a quantity for which a certain type of data (in laboratory or field) has been compiled. If n samples are collected for a quantity X, then these n values are expected to differ from one another for a variety of reasons. The quantity X may simply have varying outcomes at different times. For example, if X represents the stress at a critical location in a bridge girder, it is obvious that because of the variation in the intensity of the vehicle load applied on the bridge, the stress will have varying values at different times. There are, however, many other reasons that X might have different outcomes. Even if all conditions influencing the variation in X are identical, the error in measuring X may impose a variation in it. The n samples compiled for X actually only represent a small portion of a much larger set of possibilities for the quantity X. By definition, a sample space S for X is a set that contains all possible values of X. A specific outcome (i.e., value) for X is only one possibility and is referred to as a *sample point* (or an *element*). An *event* is defined as a subdomain or subset of the sample space S. This means that if we identify and single out a group of values in S, we have just made an event that is specific to the quantity X. The sample space S is written:

$$S : \{x_1, x_2, \ldots\} \tag{5.1}$$

which means S contains all possible values of X as described by $x_1, x_2, \ldots$, and so on. An event such as A is a subset of S and contains a portion of the sample points in S. Thus it can be said that Event A *belongs* to S. This can be shown as:

$$A \subset S \tag{5.2}$$

Thus, for example, subsets $A_1 : \{x_1\}$ and $A_2 : \{x_1, x_2\}$ are two events that belong to S. Note that, in a special case, an event can contain all the sample points in S. Such an event is the sample space itself and is referred to as one that is *certain* or *sure*. In contrary to this, an event may have none of the sample points of S. Such an event is an empty subset of S and is referred to as an *impossible* event. An impossible event is often shown with the symbol φ.

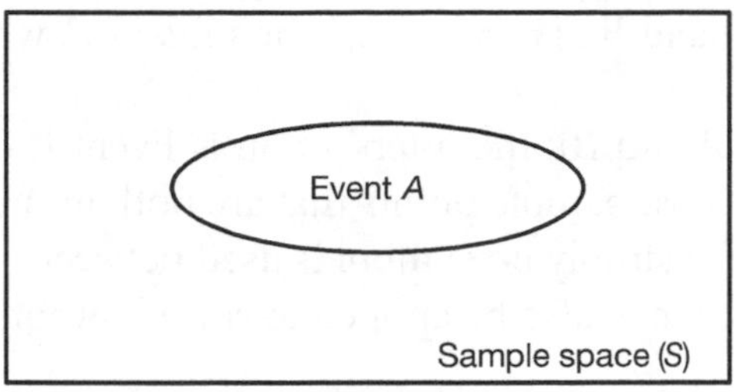

Figure 5.1 Presentation of an event using the Venn diagram

In terms of the elementary set theory, an event (which is a subset of S) can be shown with the Venn diagram representation as shown in Figure 5.1. We observe that, theoretically, many events can be defined in a sample space. Again, from the set theory, one can refer to definitions described in the following sections.

5.2.1 Complementary Event

The complementary event of A is an Event $\bar{A}$ that contains those sample points, of the space S, that do not belong to A. Thus if the space S contains n sample points designated by x_i ($i = 1, 2, \ldots, n$), and Event A within S contains only x_1 and x_2 then $\bar{A}$ will contain all x_i with the exception of x_1 and x_2. This can be described with the following series of expressions:

$$\begin{aligned} &S:\{x_1, x_2, \ldots, x_n\} \\ &A \subset S; \bar{A} \subset S \\ &A:\{x_1, x_2\} \\ &\bar{A}:\{x_3, x_4, \cdots, x_n\} \end{aligned} \tag{5.3}$$

5.2.2 Combination of Events

The *union of events* A and B is an Event C which contains those sample points that are in A or B. The symbol $\cup$, is used to describe the union. Note that union is actually an operation conducted on these two events. The union can also be conducted on several events. Considering Events A and B, one can write:

$$C = A \cup B \tag{5.4}$$

When the union of n Events $E_i(i = 1, 2, \ldots, n)$ is desired, the operation can be written:

$$C = E_1 \cup E_2 \cup \cdots \cup E_n \tag{5.5}$$

or in a more condensed form:

$$C = \bigcup_{i=1}^{n} E_i \tag{5.6}$$

For example, if $A : \{x_1, x_2, x_3\}$ and $B : \{x_1, x_2, x_4, x_6\}$ then Event C, which is the union of A and B will be $C : \{x_1, x_2, x_3, x_4, x_6\}$.

For two events (such as A and B), the intersection is Event D (known as an *intersection of events*) such that it contains those sample points that are both in A and B. The symbol $\cap$ is used to describe the intersection (or simply no symbol is used between A and B). The intersection is another operation on events. It can also be applied to several events. For Events A and B

$$D = A \cap B \quad \text{or} \quad D = AB \tag{5.7}$$

and for n events:

$$D = E_1 \cap E_2 \cap \cdots \cap E_n \quad \text{or} \quad D = E_1 E_2 \cdots E_n \tag{5.8}$$

or in a more condensed form:

$$D = \bigcap_{i=1}^{n} E_i \tag{5.9}$$

Again, considering $A : \{x_1, x_2, x_3\}$ and $B : \{x_1, x_2, x_4, x_6\}$ then Event D, which is the intersection of A and B, will be $D : \{x_1, x_2\}$.

Figure 5.2 shows the union and intersection with the help of the Venn diagram. Notice that Events C and D both belong to the sample space S. While Event C covers the entire area bounded by A and B, Event D only covers the small shaded area in common between A and B.

5.2.3 Mutually Exclusive and Collectively Exhaustive Events

If Events A and B have no common sample points, their intersection will be an impossible event (φ). In this case A and B are said to be *mutually exclusive*. As seen in Figure 5.3, two mutually exclusive events, as represented by the Venn diagram, have no common area. Also notice that an event and its complementary event are mutually exclusive.

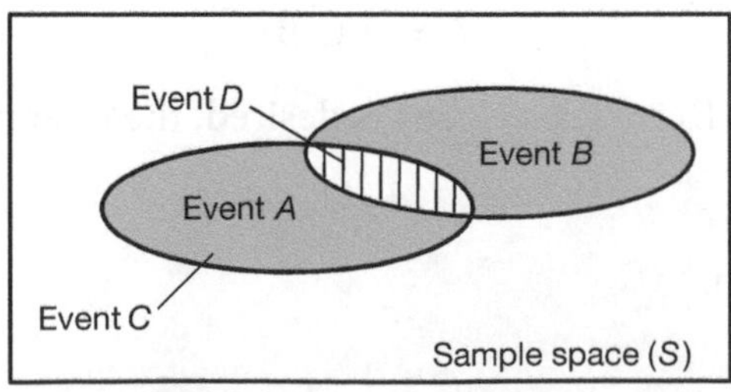

Figure 5.2 Union and intersection of events

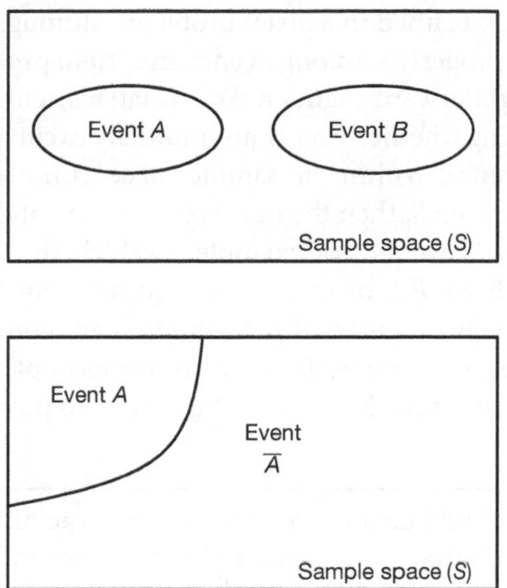

Figure 5.3 Mutually exclusive events

Two or more mutually exclusive events that make up the entire sample space are said to be *collectively exhaustive* (see Figure 5.4). With this definition, we observe that if n events $E_i(i = 1, 2, \ldots, n)$ are mutually exclusive and collectively exhaustive, then:

$$E_1 \cup E_2 \cup \cdots \cup E_n = S \tag{5.10}$$

Furthermore, notice that Event A and its complementary Event $\bar{A}$ are also collectively exhaustive. Thus:

$$A \cup \bar{A} = S \tag{5.11}$$

It is important to recognize the mutually exclusive events that make up the entire sample space and are, thus, considered collectively exhaustive. The identification of these events depends on

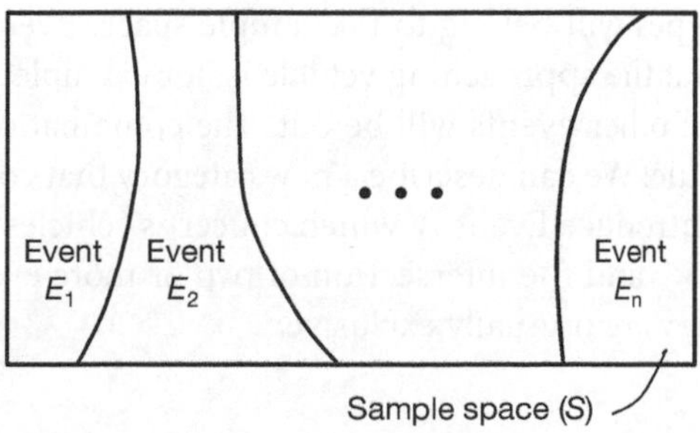

Figure 5.4 Mutually exclusive and collectively exhaustive events

the sample space and how it is defined in a given problem. Although the Venn diagram may be helpful in recognizing these properties among events, in certain problems, the diagram may not be shown in a simple, straightforward manner. As it is later discussed in the section on probability, one way of recognizing whether events are mutually exclusive and collectively exhaustive is through their occurrences within the sample space. If n events are mutually exclusive and collectively exhaustive within S, then the occurrence of one of these events will exclude the occurrence of the other $(n - 1)$ events. For example, a vehicle that is approaching an intersection will either go straight (Event E_1), or make a right turn (Event E_2) or a left turn (Event E_3). These three events are mutually exclusive. If the sample space concerns the three events only, then they are also collectively exhaustive. This means, for example, that once it is certain that the vehicle will go straight (i.e., Event E_1 occurs), the other two possibilities are out.

Example 5.1 In a series of field data observations on the type of heavy vehicles on a bridge, vehicle types are divided into seven categories as summarized in Table E5.1. Establish events describing these categories and identify mutually exclusive events and a combination of events within the sample space:

Table E5.1 Categories for vehicle types

Category	Type of vehicle	Designation
1	Buses	V_1
2	2-axle vehicles	V_2
3	3-axle trucks	V_3
4	4-axle trucks	V_4
5	5-axle trucks	V_5
6	6-axle trucks	V_6
7	Trucks with 7 or more axles	V_7

Solution Using the designation letters described in Table E5.1, we define an Event V_i to represent the event that a vehicle belongs to category i (where $i = 1$ to 7). The sample space will then describe the type of vehicles using the bridge. For a heavy vehicle approaching the bridge, the event of being one of the seven types will belong to the sample space. Events V_i are mutually exclusive. Thus, once it is known that the approaching vehicle is, for example, a 3-axle truck, Event V_3 will be certain. This means all other events will be out. The combination of Events V_i are only possible through the union rule. We can describe a new category that combines two or more of these events. For example, we introduce Event W which concerns vehicles with less than 5 axles. In this case $W = V_1 \cup V_2 \cup V_3 \cup V_4$ and the intersection of two or more events (among Events V_i) is an impossible event since they are mutually exclusive.

Example 5.2 A construction company is currently bidding for two jobs. Considering a sample space (S) containing the outcome of winning or losing these jobs:

(a) Identify all sample points in S

(b) Identify the sample points in the following events:

A = That the company will win only one job
B = That the company will win no jobs
C = That the company will win at least one job

(c) Identify sample points in the $A \cap B$ and $A \cup B$

Solution

(a) Let W represent winning a job and L is losing a job. Thus WW means winning both jobs; WL means winning the first and losing the second job, and so on. The sample points in S will then be:

$$S : \{WW, WL, LW, LL\}$$

(b) Event A will contain two sample points:

$$A : \{WL, LW\}$$

furthermore, $B : \{LL\}$ and $C : \{WL, WL, WW\}$.

(c) The intersection of A and B is an impossible event since A and B have no sample points in common. Thus $A \cap B = \varphi$ The union of A and B contains three sample points:

$$A \cup B : \{WL, LW, LL\}$$

Since events and the sample space follow the principles of the set theory, they are subject to rules that govern sets. The following equations present these rules:

$$A \cup B = B \cup A \tag{5.12}$$

$$A \cap B = B \cap A \tag{5.13}$$

$$A \cup (B \cap C) = (A \cup B) \cap (A \cup C) \tag{5.14}$$

$$A \cup (B \cap C) = (A \cup B) \cap (A \cup C) \tag{5.15}$$

5.3 PROBABILITY

We recall that within the sample space S, an Event A is a set among several other sets that belong to S. The possibility for A to occur is defined by a numerical value called *probability*. We refer to this possibility as the *probability of* A, or $P(A)$. This numerical value describes the likelihood that A will occur among all other events that can be identified within S. Mathematically speaking, $P(A)$ is bounded by 0 and 1. We also observe:

$$P(\varphi) = 0 \tag{5.16}$$

$$P(S) = 1 \tag{5.17}$$

indicating that the probability of an impossible event is zero; whereas, a sure event has a probability of 1. A logical definition for Equation 5.17 is that since S contains all possible events within the sample space, the probability of occurrence of any one event (without being specific as to which one) within S will be unity. For example, in the case of a car approaching an intersection, $P(S)$ simply means the probability that the car will either make a left turn, a right turn, or will go straight. Obviously the probability for this will be unity.

Example 5.3 Assume Table E5.3 summarizes data gathered on the number per day of each vehicle type listed in Table E5.1 as discussed in Example 5.1. The total number of data points is 192. Assuming all data points have the same weight among the total of 192, compute the probability of individual events V_i.

Table E5.3

Vehicle type	Number of veh/da
Buses	5
2-axle vehicles	15
3-axle trucks	25
4-axle trucks	30
5-axle trucks	105
6-axle trucks	6
Trucks with 7 or more axles	6
Total	192

Solution Since all data points have the same weight, the probability of an individual Events V_i will depend on the ratio of the number of data points for the specific vehicle type to the total number of data points. Thus:

$P(V_1)$ = Probability that the vehicle type will be buses = 5/192 = 0.026;
$P(V_2)$ = Probability that the vehicle type will be 2-axle trucks = 15/192 = 0.078;

Continues

Example 5.3: *Continued*

$P(V_3)$ = Probability that the vehicle type will be 3-axle trucks = 25/192 = 0.130;
$P(V_4)$ = Probability that the vehicle type will be 4-axle trucks = 30/192 = 0.156;
$P(V_5)$ = Probability that the vehicle type will be 5-axle trucks = 105/192 = 0.547;
$P(V_6)$ = Probability that the vehicle type will be 6-axle trucks = 6/192 = 0.031; and,
$P(V_7)$ = Probability that the vehicle type will be trucks with 7 or more axles = 6/192 = 0.031.

Since Events V_i are mutually exclusive and collectively exhaustive, the sum of the probabilities of the seven Events V_1 through V_7 will be equal to one because the union of the seven events makes the entire sample space, that is, $P(V_1 \cup V_2 \cup \ldots \cup V_7) = P(S) = 1$.

Knowing that an Event A and its complementary $\bar{A}$ are mutually exclusive and collectively exhaustive, as described by Equation 5.11, it can be shown that the probability of $\bar{A}$ is one minus probability of A.

$$P(A \cup \bar{A}) = P(S) = 1 \tag{5.18}$$

Thus

$$P(A) + P(\bar{A}) = 1 \tag{5.19}$$

or

$$P(\bar{A}) = 1 - P(A) \tag{5.20}$$

5.3.1 Probability of the Union of Events

In Example 5.3, the probability of the union of events can simply be written as the sum of individual probabilities because the corresponding events are mutually exclusive. However, in general, computing the probability of the union of two or more events is not as simple. Consider first the union of Events A and B. From Figure 5.2, it is obvious that if we simply add the probability of A to that of B, the result will be the probability of union of A and B plus the probability of the intersection of the two events:

$$P(A) + P(B) = P(A \cup B) + P(A \cap B) \tag{5.21}$$

Thus, from Equation 5.21, we obtain:

$$P(A \cup B) = P(A) + P(B) - P(A \cap B) \tag{5.22}$$

Equation 5.22 can be used to show that the probability of the union of three Events A, B, and C is:

$$P(A \cup B \cup C) = P(A) + P(B) + P(C) - P(A \cap B) - P(A \cap C) - P(B \cap C) + P(A \cap B \cap C) \tag{5.23}$$

An expression for the probability of the union of four or more events can also be obtained similarly. However, the expression will involve many terms. In such cases, it may be simpler if the probability of the union of several events be written in terms of the intersection of the complementary of the events following the *De Morgan* rule, which is described later.

5.3.2 Conditional Probability and Probability of Intersection of Events

To derive relations describing the probability of intersection of events, we first introduce the concept of *conditional probability*. If an Event A depends on another Event B, the occurrence of A within the domain of B is shown with $A \mid B$ and defined as the event of A given B. The corresponding probability is therefore $P(A \mid B)$ and is defined as the conditional probability of A given B. Within this definition, A depends on the occurrence of B. Thus, the conditional probability $P(A \mid B)$ assumes that B occurs; as such, B becomes a sure event. The conditional probability $P(A \mid B)$ can then be defined as the portion of Event A that is within B. As seen in Figure 5.2, this portion is the intersection of the two events. In reality Event B has a probability of occurrence, that is, $P(B)$, thus the conditional probability $P(A \mid B)$ will be:

$$P(A \mid B) = \frac{P(A \cap B)}{P(B)} \tag{5.24}$$

Similarly

$$P(B \mid A) = \frac{P(B \cap A)}{P(A)} \tag{5.25}$$

From these equations the following equations for the probability of the intersection of two events can be written:

$$P(A \cap B) = P(A \mid B)P(B) \tag{5.26}$$

and

$$P(B \cap A) = P(B \mid A)P(A) \tag{5.27}$$

Of course, the left sides of Equations 5.26 and 5.27 are equal. As explained later, this will result in Bayes' theorem that can be used to relate $P(A \cap B)$ to $P(B \cap A)$ and vice versa.

Equations 5.26 and 5.27 can be generalized to derive equations for the probability of intersection of three or more events. For example, the intersection of the three Events A, B, and C can be written:

$$P(A \cap B \cap C) = P(A \mid B \cap C)P(B \cap C) = P(A \mid B \cap C)P(B \mid C)P(C) \tag{5.28}$$

Example 5.4 Figure E5.4a shows a pipeline system made up of Links A and B. The system is used to deliver water between Points 1 and 2. The probability that during any given month either of the two links fails is 0.03. If Link A fails for any reason (say severe freezing conditions in winter months), there is 0.1 probability that Link B will also fail. Compute the probability that there will be no water delivered between Points 1 and 2 during a given month.

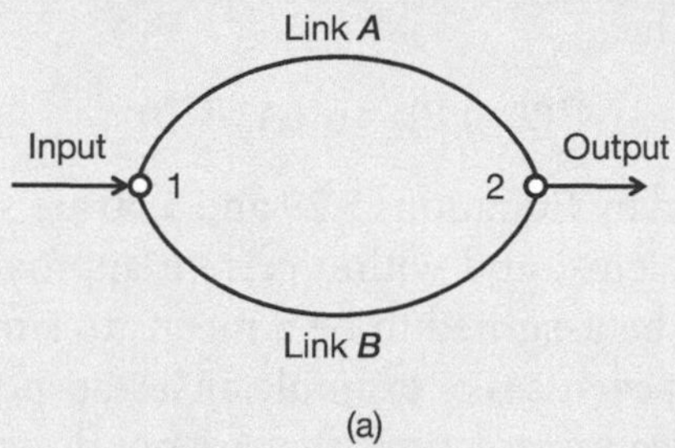

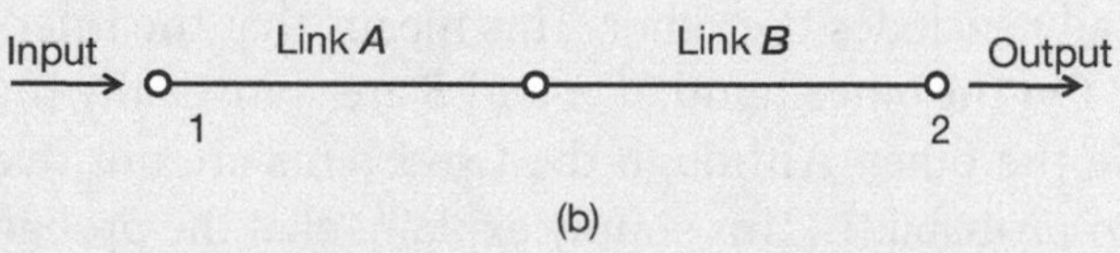

Figure E5.4 (a) Links in parallel and (b) links in series

Solution Let

$P(A)$ = Probability of failure of Link A
$P(B)$ = Probability of failure of Link B
$P(C)$ = Probability that there will be no water delivered to Point 2 from Point 1.

Since both Links A and B must fail to stop the delivery of water, Event C will be the intersection of Events A and B. According to the problem, $P(A) = P(B) = 0.03$; also. $P(B \mid A) = 0.1$. Thus:

$$P(C) = P(B \cap A) = P(B \mid A)P(A) = 0.1 \times 0.03 = 0.003$$

Example 5.5 In Example 5.4, assume Links A and B are connected in a series as shown in Figure E5.4b. With the same information for the probabilities, indicated in Example 5.4, for the failure of individual links, compute the probability that no water is delivered to Point 2 from Point 1.

Solution In this case, if either of the two, or both, links fail, water delivery at Point 2 will be disrupted. Denoting C as this event, $C = A \cup B$ and:

$$P(C) = P(A) + P(B) - P(A \cap B) = 0.03 + 0.03 - 0.003 = 0.057$$

Note that the probability of the intersection of two events can only be found when the information on the conditional probability of the occurrence of one on the other is known. If the two events are independent of each other, then the occurrence of one will not depend on the other. In this situation, one can write:

$$P(A \mid B) = P(A) \tag{5.29}$$

and, as such, Equation 5.26 will be:

$$P(A \cap B) = P(A)P(B) \tag{5.30}$$

Events A and B that are governed by Equations 5.29 and 5.30 are said to be *statistically independent.* In many engineering problems, and within certain approximations, statistical independence between two events may be assumed to be a means to simplify the computation of the probability of the intersection of events. For example, in a transportation engineering problem, the occurrence of accidents at two intersections that are not directly linked may be assumed to be independent. It is also important to distinguish the difference between statistical independence and mutually exclusive. Note that if Events A and B are mutually exclusive, the occurrence of one automatically excludes the other. This means that the intersection of A and B will be an impossible event. On the other hand, if A and B are statistically independent, occurrence of one does not exclude the other. Although the two events are not related, their intersection exists and has a nonzero probability. This simply explains that the probability of the occurrence of A and B does exist and can be computed through Equation 5.30.

Example 5.6 A motorist is driving at the posted speed along a roadway. He has two intersections ahead. The probability of encountering a red light at any of the two intersections is 0.30. Assume the event of encountering a red light at any one intersection is independent of the same event at the other intersection.

(a) Compute the probability that the motorist will encounter at least one red light.

(b) Compute the probability that the motorist will encounter red lights at both intersections.

Solution Let A be the event of encountering a red light at the first intersection and B the same event at the second intersection.

(a) In this part, the probability of encountering at least one red light means encountering a red light at either intersection or at both. Thus the union of Events A and B is desired. Defining this union as C, $P(C) = P(A) + P(B) - P(A \cap B)$. Since Events A and B are independent, then:

$$P(C) = P(A) + P(B) - P(A)P(B) = 0.3 + 0.3 - 0.3 \times 0.3 = 0.51.$$

(b) If D is defined as the event of concern in this part, then $D = A \cap B$. Thus:

$$P(D) = P(A) \times P(B) = 0.09$$

In regard to conditional probability, we further observe that the probability of the complementary event of A is written:

$$P(\bar{A} \mid B) = 1 - P(A \mid B) \tag{5.31}$$

Note that the *given* event, that is, B, does not change on both sides of the equation. That is to say that $P(\bar{A} \mid \bar{B}) \neq 1 - P(A \mid B)$ and $P(\bar{A} \mid \bar{B}) = 1 - P(A \mid \bar{B})$

5.3.3 Bayes' Theorem

In light of Equations 5.26 and 5.27, one can write:

$$P(A \mid B) = \frac{P(B \mid A)P(A)}{P(B)} \tag{5.32}$$

This relation is known as *Bayes' theorem*. In many engineering problems, Equation 5.32 may be useful as a means to compute a desired conditional probability. For example, if Event B represents delay in a construction project and A stands for the event of the shortage of certain material in the market, while $P(B \mid A)$ is the probability of delay in the project given the shortage of the material, Equation 5.32 can be used to determine the probability of the shortage of the material, if we assume there is a delay in the completion of the project. This can be interpreted that if there is a delay, the probability that the shortage of material is causing it is obtained through Equation 5.32; the new information may also be considered as an updated probability for the shortage of material.

Example 5.7 The probability of traffic congestion in the northbound lanes of a freeway during the morning rush hours in any working day is estimated as 0.02. The probability of traffic congestion during the same time period in the southbound lanes is only 0.005. There is dependence between the traffic congestion in the southbound and the northbound lanes during rush hours. If congestion in the northbound lanes occurs, there will be a 0.17 probability that the southbound lane will also experience congestion.

(a) Compute the probability that during the morning rush hours, there will be congestions in both the northbound and the southbound lanes.

(b) Compare the probability that there will be congestion in the freeway during the morning rush hours, whether in the northbound or southbound lanes.

(c) If there is congestion in the southbound lanes, what will be the probability that the northbound lanes have also experienced congestion?

Solution For simplicity, let:

A = Event of traffic congestion in the northbound lanes
B = Event of traffic congestion in the southbound lanes

Continues

Example 5.7: *Continued*

Then the following information is known from the problem:

$$P(A) = 0.02 \text{ and } P(B) = 0.005.$$

Furthermore, notice that if A occurs, there is a 0.17 probability that B will also occur. Thus:

$$P(B \mid A) = 0.17$$

(a) In this part, since the occurrence of both A and B is desired, the intersection of the two events must be formed. Calling this Event C, we can write:

$$P(C) = P(A \cap B) = P(B \mid A)P(A) = 0.17 \times 0.02 = 0.0034$$

(b) In this part, since no specific reference to the direction of the lanes has been made, the event of concern is the union of A and B. In fact, we are looking for the probability that there will be traffic congestion in either the northbound (Event A) or in the southbound lanes (Event B) or in both. Calling the event of concern D, we can write:

$$P(D) = P(A \cup B) = P(A) + P(B) - P(A \cap B) = 0.02 + 0.005 - 0.0034 = 0.0216$$

(c) In this part $P(B \mid A)$ is desired. Bayes' theorem (Equation 5.32) can be used for this purpose:

$$P(A \mid B) = \frac{P(B \mid A)P(A)}{P(B)} = \frac{0.17 \times 0.02}{0.005} = 0.68$$

Example 5.8 Derive an expression for $P(\bar{A} \mid \bar{B})$ in terms of $P(A \mid B)$, $P(A)$, and $P(B)$.

Solution Equations 5.31 and 5.32 can be used for this purpose. The following shows the sequence of operations:

$$P(\bar{A} \mid \bar{B}) = 1 - P(A \mid \bar{B}) = 1 - \frac{P(\bar{B} \mid A)P(A)}{P(\bar{B})} = 1 - \frac{[1 - P(B \mid A)]P(A)}{1 - P(B)}$$

or

$$P(\bar{A} \mid \bar{B}) = 1 - \frac{\left[1 - \dfrac{P(A \mid B)P(B)}{P(A)}\right]P(A)}{1 - P(B)} = 1 - \frac{P(A) - P(A \mid B)P(B)}{1 - P(B)}$$

Using the numerical values in Example 5.7:

$$P(\bar{A} \mid \bar{B}) = 1 - \frac{0.02 - 0.68 \times 0.005}{1 - 0.005} = 0.9833$$

Continues

Example 5.8: *Continued*

This means that if there is no traffic congestion in the southbound lanes; there will be a 0.9833 probability that the northbound lanes do not have any congestion either. We emphasize that other methods, including the use of the Venn diagram and De Morgan's rule (see next section), can also be employed in deriving an expression for $P(\bar{A} \mid \bar{B})$

5.3.4 De Morgan's Rule

De Morgan's rule applied to Events A and B is defined via the following two equations:

$$P(\overline{A \cup B}) = P(\bar{A} \cap \bar{B}) \quad (5.33)$$

$$P(\overline{A \cap B}) = P(\bar{A} \cup \bar{B}) \quad (5.34)$$

To prove these equations, the Venn diagram can be used. However, they can also be proved via equations governing the combination of events and probabilities. Considering the expression found for $P(\bar{A} \mid \bar{B})$ in Example 5.8, we can write:

$$P(\bar{A} \mid \bar{B})[1 - P(B)] = 1 - P(B) - P(A) + P(A \mid B)P(B)$$

or

$$P(\bar{A} \mid \bar{B})P(\bar{B}) = 1 - P(A \cup B)$$

The left side of the equation is $P(\bar{A} \cap \bar{B})$ and the right side is $P(\overline{A \cup B})$ Thus, Equation 5.33 is proven. Considering again the expression found in Example 5.8, we can also write:

$$P(A \mid B) = 1 - \frac{[P(\bar{A}) - P(\bar{A} \mid \bar{B})P(\bar{B})]}{1 - P(\bar{B})}$$

or

$$P(A \mid B)[1 - P(\bar{B})] = 1 - P(\bar{B}) - [P(\bar{A}) - P(\bar{A} \mid \bar{B})P(\bar{B})$$

or

$$P(A \cap B) = 1 - P(\bar{B}) - P(\bar{A}) + P(\bar{A} \cap \bar{B})P(\bar{B}) = 1 - P(\bar{A} \cup \bar{B})$$

or

$$1 - P(A \cap B) = P(\bar{A} \cup \bar{B})$$

The left side of this equation is $P(\overline{A \cap B})$ Thus, Equation 5.34 is proven.

It can easily be shown that in general forms, De Morgan's rule can be defined via the following equations:

$$P(\overline{E_1 \cup E_2 \cup E_3 \cup \ldots \cup E_n}) = P(\bar{E}_1 \cap \bar{E}_2 \cap \bar{E}_3 \cap \ldots \cap \bar{E}_n) \quad (5.35)$$

$$P(\overline{E_1 \cap E_2 \cap E_3 \cap \ldots \cap E_n}) = P(\bar{E}_1 \cup \bar{E}_2 \cup \bar{E}_3 \cup \ldots \cup \bar{E}_n) \tag{5.36}$$

The following example shows an application of De Morgan's rule and the proof of these equations through a practical engineering problem.

Example 5.9 A large engineering system is made up of 15 components. A failure in any component will result in a system malfunction. In a series of laboratory tests, the probability of failure of any one component has been estimated to be 0.0025. Assuming the conditions among individual component failures are statistically independent, compute the probability of a malfunction of the system.

Solution In general, for a system made up of n components, the probability of malfunction (failure), p_F can be written via the union rule as:

$$p_F = P(E_1 \cup E_2 \cup E_3 \cup \ldots \cup E_n)$$

in which E_i ($i = 1, 2, 3, \ldots n$) is the event of failure in component i. The probability of no malfunction in the system p_S is equal to $(1 - p_F)$ and can be written:

$$p_S = 1 - p_F = 1 - P(E_1 \cup E_2 \cup E_3 \cup \ldots \cup E_n) = P(\overline{E_1 \cup E_2 \cup E_3 \cup \ldots \cup E_n})$$

The probability p_S can also be obtained independently considering the probabilities of no failure (i.e., survival) of individual components. The system will have no malfunction if all components survive. This means that the event of no malfunction in the system is the intersection of the complementary events of E_i. Introducing $\bar{E}_i$ as the complementary of E_i (i.e., the survival of element i), then p_S can be formulated as:

$$p_S = P(\bar{E}_1 \cap \bar{E}_2 \cap \bar{E}_3 \cap \ldots \cap \bar{E}_n)$$

which proves De Morgan's rule. In this example, Events E_i are statistically independent. This means that their complimentary events are also independent. Accordingly, the probability of no malfunction will be:

$$p_S = P(\bar{E}_1)P(\bar{E}_2)P(\bar{E}_3)\ldots P(\bar{E}_n) = [1 - P(E_1)][1 - P(E_2)]$$
$$[1 - P(E_3)]\ldots[1 - P(E_n)] = (1 - 0.0025)^{15} = 0.9631$$

and the probability of malfunction is:

$$p_F = 1 - 0.9631 = 0.0369$$

5.3.5 Total Probability Theorem

If an Event A depends on n mutually exclusive and collectively exhaustive events such as E_1, E_2, E_3, ..., E_n, then the probability of A is obtained from the *total probability theorem*. As seen in

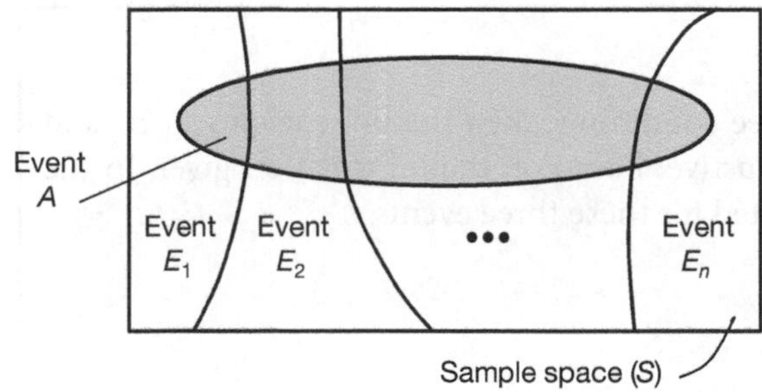

Figure 5.5 The Venn diagram presentation of total probability theorem

Figure 5.5, the intersection of Event A and any of the n Events E_i is an Event $A \cap E_i$. Since Events $A \cap E_i$ are mutually exclusive, the probability of A can be written:

$$P(A) = P(A \cap E_1) + P(A \cap E_2) + \ldots + P(A \cap E_n) \tag{5.37}$$

or

$$P(A) = P(A \mid E_1)P(E_1) + P(A \mid E_2)P(E_2) + \ldots + P(A \mid E_n)P(E_n) = \sum_{i=1}^{n} P(A \mid E_i)P(E_i) \tag{5.38}$$

In light of Equations 5.32 and 5.38, a generalized equation for Bayes' theorem can be written as:

$$P(E_i \mid A) = \frac{P(A \mid E_i)P(E_i)}{P(A)} = \frac{P(A \mid E_i)P(E_i)}{\sum_{i=1}^{n} P(A \mid E_i)P(E_i)} \tag{5.39}$$

Example 5.10 The probability of failure of a cable used in a suspension bridge depends on the tension load transferred to the cable. For simplicity, the load in the cable is divided into three groups, namely, light, moderately heavy, and heavy with relative probabilities of 3 to 2 to 1. A structural analysis of the cable shows that if the load is light, the probability of failure of the cable is only 0.0001. However, if the load is moderately heavy, the failure probability is 0.005. For a heavy load, the failure probability is 0.2.

(a) Compute the probability of failure of the cable.

(b) If the cable fails, what is the probability that the failure is caused by a light load?

Solution In this problem, let A be the event of failure of the cable for any type of load and E_i the events representing the types of loading in the cable. Thus:

E_1 = Event that the load will be light
E_2 = Event that the load will be moderately heavy
E_3 = Event that the load will be heavy

Continues

Example 5.10: *Continued*

Since the loading can only be one of these three conditions, then the three events E_1, E_2, and E_3 are mutually exclusive and collectively exhaustive. Based on the information given in the problem, the following probabilities are computed for these three events:

$$P(E_1) = \frac{3}{3+2+1} = 0.500$$

$$P(E_2) = \frac{2}{3+2+1} = 0.333$$

$$P(E_3) = \frac{1}{3+2+1} = 0.167$$

Note that the sum of these three probabilities must be equal to 1. The information on the failure probability of the cable is in the form of conditional probabilities given the three loading types. This information can be written:

$$P(A \mid E_1) = 0.0001$$
$$P(A \mid E_2) = 0.005$$
$$P(A \mid E_3) = 0.2$$

(a) From Equation 5.38, using the total probability theorem:

$$P(A) = 0.0001 \times 0.500 + 0.005 \times 0.333 + 0.2 \times 0.167 = 0.0351$$

(b) In this part, the probability of concern can be computed using Bayes' theorem (Equation 5.39). The required probability is $P(E_1 \mid A)$. From Equation 5.39:

$$P(E_1 \mid A) = \frac{0.0001 \times 0.500}{0.0351} = 0.0014$$

Example 5.11 Accidents in a roadway junction depend on the volume V of traffic (in number of cars/min) approaching the junction; and the average speed S (in km/h) of the approaching cars. For simplicity, only two possibilities for each of these factors are considered. Table E5.11 presents the probability of a traffic accident at the junction during any minute time period for various combinations of these two factors.

Table E5.11 Probability of accidents in terms of traffic volume (*V*) and average speed (*S*)

	$V \le 10$	$V > 10$
$S \le 50$	0.01	0.12
$S > 50$	0.18	0.32

Continues

Example 5.11: *Continued*

During any minute time period, there is a 50 percent chance that the average speed is less than or greater than 50 km/h. Also the traffic volume can be less than or greater than 10 cars/min with a 50 percent chance. Assume V and S are statistically independent.

(a) Compute the probability of accidents during any minute time period at the junction.

(b) If there are 8 identical such junctions in a row along a portion of the roadway, compute the probability that none of the junctions will experience accidents during any minute time period. Assume conditions between these junctions are statistically independent.

Solution Considering the possibilities listed in Table E5.11, there are 4 combinations of S and V. These combinations make 4 mutually exclusive and collectively exhaustive events as described below:

$$E_1 = \text{Event of } V \leq 10 \text{ and } S \leq 50;\ P(E_1) = 0.5 \times 0.5 = 0.25$$
$$E_2 = \text{Event of } V > 10 \text{ and } S \leq 50;\ P(E_2) = 0.5 \times 0.5 = 0.25$$
$$E_3 = \text{Event of } V \leq 10 \text{ and } S > 50;\ P(E_3) = 0.5 \times 0.5 = 0.25$$
$$E_4 = \text{Event of } V > 10 \text{ and } S > 50;\ P(E_4) = 0.5 \times 0.5 = 0.25$$

(a) From the total probability theorem, if A is the event of accidents in the junction, then $P(A)$ can be computed:

$$P(A) = P(A \mid E_1)P(E_1) + P(A \mid E_2)P(E_2) + P(A \mid E_3)P(E_3) + P(A \mid E_4)P(E_4)$$
$$P(A) = 0.01 \times 0.25 + 0.12 \times 0.25 + 0.18 \times 0.25 + 0.32 \times 0.25 = 0.1575$$

(b) The intersection rule is used in this part. If A_i is the event of accidents for junction i, the event of no accidents for the 8 junctions (say event $\bar{B}$) will be:

$$P(\bar{B}) = P(\bar{A}_1 \cap \bar{A}_2 \cap \bar{A}_3 \cap \ldots \cap \bar{A}_8)$$

and $P(\bar{B}) = (1 - 0.1575)^8 = 0.254$

Example 5.12 Life cycle analysis of bridge structures often calls for a planning study to optimize the cost associated with bridge maintenance. Maintenance includes a host of activities ranging from a minor repair to total replacement. Often the decision to repair or replace requires a careful evaluation and analysis on the part of the engineer. For simplicity, assume there are only three options considered for a bridge: (1) repair, (2) rehabilitation, and (3) replacement. Repair involves any type of work to keep the bridge in operation. Rehabilitation involves major repair along with an upgrade on the bridge condition such that an improvement in the service is achieved. Replacement means complete removal of the structure and replacement with a new structure.

An engineer is planning for the next year budget for the maintenance of the bridge that will include only one of the three options (i.e., repair, rehabilitation, or replacement). From

Continues

Example 5.12: *Continued*

past experience, the engineer estimates that the odds for similar bridges to require repair, rehabilitation, or replacement, respectively, are 1 to 2 to 3. The budget allotted to this bridge is a fixed amount. If the decision is to only repair the bridge, the probability that the cost will be within the budget will be 1. If the decision is to rehabilitate, the probability that the cost will be within the budget will be 0.99. However, if the decision is to replace the bridge, the probability that the cost will be within the budget will be 0.03. The engineer is not sure what option will finally be selected.

(a) Compute the probability that the cost associated with maintenance (that includes any of the three options) will be within the allotted budget.

(b) If there is a cost overrun (i.e., the allotted budget is exceeded by the cost), compute the probability that the decision is for the rehabilitation option.

Solution The three options are mutually exclusive and collectively exhaustive. Let:

E_1= Event that the decision will be in favor of repair
E_2= Event that the decision will be in favor of rehabilitation
E_3= Event that the decision will be in favor of replacement
A = Event that the cost of maintenance will be within the allotted budget

From the information given in the problem:

$$P(E_1) = \frac{1}{3+2+1} = 0.167$$

$$P(E_2) = \frac{2}{3+2+1} = 0.333$$

$$P(E_3) = \frac{3}{3+2+1} = 0.500$$

(a) Since it is not certain which option will be selected, then $P(A)$ can be formulated using the total probability theorem considering all possible options. From information provided by the problem:

$$P(A \mid E_1) = 1.0;\ P(A \mid E_2) = 0.99;\ P(A \mid E_3) = 0.03$$

$$\begin{aligned} P(A) &= P(A \mid E_1)P(E_1) + P(A \mid E_2)P(E_2) + P(A \mid E_3)P(E_3) \\ &= 1.0 \times 0.167 + 0.99 \times 0.333 + 0.03 \times 0.500 = 0.512 \end{aligned}$$

(b) This part is an application of Bayes' theorem. Note that the event of overrun is the complementary event of A. Thus, the required probability is $P(E_2 \mid \bar{A})$ and can be computed:

$$P(E_2 \mid \bar{A}) = \frac{P(\bar{A} \mid E_2)P(E_2)}{P(\bar{A})} = \frac{[1 - P(A \mid E_2)]P(E_2)}{1 - P(A)} = \frac{(1-0.99) \times 0.333}{1 - 0.512} = 0.007$$

Example 5.13 Occurrences of windstorms in an area are modeled as random phenomena. For simplicity, we assume that any potential storm in a given time period in the area can be either strong (Event E_s), moderate (event E_m), or weak (event E_w). These are the only three possibilities in our sample space. Note that these three events are mutually exclusive and collectively exhaustive. From past records we have the following data $P(E_s) = 0.01$, $P(E_m) = 0.20$, and $P(E_w) = 0.79$. A billboard is located in the area. If a strong storm occurs, the probability of failure of the billboard is 0.995. However, if the storm is moderate or weak, the failure probabilities are 0.056 and 0.001, respectively. Thus, denoting A as the event of failure of the billboard because of potential storms, we can write $P(A \mid E_s) = 0.995$; $P(A \mid E_m) = 0.056$; and $P(A \mid E_w) = 0.001$.

(a) What is the probability of failure of the billboard because of probable storms?

(b) If the billboard fails in a recent storm, what is the probability that the moderate storm has occurred?

Solution

(a) In general we do not know what type of storm will occur, so we must consider all possibilities. We use the total probability theorem to compute $P(A)$:

$$P(A) = P(A \mid E_s)P(E_s) + P(A \mid E_m)P(E_m) + P(A \mid E_w)P(E_w)$$
$$= 0.995 \times 0.01 + 0.056 \times 0.20 + 0.001 \times 0.79 = 0.02194$$

(b) In this case, we use Bayes' theorem:

$$P(E_m \mid A) = \frac{P(A \mid E_m)P(E_m)}{P(A)} = \frac{0.056 \times 0.20}{0.02194} = 0.510$$

Example 5.14 We are investigating the probability of failure from potential snow loads of a warehouse roof made up of a series of timber beams. The overall capacity of the roof resisting the snow load depends on the strength of the beams. Not knowing what kind of timber was used in the beams, the roof type is considered random. For simplicity, we define the roof type with only two possibilities, A and B. Type A means the roof is made up of beams from good quality materials; while Type B means the roof is made up of beams from lower quality materials. Based on our experience with other applications, we know that the odds of the roof type being A or B is 1 to 4. On the other hand, the amount of snow accumulated on the roof during a storm is also random. Again for simplicity, we assume the amount of accumulated snow is either heavy (Event H) or light (Event L) with a relative likelihood of 1 to 7. If the amount of snow accumulated on the roof is heavy, the probability of roof collapse is 0.005, for Type A and 0.075 for Type B roof, respectively. The corresponding probabilities if the snow accumulated is light are 0.001 and 0.015, respectively.

(a) Compute the probability of collapse of the roof because of potential snow loads during a future storm.

Continues

Example 5.14: *Continued*

(b) If the roof collapsed during a recent snow storm, what is the probability the roof was of Type A?

Solution

(a) This problem is again the application of the total probability and Bayes' theorem. However, in order to be able to use the total probability theorem, we must identify all possibilities for the type of roof and amount of snow accumulated during a probable storm. In this way, we will have a series of events that are mutually exclusive and collectively exhaustive based on which total probability theorem is applied. There are four such possibilities listed as denoted by Events E_1, E_2, E_3 and E_4. If A and B are the events that describe the roof types, based on the information provided, $P(A) = 1/5$ and $P(B) = 4/5$. Furthermore, $P(H) = 1/8$ and $P(L) = 7/8$. As expected, the events describing the type of roof are statistically independent of those describing the amount of snow accumulated on the roof. Thus, we can write:

Event E_1 = Roof Type A and heavy snow = $A \cap H$ and $P(E_1) = P(A \cap H) = (1/5) \times (1/8) = 0.025$

Event E_2 = Roof Type A and light snow = $A \cap L$ and $P(E_2) = P(A \cap L) = (1/5) \times (7/8) = 0.175$

Event E_3 = Roof Type B and heavy snow = $B \cap H$ and $P(E_3) = P(B \cap H) = (4/5) \times (1/8) = 0.100$

Event E_4 = Roof Type B and light snow = $B \cap L$ and $P(E_4) = P(B \cap L) = (4/5) \times (7/8) = 0.700$

Notice that the sum of the probabilities of these four events is equal to 1. Denoting C as the probability of collapse of the roof because of potential snow loads, based on the information given in the problem:

$$P(C \mid E_1) = 0.005;\ P(C \mid E_2) = 0.001;\ P(C \mid E_3) = 0.075;\ \text{and}\ P(C \mid E_4) = 0.015$$

Now by applying the total probability theorem:

$$P(C) = P(C \mid E_1)P(E_1) + P(C \mid E_2)P(E_2) + P(C \mid E_3)P(E_3) + P(C \mid E_4)P(E_4)$$

and

$$P(C) = 0.005 \times 0.025 + 0.001 \times 0.175 + 0.075 \times 0.100 + 0.015 \times 0.700 = 0.0183$$

(b) In this part, the probability of concern can be computed using Bayes' theorem (Eq. 5.39). However, we do not know whether the snow accumulation was heavy or light. Thus we must consider both possibilities in our computation. The required probability is computed through the following series of calculations:

$$P(A \mid C) = P(A \cap H \mid C) + P(A \cap L \mid C) = P(E_1 \mid C) + P(E_2 \mid C)$$

$$= \frac{P(C \mid E_1)P(E_1)}{P(C)} + \frac{P(C \mid E_2)P(E_2)}{P(C)} = \frac{0.005 \times 0.025}{0.0183} + \frac{0.001 \times 0.175}{0.0183} = 0.0164$$

SUMMARY

In this chapter, the basic rules governing events and equations related to probabilities were reviewed. Probability is a numerical measure that describes the chance of occurrence of a specific event among a number of possibilities. The need to use probability stems from the fact that there is an element of uncertainty involved in any engineering decision-making process as well as in any engineering analysis and design procedure. Essentially, the purpose of analysis and evaluation of engineering data is to provide information for a meaningful estimation of parameters that influence the decision making and system analysis and design. The basic elements of the theory of probability are especially important in understanding the underlying concepts in the evaluation of data and estimation of statistical values of parameters that are involved in engineering problems.

The review of events and probability provided in this chapter is brief. Nevertheless, the chapter provides nearly all of the essential equations needed to proceed with solving a given engineering problem. Several examples were provided to better acquaint the reader with the types of problems that an engineer may be facing. For a more complete treatment of the theory of probability as applied to engineering problems, the reader is referred to the list of references at the end of this chapter. These suggested references provide the reader with an ample number of illustrative examples that are useful for better understanding the application of the theory of probability to engineering problems.

REFERENCES

Ang, A. H.-S. and W. H. Tang (2007). Chapter 2 in *Probability Concept in Engineering: Emphasis on Applications to Civil and Environmental Engineering*, Hoboken, NJ: John Wiley & Sons.

Benjamin, J. and C. A. Cornell (1970). *Probability, Statistics and Decision-Making for Civil Engineers*, New York, NY: McGraw-Hill.

Milton, J. S. and J. C. Arnold (2002). Chapter 2 in *Introduction to Probability and Statistics: Principles and Applications for Engineering and the Computer Sciences*, 4th ed., New York, NY: McGraw-Hill.

Walpole, R. E., R. H. Myers, S. L. Myers, and K. Ye (2006). Chapter 2 in *Probability and Statistics for Engineers and Scientists*, 8th ed., Upper Saddle River, NJ: Prentice-Hall.

EXERCISES

1. Events A and B are defined with their sample points:

$$A : \{a, b, c, d, e, f\} \qquad B : \{a, c, g, h\}$$

Determine the sample points that belong to:

(a) Event C, which is the union of A and B
(b) Event D, which is the intersection of A and B

2. In designing a runway for a small airport, the wind velocity and wind direction are important factors. Assume the direction makes an angle θ with respect to the north. Possible values for θ range from 0 to 180 degrees. The wind velocity V ranges between 0 to 140 km/h.
   (a) Show possible values of θ and V in the form of a Venn diagram.
   (b) Identify and show the following events on the diagram.

$$E_1 : 20 < V \leq 140 \text{ and } 0 < \theta \leq 45$$
$$E_2 : 50 < V \leq 100 \text{ and } 0 < \theta \leq 90$$

   (c) Define and show $E_1 \cup E_2$, and $E_1 \cap E_2$

3. Events A and B contain possible values for the soil bearing capacities (kN/m^2) of three adjacent sites.

$$A = \{100,120,150,170\}$$
$$B = \{120,130,140\}$$
$$C = \{130,135\}$$

   (a) Determine the sample points in the intersection of A and B and in the union of A and B.
   (b) Determine the sample points in the intersection of B and C and in the union of B and C.
   (c) Among Events A, B, and C, which two are mutually exclusive?
   (d) Determine the sample points in the union of A, B, and C.

4. The duration of a construction project ranges between 121 to 160 da. There is an equal chance for the duration to be equal to any number of days between 121 and 160. Currently the project is scheduled to be finished in 130 da or fewer.
   (a) Compute the probability that the project will be finished on time
   (b) Compute the probability that the project will be delayed

5. A water pipeline system is made up of three links in a series. The probability of failure in each link (i.e., the probability that the pipeline link will not be able to deliver water) is 0.01 in any given year. The condition between the links is statistically independent. Compute the probability of failure of the pipeline system.

6. Repeat Exercise 5; however, assume the three links are parallel to each other.

7. Delay in a construction project depends mainly on the event of either material shortage or bad weather conditions. From past performance data, the following are known:

   - In any given season, the probability of shortage of material is 0.01; the probability of experiencing bad weather conditions is 0.05. Assume these two events are statistically independent.

- If only one of these events occurs, the probability of delay in the project is 0.10; whereas, if both events occur, the probability of project delay is 0.60.
  (a) Compute the probability of delay in the project.
  (b) If delay is certain, what is the probability that the shortage of material is causing it? Hint: Use the total probability and Bayes' theorems.

8. The water for a small town is provided from two sources A and B. In any given dry season, there is a 0.01 probability that Source A will not be able to provide the expected amount of water needed. The corresponding probability for Source B is 0.02. If Source A is not able to meet the demand, there is a 0.3 probability that B will not be able to either. If only one source cannot meet the demand, the probability of water shortage in town will be 0.40; whereas, if both sources cannot meet the demand, the probability of water shortage in the town will be 0.90.
   (a) Compute the probability of water shortage in the town in the next season.
   (b) If there is a shortage, what is the probability that only Source A is causing it?
9. Of every 100 cars approaching an interstate junction, 60 will go straight, 15 will make a left turn, and the rest will make a right turn.
   (a) What is the probability that a car approaching the junction will make a turn?
   (b) It is certain that an approaching car will make a turn, what is the probability that it will be a right turn?
10. Prove that Equation 5.23 for the union of three events is valid. Derive a similar expression for the union of four events.
11. A construction company is currently bidding on a job overseas in Country X. The chance of winning the job depends on the upcoming election results in X. If Party A wins the election, the chance of the company winning the job is 60 percent; otherwise, there is only a 20 percent chance that the company will win the job. On the other hand, from poll data, the probability that Party A will win the election is 0.85. Compute the probability of the company's success in winning the job.
12. A roof system is made up of fifteen parallel wood trusses. Proof load testing has shown that the probability of failure of any such roof truss is 0.02 if the total applied load exceeds 20 kN. If during a snowstorm, the total load transmitted to each truss exceeds 20 kN, what will be the probability of the failure of the roof system? Assume the conditions between trusses are statistically independent.
13. The on-time completion of Activity A in a construction project depends on the on-time completion of prior Activities B and C. If both B and C are completed on time, for sure A will be completed on time. If only one of the two Activities B and C are completed on time,

A will have a 50 percent probability that it will be completed on time. If neither Activities *B* and *C* are completed on time, *A* will only have a 10 percent probability that it will be completed on time. From past experience, we know that Activities *B* and *C* each have an 80 percent probability to be completed on time. Activities *B* and *C* are independent.

(a) Compute the probability that Activity *A* will be completed on time.
(b) If we allow a delay in the completion of *A*, what is the probability that *only* Activity *B* is not completed on time?

14. The probability of weather-related traffic accidents in winter months in an intersection depends on the road conditions. For simplicity, assume there are three possibilities for the road conditions: (1) dry; (2) wet; and (3) icy conditions. If, during the traffic rush hours, the roads are dry, the probability of one or more accidents at the intersection is 0.001. The corresponding probabilities for the wet and icy conditions are 0.10 and 0.30, respectively. The probability of roads being dry at any time during the winter months is 0.80. The probability for the wet condition is 0.15.

    (a) Compute the probability that the intersection will experience any number of accidents during the traffic rush hours in the winter months.
    (b) Compute the probability that three such intersections located along a road will have no accidents during the traffic rush hours in the winter months. Assume statistical independence between the intersections.

15. Three construction companies (designated *A*, *B*, and *C*) are bidding on a job. These companies have equal chances to win the job. From past performance records, if either Company *A* or *B* wins the job, there is a 0.80 probability that the job will be completed on time. The corresponding probability for Company *C* is 0.90.

    (a) Not knowing which company will win the job; compute the probability that it will be completed as scheduled
    (b) If we allow the completion time to be delayed, compute the probability that *A* wins the job
    (c) Repeat (b) for Companies *B* and *C*

16. A motorist is driving at the posted maximum speed along a roadway. If he encounters a red traffic light at an intersection, the probability that he will encounter a red light at the next intersection is 0.60. However, if he does not encounter a red light in the first intersection, there is still a 0.10 probability that he will encounter a red light at the next intersection.

    (a) The motorist has two consecutive intersections ahead. The probability of encountering a red light at the first intersection is 0.3. What is the probability that he will not encounter any red lights passing the two intersections?

(b) Based on the information given, compute the probability of encountering a red light in the second intersection.

17. A company is currently bidding on three jobs. The odds of winning any job are 0.50. The winning of a job is independent of winning the other two.

    (a) Compute the probability that the company will win all three jobs
    (b) Compute the probability that the company will win any two jobs

18. A gas distribution pipeline network consists of 24 links distributing gas to houses in a subdivision. The probability that a link will develop a leak in any given year is 0.003. Assuming the conditions between the links are statistically independent:

    (a) Compute the probability that the entire network will not develop any leak in any given year
    (b) In light of the result obtained in (a), compute the probability that the system will not develop any leak 10 yr in a row

19. A highway bridge is currently scheduled to be inspected once every two yr. There is a 3 percent probability that during an inspection a major defect will be found. Compute the probability that in the next 12 yr no defects will be found in the bridge. Assume statistical independence concerning finding defects in different inspections.

20. In Exercise 16, assume the motorist has three intersections ahead. These intersections are labeled *A*, *B*, and *C*. The motorist will reach *A* first, then *B*, and finally *C*. The probability that he will encounter a red light at *B* depends on whether or not he encountered a red light at *A*. The probability that he will encounter a red light at *C* depends on whether or not he encountered a red light at the previous intersection (i.e., *B*) only. As described in Exercise 16, if he encounters a red light at *A*, there is a 0.60 probability that he will also encounter a red light at *B*. If he does not encounter a red light at *A*, there is still a 0.10 probability that he will encounter a red light at *B*. Similarly, given a red light encounter at *B*, the probability that the motorist will encounter a red light at *C* is also 0.60. However, if no red light is encountered at *B*, there is still a 0.10 probability that he will encounter a red light at *C*. There is a 0.30 probability that he will encounter a red light at the first intersection, that is, *A*.

    (a) Compute the probability that the motorist will not encounter any red light passing through the three intersections
    (b) Compute the probability that the motorist will encounter a red light at *B*
    (c) Repeat (b) but for Intersection *C*

21. In an improvement design for an intersection, two alternatives are suggested by a traffic engineer. These are: (Alternative *A*) use a traffic light system; and (Alternative *B*) use a 4-way stop sign system. The maximum number of accidents allowed for the intersection is 20 per yr. If design Alternative *A* is selected, there is a 0.10 probability that there will be more than

the allowable 20 accidents/yr. However, if design Alternative *B* is selected, this probability will be 0.45. The odds that design Alternative *B* is selected over *A* is 3 to 1.

(a) Compute the probability that once an improvement is completed the intersection will have less than the allowable number of accidents/yr.
(b) We wish to investigate the effect of design Alternative *A* on the safety of the intersection. Assuming the accidents will exceed the allowable number after the improvement plan is implemented; compute the probability that design Alternative *A* is the choice for the improvement plan.

22. Possible values for the resistance of a type of cable used as a tension member are 70, 80, and 90 kN, respectively. These possibilities are equally likely to occur. There are also three equal possibilities for the applied load. These are 65, 75, and 85 kN, respectively.

(a) What is the probability that the resistance of such a cable selected at random will be at least equal to 80 kN?
(b) What is the possibility that the load will be at most equal to 75 kN?
(c) What is the probability of failure of the cable?

23. Prove that if Events *A* and *B* are statistically independent, their complementary events are also statistically independent.

24. A box used in a scientific lab contains an equal number of steel and aluminum beads. Each steel bead weighs 30 g; and each aluminum bead weighs 10 g. If three beads are selected at random, what will be the probability that the total weight of the three beads is exactly equal to 50 g?

Principles of Probability: Part II—Random Variables and Probability Distributions

6

6.1 RANDOM VARIABLES

Nearly all events can be defined by *random variables*. Similar to any other variable, a random variable such as X can accept different values; however, these values follow an uncertain or random pattern. If specific values or ranges are assigned to a random variable, then events are formed. Thus, a random variable offers a mathematical tool that can be used to introduce events within a defined sample space. The sample space is simply defined by a lower bound and an upper bound that specify the range of all possible values that the random variable can accept. If no lower and upper bounds are specified, then the range for the random variable is within $\pm\infty$. For example, if X is used to define the vertical acceleration in an aircraft during a flight, the ranges are within a negative acceleration (e.g., $-3g$, where g is the gravity acceleration) to a maximum positive acceleration (e.g., $+5g$). Any specific range assigned to X will define an event. For example, the expression $-2g < X \leq 2g$ indicates the event that the acceleration at any given time period will be within $\pm 2g$. In this example, $-3g < X \leq 5g$ covers the entire sample space and, as such, defines a sure event. Note that within this range many events are possible and can be defined as either a one- or two-sided inequality.

6.1.1 Discrete Random Variables

In some engineering problems, the possibilities within the lower and upper bound of a random variable are countable. Such a random variable is *discrete*. Problems dealing with a specific number of occurrences of a quantity are often defined with discrete random variables. For example, the number of defective products in any given day in the production line of a manufacturing firm can be defined as a discrete random variable. The possibilities are countable. If X is used to define the number of defective products, then possibilities are $X = 0, X = 1 \ldots, X = x$. Note that each possibility is an event and that these individual events are mutually exclusive. Furthermore, events can also be formed by assigning a range for X. For example, $2 < X \leq 6$ defines an event that the number of defective units will be ≤ 6 and > 2. This event contains four sample points, that is, $X = 3$, $X = 4$, $X = 5$, and $X = 6$.

Several other examples of quantities that can be defined by discrete random variables are:

- Number of projects a construction firm wins in a given year.
- Occurrences of natural hazards such as earthquakes or tornadoes. The number of such events in any given time period can be defined via discrete random variables.
- Number of component failures in a multicomponent engineering system.
- Number of accidents in a given time period at a busy intersection.
- Number of airplanes taking off and landing during a given time period in an airport.
- Number of cars approaching a toll booth in a tollway during any given hour.

6.1.2 Continuous Random Variables

If the number of possibilities defined by a random variable is infinite, then the variable is said to be *continuous*. For example, the strength of a given material can be considered a continuous random variable. Theoretically, the strength can be any positive value. Denoting this random variable as X, events can be formed via specific ranges assigned to X. These ranges are defined using inequality expressions. An inequality expression such as $a < X \leq b$ describes the event that X will be within the two limits a and b; whereas, an inequality $X > c$, for example, describes the event that X will be within c and infinity (∞). With continuous random variables, the events are always defined with ranges even if these ranges are infinitesimally small. Thus, although $X = x$ is practically meaningful, theoretically, it can define an event only if it is shown with the inequality expression $x < X \leq x + dx$ in which dx is an infinitesimally small range of X.

The following is an example of several other engineering quantities that are defined via continuous random variables:

- The magnitude of load applied on a structural system
- Duration of a construction project
- The length of a crack in metals
- The time intervals between subsequent breakdowns of a mechanical system
- The amount of a certain raw material available in a manufacturing unit
- The amount of rainfall (using mm) during a future rainstorm

It is noted that the above quantities can also be defined via discrete random variables. This will be possible only if a series of definite discrete ranges are assigned to each quantity as demonstrated in Example 6.1.

Example 6.1 The stress data for a main girder in a timber bridge have been compiled as shown in Table E6.1. The stress has been defined via six ranges in the megapascal (MPa) unit. Each range is a discrete possibility within the total of six possibilities. The random variable describing these possibilities is X. There are no occurrences for stresses larger than 6 MPa.

Continues

Example 6.1: *Continued*

(a) Compute the probability of each of the six events described by values of X.

(b) Compute the probability of $X \leq 2$.

(c) Compute the probability of $1 < X \leq 5$.

Table E6.1 Stress range data values

Values of X	Stress range (MPa)	Occurrences/da	Relative frequency
1	0.00 to 1.00	4234	4234/7017 = 0.603
2	1.01 to 2.00	1432	1432/7017 = 0.204
3	2.01 to 3.00	850	850/7017 = 0.121
4	3.01 to 4.00	411	411/7017 = 0.059
5	4.01 to 5.00	85	85/7017 = 0.012
6	5.01 to 6.00	5	5/7017 = 0.001
		Total = 7017	1.000

Solution

(a) The six possibilities make up the entire sample space; they are mutually exclusive and collectively exhaustive. Thus, for the discrete random variable X, the probabilities are the relative frequencies computed and shown in Table E6.1, that is,

$P(X = 0) = 0$
$P(X = 1) = 0.603$
$P(X = 2) = 0.204$
$P(X = 3) = 0.121$
$P(X = 4) = 0.059$
$P(X = 5) = 0.012$
$P(X = 6) = 0.001$
$P(X > 6) = 0$

(b) From the results in (a):

$$P(X \leq 2) = P(X = 1) + P(X = 2) = 0.807$$

(c) Again from results in (a):

$$P(1 < X \leq 5) = P(X = 2) + P(X = 3) + P(X = 4) + P(X = 5) = 0.396$$

Note that if no discrete ranges are specified in Example 6.1, and that the strain can accept any value within 0 to 6 MPa, then the random variable describing the stress will be a continuous one.

Example 6.2 The error associated with the outcome of a laboratory measurement device is a continuous random variable, X, and ranges uniformly between −6 percent to +6 percent.

(a) Compute the probability that the error will be between −2 percent and +2 percent.

(b) Compute the probability that the error will be a positive value.

(c) Compute the probability that the random variable X will be within a small range dx.

Solution

(a) Since the random variable ranges uniformly between −6 and +6, the probability that the variable is between a given range will be the ratio of the range to the overall range of the random variable. The overall range of the variable is 6 − (−6) = 12. Thus:

$$P(-2 < X \leq +2) = 4/12 = 0.333$$

(b) In this part, the probability of the Event $X > 0$ is required. Thus:

$$P(X > 0) = 6/12 = 0.500$$

(c) In this part, the probability of the Event $x < X \leq x + dx$ is desired. Since the range for X is dx:

$$P(x < X \leq x + dx) = dx/12$$

6.2 PROBABILITY FUNCTIONS

Probability functions are used to define probabilities of events associated with a random variable. Most such functions are in mathematical forms. However, there may be cases in which a probability function is given in a tabular or graphical format. Probabilities of events defined by a discrete random variable are described by means of a *probability mass function* and a *cumulative distribution function*. For a continuous random variable, the corresponding functions are the *probability density function* and *probability distribution function*, respectively.

6.2.1 Probability Mass Function and Cumulative Distribution Function

The probability mass function of a discrete random variable X is defined with a function $p(x)$. Notice that p is written as a function of x (and not X). This is because x describes the specific values that X can take; that is, $x \subset X$. Furthermore, $p(x)$ describes the probability of $X = x$. Thus:

$$p(x) = P\,(X = x) \tag{6.1}$$

Note that since X is a discrete random variable, the values of x can only be whole integer numbers such as 0, ±1, ±2. . . . Furthermore, note that the function $p(x)$ is also discrete. The cumulative distribution function is shown as $F(x)$ and describes the probability of the Event $X \leq x$. Since the Events $X = x$ are mutually exclusive, then $F(x)$ can be written:

$$F(x) = P(X \leq x) = \sum_{\text{all } x_i \leq x} p(x_i) \tag{6.2}$$

In Equation 6.2, the summation is overall x_i values that are $\leq x$. When plotted, the function $F(x)$ appears in the form of a step graph. Since this function defines a probability, it is then bounded by 0 and 1:

$$F(-\infty) = P(X \leq -\infty) = 0 \tag{6.3}$$

and

$$F(+\infty) = P(X \leq +\infty) = 1 \tag{6.4}$$

The latter equation covers the entire sample space. Thus, Equation 6.4 can also be written:

$$F(+\infty) = \sum_{\text{all } x_i} p(x_i) \tag{6.5}$$

In Equation 6.5, the summation covers all values of the random variable X. Considering relations between events, as discussed in Chapter 5, one can also write:

$$F(X > x) = 1 - P(X \leq x) = 1 - F(x) \tag{6.6}$$

$$P(a < X \leq b) = F(b) - F(a) = \sum p(x_i) \quad \text{for } a < x_i \leq b \tag{6.7}$$

Figure 6.1 shows an example of a probability mass function, $p(x)$, and the corresponding cumulative distribution function, $F(x)$.

6.2.2 Probability Density and Distribution Functions

Probability density and distribution functions are specific to continuous random variables. The probability density function is defined by $f(x)$, where:

$$f(x) = \frac{P(x < X \leq x + dx)}{dx} \tag{6.8}$$

or

$$P(x < X \leq x + dx) = f(x)\,dx \tag{6.9}$$

Again, notice that f is written as a function of x (and not X). As seen in Figure 6.2, $P(x < X \leq x + dx) = f(x)dx$ is the small shaded area, that is, $f(x)dx$, under the probability density function curve in the x, $x + dx$ interval. Thus, the probability of the event $(-\infty < X \leq x)$ is the sum of an infinite number of small areas between $-\infty$ and x. This means that:

$$P(-\infty < X \leq x) = P(X \leq x) = \int_{-\infty}^{x} f(x)\,dx \tag{6.10}$$

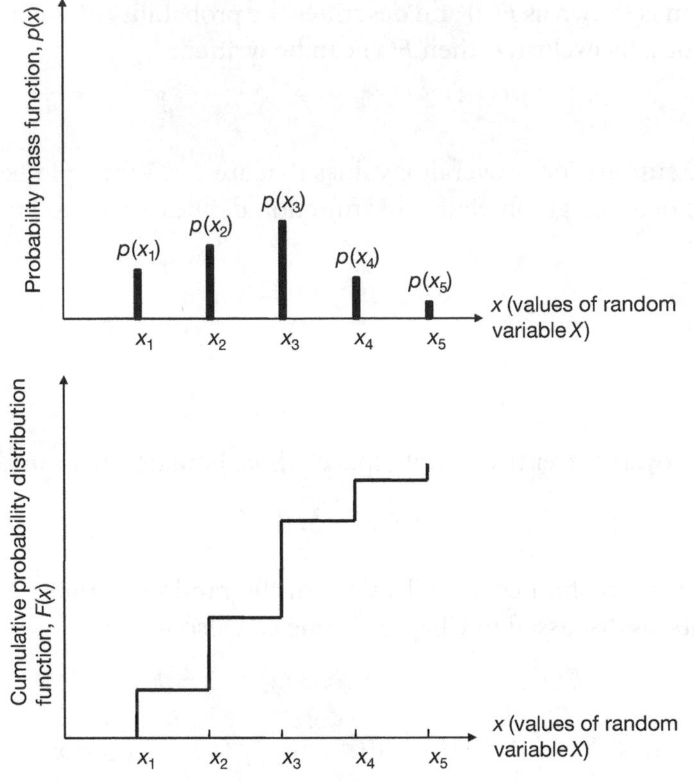

Figure 6.1 Probability mass function $p(x)$ and cumulative distribution function $F(x)$

The probability distribution function is shown with $F(x)$ and is defined:

$$F(x) = P(X \le x) \tag{6.11}$$

In light of Equations 6.10 and 6.11:

$$F(x) = \int_{-\infty}^{x} f(x)\,dx \tag{6.12}$$

or

$$f(x) = \frac{dF(x)}{dx} \tag{6.13}$$

Equation 6.12 implies that the probability distribution function is the area under the density function curve (see Figure 6.2). Furthermore, one can write:

$$P(a < X \le b) = \int_{a}^{b} f(x)\,dx = F(b) - F(a) \tag{6.14}$$

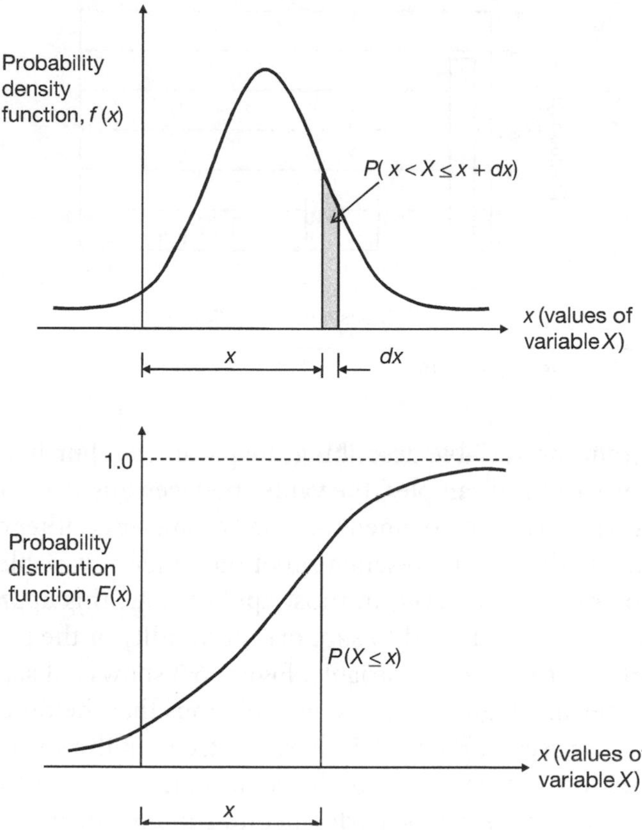

Figure 6.2 Probability density and distribution functions

and

$$F(-\infty) = 0 \tag{6.15}$$

and

$$F(+\infty) = 1 \tag{6.16}$$

The latter equation also indicates that:

$$\int_{-\infty}^{+\infty} f(x)\,dx = 1 \tag{6.17}$$

This simply means that the area under the entire probability density function is equal to 1.

In real situations, engineering data are compiled and presented in discrete forms. A strain in a critical component of a structure, for example, is measured using a specific sampling rate. If n samples, that is, $x_1, x_2 \ldots, x_n$ are collected, the data compiled represent a small subdomain of all

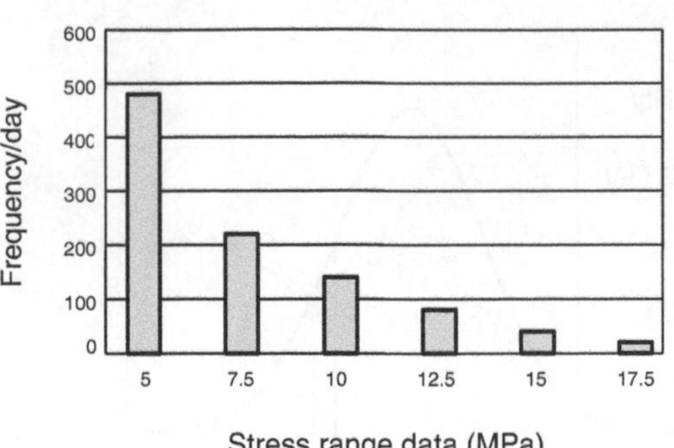

Figure 6.3 Discrete plot of stress range data

possible values of the random variable describing the parameter. Furthermore, it is noted that due to the discrete nature of the n samples, the values between any two consecutive values (say x_i and x_{i+1}) are not known (it is often assumed that the parameter is either constant or varies linearly between x_i and x_{i+1}). Thus the representation of the random variable as a continuous one will only be an approximation. However, in most applications, this approximation is acceptable if additional analyses are conducted to support the validity of the probability distribution model selected to represent the random variable. Figure 6.3 shows a discrete plot of stress data compiled in a main girder in a highway bridge. It is obvious that the random variable describing stress is continuous. As seen in Figure 6.3, there is a trend in the stress data in the form of a probability density function. Once the type of distribution model that can be used to represent the data is selected, statistical tests can be performed to support the validity of the model. This subject is described in subsequent chapters.

6.3 DESCRIBING PARAMETERS OF A RANDOM VARIABLE

The probability functions described in Section 6.2 provide a complete description on random variables. In addition, there are also several parameters that are considered descriptors of a random variable. The descriptors are parameters that can be found using the probability mass or density functions. These parameters are discussed in this section.

6.3.1 Mathematical Expectation

By definition, the mathematical expectation of any function $g(X)$ of the random variable X is shown as $E[g(X)]$ and is mathematically described by:

$$E[g(X)] = \sum_{\text{all } x_i} g(x_i)p(x_i) \tag{6.18}$$

for discrete random variables, where the summation covers all possible values of x_i; and:

$$E[g(X)] = \int_{-\infty}^{+\infty} g(x)f(x)\,dx \tag{6.19}$$

for continuous random variables. The mathematical expectation is a deterministic parameter and is in fact a weighed average. Considering a discrete random variable X, the average value of all $g(x_i)$ values using $p(x_i)$ as the weights can be written:

$$\text{Weighed average} = \frac{\sum_{\text{all } x_i} g(x_i)p(x_i)}{\sum_{\text{all } x_i} p(x_i)}$$

in which the summations are for all values of x_i, as indicated. Since $\sum p(x_i) = 1$ then the average so obtained is $E[g(X)]$. A similar logic can also be slated for a continuous random variable.

In a special case where the function $g(X)$ is simply X, the mathematical expectation, that is, $E(X)$, is called the *expected value* of X, or the *mean value* of X (as described in the following section). If the function $g(X)$ is equal to X^2, the mathematical expectation, that is, $E(X^2)$, is called the *mean square* of X.

6.3.2 Mean or Expected Value

The mean value μ of the random variable X is the mathematical expectation of X. Thus:

$$\mu = E(X) = \sum_{\text{all } x_i} x_i p(x_i) \tag{6.20}$$

for a discrete random variable and:

$$\mu = E(X) = \int_{-\infty}^{+\infty} xf(x)\,dx \tag{6.21}$$

for a continuous random variable. Note that the mean value is simply a weighed average of all values of the random variable X.

6.3.3 The Variance and Standard Deviation

The variance is the mathematical expectation of the function $(X - \mu)^2$. Thus, if function $g(X)$ is taken equal to $(X - \mu)^2$ then:

$$E(X - \mu)^2 = \sum_{\text{all } x_i} (x_i - \mu)^2 p(x_i) \tag{6.22}$$

for the case where X is discrete and:

$$E(X - \mu)^2 = \int_{-\infty}^{+\infty} (x - \mu)^2 f(x)\,dx \tag{6.23}$$

for the case where X is continuous. In an alternative form the variance can be written in terms of the mean square, that is, $E(X^2)$, as demonstrated:

$$E[(X-\mu)^2] = E[X^2 - 2\mu X + \mu^2] = E(X^2) - 2\mu E(X) + \mu^2 = E(X^2) - \mu^2 \quad (6.24)$$

Equation 6.24 can also be proven by expanding the right-hand side of either Equation 6.22 or Equation 6.23.

The square root of the variance is the standard deviation and is shown with σ, that is:

$$\sigma = \sqrt{E(X-\mu)^2} \quad (6.25)$$

From Equation 6.24, it can be shown that:

$$E(X^2) = \sigma^2 + \mu^2 \quad (6.26)$$

Note that σ and μ have the same unit as does the random variable. The mean may be a positive, negative, or zero value. However, the standard deviation for a random variable is always a non-zero positive value. In the case in which the standard deviation is zero, the variable will not be random. Dividing the standard deviation by the mean value results in a dimensionless quantity called the *coefficient of variation* (COV). The COV is shown with δ and describes the standard deviation as a fraction of the mean value. The COV is then computed:

$$\text{COV} = \delta = \frac{\sigma}{\mu} \quad (6.27)$$

The COV can also be presented as a percentage.

6.3.4 Median and Mode

The median is a specific value of the random variable X at which the 50 percentile probability is obtained. The median (x_m) is obtained from the following:

$$P(X > x_m) = 0.5 \quad (6.28)$$

The mode of a random variable is a specific value of the variable at which the density function has its peak.

6.3.5 Moments of a Random Variable

The moments of a random variable are mathematical expectations of X^n in which n is a numerical value. According to this definition, the mean value is the first moment (since $n = 1$); whereas, the variance is a second moment. The third moment is used to describe the skewness in the probability density (or mass) function. The skewness factor, θ, is defined:

$$\theta = \frac{E(X-\mu)^3}{\sigma^3} \quad (6.29)$$

If the density (or mass) function is symmetric, then $\theta = 0$.

Example 6.3 The tensile strength of a cable (X) is defined as a random variable following a triangular probability density function as shown in Figure E6.3. The values of x (the tensile strength) given in the figure using the kilonewton (kN) unit.

(a) Compute h and write an equation for the probability density function of X.

(b) Compute the mean and standard deviation of X.

(c) If the load applied to the cable is 25 kN, compute the probability of the failure of the cable.

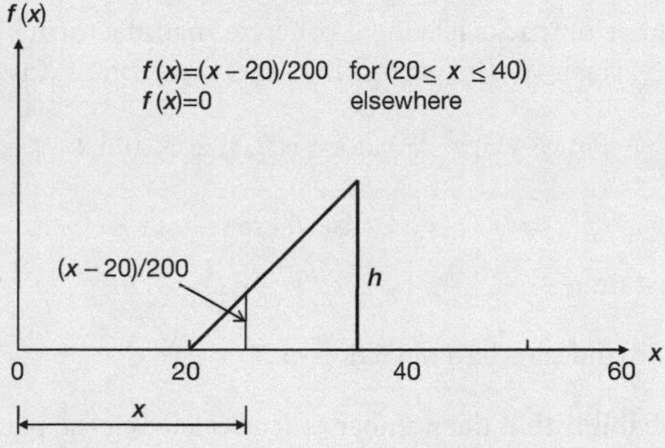

Figure E6.3 Triangular probability density function

Solution

(a) In light of Equation 6.17, $h \times (40 - 20)/2 = 1$, $h = 1/10$. The equation for $f(x)$ is written:

$$f(x) = \frac{x - 20}{200} \quad \text{for } 20 \le x \le 40$$

$$f(x) = 0 \quad \text{elsewhere}$$

(b) In this part we use Equations 6.21 and 6.23:

$$\mu = E(x) = \int_{-\infty}^{+\infty} xf(x)\,dx = \int_{20}^{40} x\frac{x - 20}{200}dx = 33.3$$

$$\sigma^2 = E(X - \mu)^2 = \int_{-\infty}^{+\infty} (x - \mu)^2 f(x)\,dx = \int_{20}^{40} (x - 33.3)^2 \frac{x - 20}{200}dx = 22.2$$

and the standard deviation (σ) = 4.71.

(c) The probability of failure is defined by event ($X \le 25$); because the failure occurs when the load exceeds the resistance of the cable. To compute this probability we can utilize Equation

Continues

Example 6.3: *Continued*

6.14. Alternatively, the area under the $f(x)$ between $x = 20$ and 25 can be computed as the probability of concern. Using Equation 6.14:

$$P(\text{cable failure}) = P(X \leq 25) = \int_{20}^{25} \frac{x-20}{200}dx = 0.063$$

Example 6.4 The number of trucks leaving a concrete manufacturing plant in any hour is defined with a random variable X. The probability mass function of X is proposed:

$$p(x) = \frac{c}{x^2+1} \quad \text{for } x = 0, 1, 2, 3, \text{and } 4$$

$$p(x) = 0 \quad \text{elsewhere}$$

(a) Compute the constant c.

(b) Compute the mean and standard deviation of X.

(c) Compute the probability that the number of trucks leaving the plant in any hour is ≤ 2.

Solution

(a) From Equation 6.5:

$$c\left(\frac{1}{x_1^2+1} + \frac{1}{x_2^2+1} + \frac{1}{x_3^2+1} + \ldots\right) = c\left(\frac{1}{0+1} + \frac{1}{1+1} + \frac{1}{4+1} + \frac{1}{9+1} + \frac{1}{16+1}\right) = 1$$

This results in $c = 0.538$.

(b) The mean and standard deviation are computed from Equations 6.20 and 6.22:

$$\mu = c\left[x_1\frac{1}{x_1^2+1} + x_2\frac{1}{x_2^2+1} + x_3\frac{1}{x_3^2+1} + \ldots\right]$$

$$= c\left[0\times\frac{1}{0+1} + 1\times\frac{1}{1+1} + 2\times\frac{1}{4+1} + 3\times\frac{1}{9+1} + 4\times\frac{1}{16+1}\right] = 0.772$$

and

$$\sigma^2 = c\left[(x_1-\mu)^2\frac{1}{x_1^2+1} + (x_2-\mu)^2\frac{1}{x_2^2+1} + (x_3-\mu)^3\frac{1}{x_3^2+1} + \ldots\right]$$

$$= c\left[(0-0.772)^2\times\frac{1}{0+1} + (1-0.772)^2\times\frac{1}{1+1} + \ldots\right] = 1.094$$

Continues

Example 6.4: *Continued*

This results in $\sigma = 1.05$. Of course, the mean value in this case is a symbolic measure since the number of trucks can only be a whole number.

(c) In this case $P(X \leq 2)$ is desired. This probability from Equation 6.2:

$$P(X \leq 2) = P(X = 0) + P(X = 1) + P(X = 2)$$

$$= c\left[\frac{1}{x_1^2 + 1} + \frac{1}{x_2^2 + 1} + \frac{1}{x_3^2 + 1}\right] = c\left[\left(\frac{1}{0+1} + \frac{1}{1+1} + \frac{1}{4+1}\right)\right] = 0.915$$

6.4 SEVERAL USEFUL PROBABILITY FUNCTIONS

In this section, several probability functions are described. Additional functions are introduced throughout the book whenever a particular discussion necessitates the introduction of a specific type of distribution model.

6.4.1 Normal (Gaussian) Probability Density Function

The normal probability density function is widely used to describe continuous random variables in many applications. The normal probability density function is:

$$f(x) = \frac{1}{\sigma\sqrt{2\pi}} \exp\left[-\frac{1}{2}\left(\frac{x-\mu}{\sigma}\right)^2\right] \quad -\infty \leq x \leq +\infty \tag{6.30}$$

where μ and σ are the parameters of the distribution function. It can be shown that the expected value of X using Equation 6.30 will result in the parameter μ; whereas the variance will be the square of the parameter σ (i.e., the standard deviation). The normal function has the familiar bell shape. It is symmetric with respect to the mean value. As shown in Figure 6.4, the standard

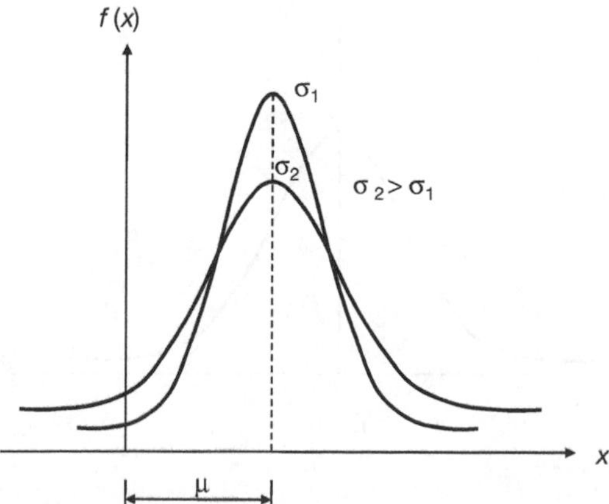

Figure 6.4 Normal probability density function

deviation determines the shape of the curve. A large standard deviation indicates that the random variable contains values that are scattered with less concentration of values around the mean. A small standard deviation, on the other hand, indicates that values of the random variable are mainly concentrated around the mean.

A special form of the normal probability density function has a zero mean and a *standard deviation* equal to unity (i.e., $\sigma = 1$). This is called the *standard normal* probability *density function*. Using S as the random variable represented by the standard normal function, $f(s)$ can be written:

$$f(s) = \frac{1}{\sqrt{2\pi}} \exp\left(\frac{s^2}{2}\right) \quad -\infty \leq s \leq +\infty \tag{6.31}$$

The standard normal function is shown in Figure 6.5. The probability distribution function of the standard normal random variable S is the integral of Equation 6.31. Rather than the usual form $F(s)$, the standard normal probability distribution function is shown with $\Phi(s)$; that is:

$$\Phi(s) = P(S \leq s) = \frac{1}{\sqrt{2\pi}} \int_{-\infty}^{s} \exp\left(-\frac{s^2}{2}\right) ds \tag{6.32}$$

The function $\Phi(s)$ cannot be obtained in closed form. It can be computed numerically for various values of s. Table A.1 in Appendix A presents the values of $\Phi(s)$. Since the standard normal function is symmetric with respect to $s = 0$ (i.e., the mean), the following relations can be written:

$$\Phi(0) = 0.5, \Phi(-\infty) = 0 \text{ and, } \Phi(+\infty) = 1 \tag{6.33}$$

$$\Phi(-s) = 1 - \Phi(s) \tag{6.34}$$

A linear transformation can be used to convert any normal function into the standard normal form. This can be proven using the following sequence of computations.

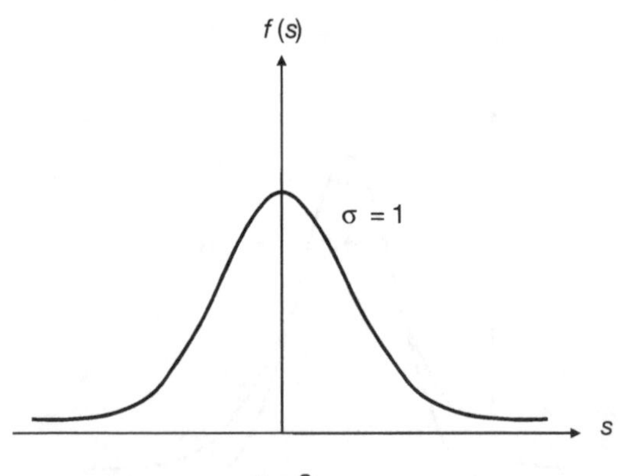

Figure 6.5 Standard normal probability density function

Let $S = (X - \mu)/\sigma$ or $X = \mu + S\sigma$ If s and x are specific values of the random variables S and X, respectively, then, $x = \mu + s\sigma$ or $dx = \sigma\, ds$. Now recall that the probability distribution function of X is the integral of $f(x)$. Thus:

$$F(x) = \frac{1}{\sigma\sqrt{2\pi}} \int_{-\infty}^{x} \exp\left[-\frac{1}{2}\left(\frac{x-\mu}{\sigma}\right)^2\right] dx \tag{6.35}$$

Since $dx = \sigma\, ds$ and $s = (x - \mu)/\sigma$ then:

$$F(x) = P(X \le x) = \frac{1}{\sqrt{2\pi}} \int_{-\infty}^{(x-\mu)/\sigma} \exp\left[-\frac{s^2}{2}\right] ds \tag{6.36}$$

This equation is identical to the standard normal function (Equation 6.32) with the only difference being the upper limit of the integral. Thus:

$$F(x) = P(X \le x) = \Phi\left(\frac{x-\mu}{\sigma}\right) \tag{6.37}$$

In light of Equations 6.14, 6.36, and 6.37, one can also write:

$$P(a < X \le b) = P(X \le b) - P(X \le a) = \Phi\left(\frac{b-\mu}{\sigma}\right) - \Phi\left(\frac{a-\mu}{\sigma}\right) \tag{6.38}$$

Example 6.5 The target time for the completion of a project is 120 da. The project duration is a normal random variable X with a mean of 115 and standard deviation of 20 da.

(a) Compute the probability of project delay (i.e., the project will not be finished within the target time).

(b) If a probability of delay equal to 0.10 is accepted for the project, compute the target time that corresponds to this probability.

Solution

(a) In this part, $P(X > 120)$ is desired. This is computed using Equation 6.37:

$$P(X > 120) = 1 - P(X \le 120) = 1 - \Phi\left(\frac{120-115}{20}\right) = 1 - \Phi(0.25)$$

From the table of normal probability values in the Appendix, $\Phi(0.25) = 0.5987$, thus:

$$P(X > 120) = 1 - 0.5987 = 0.4013$$

Continues

Example 6.5: *Continued*

(b) In this part, the target time is to be determined. Taking the target time as t, we can write $P(X > t) = 0.10$; or $P(X \le t) = 0.90$; or:

$$\Phi\left(\frac{t-115}{20}\right) = 0.90$$

This expression is also shown:

$$\left(\frac{t-115}{20}\right) = \Phi^{-1}(0.90)$$

From the normal probability table $\Phi(1.28) \cong 0.90$; or $\Phi^{-1}(0.90) = 1.28$. Thus:

$$\frac{t-115}{20} = 1.28 \quad \text{or} \quad t \cong 141\,da$$

Example 6.6 In Example 6.5, (a), assume that rather than 120 da, there is an uncertainty involved in the target time. In fact, there is 80 percent probability that the target time $t = 120$ da and 20 percent probability that it will be equal to100 da. These are the only two possibilities for the target time. Now recompute the probability of the project delay.

Solution The sample space for the target time has only two possibilities, that is, $t = 120$ da and $t = 100$ da. These are mutually exclusive and collectively exhaustive. Thus the total probability theorem can be used as follows:

$$P(X > t) = P(X > t \mid t = 120)P(t = 120) + P(X > t \mid t = 100)P(t = 100)$$

The individual terms in the above equation are computed:

$$P(X > t \mid t = 120) = 1 - P(X \le t \mid t = 120)$$

$$= 1 - \Phi\left(\frac{120-115}{20}\right) = 1 - \Phi(0.25) = 1 - 0.5987 = 0.4013$$

$$P(X > t \mid t = 100) = 1 - P(X \le t \mid t = 100)$$

$$= 1 - \Phi\left(\frac{100-115}{20}\right) = 1 - \Phi(0.75) = \Phi(0.75) = 0.7734$$

$$P(t = 120) = 0.8 \text{ (given by the problem)}$$

$$P(t = 100) = 0.20 \text{ (also given by the problem)}$$

Thus:

$$P(X > t) = 0.4013 \times 0.8 + 0.7734 \times 0.20 = 0.4757$$

Example 6.7 In Example 6.5, (a), assume the target time is 120 da only and that the standard deviation for the duration of the project is 10 da rather than 20 da. This simply means that there is now less uncertainty regarding the duration of the project. Compute the probability of delay.

Solution

$$P(X > 120) = 1 - P(X \leq 120) = 1 - \Phi\left(\frac{120 - 115}{10}\right) = 1 - \Phi(0.5) = 1 - 0.6915 = 0.3085$$

Notice that in Example 6.5 the mean duration time was shorter than the target time. Nevertheless, there was a probability that a delay would take place. This is because of the randomness in the project duration time (i.e., uncertainty in the duration time). In Example 6.6 there was also an uncertainty in the target time. Because of this uncertainty, a larger probability of delay than that in Example 6.5 was obtained. Example 6.7 had a smaller standard deviation (i.e., less uncertainty) for the project duration. This resulted in a smaller probability of delay than that in Example 6.5.

In Examples 6.5-6.7, the project duration time was assumed to be a normal random variable. Theoretically, this random variable can range between $-\infty$ to $+\infty$. However, it is obvious that the duration time cannot accept negative values. To overcome this difficulty, a different distribution such as *logarithmic normal* (or *lognormal*) or *Rayleigh* can be used. Both of these models only accept positive values and are more suitable to model variables that are always positive. These types of probability density functions are explained in the following sections.

6.4.2. Rayleigh Probability Density Function

The Rayleigh probability density function is in the following form:

$$f(x) = \frac{x}{a^2}\exp\left(-\frac{x^2}{2a^2}\right) \quad 0 \leq x < \infty \tag{6.39}$$

in which a is the parameter of the Rayleigh probability density function. Probability of events $(X \leq x)$ can be found through Equation 6.10. The advantage of the Rayleigh probability density function over the normal probability density function is that Equation 6.39, used along with Equation 6.10, will result in a closed form solution for $P(X \leq x)$. Using Equations 6.21 and 6.23, one can find the following equations between the mean and standard deviation of X (i.e., μ and σ) and the parameter a (Ang and Tang, 2007).

$$\mu = E(X) = a\sqrt{\frac{\pi}{2}} \text{ and } \sigma = a\sqrt{2 - \frac{\pi}{2}} \tag{6.40}$$

6.4.3 Logarithmic Normal Probability Density Function

The logarithmic normal (or lognormal) probability density function is defined via the following function:

$$f(x) = \frac{1}{\nu\, x\sqrt{2\pi}} \exp\left[-\frac{1}{2}\left(\frac{\ell n\, x - \lambda}{\nu}\right)^2\right] \quad 0 < x \leq +\infty \tag{6.41}$$

in which λ and ν are the parameters of the logarithmic distribution and are the mean and standard deviation of $\ell n(X)$, respectively. These parameters are written as:

$$\lambda = E[\ell n(X)] \quad \text{and} \quad \nu = \sqrt{Var[\ell n(X)]}$$

Figure 6.6 shows the logarithmic normal probability density function. As seen in this figure, smaller values of the parameter ν result in a more condensed variation. This indicates less variability in the random variable. The logarithmic normal distribution is related to the normal distribution. It can be shown that any probability associated with the logarithmic normal distribution can easily be computed from the standard normal probability distribution function Φ using the following equation:

$$P(a < X \leq b) = \Phi\left(\frac{\ell n\, b - \lambda}{\nu}\right) - \Phi\left(\frac{\ell n\, a - \lambda}{\nu}\right) \tag{6.42}$$

Furthermore, it can be shown that:

$$\nu^2 = \ell n(1 + \delta^2) \text{ and } \lambda = \ell n\, \mu - \frac{\nu^2}{2} \tag{6.43}$$

The proof of Equations 6.42 and 6.43 is left to the reader as an exercise problem.

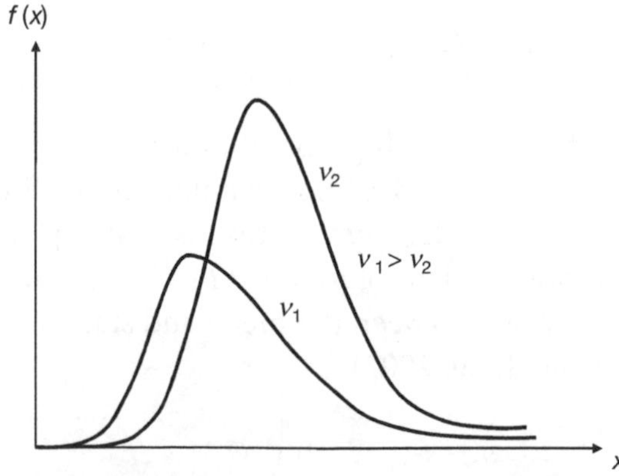

Figure 6.6 Lognormal probability density function

Example 6.8 The resistance of a tension member used in a truss is a logarithmic normal random variable with a mean value of 60 kN and a standard deviation of 15 kN.

(a) Compute the probability that the resistance at any given time will be between 40 and 70 kN.

(b) If a load equal to 30 kN is transferred to this member, compute the probability of failure of the member.

Solution

(a) In this part we are interested in $P(40 < X \leq 70)$. Using Equation 6.42:

$$P(40 < X \leq 70) = \Phi\left(\frac{\ell n\, 70 - \lambda}{\nu}\right) - \Phi\left(\frac{\ell n\, 40 - \lambda}{\nu}\right)$$

The parameters λ and ν are computed from Equation 6.43:

$$\text{COV} = \delta = 15/60 = 0.25, \text{ and } \nu^2 = \ell n\,[1 + (0.25)^2] = 0.06 \text{ and } \nu = 0.25$$

$$\lambda = \ell n\, 60 - (0.25)^2 = 4.06$$

Thus:

$$P(40 < X \leq 70) = \Phi(0.75) - \Phi(-1.48) = 0.704$$

(b) In this part, we observe that the failure will occur when the resistance X will be less than the applied load. Thus, we are interested in $P(X \leq 30)$. Although the mean value of the resistance is much larger than the load, there still exists a chance for failure.

$$P(X \leq 30) = \Phi\left(\frac{\ell n\, 30 - \lambda}{\nu}\right) = \Phi(-2.64) = 0.004$$

6.4.4 The *t*-Distribution

The *t*-distribution (also called the student *t*-distribution) is also used for continuous random variables and is similar to the standard normal in shape. The difference is that it is only defined with one parameter called the degree of freedom (ν). The probability density function for the *t*-distribution is symmetric with respect to 0. Using T as the random variable described by the *t*-distribution, the probability density function, $f(t)$, is written as:

$$f(t) = \frac{\Gamma\left(\frac{\nu+1}{2}\right)}{\Gamma\left(\frac{\nu}{2}\right)\sqrt{\pi\nu}}\left(1 + \frac{t^2}{\nu}\right)^{-\frac{\nu+1}{2}} \quad -\infty < t < +\infty \tag{6.44}$$

in which $\Gamma(\alpha)$ is the *gamma* function described with the following integral:

$$\Gamma(\alpha) = \int_0^\infty x^{\alpha-1} e^{-x} dx \quad \alpha > 0 \tag{6.45}$$

and with the following properties:

$$\Gamma(\alpha) = (1-\alpha)\Gamma(1-\alpha) \text{ and } \Gamma(1) = 1 \tag{6.46}$$

The degree of freedom (ν) is a positive integer value. It is taken as $n - 1$ in which n is the number of sample data points collected in a data acquisition process for a variable. The parameter ν is in fact a shape parameter. As the value of ν increases, the variance of the random variable T decreases. This means that larger ν values are associated with more compact density functions (see Figure 6.7). The t-distribution density function approaches the standard normal function as ν approaches ∞. The t-distribution tables are summarized for different t values for various probability levels and degrees of freedom (see Appendix A). For example, at $\nu = 10$ and 0.95 probability, $t = 1.812$. This means that with a degree of freedom equal to 10, $P(T \leq 1.812) = 0.95$. At $\nu = \infty$, and 0.95 probability, $t = 1.645$. This value is identical with the corresponding value from the normal probability table, that is, $P(T \leq 1.645) = 0.95$, both from the t-distribution table and the normal probability table. The application of the t-distribution is mainly in establishing confidence intervals for a mean value that is estimated from compiled data and also in the analysis of variance when means from two data sets (two populations) are compared. This is discussed in subsequent chapters.

6.4.5 Binomial Distribution Function

The binomial distribution function (also known as the Bernoulli sequence) is used for discrete random variables. The function is suitable in modeling random variables that concern the occurrence of x trials among a total of n trials. The basic assumptions in the binomial distribution are

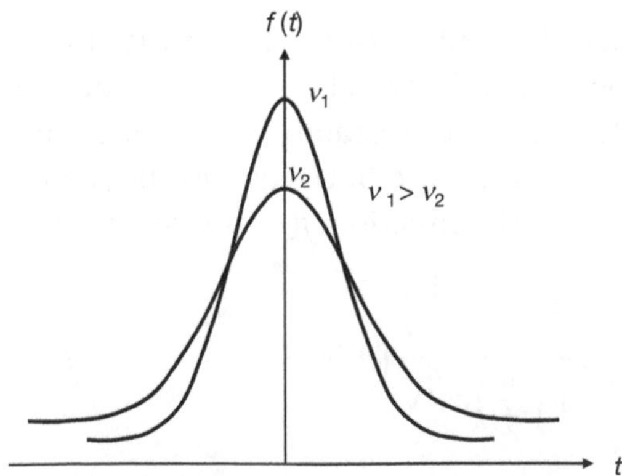

Figure 6.7 *t*-distribution

that: (1) each trial has only two outcomes, namely, either occurrence or nonoccurrence; (2) the probability of occurrence of any one trial is p, which is the same for all trials; and (3) the trials are statistically independent. A classic example of binomial distribution in engineering is the problem of a system that is made up of n identical components. Assuming that the probability of failure of any one component is p, then the binomial distribution can be used to estimate the probability that $X = x$ component failures (out of a total of n) will occur. For example, if $n = 3$ and the probability of exactly two component failures ($X = 2$) is desired, this probability can be computed by considering all possible cases that two trials (out of three) will take place. It is obvious that there will be three possibilities for two component failures (and one no-failure). Thus $P(X = 2) = 3p^2(1 - p)$. Notice that each possibility term is obtained as the intersection of two occurrence events and one nonoccurrence event, that is, $p^2(1 - p)$. The intersection is simply written as the product of the three probabilities p, p, and $(1 - p)$ because the respective events are statistically independent of one another. In a general case, when x trials among a total number of n are desired, the number of possibilities n, can be computed by considering all possible combinations of x trials in the set of n total trials. Thus:

$$P(X = x) = \frac{n!}{x!(n - x)!} p^x (1 - p)^{n-x} \quad x = 0, 1, 2, \ldots, n \tag{6.47}$$

Notice that Equation 6.47 describes the probability mass function for the discrete random variable X and that p is the parameter of the function.

Example 6.9 An engineering system is made up of 11 components each with a probability of failure $p = 0.0004$. Compute the probability of failure of the system. Component failures are independent of one another.

Solution Assuming that any component failure will result in the system failure, the binomial distribution can be used to compute the system failure probability. In this case, a trial is the failure of any one component. This failure may or may not occur.

$$P(\text{system failure}) = P(X \geq 1) = 1 - P(X < 1) = 1 - P(X = 0)$$

$$= \frac{11!}{0! \times 11}(0.0004)^0(1 - 0.0004)^{11-0} = 1 - 0.9956 = 0.0044$$

The binomial distribution can easily be applied to time-dependent problems also. In such a case, time is first divided into identical intervals. It is then assumed that within each interval only one occurrence of the event of concern is possible. In electronic systems a time unit equal to 1 hr is often used. The component probabilities are then defined via a *failure rate* that is the probability of failure per hour. The probability that the failure will happen after t time intervals can also be obtained from the binomial distribution. However, in such a case the distribution is called the *geometric distribution*. In a more general case, the probability that the failure occurs after t intervals for the nth time can also be computed. The corresponding distribution is then referred to as the *hypergeometric distribution*. The reader is referred to references at the end of this chapter for more information on these distributions. Considering the failure rate of electronic components, using the geometric and hypergeometric distributions, it can be shown that the average time between failures is equal to $1/p$. The average time between failures is often used as a measure of the reliability of a system.

Example 6.10 A flight data recorder is made up of 267 electronic components. The reliability of each component is 0.999999 per hr. Assuming that failures among components are independent and that the component failure rates are identical:

(a) Compute the failure rate and average time between failures of any given component.

(b) Compute the failure rate of the recorder per hour.

(c) Compute the reliability of the recorder and its average time between failures.

Solution

(a) The failure rate is the probability of failure per hour (p) which is:

$$p = 1 - 0.999999 = 10^{-6} \text{ per hr}$$

The average time between failures for any one component is $1/10^{-6} = 1{,}000{,}000$ hr.

(b) The failure rate (failure probability) of the recorder is obtained using the binomial distribution. The assumption is that the failure in any one component causes the failure of the recorder.

$$P(\text{system failure}) = P(X \geq 1) = 1 - P(X = 0) = (1 - 0.000001)^{267} = 0.000267 \text{ per hr}$$

(c) The reliability of the recorder =1 − 0.000267 = 0.999733; and the average time between failure is 1/0.000267 = 3,745 hr.

6.4.6 Poisson Distribution Function

The Poisson distribution function (also known as the Poisson process) is often used in time- or space-related problems. It is suitable for discrete random variables and defines the probability of occurrence of a desired number of identical events in a desired time or space interval. Several examples of events used in conjunction with the Poisson distribution function are provided:

- Occurrence of at least one traffic accident in a week at an intersection
- Occurrence of no accidents in a construction site during the next 30 mo

- Occurrence of any number of earthquakes (i.e., at least one earthquake) in a certain county in the central United States in the next 25 yr
- Occurrence of exactly five airplanes arriving at an airport in the next 20 min
- Occurrence of discovering at least two cracks while inspecting a 2-meter long segment of an aluminum structural component

The latter example deals with space; all the others deal with time.

The key parameter in the Poisson distribution function is λ, which is called the *occurrence rate* or the *activity rate*. The parameter λ is the average number of occurrences of the event of concern in unit time or unit space. Obviously, the information on λ must be compiled from past occurrences or test data, depending on the type of problem. The probability mass function of a discrete random variable X that follows the Poisson distribution function is defined:

$$P(X = x) = \frac{(\lambda t)^x}{x!} e^{-\lambda t} \quad x = 0, 1, 2, \ldots, n \tag{6.48}$$

Notice that $X = x$ is the event of x occurrences within time t. Equation 6.48 can be found directly or from the binomial distribution. In the latter form, it can be shown that in a time-related problem, for example, when the number of time intervals approaches infinity, the binomial distribution approaches the Poisson. This can be shown with the aid of the *Maclaurin* series defined:

$$e^r = 1 + r + \frac{r^2}{2!} + \frac{r^3}{3!} + \frac{r^4}{4!} + \ldots \tag{6.49}$$

The proof is beyond the scope of this book and is left to the reader as an exercise (see Ang and Tang, 2007 for a complete discussion).

As in the binomial distribution function, the individual events associated with the Poisson distribution function are assumed to be independent. However, the Poisson model offers several advantages over the binomial when time-related problems are involved. In the Poisson model there is no need to divide the time into intervals. Furthermore, any number of occurrences within the unit time is possible. As you recall, in the binomial model a basic assumption was that within each interval only one occurrence is possible.

Example 6.11 The data compiled at an intersection indicate that in the past 3 yr nine major traffic accidents have occurred at the intersection. Compute the following:

(a) The probability that next month the intersection will be accident free

(b) The probability that in the next 3 mo exactly two such accidents will occur at the intersection

(c) The probability that the intersection will have at most two such accidents in the next 3 mo

Continues

Example 6.11: *Continued*

Solution The Poisson distribution function will be used to model the number of accidents. The parameter λ is estimated from the available data. A time unit equal to 1 mo will be considered. Thus:

$$\lambda = 9/36 = 0.25 \text{ accidents/mo}$$

Note that theoretically the parameter λ can accept a value as a fraction of 1.

(a) In this part $t = 1$ mo and $\lambda t = 0.25 \times 1 = 0.25$. The desired probability is $P(X = 0)$. Thus

$$P(X = 0) = \frac{(\lambda t)^0}{0!} e^{-\lambda t} = e^{-0.25} = 0.779$$

(b) In this case $\lambda t = 0.25 \times 3 = 0.75$; and the desired probability is $P(X = 2)$. Thus:

$$P(X = 2) = \frac{(0.75)^2}{2!} e^{-0.75} = 0.113$$

(c) In this case $\lambda t = 0.75$; however, the desired probability is $P(X \leq 2)$.

$$P(X \leq 2) = P(X = 0) + P(X = 1) + P(X = 2) = \frac{(0.75)^0}{0!} e^{-0.75} + \frac{(0.75)^1}{1!} e^{-0.75} + \frac{(0.75)^2}{2!} e^{-0.75}$$

$$= 0.472 + 0.354 + 0.133 = 0.959$$

6.4.7 Exponential Distribution Function

For a series of events that follow the Poisson distribution function, we introduce the random variable T as the time until the first occurrence of such events. The probability $P(T > t)$ can be obtained from Equation 6.48. Note that the first occurrence is after t, thus, during the time interval 0 to t, there is no occurrence of the event. Thus:

$$P(T > t) = P(X = 0) = \frac{(\lambda t)^0}{0!} e^{-\lambda t} = e^{-\lambda t} \tag{6.50}$$

Introducing the probability distribution function of T as $F(t)$, then:

$$F(t) = P(T \leq t) = 1 - P(X = 0) = 1 - e^{-\lambda t} \tag{6.51}$$

and the probability density function of T is:

$$f(t) = \frac{dF(t)}{dt} = \lambda e^{-\lambda t} \tag{6.52}$$

The function described by Equation 6.52 is called the *exponential probability density function*; and the one described by Equation 6.51 is called the *exponential distribution function*. Although they are used in conjunction with the Poisson distribution function, they can also be used independently. The exponential distribution function is suitable for modeling the useful life (life to failure) of many engineering systems. In such a case, the random variable T will be the system useful life. Using Equation 6.52, the expected value of T can be shown:

$$E(T) = \frac{1}{\lambda} \tag{6.53}$$

This is the mean time between failures, or simply the average life of a system, and is also called the *return period.*

For events modeled with the Poisson distribution function, it is also possible to compute the probabilities associated with a time after which an event of concern will occur for the second, third, or nth time. The corresponding function is called the *gamma distribution function*. The discussion on this type of distribution function is provided in Ang and Tang (2007).

Example 6.12 An electronic device has an average useful life of 1000 operating hr. Compute the probability that the device will fail in the first 500 hr of its operation.

Solution From Equation 6.53, the average failure rate of the device $\lambda = 1/1000 = 0.001$. Using Equation 6.51 we find:

$$P(T \leq 500) = 1 - e^{-\lambda t} = 1 - e^{-0.001 \times 500} = 0.393$$

Of course, this probability could have been directly found from Equation 6.48 by computing $P(X = 0)$ for $t = 500$ hr.

SUMMARY

In this chapter the definition of a random variable and distribution functions used for computing various events associated with it were provided. Furthermore, mathematical parameters known as descriptors of a random variable were introduced and expressions for computing them were presented. Several probability distribution functions were described to familiarize the reader with the underlying concept in using such functions in computing the probability of a desired event. Additional probability functions are introduced in subsequent chapters wherever there is a need to use a specific type of probability distribution function. Several additional topics related to random variables are also important in the application of the theory of probability. Among these are the joint probability distribution functions and the mathematical expressions as functions of one or more random variables. The joint probability distribution functions are used wherever an event must be defined with two (or more) random variables that are correlated. This subject is important in understanding the statistical correlation between two (or more) variables and is discussed in subsequent chapters. The functions of random variables are mathematical expressions that are written in terms of one or more random variables. These functions are used in a variety of engineering problems when the formulation of an event of concern involves a relation between several random variables. Computation of the probability of the event would then require information on the individual random variable probability distribution functions and their respective joint distributions. This subject is discussed in Chapter 7.

REFERENCES

Ang, A. H.-S. and W. H. Tang (2007). Chapters 3 and 4 in *Probability Concept in Engineering: Emphasis on Applications to Civil and Environmental Engineering*, Hoboken, NJ: John Wiley & Sons.

Benjamin, J. and C. A. Cornell (1970). *Probability, Statistics and Decision-Making for Civil Engineers*, New York, NY: McGraw-Hill.

Milton, J. S. and J. C. Arnold (2002). Chapters 3-5 in *Introduction to Probability and Statistics: Principles and Applications for Engineering and the Computer Sciences*, 4th ed., New York, NY: McGraw-Hill.

EXERCISES

1. The maximum tensile force in a cable follows a normal probability density function with a mean = 55 kN and a COV = 0.25. The resistance of the cable is 75 kN (the resistance is not a random variable).
   (a) Compute the probability of failure of the cable.
   (b) Two such cables are used in parallel in carrying the load of an elevator. Assuming the conditions between the cables are statistically independent, compute the probability of failure of the two-cable system.
   (c) Considering one cable again, if the desired probability of failure is 0.01, determine what the resistance should be.
2. The waiting time (T) for a car before it can make a left turn at an intersection is modeled with the following probability density function:

$$f(t) = h\left(2 - \frac{t}{30}\right) \quad \text{for } 30 \leq t \leq 60 \text{ sec}$$

$$f(t) = 0 \quad \text{elsewhere}$$

   (a) If h is constant, compute h.
   (b) Draw this probability density function and show (on the graph) what portion of the area under the curve represents the probability that the waiting time is less than 45 sec. Compute this probability.
   (c) Compute the mean and standard deviation of T.

3. Leaks in a 2-mile gas delivery pipeline occur randomly. In the past 36 mo there were five leaks reported along the pipeline.
   (a) What type of distribution model is suitable to represent the number of leaks at any time period along the pipeline?
   (b) Compute the probability that in the next 6 mo there will be no leaks occurring along the pipeline.
   (c) If a leak occurs, there is a 0.15 probability that the gas delivery will be disrupted so that a repair can be done. What is the probability that there will be no disruption in gas delivery due to probable leaks in the next 6 mo?
4. The maximum bending moment in a beam is a normal random variable with a mean = 50,000 newton.meter (N.m) and a COV = 0.3. The resistance of the beam is 85,000 N.m (the resistance is not a random variable).
   (a) Compute the probability of failure of the beam due to bending.
   (b) Twelve such beams are used in a roof system. Failure in at least two beams will result in the failure of the entire roof system. If conditions between beams are statistically independent, compute the failure probability of the roof system.
5. Accidents in a construction site occur following a Poisson distribution model. Data from similar sites indicate that on the average two accidents occur every year.
   (a) If the duration of the construction project is expected to be 36 mo, estimate the probability that the site will be accident free for the entire duration of the project
   (b) Compute the probability that exactly two accidents will occur during the 36-mo period
6. A motorist drives at the posted speed on a roadway. The motorist will pass seven intersections in a 6-km stretch of the roadway. The probability that the motorist will encounter a red light at an intersection is 0.3. This probability is constant for all seven intersections. Assume the event of encountering a red light at an intersection is independent of the same event for all the other intersections.
   (a) Compute the probability that the motorist will not encounter any red light for all seven intersections
   (b) Compute the probability that the motorist will only encounter one red light within the 6 km of travel
7. Derive Equations 6.42 and 6.43.
8. Repeat Exercise 4 but use a logarithmic normal distribution function instead of normal.
9. Repeat Exercise 1 but use a logarithmic normal distribution function instead of normal.

10. The useful life of an electrical component used in a flight data recorder follows an exponential probability density function. The mean useful time is specified as 20,000 hr by the manufacturer.
    (a) Using T as the random variable describing the useful life of the component, establish the equation for the probability density function of T.
    (b) If a target useful time is specified as 1000 hr, compute the probability that the component will actually surpass this target time.
    (c) The flight data recorder is made up of 20 such components. Compute the probability that the recorder will survive the first 1000 hr of operation. Assume the conditions between the components are independent. Furthermore, assume that the failure in any one component will cause the failure of the recorder.
    (d) Assume a specification requires 0.8 probability of survival for the recorder for the first 1000 hr of operation. Compute the desired mean useful life for the electrical component.

11. The occurrences of strong motion earthquakes in an area are modeled using a Poisson distribution function. The historical data shows that in the past 40 yr there were two such earthquakes that occurred in the area.
    (a) Compute the probability that there will be no such earthquakes in the area in the next 10 yr.
    (b) Compute the probability that such an earthquake may occur next year.
    (c) If a strong motion earthquake occurs in the area, there is a 0.01 probability that a dam located in the area will suffer damage. Compute the probability of damage to the dam in the next 10 yr due to earthquakes.

12. In studying the possibility of foundation settlement (using mm as the unit) in piers supporting a highway bridge, the data from similar bridges indicate the following probability values:

    $P(X = 0) = 0.20$
    $P(X = 5) = 0.30$
    $P(X = 10) = 0.20$
    $P(X = 15) = 0.15$
    $P(X = 20) = 0.10$
    $P(X = 25) = 0.05$

    In which X is a discrete random variable describing the foundation settlement in millimeters (mm). The data given constitute the only possibilities for X.

    (a) Plot the probability mass function of the foundation settlement
    (b) Compute the mean and standard deviation of the foundation settlement
    (c) Compute the probability that the foundation settlement will exceed 15 mm

13. The duration of a construction project is modeled as a normal random variable. The mean duration is 90 da and the corresponding COV 0.25.

    (a) Compute the probability that the project will be delayed, if the target time for the completion of the project is 100 da.
    (b) With improved workmanship, the COV can be reduced to 0.15. With the same duration, compute the target time if the accepted probability of delay is 0.10.

14. On average, an intersection experiences five major accidents per year. If an accident occurs, there is a 0.02 probability that the traffic will need to be rerouted to avoid the accident site.

    (a) What is the probability that there will be no accidents in the next 3 mo?
    (b) What is the probability that no traffic needs to be rerouted due to potential accidents in the next 3 mo?

15. Cars approach a toll bridge according to a Poisson distribution function. Currently, according to design specifications, the bridge can handle 2 cars per min with a 0.20 probability of delay every minute (delay means traffic exceeding this capacity). If the bridge is improved, the capacity can be increased to 5 cars per min. Estimate the probability of delay when the bridge is improved.

16. In the design of a dam, the occurrence of floods is modeled with the Poisson distribution function. The occurrence rate for the floods is mentioned as 0.01. However, further investigations show that the occurrence rate may be as high as 0.02. Since there are merits with both estimates, we decide that the two occurrence rates are both likely with 50-50 chances.

    (a) Compute the probability that the dam will not experience any floods in the next 10 yr
    (b) Compute the probability that the dam will experience no more than two floods in the next 10 yr

17. Within the past 20 yr, a construction company has been consistently bidding on five jobs every year. The company's performance shows that it was successful to win 29 jobs in the 20-yr period.

    (a) Based on data provided, what is the probability of winning any job per year?
    (b) The company is bidding on five jobs this year; what is the probability that it will win all five jobs?
    (c) Repeat (b) but compute the probability that the company will only win one job.

18. Two different sources publish different values for the rate of tornado activities in a certain county in the Midwest. Source *A* indicates 0.2 tornadoes per yr, whereas Source *B* has

indicated a value equal to 0.4 per yr. To utilize both sources, we assume they are both credible and the activity rate may be either value with 50-50 chances.

(a) Based on data given, estimate the probability that next year there will be no tornado in the county
(b) Estimate the probability that there will be exactly one tornado in the county next year

19. The past performance of a construction company indicates the following probabilities of success in winning jobs every year (X = number of jobs won):

$P(X = 0) = 0.15$
$P(X = 1) = 0.40$
$P(X = 2) = 0.20$
$P(X = 3) = 0.20$
$P(X = 4) = 0.05$

(a) Plot the probability mass function of X
(b) Compute the mean and standard deviation of X
(c) Compute the probability that the company will win at least two jobs next year

20. The resistance of a cable follows a lognormal distribution. The cable has a 90 percent chance of survival carrying a 50 kN load. If the COV of the cable resistance is 0.15, estimate the probability that the cable will survive a 70 kN load.

21. In Exercise 12, provide an estimate for the median of the foundation settlement.

22. An antenna system is designed for a wind velocity of 120 mph. The metrological data forecasts steady winds with a mean velocity of 90 mph and a corresponding standard deviation of 25 mph for tomorrow.

(a) Using a normal probability density function for the wind velocity, compute the probability of failure of the antenna.
(b) If we wish to limit the probability of failure of the antenna to only 0.01, the structure must be strengthened to withstand a higher wind velocity. Compute this new higher wind velocity.

23. The duration of an activity in a construction project is a random variable following a lognormal density function. The mean value for the duration is 130 da. The corresponding COV is 0.20. The target time to finish this activity is estimated to be either 120 or 140 da with 1 to 2 chances. Compute the probability of delay in finishing the activity.

Principles of Probability: Part III—Joint Probability Functions and Correlated Variables 7

7.1 INTRODUCTION

In Chapters 5 and 6 the basic concepts underlying the theory of probability and probability distribution functions were presented. The discussion focused on a single random variable. Many engineering problems, however, may involve two or more random variables that can be correlated with one another via a mathematical function. For example, the amount of toll collected on any given day at a toll booth depends on the number of vehicles approaching the booth in a day. Assume Y describes the amount of toll collected in a day and X_i ($i = 1, 2 \ldots, n$) presents the random variable describing the number of vehicles of type i that approach the booth. With n different types of vehicles identified, Y (*dependent variable*) will be a function of n random variables $X_1, X_2 \ldots, X_n$ (*independent variables*). In this example, the relationship between Y and X_i ($i = 1, 2 \ldots, n$) is obtained by observation using collected data. However, there are other examples in engineering in which the relationship between the dependent variable and independent variables can be established through basic principles governing engineering problems. For example, consider the stress at a location in a girder used in a bridge. The stress (dependent variable) is theoretically related to the applied load and the cross-sectional properties of the girder through equations governing laws in the strength of materials. However, no matter what the base of the relationship between Y and X_i ($i = 1, 2 \ldots, n$) is (whether theoretical or empirical), the relationship can be written in a general form:

$$Y = g(X_1, X_2, \ldots X_n) \tag{7.1}$$

in which g is the function describing the relationship between Y and $X_1, X_2, \ldots, X_n$. In a typical probabilistic problem, we will be interested in the expected value and the variance of Y and probabilities associated with Y. These can be computed based on either the expected values or variances of $X_1, X_2, \ldots, X_n$, or the probability distribution functions describing them. In this chapter we focus on probability concepts governing functions between two or more random variables. We start out with the basic assumption that the function g is known and the dependent variable can be written in the form of Equation 7.1. We then treat the problem of correlation between two random variables for which joint probability functions governing probability values associated with the combined effects of the random variables are known. This is followed by a discussion on methods used for approximate estimation of the expected value, variance, and probabilities associated with a dependent variable (such as Y in Equation 7.1).

7.2 JOINT PROBABILITY FUNCTIONS

In many engineering problems an event of concern depends on a combination of two or more other events. In such problems, more than one random variable will be needed to formulate the event of concern. If these random variables are correlated among themselves, then a new set of probability functions (e.g., probability mass function or probability density function) need to be defined to describe the combined or joint occurrence of events represented by these individual random variables. An example will clarify this problem. Suppose a traffic engineer is interested in knowing the probability that the rush hour traffic volumes in the northbound and southbound directions of an urban toll way each are less than a certain value (say x_1 and x_2). Using X_1 and X_2 as random variables describing the rush hour traffic volumes in the northbound and southbound directions, respectively, the event of concern can be described with the combined events of $X_1 \le x_1$ and $X_2 \le x_2$. The corresponding probability will be $P(X_1 \le x_1 \cap X_2 \le x_2)$ or simply $P(X_1 \le x_1;\ X_2 \le x_2)$. If the two events $(X_1 \le x_1)$ and $(X_2 \le x_2)$ are independent, the probability can simply be written $P(X_1 \le x_1) \times P(X_2 \le x_2)$ These individual probabilities can be computed knowing the probability density functions of X_1 and X_2. However, if X_1 and X_2 are correlated, the probability $P(X_1 \le x_1;\ X_2 \le x_2)$ can only be computed if a probability function describing the combined or joint occurrences of the events $(X_1 \le x_1)$ and $(X_2 \le x_2)$ is known. A probability density function (or a mass function if random variables are discrete) corresponding to such a probability function is called a *combined* or *joint* probability density (or mass) function.

7.2.1 Joint Probability Mass Function

If X_1 and X_2 are two correlated random variables, their joint probability mass function, $p_{1,2}(x_1, x_2)$, is defined:

$$p_{1,2}(x_1,x_2) = P(X_1 = x_1; X_2 = x_2) \tag{7.2}$$

The joint cumulative probability function $F_{1,2}(x_1, x_2)$ is defined as:

$$F_{1,2}(x_1,x_2) = P(X_1 \le x_1; X_2 \le x_2) = \sum_i \sum_j p_{1,2}(x_{1i},x_{2j}) \tag{7.3}$$

in which x_{1i} (i = 1, 2, 3, . . .) are values of the random variable X_1 and x_{2i} (j = 1, 2, 3, . . .) are values of the random variable X_2. The double summation in Equation 7.3 is for all $x_{1i} \le x_1$, and all $x_{2i} \le x_2$. Notice that:

$$F_{1,2}(+\infty,+\infty) = 1 \tag{7.4}$$

In light of Equation 7.2, we can also write:

$$p_{1,2}(x_1,x_2) = P(X_1 = x_1 \mid X_2 = x_2)P(X_2 = x_2) \tag{7.5}$$

Denoting the conditional probability of x_1 given x_2 as $p_{1|2}$ we have:

$$p_{1,2}(x_1,x_2) = P(X_1 = x_1 \mid X_2 = x_2) \tag{7.6}$$

Combining Equations 7.5 and 7.6 will result in:

$$p_{1,2}(x_1,x_2) = p_{1|2}(x_1 \mid x_2)p_2(x_2) \tag{7.7}$$

The function $p_{1|2}(x_1 \mid x_2)$ is referred to as the *conditional probability mass function* of X_1 given X_2; whereas $p_2(x_2)$ is the *marginal probability mass function* of X_2. Notice that $p_{1|2}(x_1 \mid x_2)$ describes the conditional probability or $X_1 = x_1$ given $X_2 = x_2$. From Equation 7.7:

$$p_{1|2}(x_1 \mid x_2) = \frac{p_{1,2}(x_1,x_2)}{p_2(x_2)} \tag{7.8}$$

Equation 7.3 can also be written:

$$F_{1,2}(x_1,x_2) = \sum_i \sum_j p_{1|2}(x_{1i} \mid x_{2j})p_2(x_{2j}) \tag{7.9}$$

If $x_1 = +\infty$, then:

$$F_{1,2}(+\infty,x_2) = \sum_j p_2(x_{2j}) \tag{7.10}$$

which indicates that:

$$\sum_j p_2(x_{2j}) = \sum_i \sum_j p_{1,2}(x_{1i},x_{2j}) \quad \text{for all } x_{1i} \tag{7.11}$$

or

$$p_2(x_2) = \sum_i p_{1,2}(x_{1i},x_2) \quad \text{for all } x_{1i} \tag{7.12}$$

This indicates that the marginal probability mass function of the random variable X_2 can be found by the summation of the values of the joint probability mass function for all values of x_1. Similarly:

$$p_1(x_1) = \sum_j p_{1,2}(x_1,x_{2j}) \quad \text{for all } x_{2j} \tag{7.13}$$

If X_1 and X_2 are statistically independent of one another, then:

$$p_{1,2}(x_1,x_2) = P(X_1 = x_1)P(X_2 = x_2) = p_1(x_1)p_2(x_2) \tag{7.14}$$

and as such $p_{1|2}(x_1 \mid x_2) = p_1(x_1)$ and $p_{2|1}(x_2 \mid x_1) = p_2(x_2)$.

Example 7.1 Two random variables X_1 and X_2, respectively, define the number of trucks entering and leaving a concrete mixing plant every hour. The joint probability mass function of X_1 and X_2 is assumed to be:

$$p_{1,2}(x_1,x_2) = \frac{c}{1 + x_1 + x_2} \quad \text{for } x_1 = 0, 1 \text{ and } 2 \text{ and } x_2 = 0, 1, \text{ and } 2$$

$$p_{1,2}(x_1,x_2) = 0 \quad \text{elsewhere}$$

(a) Compute c.

(b) Compute the probability that the number of trucks entering the plant every hour is less than two and no truck is leaving the plant during the same hour.

(c) Compute the marginal probability mass functions $p_1(x_1)$ and $p_2(x_2)$.

(d) Compute the conditional probability mass functions $p_{1|2}(x_1|x_2)$ and $p_{2|1}(x_2|x_1)$.

Solution

(a) From Equation 7.4:

$$\sum_{i=1}^{3}\sum_{j=1}^{3} p_{1,2}(x_{1i},x_{2j}) = 1.0$$

Thus:

$$c\left[\frac{1}{1 + x_{11} + x_{21}} + \frac{1}{1 + x_{11} + x_{22}} + \frac{1}{1 + x_{11} + x_{23}} + \frac{1}{1 + x_{12} + x_{21}} + \frac{1}{1 + x_{12} + x_{22}} + \ldots\right] = 1$$

in which x_{11}, x_{12}, and x_{13} are values of X_1 and are equal to 0, 1, and 2, respectively. Similarly, x_{21}, $x_{22,}$ and x_{23} are values of X_2 and are also equal to 0, 1, and 2, respectively, using these values, 3.7c = 1.0. This results in $c = 0.27$.

(b) In this part, $P(X_1 < 2; X_2 = 0)$ is desired:

$$p(X_1 < 2; X_2 = 0) = P(X_1 \leq 1; X_2 = 0) = p_{1,2}(0,0) + p_{1,2}(1,0)$$

$$= 0.27\left[\frac{1}{1 + 0 + 0} + \frac{1}{1 + 1 + 0}\right] = 0.405$$

(c) From Equation 7.13:

$$p_1(x_1) = 0.27\left[\frac{1}{1 + x_1 + 0} + \frac{1}{1 + x_1 + 1} + \frac{1}{1 + x_1 + 2}\right]$$

The values of $p_1(x_1)$ are p_1 (0) = 0.495, $p_1(1) = 0.293$, and $p_1(2) = 0.212$. Similarly:

$$p_2(x_2) = 0.27\left[\frac{1}{1 + 0 + x_1} + \frac{1}{1 + 1 + x_1} + \frac{1}{1 + 2 + x_2}\right]$$

Continues

Example 7.1: *Continued*

(d) From Equation 7.8:

$$p_{1|2}(x_1 \mid x_2) = \frac{\dfrac{0.27}{1 + x_1 + x_2}}{\dfrac{0.27}{1 + x_2} + \dfrac{0.27}{2 + x_2} + \dfrac{0.27}{3 + x_2}} = \frac{1}{(1 + x_1 + x_2)\left(\dfrac{1}{1 + x_2} + \dfrac{1}{2 + x_2} + \dfrac{1}{3 + x_2}\right)}$$

For example, given $x_2 = 0$, the values of the conditional probability mass function $p_{1/2}$ are $p_{1|2}(0|0) = 0.5455$, $p_{1|2}(1|0) = 0.2727$, and $p_{1|2}(2|0) = 0.1818$. Since the function $p_{1,2}(x_1, x_2)$ is symmetric with respect to x_1 and x_2, the function $p_{2|1}(x_2|x_1)$ can be obtained from $p_{1|2}(x_1|x_2)$ by changing x_1 to x_2 and x_2 to x_1.

Example 7.2 Assume X_1 and X_2 are discrete random variables describing, respectively, the number of accidents per hour in the northbound and the southbound lanes of an urban freeway during rush hours. Assume hypothetically that the joint probability mass function of X_1 and X_2 are available by the values in Table E7.2 for $0 \leq x_1 \leq 3$ and $0 \leq x_2 \leq 3$. For any other x_1 and x_2, the values of the joint probability mass function are zero.

(a) Verify the validity of the numbers given in Table E7.2 as joint probability mass function valucs.

(b) Compute the probability that the number of accidents in the northbound lanes and the southbound lanes each is less than two.

(c) Compute and tabulate the values of marginal probability functions $p_1(x_1)$ and $p_2(x_2)$.

(d) Compute and tabulate the values of the conditional probability mass functions $p_{1|2}(x_1|x_2)$ and $p_{2|1}(x_2|x_1)$.

Table E7.2 Values of $p_{1,2}(x_1, x_2)$

	$x_1 = 0$	$x_1 = 1$	$x_1 = 2$	$x_1 = 3$
$x_2 = 0$	0.16	0.14	0.08	0.06
$x_2 = 1$	0.14	0.07	0.05	0.03
$x_2 = 2$	0.08	0.05	0.03	0.01
$x_2 = 3$	0.06	0.03	0.01	0

Solution

(a) From Equation 7.4:

$$\sum_{i=0}^{3}\sum_{j=0}^{3} p_{1,2}(x_{1i}, x_{2j}) = 0.16 + 0.14 + 0.08 + 0.06 + 0.014 + \ldots = 1.0$$

Continues

Example 7.2: *Continued*

Thus the numbers listed in Table E7.2 are the values of the joint probability mass function within the limits listed for the two random variables X_1 and X_2.

(b) In this part $P(X_1 < 2; X_2 < 2)$ is desired. This probability can be computed from Equation 7.3:

$$P(X_1 < 2; X_2 < 2) = P(X_1 \leq 1; X_2 \leq 1) = 0.16 + 0.14 + 0.14 + 0.07 = 0.51$$

(c) From Equation 7.13, the values of the marginal probability mass function $p_1(x_1)$ can be obtained by summation of the values of $p_{1,2}(x_1, x_2)$ in individual columns of Table E7.2. Thus:

$$p_1(0) = 0.16 + 0.14 + 0.08 + 0.06 = 0.44$$
$$p_1(1) = 0.14 + 0.07 + 0.05 + 0.03 = 0.29$$
$$p_1(2) = 0.08 + 0.05 + 0.03 + 0.01 = 0.17$$
$$p_1(3) = 0.06 + 0.03 + 0.01 + 0 = 0.10$$

Similarly, from Equation 7.12, the values of $p_2(x_2)$ are obtained by summation of the $p_{1,2}(x_1, x_2)$ in individual rows of Table E7.2. Thus:

$$p_2(0) = 0.16 + 0.14 + 0.08 + 0.06 = 0.44$$
$$p_2(1) = 0.29$$
$$p_2(2) = 0.17$$
$$p_2(3) = 0.10$$

(d) In this part, for example, given $x_2 = 0$, the values of the conditional probability mass function $p_{1|2}(x_1|x_2)$ can be computed:

$$p_{1|2}(0 \mid 0) = \frac{p_{1,2}(0,0)}{p_2(0)} = \frac{0.16}{0.44} = 0.36$$

$$p_{1|2}(1 \mid 0) = \frac{p_{1,2}(1,0)}{p_2(0)} = \frac{0.14}{0.44} = 0.32$$

$$p_{1|2}(2 \mid 0) = \frac{p_{1,2}(2,0)}{p_2(0)} = \frac{0.08}{0.44} = 0.18$$

$$p_{1|2}(3 \mid 0) = \frac{p_{1,2}(3,0)}{p_2(0)} = \frac{0.06}{0.44} = 0.14$$

Identical results are also found for $p_{2|1}$ values.

7.2.2 Joint Probability Density Function

The probabilities of combined events of two or more continuous random variables are obtained through joint probability density functions. If only two random variables are involved, a joint probability density function is shown with $f_{1,2}(x_1, x_2)$ and is described in the following equation:

$$P[(x_1 < X_1 \leq x_1 + dx_1) \cap (x_2 < X_2 \leq x_2 + dx_2)] = f_{1,2}(x_1, x_2)\, dx_1\, dx_2 \tag{7.15}$$

Using Equation 7.15, the joint distribution function $F_{1,2}(x_1, x_2)$ of random variables X_1 and X_2 is:

$$F_{1,2}(x_1, x_2) = \int_{-\infty}^{x_1} \int_{-\infty}^{x_2} f_{1,2}(x_1, x_2)\, dx_1\, dx_2 = P(X_1 \leq x_1; X_2 \leq x_2) \tag{7.16}$$

and as such:

$$F_{1,2}(+\infty, +\infty) = 1 \tag{7.17}$$

Introducing $f_{1|2}(x_1|x_2)$ and $f_{2|1}(x_2|x_1)$ as conditional probability density functions and $f_1(x_1)$ and $f_2(x_2)$ as the marginal probability density functions, we can write:

$$P(X_1 \leq x_1; X_2 \leq x_2) = \int_{-\infty}^{x_2} P[(X_1 \leq x_1) \mid (x_2 < X_2 \leq x_2 + dx_2)]$$
$$f_2(x_2)\, dx_2 = \int_{-\infty}^{x_1} \int_{-\infty}^{x_2} f_{1|2}(x_1 \mid x_2) f_2(x_2)\, dx_1\, dx_2 \tag{7.18}$$

Comparing Equations 7.16 and 7.18:

$$f_{1,2}(x_1, x_2) = f_{1|2}(x_1|x_2) f_2(x_2) \tag{7.19}$$

Similarly:

$$f_{1,2}(x_1, x_2) = f_{2|1}(x_2|x_1) f_1(x_1) \tag{7.20}$$

From Equations 7.19 and 7.20

$$f_{1|2}(x_1 \mid x_2) = \frac{f_{1,2}(x_1, x_2)}{f_2(x_2)} \tag{7.21}$$

and

$$f_{2|1}(x_2 \mid x_1) = \frac{f_{1,2}(x_1, x_2)}{f_1(x_1)} \tag{7.22}$$

When $x_1 = +\infty$

$$P(X_1 \leq x_1; X_2 \leq x_2) = \int_{-\infty}^{x_2} \left[\int_{-\infty}^{+\infty} f_{1,2}(x_1, x_2)\, dx_1 \right] dx_2$$
$$= \int_{-\infty}^{x_2} \left[\int_{-\infty}^{+\infty} f_{1|2}(x_1 \mid x_2)\, dx_1 \right] f_2(x_2)\, dx_2 = \int_{-\infty}^{x_2} f_2(x_2)\, dx_2 \tag{7.23}$$

From Equation 7.23:

$$f_2(x_2) = \int_{-\infty}^{+\infty} f_{1,2}(x_1,x_2)\,dx_1 \tag{7.24}$$

and similarly:

$$f_1(x_1) = \int_{-\infty}^{+\infty} f_{1,2}(x_1,x_2)\,dx_2 \tag{7.25}$$

In a special case where the random variables X_1 and X_2 are statistically independent

$$f_{1|2}(x_1|x_2) = f_1(x_1) \quad \text{and} \quad f_{2|1}(x_2|x_1) = f_2(x_2) \tag{7.26}$$

$$P(X_1 \leq x_1 : X_2 \leq x_2) = \int_{-\infty}^{x_1} f_1(x_1)\,dx_1 \int_{-\infty}^{x_2} f_2(x_2)\,dx_2 \tag{7.27}$$

Example 7.3 The live loads on any two floors in an office building are correlated. Assume for simplicity the live loads on two given floors are random variables X_1 and X_2 and are defined by the following joint probability density functions:

$$f_{1,2}(x_1, x_2) = c(e^{-x_1} + e^{-x_2}) \quad \text{for } 0 \leq (x_1 \text{ and } x_2) \leq 2{,}400 \text{ N/m}^2$$

$$f_{1,2}(x_1, x_2) = 0 \quad \text{elsewhere}$$

(a) Compute c.

(b) Compute the probability that both X_1 and X_2 are not more than 1000 N/m^2.

(c) Derive expressions for the marginal probability density functions of X_1 and X_2.

(d) Derive expressions for the conditional probability density functions $f_{1|2}(x_1|x_2)$ and $f_{2|1}(x_2|x_1)$.

Solution

(a) From Equation 7.17:

$$\int_0^{2400}\int_0^{2400} c(e^{-x_1} + e^{-x_2})\,dx_1\,dx_2 = 1.0$$

This results in $c = 1/4800$.

(b) In this part P($X_1 \leq 1000$; $X_2 \leq 1000$) is desired. Using 7.16:

$$P(X_1 \leq 1000; X_2 \leq 1000) = \frac{1}{4800}\int_0^{1000}\int_0^{1000} (e^{-x_1} + e^{-x_2})\,dx_1\,dx_2 = 0.417$$

Continues

Example 7.3: *Continued*

(c) In this part, we apply Equation 7.25:

$$f_1(x_1) = \int_0^{2400} f_{1,2}(x_1, x_2)\,dx_2 = \frac{e^{-x_1}}{2} + \frac{1}{4{,}800} \qquad \text{for } 0 \le x_1 \le 2400$$

$$f_1(x_1) = 0 \qquad \text{elsewhere}$$

Similarly, using Equation 7.24:

$$f_2(x_2) = \int_0^{2400} f_{1,2}(x_1, x_2)\,dx_1 = \frac{e^{-x_2}}{2} + \frac{1}{4{,}800} \qquad \text{for } 0 \le x_2 \le 2400$$

$$f_2(x_2) = 0 \qquad \text{elsewhere}$$

(d) From Equation 7.21:

$$f_{1|2}(x_1 \mid x_2) = \frac{e^{-x_1} + e^{-x_2}}{2400e^{-x_2} + 1}$$

Again, the function is valid for x_1 and x_2 between 0 and 2400. Similarly:

$$f_{2|1}(x_2 \mid x_1) = \frac{e^{-x_1} + e^{-x_2}}{2400e^{-x_1} + 1}$$

Notice that the joint probability density function forms a surface in the three-dimensional space. Any probability value represented by Equation 7.16 will be a portion of the volume under this surface. The total volume under the surface will be equal to unity. Figure 7.1 shows an example of a joint probability density function. This particular one is represented by the joint normal probability density function (bivariate normal) that is described in Section 7.3.3.

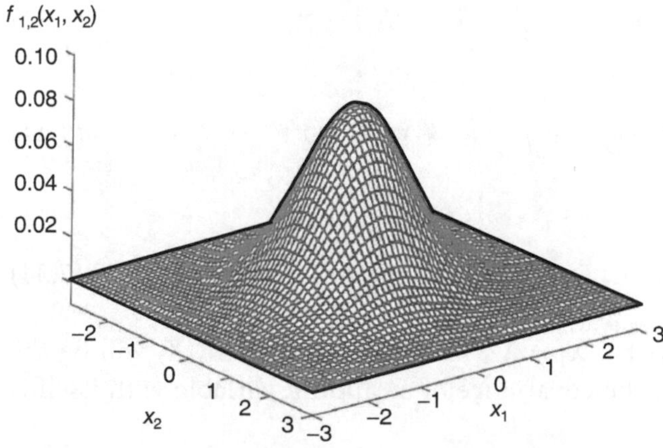

Figure 7.1 Illustration of a joint probability density function

7.3 COVARIANCE AND CORRELATION

7.3.1 Covariance

As discussed in Chapter 6, the moments (or mathematical expectation) of a random variable can be computed using the generalized relationship of Equations 6.18 and 6.19. The same concept can also be applied to two random variables that are defined with a joint probability density function or a joint probability mass function. By definition, the mathematical expectation of a function $g(x_1, x_2)$ is computed using the following expression:

$$E[g(X_1, X_2)] = \sum_i \sum_j g(x_{1i}, x_{2j}) p_{1,2}(x_{1i}, x_{2j}) \tag{7.28}$$

for discrete random variables:

$$E[g(X_1, X_2)] = \int_{-8}^{+\infty} \int_{-\infty}^{+\infty} g(x_1, x_2) f_{1,2}(x_1, x_2)\, dx_1\, dx_2 \tag{7.29}$$

for continuous random variables. In Equation 7.28 the double summation is for all values of the random variable X_1 (i.e., $x_{11}, x_{12}, x_{13}. \ldots$), and all values of the random variable X_2 (i.e., $x_{21}, x_{22}, x_{23} \ldots$). In a special case where function $g = (X_1 - \mu_1)(X_2 - \mu_2)$, with μ_1 and μ_2, being the mean values of the random variables X_1 and X_2, respectively, the mathematical expectation is called the *covariance* of X_1 and X_2 and is identified with Cov(X_1, X_2). Using Equations 7.28 and 7.29, the covariance of X_1 and X_2 is:

$$\mathrm{Cov}(X_1, X_2) = E[(X_1 - \mu_1)(X_2 - \mu_2)] = \sum_i \sum_j (x_{1i} - \mu_1)(x_{2j} - \mu_2) p_{1,2}(x_{1i}, x_{2j}) \tag{7.30}$$

for discrete random variables:

$$\mathrm{Cov}(X_1, X_2) = \int_{-\infty}^{+\infty} \int_{-\infty}^{+\infty} (x_1 - \mu_1)(x_2 - \mu_2) f_{1,2}(x_1, x_2)\, dx_1\, dx_2 \tag{7.31}$$

for continuous random variables. Another form for Cov(X_1, X_2) can be written:

$$\mathrm{Cov}(X_1, X_2) = E[(x_1 - \mu_1)(x_2 - \mu_2)] = E(X_1\, X_2) - \mu_1 \mu_2 \tag{7.32}$$

in which (by virtue of Equations 7.28 and 7.29):

$$E(X_1\, X_2) = \sum_i \sum_j x_{1i}\, x_{2j} p_{1,2}(x_{1i}, x_{2j}) \tag{7.33}$$

for discrete random variables:

$$E(X_1\, X_2) = \int_{-\infty}^{+\infty} \int_{-\infty}^{+\infty} (x_1, x_2) f_{1,2}(x_1\, x_2)\, dx_1\, dx_2 \tag{7.34}$$

for continuous random variables. Notice that if $X_1 = X_2$, the covariance of X_1, X_2 will be the same as the variance of X_1. This means that the covariance of a random variable with itself is the same as the variance of that random variable. In a special case where X_1, X_2 are statistically independent of each other, Cov(X_1,X_2)= 0. This can be proven easily by virtue of the fact that for two random variables that are statistically independent of each other, $p_{1,2}(x_1, x_2) = p_1(x_1)\, p_2(x_2)$ and $f_{1,2}(x_1, x_2) = f_1(x_1)\, f_2(x_2)$. For example, for two continuous random variables from Equation 7.34:

$$E(X_1 X_2) = \int_{-\infty}^{+\infty}\int_{-\infty}^{+\infty} (x_1 x_2) f_1(x_1) f_2(x_2)\, dx_1 dx_2 = \mu_1 \mu_2 \tag{7.35}$$

Substituting Equation 7.35 in Equation 7.32 will result in Cov(X_1, X_2) = 0.

Example 7.4 In Example 7.1, compute $E(X_1, X_2)$ and Cov(X_1, X_2).

Solution From Equation 7.33:

$$E(X_1, X_2) = c\left[\frac{x_{11}x_{21}}{1 + x_{11} + x_{21}} + \frac{x_{11}x_{22}}{1 + x_{11} + x_{22}} + \frac{x_{11}x_{23}}{1 + x_{11} + x_{23}} + \frac{x_{12}x_{21}}{1 + x_{12} + x_{21}} + \ldots\right] = 213c = 0.576$$

The mean values of X_1 and X_2 can be computed from $p_1(x_1)$ and $p_2(x_2)$, respectively.

$$\mu_1 = \sum_i x_{1i} p_1(x_{1i}) \quad \text{and} \quad \mu_2 = \sum_j x_{2j} p_2(x_{2j})$$

For example, since:

$$p_1(x_1) = 0.27\left(\frac{1}{1 + x_1} + \frac{1}{2 + x_1} + \frac{1}{3 + x_1}\right)$$

$\mu_1 = 0.716$. Similarly, $\mu_2 = 0.716$. Thus, Cov(X_1,X_2) = 0.576 − 0.716 × 0.716 = 0.064.

7.3.2 Correlation Coefficient

By definition, the correlation coefficient (ρ) of two random variables X_1 and X_2 is computed by dividing the covariance of X_1 and X_2 by their respective standard deviations. This means that

$$\rho = \frac{Cov(X_1, X_2)}{\sigma_1 \sigma_2} = \frac{E(X_1 X_2) - \mu_1 \mu_2}{\sigma_1 \sigma_2} \tag{7.36}$$

in which σ_1 and σ_2 are the standard deviations of X_1 and X_2, respectively. Notice that if the two random variables X_1 and X_2 are statistically independent, $\rho = 0$. On the other hand if X_1 and X_2 are perfectly correlated, and the relationship between X_1 and X_2 is linear, one of the two random variables, say X_2, can be fully represented by the other, i.e., X_1. This means that $\sigma_1 = \sigma_2$ and also $E(X_1 X_2) = E(X_1^2)$ and $\mu_1 = \mu_2$; and as such $\rho = 1$. It is noted that ρ may be negative. In such a case, a perfect *negative* correlation will be represented by $\rho = -1$. It is also noted that ρ is a measure of a linear correlation between X_1 and X_2.Thus, if a nonlinear correlation between X_1

and X_2 exists, Equation 7.36 will result in $\rho < 1$. Figure 7.2 illustrates a few examples of values of ρ for several types of relationships between X_1 and X_2. In Figures 7.2a and 7.2b, the correlation is perfect (linear relationship). Figures 7.2c and 7.2d show no correlation. In these cases, changes in one variable will not affect the other. In Figures 7.2e and 7.2f, although in each case a relationship between X_1 and X_2, exists, ρ will not be equal to unity because these relationships are not linear. It can be said that in these cases, the linear correlation between X_1 and X_2 is only partial. To further clarify on cases where two random variables may only be partially correlated, consider the following example.

Let's define X_1 and X_2 as the volume of concrete and reinforcing steel, respectively, delivered to a construction site each day. The delivery of the two materials to the site will depend on several factors such as the means of transportation, source of each material, delays in roadways, shortage of raw materials in the market, etc. If steel and concrete are provided through two different sources, factors that affect one variable may not necessarily affect the other. However, at any given time, there may be several other factors that would affect both variables, i.e., X_1 and X_2. Figure 7.3 presents three different

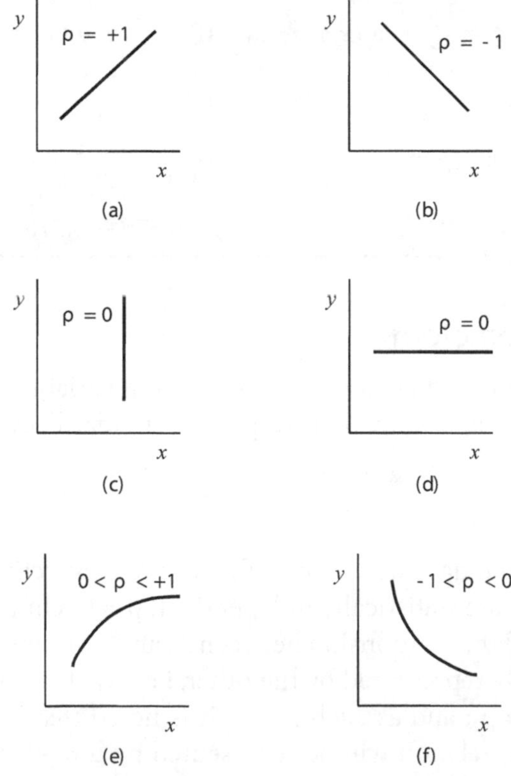

Figure 7.2 Values of correlation coefficient for several cases of *y* vs. *x*

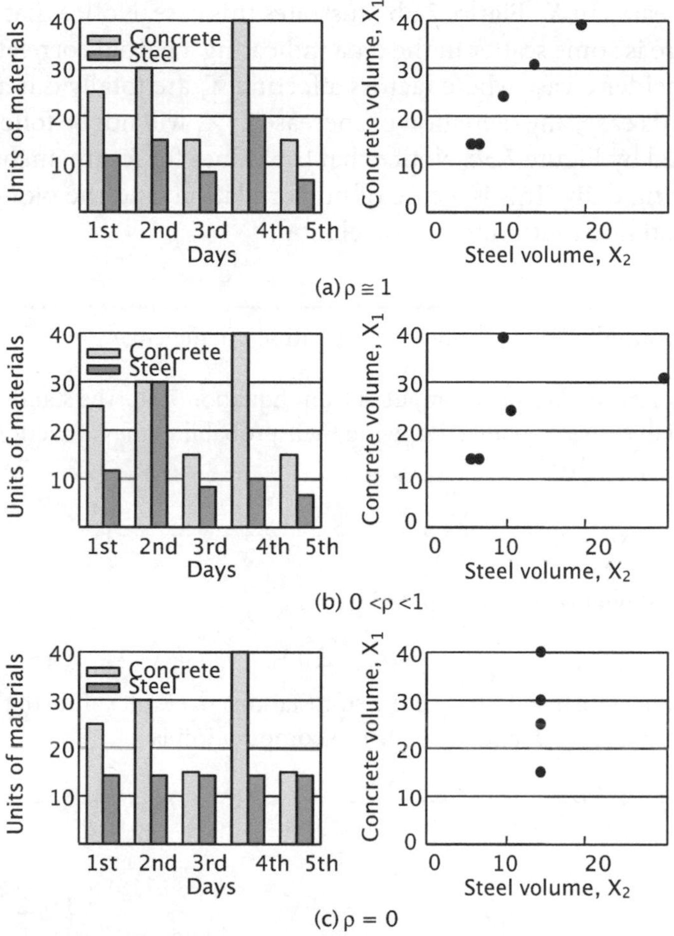

(a) ρ ≅ 1

(b) 0 <ρ <1

(c) ρ = 0

Figure 7.3 Examples of correlation coefficient in an engineering problem

cases for values of X_1 and X_2 in a five-day period. If in a special case, both materials are supplied through the same source; and exactly the same factors are affecting X_1 and X_2, the two variables are likely to be perfectly correlated. Figure 7.3a is an example of this case. This means that if there is any reduction or increase in the value of X_1, it is very likely that X_2 will also experience the same changes and will follow X_1. In Figure 7.3a, notice that the ratio of the volume of steel to that of concrete delivered daily to the site for the five-day period was constant. Furthermore, in this case, the plot of X_1 versus X_2 shows a perfect linear relationship. Now assume only some of the factors are commonly affecting X_1 and X_2. In this case, the correlation between the two random variables will be only partial. This means that if there are, for example, reductions in X_1 there will be reductions in X_2 only on some days; while on other days there are no changes or there may even be increases in X_2. Figure 7.3b illustrates this case. Notice that when X_1 versus X_2 is plotted, there is some scatter in the data indicating a partial correlation between X_1 and X_2. Now consider a case where factors affecting X_1 are totally different from those affecting X_2. In this case, any reduction or increase in X_1 will not be followed by X_2. This case is represented by Figure 7.3c. Notice that the values for X_2 are unchanged, while X_1 values are changing daily. This is a case of no correlation; and the plot of X_1 versus X_2, results in a vertical line indicating no correlation.

Example 7.5 In Example 7.4, compute the correlation coefficient.

Solution The correlation coefficient is computed from Equation 7.36. The standard deviations of the two random variables are computed knowing their probability mass functions. For example, for X_1:

$$p_1(x_1) = 0.27\left(\frac{1}{1+x_1} + \frac{1}{2+x_1} + \frac{1}{3+x_1}\right)$$

and the standard deviation σ_1 is computed from:

$$\sigma_1^2 = E(X_1 - \mu_1)^2 = \sum (x_{1i} - \mu_1)^2 p_1(x_{1i})$$

in which the mean value $\mu_1 = 0.716$ and the summation covers all values of the random variable X_1 (which are $x_{11} = 0$, $x_{12} = 1$, and $x_{13} = 2$). The computation is:

$$\sigma_1^2 = 0.27\left[(0-0.716)^2\left(\frac{1}{1+0} + \frac{1}{2+0} + \frac{1}{3+0}\right) + (1-0.716)^2\left(\frac{1}{1+1} + \frac{1}{2+1} + \frac{1}{3+1}\right) + (2-0.716)^2\left(\frac{1}{1+2} + \frac{1}{2+2} + \frac{1}{3+2}\right)\right]$$

or $\sigma_1 = 0.791$. Similarly $\sigma_2 = 0.791$ and:

$$\rho = \frac{0.064}{0.791 \times 0.791} = 0.102$$

This indicates that the linear correlation between the two random variables X_1 and X_2 is weak.

Example 7.6 Figure E7.6 shows a beam that provides a support for a crane system. At any given time, the magnitude of the load Q and its location X from the left support are random variables. Random variable Q follows a uniform distribution ranging between 0 and 160 kN. Random variable X also follows a uniform distribution ranging between 0 and 12 meters.

Continues

Example 7.6: *Continued*

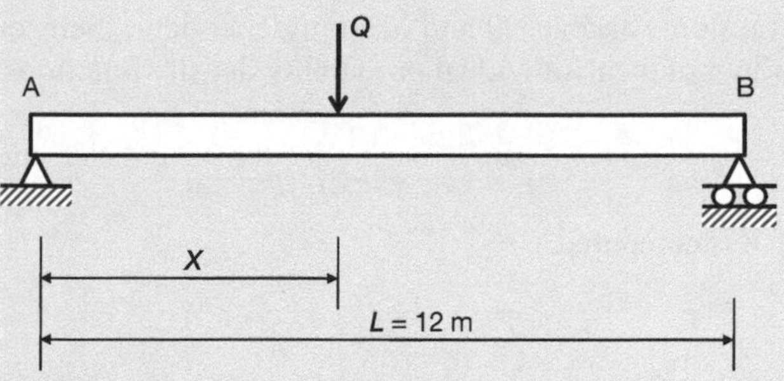

Figure E7.6 Simply supported beam with random load location

(a) At the location of the load, compute shear V (immediately to the left of the load) and bending moment M.

(b) Compute the mean and standard deviation of Q and X.

(c) Compute the mean value of V and M, and $E(VM)$.

(d) Compute the correlation coefficient between V and M.

Solution

(a) The shear and moment at the location of the load are:

$$V = \left(1 - \frac{X}{12}\right)Q \quad \text{and} \quad M = X\left(1 - \frac{X}{12}\right)Q$$

(b) The uniform probability density function for a random variable X ranging between a and b is defined:

$$f(x) = \frac{1}{b - a} \quad \text{for } a \le x \le b$$

$$f(x) = 0 \quad \text{elsewhere}$$

This results in a mean value μ and standard deviation σ defined via the following equations:

$$\mu = \frac{a + b}{2} \quad \text{and} \quad \sigma = \frac{b - a}{2\sqrt{3}}$$

Applying these equations to the two random variables Q and X:

$$\mu_Q = 80 \text{ kN}, \sigma_Q = 46.2 \text{ kN}, \text{ and } \mu_X = 6 \text{ m and } \sigma_X = 3.46 \text{ m}$$

Continues

Example 7.6: *Continued*

(c) Since the two random variables Q and X are independent, their joint probability density function is the product of their individual probability density functions. This means:

$$f_{Q,X}(q, x) = f_Q(q)f_X(x) = \frac{1}{160} \times \frac{1}{12} = \frac{1}{1920} \quad \text{for } 0 \le q \le 160; \text{ and } 0 \le x \le 12$$

The mean value of V is computed:

$$\mu_V = E(V) = E\left[\left(1 - \frac{X}{12}\right)Q\right]$$

or

$$\mu_V = \int_0^{160}\int_0^{12}\left[\left(1 - \frac{x}{12}\right)q\right]f_{Q,X}(q, x)\,dqdx = 40.$$

and

$$\sigma_V^2 = E(V^2) - \mu_V^2 = E\left[\left(1 - \frac{X}{12}\right)^2 Q^2\right] - 40^2 = \int_0^{160}\int_0^{12}\left[\left(1 - \frac{x}{12}\right)^2 q^2\right]f_{Q,X}(x, q)\,dqdx - (40)^2$$

$$\sigma_V^2 = \int_0^{160}\int_0^{12}\left[\left(1 - \frac{x}{12}\right)^2 q^2\right] \times \frac{1}{1920}\,dqdx - (40)^2 = 1246$$

which results in $\sigma_V = 35.3$. Similarly, the mean and standard deviation of M can be computed:

$$\mu_M = E\left[\left(1 - \frac{X}{12}\right)XQ\right] = \int_0^{160}\int_0^{12}\left[\left(1 - \frac{x}{12}\right)xq\right]f_{Q,X}(q, x)\,dqdx = 160.$$

and

$$\sigma_M^2 = E(M^2) - \mu_M^2 = \int_0^{160}\int_0^{12}\left[\left(1 - \frac{x}{12}\right)^2 x^2 q^2\right]f_{Q,X}(x, q)dqdx - (160)^2 = 15{,}360.$$

which results in $\sigma_M = 123.9$. To compute the correlation coefficient, we also need to compute $E(VM)$. This is computed:

$$E(VM) = E\left[\left(1 - \frac{X}{12}\right)Q\left(1 - \frac{X}{12}\right)XQ\right] = \int_0^{160}\int_0^{12}\left[\left(1 - \frac{x}{12}\right)^2 xq^2\right]f_{Q,X}(x, q)dqdx = 8533$$

Using Equation 7.36, the correlation coefficient of V and M will be:

$$\rho = \frac{(8{,}533 - 40 \times 160)}{35.3 \times 123.9} = 0.49$$

Notice that although Q and X are independent, there still exists some correlation between V and M. This is because the equations for V and M depend on the same random variables.

In cases where n random variables $X_1, X_2, \ldots, X_n$ are involved, the expected value of any two variables (say X_i and X_j) is:

$$E(X_iX_j) = \int_{-\infty}^{+\infty}\int_{-\infty}^{+\infty} x_i x_j f_{i,j}(x_i, x_j)\,dx_i\,dx_j \quad \text{if } i \neq j \tag{7.37}$$

and

$$E(X_iX_j) = E(X_i^2) = \mu_{X_i}^2 + \sigma_{X_i}^2 \quad \text{if } i = j \tag{7.38}$$

These expected values are arranged in a symmetric matrix such as:

$$\begin{bmatrix} E(X_1^2) & E(X_1X_2) & \ldots & E(X_1X_n) \\ E(X_2X_1) & E(X_2^2) & \ldots & E(X_2X_n) \\ \ldots & \ldots & \ldots & \ldots \\ E(X_nX_1) & E(X_nX_2) & \ldots & E(X_n^2) \end{bmatrix} \tag{7.39}$$

and in terms of correlation coefficient as:

$$\begin{bmatrix} 1 & \rho_{1,2} & \ldots & \rho_{1,n} \\ \rho_{2,1} & 1 & \ldots & \rho_{2,n} \\ \ldots & \ldots & \ldots & \ldots \\ \rho_{n,1} & \rho_{n,2} & \ldots & 1 \end{bmatrix} \tag{7.40}$$

Notice that the principal diagonal elements of matrix in Equation 7.40 are all equal to one. This is because the correlation coefficient of a random variable with itself is equal to unity.

In practice, the correlation coefficient is rarely computed from Equation 7.36. Instead, an estimate of ρ (which is often shown with r) is computed from data obtained for X_1 and X_2. This subject is discussed in Chapter 9.

7.3.3 Bivariate Normal Distribution

The joint probability density function of two normally distributed random variables X and Y with a correlation coefficient ρ is defined by the bivariate normal function:

$$f_{X,Y}(x, y) = \frac{1}{2\pi\sigma_X\sigma_Y\sqrt{1-\rho^2}} \exp\left\{-\frac{1}{2(1-\rho^2)}\left[\left(\frac{x-\mu_X}{\sigma_X}\right)^2 - 2\rho\left(\frac{x-\mu_X}{\sigma_X}\right)\left(\frac{y-\mu_X}{\sigma_y}\right) + \left(\frac{y-\mu_Y}{\sigma_Y}\right)^2\right]\right\} \tag{7.41}$$

for $-\infty < x$ and $y < +\infty$.

In Equation 7.41, if the correlation coefficient is zero (i.e., the random variables are statistically independent), the joint density function $f_{X,Y}$ will become the product of the individual probability density functions f_X and f_Y. Equation 7.41 is plotted as shown in Figure 7.1. It appears in the shape of a bell in the three-dimensional space.

7.4 FUNCTIONS OF A SINGLE RANDOM VARIABLE

As discussed in Section 7.1, a quantity may be related to a random variable through a definite function. Defining this function as g, a quantity Y can be written in terms of a random variable X as:

$$Y = g(X) \tag{7.42}$$

This is a simple form of Equation 7.1. The dependent variable Y follows the independent random variable X within the rule governed by the function g. If the probability function (i.e., probability mass function or density function) of X is available, any probability associated with the random variable Y can be computed by virtue of the fact that for any given value of X, such as x, there is a value for Y, such as y. The relationship between y and x is:

$$y = g(x) \tag{7.43}$$

Furthermore, moments of Y (e.g., the mean and variance) can also be computed if the probability function of X is available.

7.4.1 Discrete Random Variable Case

When random variables X and Y are discrete, any probability associated with Y can be written:

$$P(Y = y) = P(X = x) = p_X(x) \tag{7.44}$$

in which $p_X(x)$ is the probability mass function of X. Equation 7.44 can be evaluated only if x can be solved in terms of y from Equation 7.43. Solving for x from Equation 7.43:

$$x = g^{-1}(y) \tag{7.45}$$

where g^{-1} is the inverse function of g. Furthermore, we can write:

$$P(Y \le y) = P(X \le x) = \Sigma p_X(x_i) \tag{7.46}$$

in which the summation includes all $x_i \le x$ or all $x_i \le g^{-1}(y)$. The expected value and variance of Y can also be computed in terms of $p_X(x)$ as:

$$E(Y) = \mu_Y = E[g(X)] = \Sigma\,[g(x_i)p_X(x_i)] \tag{7.47}$$

$$\mathrm{Var}(Y) = E(Y^2) - \mu_Y^2 = \Sigma\,[g^2(x_i)p_X(x_i)] - \mu_Y^2 \tag{7.48}$$

In both Equations 7.47 and 7.48, the summation includes all x_i.

Example 7.7 Figure E7.7 shows a cantilever beam with a uniformly distributed load q acting over the length X. Length X is a discrete random variable. Possible values for X are L, $3L/4$, $L/2$, and $L/4$. The probability mass function of X is defined:

$$p_X(x) = \frac{cx}{L} \quad \text{for } x = L/4, L/2, 3L/4 \text{ and } L$$

$$p_X(x) = 0 \quad \text{elsewhere}$$

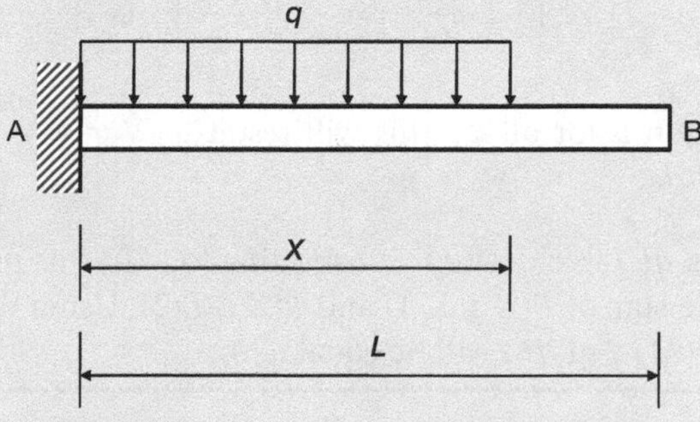

Figure E7.7 Cantilever beam with load area

Assume q and L are deterministic.

(a) Compute constant c, and mean and standard deviation of X.

(b) Write an expression for moment M_A at A in terms of X.

(c) Compute the mean and standard deviation of M_A.

(d) Compute the probability that M_A will be less than or equal to $qL^2/8$.

Solution

(a) Since $\Sigma p_X(x_i) = 1$, using $x_1 = L/4$, $x_2 = L/2$, $x_3 = 3L/4$ and $x_4 = L$:

$$c = 0.4,\ \mu_X = 3L/4 \text{ and } \sigma_X = L/4.$$

(b) The moment at A is $M_A = qX^2/2$.

(c) From Equation 7.47:

$$E(M_A) = E\left(\frac{qX^2}{2}\right) = \sum\left[\frac{qx_i^2}{2} p_X(x_i)\right] \quad \text{for all } x_i$$

Since q is deterministic:

$$E(M_A) = q\sum\left[\left(\frac{x_i^2}{2}\right)p_X(x_i)\right]$$

Continues

Example 7.7: *Continued*

Thus, using $x_1 = L/4$, $x_2 = L/2$, $x_3 = 3L/4$ and $x_4 = L$ will result in:

$$E(M_A) = \frac{5qL^2}{16}$$

The variance and standard deviation are computed from Equation 7.48:

$$\text{Var}(M_A) = E\left[M_A^2\right] - \left(\frac{5qL^2}{16}\right)^2 = \sum\left[\left(\frac{qx_i^2}{2}\right)p_X(x_i)\right] - \left(\frac{5qL^2}{16}\right)^2$$

In which, the summation is for all x_i. This will result in $\text{Var}(M_A) = \frac{15q^2L^4}{512}$, or a standard deviation equal to $0.17qL^2$.

(d) In this part, $P(M_A \le qL^2/8)$ is desired. Substituting for M_A, this probability will be equal to $P(X \le L/2)$ which is the sum of $P(X \le L/4)$ and $P(X \le L/2)$. Using $c = 0.4$ and the probability mass function given, $P(M_A \le qL^2/8)$ will be equal to 0.3.

7.4.2 Continuous Random Variable Case

In this case, the probability density function of the independent random variable X is defined by $f_X(x)$. Again, the dependent random variable Y is related to X through Equation 7.42. Furthermore, in light of Equation 7.43, for any given value of X, such as x, there exists a value for Y (such as y). It is assumed that an expression for x can be obtained by solving Equation 7.45. The expected value and variance of Y can be computed:

$$\mu_X = E(Y) = E[g(X)] = \int_{-\infty}^{+\infty} g(x)f_X(x)\,dx \tag{7.49}$$

$$\text{Var}(Y) = E(Y^2) - \mu_Y^2 = \int_{-\infty}^{+\infty} g^2(x)f_X(x)\,dx - \mu_Y^2 \tag{7.50}$$

In addition, any probability associated with the random variable Y can be computed in terms of $f_X(x)$:

$$P(Y \le y) = P(X \le x) = \int_{-\infty}^{+\infty} f_X(x)\,dx = \int_{-\infty}^{g^{-1}(y)} f_X(x)\,dx \tag{7.51}$$

In light of Equation 7.51, it can be shown that the probability density function of Y is:

$$f_Y(y) = f_X(g^{-1})\left|\frac{dg^{-1}}{dy}\right| \tag{7.52}$$

The proof of this equation is left to the reader (the absolute value used in the equation is because the function $f_Y(y)$ is positive for all values of y). Usually there is no need to use Equation 7.52 in computing the probabilities associated with the random variable Y since these probabilities can readily be computed from Equation 7.51.

Example 7.8 In a special case of Equation 7.42, Y is a linear function of X in the form of $Y = a + bX$, in which a and b are constants and X is a continuous random variable following a normal probability density function.

(a) Compute the mean and standard deviation of Y in terms of the mean and standard deviation of X.

(b) Show that Y will also be a normally-distributed random variable.

Solution

(a) From Equations 7.49 and 7.50:

$$\mu_Y = \int_{-\infty}^{+\infty} (a + bx) f_X(x)\,dx = a \int_{-\infty}^{+\infty} f_X(x)\,dx + b \int_{-\infty}^{+\infty} x f_X(x)\,dx = a + b\mu_X$$

and

$$\mathrm{Var}(Y) = E(Y^2) - \mu_Y^2 = \int_{-\infty}^{+\infty} (a + bx)^2 f_X(x)\,dx - \mu_Y^2$$

$$= \int_{-\infty}^{+\infty} (a^2 + 2abx + b^2x^2) f_X(x)\,dx - (a + b\mu_X)^2$$

$$\mathrm{Var}(Y) = a^2 + 2ab\mu_X + b^2E(X^2) - a^2 - 2ab\mu_X - b^2\mu_X^2 = b^2\left[E(X^2) - \mu_X^2\right] = b^2\sigma_X^2$$

Thus the standard deviation of Y, $\sigma_Y = |b\sigma_X|$. Notice that even if b may be negative value, σ_Y will always be positive.

(b) In this part, we start out by computing $P(Y \le y)$. Since $y = a + bx$, $g^{-1}(y) = (y - a)/b$. Then from Equation 7.51:

$$P(Y \le y) = P(X \le x) = P\left(X \le \frac{y-a}{b}\right) = \Phi\left(\frac{\frac{y-a}{b} - \mu_X}{\sigma_X}\right) = \Phi\left[\frac{y - (a + b\mu_X)}{b\sigma_X}\right] = \Phi\left(\frac{y - \mu_Y}{\sigma_Y}\right)$$

This result indicates that Y is a normally-distributed random variable with a mean equal to $a + b\mu_X$ and a standard deviation $b\sigma_X$. Alternatively, Equation 7.52 can be used to prove that Y is a normally-distributed random variable.

Example 7.9 In Example 7.7, assume X is a continuous random variable with the following probability density function:

$$f_X(x) = c(1 - e^{-x/L}) \quad \text{for } 0 \leq x \leq L$$

$$f_X(x) = 0 \qquad \text{elsewhere}$$

(a) Compute c.

(b) Compute the mean and standard deviation of X.

(c) Compute the expected value, variance and standard deviation of M_A, the moment at A.

(d) Compute $P(M_A \leq qL^2/8)$.

Solution

(a) Since the integral of the probability density function for the entire range of the random variable X is equal to one, we can write:

$$c\int_0^L (1 - e^{-x/L})\,dx = 1$$

This will result in $c = 2.718/\,L$.

(b) The mean and standard deviation of X are computed as:

$$\mu_X = \int_0^L x f_x(x)\,dx = \frac{2.718}{L}\int_0^L x(1 - e^{-x/L})\,dx = 0.641L$$

$$\sigma_X^2 = E(X^2) - \mu_X^2 = \int_0^L x^2 f_x(x)\,dx - (0.641L)^2 = \frac{2.718}{L}\int_0^L x^2(1 - e^{-x/L})\,dx - (0.641L)^2 = 0.0586L^2.$$

or $\sigma_X = 0.242L$.

(c) The moment at A is defined as $M_A = qX^2/2$. The expected value and variance of M_A are computed using Equations 7.49 and 7.50.

$$E(M_A) = \frac{q}{2}E(X^2) = \frac{q}{2}\int_0^L x^2 f_X(x)\,dx = 0.235qL^2$$

$$\text{Var}(M_A) = E(M_A^2) - \mu_A^2 = E\left(\frac{q^2X^4}{4}\right) - (0.235qL^2)^2 = \frac{q^2}{4}\int_0^L x^4 f_X(x)\,dx - (0.235qL^2)^2 = 0.021q^2L^4$$

This results in a standard deviation equal to $0.145qL^2$.

(d) In this part, we observe that Event $M_A \leq qL^2/8$ is identical with Event $X \leq L/2$. Thus:

$$P(M_A \leq qL^2/8) = P(X \leq L/2) = \int_0^{L/2} f_X(x)\,dx = 0.29$$

7.5 FUNCTIONS OF MULTIPLE RANDOM VARIABLES

In certain engineering problems, the dependent variable Y is a function of two or more random variables $X_1, X_2, \ldots, X_n$, that is:

$$Y = g(X_1, X_2 \ldots, X_n) \tag{7.53}$$

In a special case where Y is a function of two discrete random variables (say X_1 and X_2), the expected value and variance of Y can be computed once the joint probability mass function of X_1 and X_2 is available as given:

$$E(Y) = E[g(X_1, X_2)] = \sum_i \sum_j g(x_{1i}, x_{2j}) p_{1,2}(x_{1i}, x_{2j}) \tag{7.54}$$

$$\text{Var}(Y) = E(Y^2) - \mu_Y^2 = E[g(X_1, X_2)] - \mu_Y^2 = \sum_i \sum_j g^2(x_{1i}, x_{2j}) p_{1,2}(x_{1i}, x_{2j}) - \mu_Y^2 \tag{7.55}$$

In this case x_{1i} ($i = 1, 2, \ldots$) are all values belonging to the random variable X_i and x_{2j} ($j = 1, 2, \ldots$) are all values belonging to the random variable X_2. The double summation is for all x_{1i} ($i = 1, 2, \ldots$) and all x_{2j} ($j = 1, 2, \ldots$). Furthermore, any probability associated with Y, such as $P(Y = y)$ will be:

$$P(Y = y) = P[g(X_1, X_2) = y] \tag{7.56}$$

Since for any given x_1 and x_2 there is a y that satisfies Equation 7.53, solving x_1, in terms of y and x_2 will result in:

$$x_1 = g^{-1}(y, x_2) \tag{7.57}$$

and Equation 7.56 becomes:

$$P(Y = y) = P(X_1 = x_1) = p_1(x_1) = \sum_{\text{all}x_{2j}} p_{1,2}(g^{-1}, x_{2j}) \tag{7.58}$$

in which $g^{-1} = g^{-1}(y, x_{2j})$. The problem becomes more complicated when Y is a function of three or more discrete random variables. However if $X_1, X_2, \ldots, X_n$ are independent of each other, the computation of the expected value and variance of Y and probabilities associated with Y becomes simpler since the joint probability mass function can be written as a product of the individual probability mass functions of $X_1, X_2, \ldots, X_n$.

Example 7.10 In Example 7.7, assume the load is also a discrete random variable Q that is statistically independent of X. There are only two possible values for Q; namely $0.5\bar{q}$ and $\bar{q}$. The probability mass function of Q is defined:

$$p_Q(q) = c_1 q / \bar{q} \quad \text{for } q = 0.5\bar{q} \text{ and } \bar{q}$$

$$p_Q(q) = 0 \qquad \text{elsewhere}$$

Continues

Example 7.10: *Continued*

The parameter $\bar{q}$ is a constant.

(a) Compute $E(M_A)$ and $\mathrm{Var}(M_A)$.

(b) Compute $P(M_A \leq \bar{q}L^2/8)$.

Solution

(a) We first compute the constant c_1.

$$c_1\left(\frac{0.5\bar{q}}{\bar{q}} + \frac{\bar{q}}{\bar{q}}\right) = 1$$

This results in $c_1 = 2/3$. Since Q and X are independent:

$$p_{Q,x}(q,x) = p_Q(q)p_X(x) = \frac{2q}{3\bar{q}} \times \frac{0.4x}{L} = \frac{4qx}{15\bar{q}L}$$

From Equation 7.54:

$$\mathrm{Var}(M_A) = \mu_{M_A} = \sum_{\text{all}q_i}\sum_{\text{all}x_j} g(q_i,x_j)p_{Q,X}(q_i,x_j) = \sum_{\text{all}q_i}\sum_{\text{all}x_j}\frac{q_i x_j^2}{2}\frac{4q_i x_j}{15\bar{q}L} = \frac{25\bar{q}L^2}{96}$$

The variance is:

$$\mathrm{Var}(M_A) = E(M_A^2) - (\mu_{M_A})^2 = \sum_{\text{all}q_i}\sum_{\text{all}x_j} g^2(q_i,x_j)p_{Q,X}(q_i,x_j) - (\mu_{M_A})^2$$

$$\mathrm{Var}(M_A) = \sum_{\text{all}q_i}\sum_{\text{all}x_j}\left(\frac{q_i x_j^2}{2}\right)\left(\frac{4q_i x_i}{15\bar{q}L}\right) - \left(\frac{25\bar{q}L^2}{96}\right)^2 = 0.0274(\bar{q})^2 L^4$$

The standard deviation of M_A is $0.17\bar{q}L^2$.

(b) In this part, we first identify all possible values for the random variable M_A.These are $\bar{q}L^2/64$, $2\bar{q}L^2/64$, $4\bar{q}L^2/64$, $8\bar{q}L^2/64$, $9\bar{q}L^2/64$, $18\bar{q}L^2/64$, $16\bar{q}L^2/64$, and $32\bar{q}L^2/64$. The probabilities associated with these possibilities are computed from the probability mass functions of Q and X. These probabilities are, respectively, 1/30, 2/30, 2/30, 4/30, 3/30, 6/30, 4/30, and 8/30. For example:

$$P(M_A = \bar{q}L^2/64) = P(Q = 0.5\bar{q})P(X = L/4) = [(2/3)\times 0.5]\times[0.4\times(L/4)/L] = 1/30$$

and so on; and as such, the probability $P(M_A \leq \bar{q}L^2/8)$ will then be:

$$1/30 + 2/30 + 2/30 + 4/30 = 9/30 = 0.30$$

If the n random variables $X_1, X_2, \ldots, X_n$ are continuous, $E(Y)$, Var(Y), and any desired probability associated with the dependent variable Y can be computed using the joint probability density function of $X_1, X_2, \ldots, X_n$:

$$E(Y) = E[g(X_1,X_2,\ldots X_n)] = \int_n g(x_1,x_2,\ldots x_n) f_n(x_1,x_2,\ldots x_n)\, dx_1, dx_2,\ldots dx_n \tag{7.59}$$

$$\text{Var}(Y) = \int_n g^2(x_1,x_2,\ldots x_n) f_n(x_1,x_2,\ldots x_n)\, dx_1, dx_2,\ldots dx_n - (\mu_Y)^2 \tag{7.60}$$

$$P(Y \leq y) = \int_{n,g\leq y} f_n(x_1,x_2,\ldots x_n)\, dx_1, dx_2,\ldots dx_n \tag{7.61}$$

in which $\int_n \ldots$ is an n-fold integral with $\pm\infty$ limits, $\int_{n,g\leq y} \ldots$ is an n-fold integral computed over the region specified by the inequality $g(x_1, x_2, \ldots, x_n) \leq y$ in the n-dimensional space, and f_n is the joint probability density function of the random variables $X_1, X_2, \ldots, X_n$. If the random variables $X_1, X_2, \ldots, X_n$ are statistically independent, the joint probability density function f_n will simply be equal to the product of the individual probability density functions of $X_1, X_2, \ldots, X_n$.

The computation of $E(Y)$, $Var(Y)$, and $P(Y \leq y)$ from Equations 7.59-7.61 is complicated. In general, in many real application problems, the joint density function f_n is not known; and at best, only the individual density functions for $X_1, X_2, \ldots, X_n$ are available. In such a case, one may have to assume that $X_1, X_2, \ldots, X_n$ are independent so that the computation of $E(Y)$, Var(Y), and $P(Y \leq y)$ may become simpler.

Example 7.11 In Example 7.7, assume X and Q are independent continuous random variables with the density functions given:

$$f_X(x) = c(1 - e^{-x/L}) \quad \text{for } 0 \leq x \leq L$$
$$f_X(x) = 0 \quad \text{elsewhere}$$

$$p_Q(q) = c_1 q/\bar{q} \quad \text{for } 0 \leq q \leq \bar{q}$$
$$p_Q(q) = 0 \quad \text{elsewhere}$$

In which $\bar{q}$, c, and c_1 are constants.

(a) Compute c_1.

(b) Compute the mean and standard deviation of the moment at A (i.e., M_A).

(c) Compute the probability that M_A is less than or equal to $\bar{q}L^2/8$.

Solution

(a)

$$\int_0^{\bar{q}} f_Q(q)\, dq = 1, \therefore c_1 = 2/\bar{q}$$

Continues

Example 7.11: *Continued*

(b) Since Q and X are independent, the joint probability density function of them is simply written as the product of their respective probability density functions, that is:

$$f_Q(q,x) = f_Q(q)f_X(x) = \left(2q/\bar{q}\right)\left[\frac{2.718}{L}\left(1 - e^{-x/L}\right)\right]$$

The moment at A is $M_A = QX^2/2$, thus:

$$E(M_A) = \int_0^{\bar{q}}\int_0^{L}\left(\frac{qx^2}{2}\right)f_{Q,X}(q,x)\,dqdx = 0.156\bar{q}L^2$$

Also:

$$\mathrm{Var}(M_A) = E(M_A^2) - (\mu_{M_A})^2 = \int_0^{\bar{q}}\int_0^{L}\left(\frac{qx^2}{2}\right)f_{Q,X}(q,x)\,dqdx - (0.156\bar{q}L^2)^2 = 0.0137\,(\bar{q}L^2)^2.$$

This results in a standard deviation equal to $0.117\bar{q}L^2$.

(c) Let, $m = \bar{q}L^2/8$, then, the desired probability is $P(M_A \leq m)$. Since $M_A = g(Q,X) = QX^2/2$, the domain of integration in Equation 7.61 is $qx^2/2 \leq m$. Solving for x, $x \leq \sqrt{(2m/q)}$ and:

$$P(M_A \leq m) = \int_{q=0}^{\bar{q}}\int_{x=0}^{\sqrt{2m/q}} f_{Q,X}(q,x)\,dqdx = 0.516.$$

In solving this integral, we set $\bar{q}$ and L each equal to one and use commercial math solver software.

7.5.1 Special Case: Linear Functions of Independent Normal Random Variables

In a special case where Y in Equation 7.53 is a linear combination of n independent normally distributed random variables, Y will also be a normal random variable. This is presented without proof in this book. The reader is referred to the references cited at the end of this chapter for the proof. In the special case:

$$Y = a_1X_1 + a_2X_2 + \ldots + a_nX_n \tag{7.62}$$

$a_1, a_2, \ldots, a_n$ are constants and $X_1, X_2, \ldots, X_n$ are statistically independent random variables. The mean and standard deviation of Y are:

$$\mu_Y = a_1\mu_1 + a_2\mu_2 + \ldots + a_n\mu_n \tag{7.63}$$

$$\sigma_Y = \sqrt{a_1^2\sigma_1^2 + a_2^2\sigma_2^2 + \ldots + a_n^2\sigma_n^2} \tag{7.64}$$

where μ_i and σ_i ($i = 1, 2, \ldots n$) are the mean and standard deviation of the random variable X_i, respectively.

7.5.2 Special Case: The Product (or Quotient) of Independent Lognormal Random Variables

In this special case, Y in Equation 7.53 is written:

$$Y = X_1 X_2 \ldots X_n \tag{7.65}$$

or, for example:

$$Y = \frac{X_1 X_2}{X_3} \tag{7.66}$$

If the random variables X_i ($i = 1, 2, \ldots, n$) are independent lognormal random variables, then the random variable Y will also be a lognormal with parameters λ and ν. In Equation 7.65 these parameters are:

$$\lambda_Y = \lambda_1 + \lambda_2 + \ldots + \lambda_n \tag{7.67}$$

and

$$\nu_Y = \sqrt{\nu_1^2 + \nu_2^2 + \ldots + \nu_n^2} \tag{7.68}$$

In Equation 7.66, they are:

$$\lambda_Y = \lambda_1 + \lambda_2 - \lambda_3 \tag{7.69}$$

and

$$\nu_Y = \sqrt{\nu_1^2 + \nu_2^2 + \nu_3^2} \tag{7.70}$$

Again, the above equations are presented without proof. The reader is referred to the references cited at the end of this chapter for the proof.

Example 7.12 An elevator cable is subject to a random load (Q) with a mean and standard deviation of 12,000 and 750 N, respectively. The ultimate strength of the cable (R) is also a random variable with a mean and standard deviation of 16,000 and 1,500 N, respectively.

(a) Assume Q and R are independent normal variables and compute the probability of failure of the cable.

(b) Assume Q and R are independent lognormal random variables. Recompute the probability of failure of the cable.

Solution

(a) In this part, we observe that failure will occur if $Q > R$. Although the mean of Q is less than the mean of R, there is a chance that Q may actually be larger than R. This is shown in Fig. E7.12. The shaded area represents the region where $Q > R$ and thus is equal to the probability

Continues

Example 7.12: *Continued*

of failure. A simple method for solving the probability of failure is by using Equations 7.63 and 7.64. Let $Y = R - Q$, thus the failure probability will be defined via Event $Y \leq 0$. Because R and Q are independent normal variables, Y will also be a normal random variable with a mean μ_Y and a standard deviation σ_Y. These are:

$$\mu_Y = \mu_R - \mu_Q = 16{,}000 - 12{,}000 = 4{,}000 \text{ N}$$

$$\sigma_Y = \sqrt{\sigma_R^2 + \sigma_Q^2} = \sqrt{(1{,}500)^2 + (750)^2} = 1{,}677 \text{N}$$

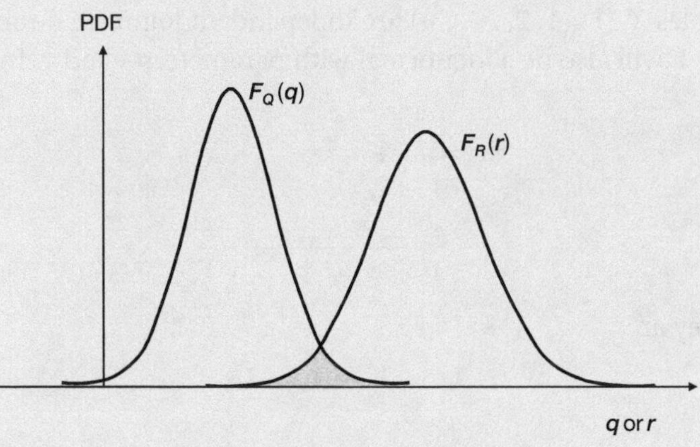

Figure E7.12 Presentation of failure probability of cable in tension (shaded area = probability of failure)

The probability of failure is then computed as follows:

$$P_F = P(Y \leq 0) = \Phi\left(\frac{0 - 4{,}000}{1{,}677}\right) = 0.0084$$

(b) In this part, because Q and R are independent lognormal random variables, we use $Y = R/Q$. Thus, $P_F = P(Y \leq 1)$. Based on the data given in the problem, $\delta_R = 1{,}500/16{,}000 = 0.094$ and $\delta_Q = 750/12{,}000 = 0.063$, and since $v_R^2 = \ell n(1 + \delta_R^2)$, $v_R = 0.094$. Similarly, $v_Q^2 = \ell n(1 + \delta_Q^2)$, $v_Q = 0.063$. Also, $\lambda_R = \ell n(\mu_g) - v_R^2/2 = 9.68$. Similarly, $\lambda_Q = \ell n(\mu_Q) - v_Q^2/2 = 9.39$. Now applying Equations 7.69 and 7.70, we get:

$$\lambda_Y = \lambda_R - \lambda_Q = 9.68 - 9.39 = 0.29 \text{ and } v_Y = \sqrt{v_R^2 + v_Q^2} = \sqrt{(0.094)^2 + (0.063)^2} = 0.11$$

Thus:

$$P_F = P(Y \leq 1) = \Phi\left(\frac{\ell n 1 - 0.29}{0.11}\right) = 0.0041$$

7.6 APPROXIMATE METHODS FOR COMPUTING MEAN, VARIANCE, AND PROBABILITY

As it is evident in Examples 7.10 and 7.11, computing the expected value, variance and probabilities associated with the random variable Y in Equation 7.53 can become complicated when more than two random variables are involved. Frequently, we encounter problems in engineering in which Equation 7.53 forms a nonlinear function of two or more random variables that some of which may be interrelated. Consider the following examples.

In materials engineering, a mode of metal failure is fatigue. Fatigue is a sudden failure resulting from rupture in the material. The failure occurs when the material is subject to a cyclical stress application over an extended period of time. The relationship between the number of cycles of stress until failure to occur and the intensity of stress range (stress range is the rise and fall of stress in one cycle of load application) is empirically obtained in the following form:

$$N = \frac{C}{S^M} \tag{7.71}$$

in which N = number of cycles of stress range S to failure and C and M are constants that depend on the type of material. Clearly N (which plays the same role as Y in Equation 7.53) is a nonlinear function of three random variables C, S, and M (these are similar to X_1, X_2, and X_3 in Equation 7.53). In Equation 7.71, we are often interested in computing the expected value and variance of N. If C and M are assumed to be deterministic, knowing the probability density function of S, $E(N)$, and $\mathrm{Var}(N)$ can easily be computed from Equations 7.49 and 7.50. However, if C and M are treated as random variables (which may also be intercorrelated), computation of $E(N)$ and $\mathrm{Var}(N)$ may become complicated.

In strength of materials, the relationship between bending stress (S) and bending moments M_X and M_Y (acting in two perpendicular directions on the cross section of a beam) is defined:

$$S = \frac{M_X C_Y}{I_X} + \frac{M_Y C_X}{I_Y} \tag{7.72}$$

in which I_X and I_Y are the moments of inertia of the beam cross section with respect to x and y axes of the cross section, respectively. C_Y is the distance of the extreme top (or bottom) fiber of the cross section from the x axis; and C_X is the distance of the extreme right (or left) fiber of the cross section from the y axis. In Equation 7.72, S is the dependent variable. If all independent variables are considered random, there will be at least six variables involved. In this case, the computation of the mean and variance of S will be very complicated.

In transportation planning, when dealing with the life-cycle cost of a facility (such as a highway bridge), the cost (C) of action items such as repair, rehabilitation, or reconstruction is written in terms of several variables as:

$$C = a_1C_1 + a_2C_2 + \ldots + a_nC_n + C_I \quad (7.73)$$

in which C_i ($i = 1, 2, \ldots, n$) = cost of action item i in one life cycle, C_I = initial cost or users' cost, and a_i ($i = 1, 2, \ldots, n$) are factors that depend on the structural condition of the facility. There are often uncertainties involved in all variables in Equation 7.73; and as such, they are, more appropriately, treated as random variables. The uncertainties arise from the fact that generally the capital budget available for repair, rehabilitation, or reconstruction of a given facility is not known. Furthermore, when planning for future years is being considered, it is not known exactly what action items would best suit the facility. Accordingly, the dependent variable C will be a function of several random variables that may be intercorrelated. The computation of the expected value and variance of C will then become complicated.

In each of the examples cited previously, approximate methods are used in computing the expected value and the variance of, and any probabilities associated with, the dependent random variable. There are two approximate methods via which these quantities can be computed: (1) the simulation method and (2) the method based on the first- or second-order approximation of function g in Equation 7.53.

7.6.1 Simulation Method

The simulation method is generally used for estimating the probabilities associated with a dependent variable Y. However, the method can also be used if estimates for the expected value and the variance of Y are desired. The procedure can be conducted in a variety of ways. A necessity in conducting simulation is a means by which random numbers can be generated, because the method is based on artificially generated values that belong to a specific random variable such as X_i. The method becomes quite efficient when the probability density functions of individual independent random variables X_i in Equation 7.53 are available and can be used for generating values that belong to X_i. To clarify this method, assume we have a means available that can be used to generate values belonging to random variables $X_1, X_2, \ldots, X_n$. We start the process by generating a random value for each of the n random variables $X_1, X_2, \ldots, X_n$. Assume these values are, respectively, $x_{11}, x_{21}, \ldots, x_{n1}$. Notice that the first subscript identifies the random variable; whereas, the second subscript indicates the first round of generating random values. The random values in the second round will be $x_{12}, x_{22}, \ldots, x_{n2}$, and in general, in the round k they are $x_{1k}, x_{2k}, \ldots, x_{nk}$. Upon generating random values in each round, we substitute them in Equation 7.53 and compute one value for the dependent variable Y. Thus if the simulation process is conducted in k rounds, there will be k values for Y. These will be $y_1, y_2, \ldots,$ and y_k. We can state that:

$$y_1, y_2, \ldots, \text{ and } y_k \subset Y$$

Now any probability associated with the random variable Y can be computed if k is sufficiently large. Suppose we are interested in estimating the probability $P(Y \leq y)$. Furthermore, suppose that among those k values obtained for Y, a total of k_1 of them are less than or equal to y. The probability of concern is then approximated as:

$$P(Y \leq y) = \frac{k_1}{k}$$

Although this process seems to be rather simple, it can effectively be used to arrive at a reasonable estimate for $P(Y \leq y)$. In practice, a computer program with capabilities to generate random numbers is needed to conduct the process. Usually a very large k (i.e., a large number of simulations or trials) is required to come up with a reasonable estimate of the probability of concern. This is especially true if the probability of concern is very small. The simulation process can also be conducted in several batches. Say for example, the simulation is conducted in 15 batches and each batch includes 10,000 trials. Each batch of random numbers will result in a different estimate for $P(Y \leq y)$. This means there will be 15 different estimates for this probability. The average of these estimates is then used to represent a single value for $P(Y \leq y)$. The details of the simulation process can be found in Ang and Tang (1984).

As it was discussed earlier, the simulation process can also be used to arrive at estimates for $E(Y)$ and $\mathrm{Var}(Y)$. This can simply be achieved by treating $y_1, y_2, \ldots,$ and y_k as any other observed set of data collected for Y and use the statistical methods of estimating mean and variance as explained in Chapter 8.

The process of generating random values for a random variable X_i starts by generating a random number between 0 and 1. If we call this random number u, as it is described in Ang and Tang (1984), a random value for a variable X is computed as:

$$x_i = F^{-1}(u) \tag{7.74}$$

in which F^{-1} *is* the inverse of the probability distribution function $F(x)$ of the random variable X. As an example, assume using a random number generator, u becomes equal to 0.61. If X follows a standard normal probability distribution function, then a random value for X (i.e., x_i) can be obtained from the table of standard normal probability values. This means:

$$x_i = \Phi^{-1}(0.61) = 0.28$$

If the random variable X is a normal variable with a mean value equal to μ and standard deviation equal to σ with $u = 0.61$, the random value x_i will be:

$$\frac{(x_i - \mu)}{\sigma} = \Phi^{-1}(0.61) = 0.28 \quad \text{or} \quad x_i = \mu + 0.28\sigma$$

As evident, generating random numbers using Equation 7.74 is often complicated since the computation of the inverse of the function $F(x)$ will be required for each round of simulation. If the random variable X is normal or lognormal, a simpler method to generate X, is available. If u_1 and u_2 are a pair of random numbers between 0 and 1, then the following are two *independent* standard normal random values:

$$s_1 = \sqrt{-2\ell n(u_1)}\cos(2\pi u_2) \tag{7.75}$$

$$s_2 = \sqrt{-2\ell n(u_1)}\sin(2\pi u_2) \tag{7.76}$$

In which s_1 and s_2 are two random values that belong to independent standard normal variables S_1 and S_2, respectively. The proof of this theorem can be found in Ang and Tang (1984) and Au et al. (1972). Notice that if X_1 and X_2 are two independent normal random variables, random values that belong to these two random variables can be generated using Equations 7.75 and 7.76 in terms of the means and standard deviations of X_1 and X_2. If these random values are designated with x_1 and x_2 then:

$$x_1 = \mu_1 + \sigma_1 s_1 \tag{7.77}$$

$$x_2 = \mu_2 + \sigma_2 s_2 \tag{7.78}$$

in which μ_1, and σ_1 are the mean and standard deviation of the random variable X_1, respectively; and μ_2, and σ_2 are those of the random variable X_2. Similarly if X_1 and X_2 are independent lognormal, random numbers belonging to these two random variables can be computed as:

$$x_1 = \exp(\lambda_1 + \nu_1 s_1) \tag{7.79}$$

$$x_2 = \exp(\lambda_2 + \nu_2 s_2) \tag{7.80}$$

in which λ_1 and ν_1 are the parameters of the random variable X_1; and λ_2 and ν_2 are the parameters of the random variable X_2, respectively.

Example 7.13 In a simplified version of Equation 7.72, when only one bending moment component exists, $S = M_X C_Y / I_X$. Assume we are computing the bending stress in a timber beam. Due to the uncertainties in the cross-sectional properties of a timber beam, I_X is treated as a random variable. For a given beam, this random variable is uniformly distributed between the two values 10.5×10^8 and 12.2×10^8 mm^4. Furthermore, assume also that both M_X and C_Y are uniform random variables. M_X varies between 110 and 135 kN·m; and C_Y between 190 to 210 mm.

(a) Using a simulation method, compute the probability of failure of the beam. Failure happens when stress S exceeds the beam resistance that is 23 MPa. Use 20 samples in the simulation process.

Continues

Example 7.13: *Continued*

(b) In this part, assume the resistance is also a uniform random variable ranging between 18 and 28 MPa. Repeat (a) and compute the probability of failure using 20 samples in the simulation process.

Solution

(a) We will use a scientific calculator to generate random numbers. Most such calculators have a built-in function that can be used to generate a random number between 0 and 1. If this random number is u, then from Equation 7.74, a random value x_i that belongs to the uniform random variable X is:

$$x_i = a + (b - a)u$$

in which a and b are the limits of the uniform random variable X. For example, for M_X, $a = 110$ and $b = 135$. Using this equation, twenty random values for each of the three random variables M_X, C_Y, and I_X are generated as shown in Table E7.13a. The reader is reminded that depending on the calculator, the sequence of random numbers will be different from those listed in Table E7.13a.

Table E7.13a

Trial	M_X (kN·m)	C_Y (mm)	I_X (mm^4)	S (MPa)	Is $S > 23$?, Y or N
1	121.6	203.3	11.10	22.3	N
2	125.8	200.8	11.21	22.5	N
3	112.9	202.5	11.18	20.5	N
4	124.4	194.6	11.08	21.8	N
5	133.1	195.5	11.11	23.4	Y
6	110.4	196.5	11.39	19.0	N
7	131.3	202.3	10.76	24.7	Y
8	134.0	191.3	11.32	22.6	N
9	132.4	201.6	11.07	24.1	Y
10	120.6	204.1	10.92	22.5	N
11	118.8	192.3	11.62	19.7	N
12	118.4	205.0	10.96	22.1	N
13	113.2	208.1	11.89	19.8	N
14	123.1	190.1	10.75	23.4	Y
15	126.1	206.2	11.62	22.4	N
16	114.3	209.1	12.10	19.8	N
17	130.9	194.0	11.59	21.9	N
18	122.5	200.3	12.01	20.4	N
19	120.6	204.4	11.36	21.7	N
20	132.5	203.2	10.88	24.8	Y

Continues

Example 7.13: *Continued*

However, the final outcome of the simulation process will be close to that presented here. For each set of simulated values for M_X, C_Y, and I_X, a value for S is computed. If this value of S is larger than 23, a Y (Y = Yes) is entered in the table; otherwise an N (N = No) is listed. The probability of failure is then estimated as the ratio of the total number of Ys to the total number of trials (i.e., 20 in this example). Since there are only five cases where $S > 23$, the probability of failure $\cong 5/20 = 0.25$. Notice that with a limited 20 trials we are only able to obtain an estimate of the probability with a limited accuracy. More trials will result in a more refined estimate of the probability.

(b) The event of failure in this case is represented by $S > R$, where R is the resistance that is also a uniform random variable. We randomly generate 20 values for R within the specified range for R. Then we count the number of times $S > R$. As seen in Table E7.13b, this occurs nine times. Thus, the probability of failure $\cong 9/20 = 0.45$.

Again, the answer is approximate based on the 20 trials used. If the probability of failure is expected to be small, then more trials will be needed. For example, if the probability of failure is expected to be about 0.001, then at least 1000 trials will be needed, since in every 1000 trials on the average only once S will be larger than R. Of course, the trials can be conducted in batches and these results can be averaged to obtain an estimate for the probability of concern as described earlier.

Table E7.13b

S (MPa)	*R* (MPa)	Is *S* > *R*? Y or N
22.3	21.0	Y
22.5	18.9	Y
20.5	19.1	Y
21.8	22.1	N
23.4	23.8	N
19.0	21.9	N
24.7	25.2	N
22.6	22.4	Y
24.1	19.3	Y
22.5	23.8	N
19.7	26.1	N
22.1	19.8	Y
19.8	19.9	N
23.4	18.1	Y
22.4	23.3	N
19.8	19.0	Y
21.9	25.1	N
20.4	21.3	N
21.7	20.7	Y
24.8	25.4	N

7.6.2 Approximate First- and Second-order Estimates for the Mean and Variance

In this method, Equation 7.53 is expanded into a polynomial using the Taylor series. The expansion is carried around the mean values of the random variables X_i. In the first-order approximation, only the linear terms in the Taylor series are selected; whereas, in the second-order approximation both the linear and parabolic terms are selected. To describe this method further, consider the Taylor expansion of Equation 7.53:

$$Y = g(\mu_1, \mu_2, \dots \mu_n) + \sum_{i=1}^{n} (X_i - \mu_i)\frac{\partial g}{\partial X_i} + \frac{1}{2!}\sum_{i=1}^{n}\sum_{j=1}^{n} (X_i - \mu_i)(X_j - \mu_j)\frac{\partial^2 g}{\partial X_i \partial X_j} + \dots \tag{7.81}$$

in which μ_i is the mean value of X_i and all partial differential terms are evaluated at mean values μ_i and μ_j. This means they are all constants. Adopting the first-order approximation and taking $c_i = \partial g/\partial X_i$:

$$Y = g(\mu_1, \mu_2, \dots \mu_n) + \sum_{i=1}^{n} (X_i - \mu_i)c_i \tag{7.82}$$

The expected value and variance of Y can easily be computed from the linear Equation 7.82:

$$E(Y) = \mu_Y = g(\mu_1, \mu_2, \dots \mu_n) \tag{7.83}$$

$$\text{Var}(Y) = E(Y^2) - (\mu_Y)^2 = E\left[g^2(\mu_1, \mu_2, \dots \mu_n) + \sum_{i=1}^{n}\sum_{j=1}^{n} (X_i - \mu_i)(X_j - \mu_j)c_i c_j\right] - (\mu_Y)^2 \tag{7.84}$$

or

$$\text{Var}(Y) = E\left[\sum_{i=1}^{n}\sum_{j=1}^{n} (X_i - \mu_i)(X_j - \mu_j)c_i c_j\right] = \sum_{i=1}^{n}\sum_{j=1}^{n} c_i c_j \text{Cov}(X_i, X_j). \tag{7.85}$$

In Equation 7.85, when $i = j$, the covariance will become the variance, and the corresponding term can be written as $c_i^2 \sigma_i^2$ Equation 7.85, can also be written in terms of ρ_{ij}, i.e., the correlation coefficient between the two random variable X_i and X_j:

$$\text{Var}(Y) = \sum_{i=1}^{n}\sum_{j=1}^{n} c_i c_j \sigma_i \sigma_j \rho_{ij} \tag{7.86}$$

Notice that in these equations when $i = j$, $\rho_{ij} = 1$, since the correlation coefficient of a random variable with itself is equal to 1. Finally, if the random variables X_i are statistically independent of one another, Equation 7.86 will be written in the following simple form:

$$\text{Var}(Y) = \sigma_Y^2 = \sum_{i=1}^{n} c_i^2 \sigma_i^2 \tag{7.87}$$

The first-order approximation works well when the coefficients of variation (δ_i) of the random variables X_i are small. The approximate expected value and variance of Y can be improved by using both the linear and parabolic terms in the Taylor series of Equation 7.81. In this case, the computation becomes more involved and includes higher moments of the random variables X_i. For example, if Y is only a function of one random variable X, the approximate expected value and variance of Y can be shown to be (Ang and Tang, 2007):

$$E(Y) = g(\mu_X) + \frac{1}{2}\sigma_X^2 \frac{d^2 g}{dX^2} \tag{7.88}$$

$$\mathrm{Var}(Y) = \sigma_X^2 \left(\frac{dg}{dX}\right)^2 - \frac{1}{4}\sigma_X^4 \left(\frac{d^2 g}{dX^2}\right)^2 + E(X-\mu_X)^3 \frac{dg}{dX} \times \frac{d^2 g}{dX^2} + \frac{1}{4}E(X-\mu_X)^4 \left(\frac{d^2 g}{dX^2}\right)^2 \tag{7.89}$$

Again, the derivatives are all evaluated at the mean values.

Example 7.14 In Example 7.13, use the first-order approximation and compute the mean and standard deviation of the dependent variable S.

Solution Notice that the probability density functions for the three random variables M_X, C_Y, and I_X are uniform. For a random variable with uniform distribution bounded by limits a and b, the mean and variance are $(a + b)/2$ and $(b - a)^2/12$, respectively. Using the information given in Example 7.13 for the three independent random variables, we obtain the mean and standard deviations summarized in Table E17.4:

Table E17.4

Random variable	Mean	Standard deviation
M_X	$\mu_M = 122.5$	$\sigma_M = 7.22$ kN·m
C_Y	$\mu_C = 200$	$\sigma_C = 5.77$ mm
I_X	$\mu_I = 11.35 \times 10^8$	$\sigma_I = 0.49 \times 10^8$ mm^4

Using Equation 7.83, the mean of S is:

$$\mu_S = \frac{\mu_M \mu_C}{\mu_I} = \frac{(122.5 \times 10^6)(200)}{11.35 \times 10^8} = 21.6 \quad \text{MPa}$$

and from Equation 7.87, the variance is:

$$\mathrm{Var}(S) = \sum_{i=1}^{3} c_i^2 \sigma_i^2 = c_M^2 \sigma_M^2 + c_C^2 \sigma_C^2 + c_I^2 \sigma_I^2$$

in which:

$$c_M = \frac{\partial S}{\partial M_X} = \frac{C_Y}{I_X}\bigg|_{\text{At mean value}} = \frac{\mu_C}{\mu_I} = \frac{200}{11.35 \times 10^8} = 17.62 \times 10^{-8}$$

Continues

Example 7.14: *Continued*

and

$$c_C = \frac{\partial S}{\partial C_Y} = \frac{\mu_M}{\mu_I} = \frac{122.5 \times 10^6}{11.35 \times 10^8} = 0.108$$

and

$$c_I = \frac{\partial S}{\partial I_X} - \frac{\mu_M \mu_C}{\mu_I^2} = -\frac{122.5 \times 10^6 \times 200}{(11.35 \times 10^8)^2} = -190.2 \times 10^{-10}$$

Thus

$$\text{Var}(S) = 2.87 \quad \text{and} \quad \sigma_S = 1.69 \text{ MPa}$$

SUMMARY

In this chapter, functions of random variables, probability functions for multiple random variables, and correlated random variables were discussed. In particular, the procedure for computing important parameters (e.g., mean and variance) of a dependent variable in terms of the means and variances of independent variables was described. We noticed that in many practical problems it is difficult to use the mathematical formulations presented in this chapter in computing the probabilities associated with a dependent variable. Certain approximations may result in simplifications of these mathematical expressions. One of these approximations was to assume that the random variables are statistically independent of one another. This assumption may be helpful in cases where the intercorrelation among variables is rather weak. Other approximations included the use of the first-order simplification in computing the mean and variance of the dependent variable and simulation for estimating probabilities. The first-order simplification works relatively well where the variances of the variables are small. The simulation is simple, but its accuracy depends on the magnitude of the probability value that is being estimated. If the probability of concern is expected to be small, a large number of trials may be necessary to arrive at a reasonable estimate of the probability. Usually simulations are conducted via a computer program using a random number generator.

REFERENCES

Ang. A. H.-S. and W. H. Tang (1984). Chapter 6 in *Probability Concept in Engineering Planning and Design, Volume II*, New York, N.Y: John Wiley & Sons.

———(2007). Chapter 4 in *Probability Concept in Engineering: Emphasis on Applications to Civil and Environmental Engineering*, Hoboken, NJ: John Wiley & Sons.

Au, T., R. M. Shang, and L. A. Hoel (1972). Chapters 3 and 10 in *Fundamentals of System Engineering, Probabilistic Models*, Reading, MA: Addison-Wesley Publishing.

Walpole, R. E., R. H. Myers, S. L. Myers, and K. Ye (2006). Chapter 7 in *Probability and Statistics for Engineers and Scientists*, 8th ed., Upper Saddle River, NJ: Prentice-Hall.

EXERCISES

1. Figure P7.1 shows a simply supported beam with a load q applied along the length X. The length X is a discrete random variable with possible values equal to $0.2L$, $0.4L$, $0.6L$, $0.8L$ and L, respectively. The probability mass function of X is defined:

$$p(x) = cx/L$$

in which c is a constant.

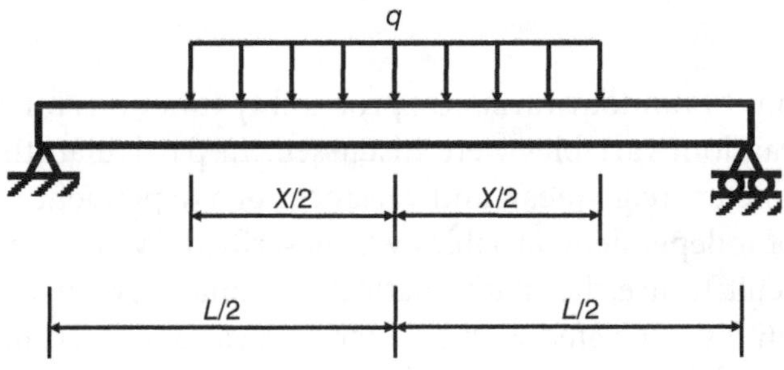

Figure P7.1 Simply supported beam with random load area

(a) Write an expression for the moment at the midspan of the beam.
(b) Compute the constant c.
(c) Compute $E(X)$ and $\text{Var}(X)$.
(d) Compute the expected value and the variance of the moment al the midspan.
(e) Compute the probability that the moment at the midspan will be less than $qL^2/8$.

2. Loads X_1 and X_2 are independent random variables applying on a cantilever beam as shown in Fig. P7.2. The probability density functions for these random variables are f_1 and f_2, respectively, and described as:

$$f_1(x_1) = c_1 \frac{x_1}{70} \quad 0 \le x_1 \le 70 \text{ kN}$$

$$f_2(x_2) = c_2 \frac{x_2}{90} \quad 0 \le x_2 \le 90 \text{ kN}$$

in which c_1 and c_2 are constants.

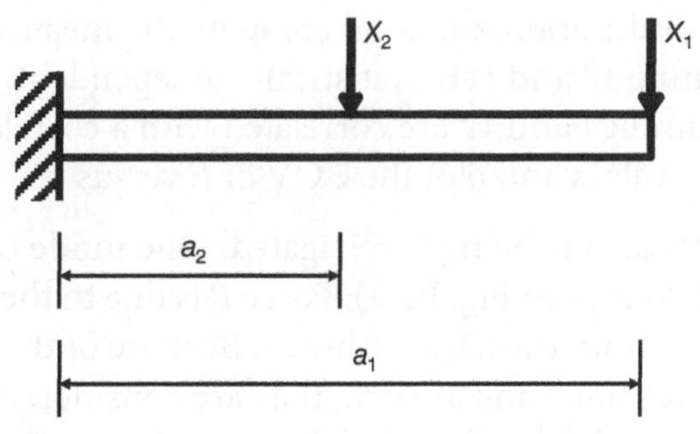

Figure P7.2 Cantilever beam with random loads

(a) Compute the constants c_1 and c_2

(b) Form the joint probability density function of X_1 and X_2, and compute the mean and standard deviation of the moment that the fixed end of the cantilever beam

(c) Compute the probability that the moment that the fixed end of the beam exceeds the beam resistance that is 350 kN. Use $a_1 = 4$ m and $a_2 = 2$ m

3. The axial load in an interior column in a two-story building is a random variable Y described as the sum of the floor loads X_1 and X_2. The floor load each is a normally distributed random variable with a mean value of 90 kN·m and a coefficient of variation (COV) of 0.20. The random variables X_1 and X_2 are independent.

   (a) Compute the mean and standard deviation of Y

   (b) If the resistance of the column (R) is a normal random variable with a mean of 270 kN and a COV of 0.20, compute the probability of failure of the column assuming that Y and R are statistically independent

4. In Exercise 3, (b), compute the probability of failure using the simulation process. Use five rounds of 100 samples in each round. Then average the results computed for the five rounds.

5. In Example 7.2:

   (a) Compute $E(X_1, X_2)$

   (b) Compute $\mathrm{Cov}(X_1, X_2)$

   (c) Compute the correlation coefficient between X_1 and X_2

6. The modulus of elasticity of concrete E depends on the density W and the compressive strength of concrete f. The relationship is written:

$$E = 0.043\,W^{1.5}\sqrt{f}$$

in which E and f are in MPa and W is in kg/m^3. Assume W and f are random variables with mean values equal to 2350 kg/m^3 and 28 MPa, respectively. The COV for these random variables each is equal to 0.20.

(a) Using the first-order approximation compute the mean value and standard deviation of E assuming W and f are statistically independent.
(b) Repeat (a) assuming f and W are correlated with a correlation coefficient ρ. Take ρ equal to several values and plot the COV of E versus ρ.

7. The stability of an earth dam is being investigated. One mode of failure is the dam overturning about the toe (Point A in Fig. P7.7). Force P is due to the hydrostatic pressure and is applied at a distance X from the base as shown. Because of the variation in water depth, both P and X change with time; and as such, they are considered random variables. There is also a small variation in the weight W of the dam. Random variables P and X are correlated with a correlation coefficient $\rho = 0.8$; whereas, there is no correlation between P and W nor between W and X. The means and standard deviations for the three random variables are:

Variable P: Mean = 450; Standard deviation = 67 kN
Variable W: Mean = 1,125; Standard deviation = 36 kN
Variable X: Mean = 4.5; Standard deviation = 0.9 meters

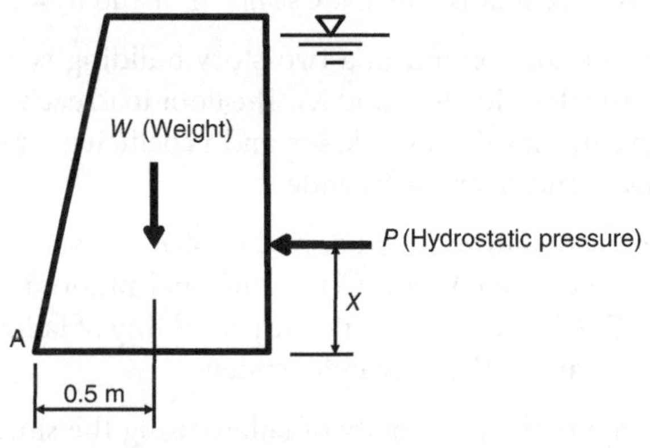

Figure P7.7 Loads applied on an earth dam

(a) Write a function (Z) describing the overturning mode of failure of the dam (Z = resisting moment – overturning moment). Compute the mean and standard deviation of Z assuming the first-order approximation.
(b) If Z is assumed to be normal, compute the failure probability of the dam because of overturning.

8. Using the probability of the dependent random variable Y given in Equation 7.43, prove that the probability density function of Y can be computed from Equation 7.52.

9. Prove that the random values described by Equations 7.75 and 7.76 belong to two independent standard normal random variables (such as S_1 and S_2).

10. In Equation 7.71, assume that C and M are deterministic and that S is the only random variable. If S is a normally-distributed random variable with a mean and standard deviation equal to μ and σ, respectively, derive an expression for the probability density function of N.

11. In a simple form, the productivity in a construction project is modeled in terms of the volume of work scheduled to be completed in a day (X_1) and the work shift hours per da (X_2). In a particular project, the possible values for X_1 are 1, 2, and 3 units; and the possible values for X_2 are 8, 9, and 10. Furthermore, in this project, the productivity Y is defined as the ratio of work completed to the work planned to be completed. Assume the following relationship is given for Y:

$$Y = \frac{0.2}{X_1} + \frac{6.0}{X_2}$$

Furthermore, the values of the probability mass function of X_1 are:

$p_1(1) = P(X_1 = 1) = 0.6$
$p_1(2) = P(X_1 = 2) = 0.3$
$p_1(3) = P(X_1 = 3) = 0.1$

Also, the values of the probability mass function of X_2 are:

$p_2(8) = P(X_2 = 8) = 0.50$
$p_2(9) = P(X_2 = 9) = 0.25$
$p_2(10) = P(X_2 = 10) = 0.25$

(a) Assuming X_1 and X_2 being statistically independent, compute the expected value and the standard deviation of Y

(b) Compute the probability that the productivity will be greater than 0.8

12. Using Equation 7.72, the maximum stress in a beam under a biaxial moment action is computed. In a particular beam, the following information for the mean and standard deviation of the six independent variables in Equation 7.72 is available (Table P7.12):

Table P7.12

Random variable	Mean	Standard deviation
M_X	150	7.00 kN
C_Y	180	7.00 mm
I_X	11.35×10^8	0.5×10^8 mm^4
M_Y	50	3.00 kN
C_X	80	2.00 mm
I_Y	2.5×10^8	0.24×10^8 mm^4

Using the first-order approximation, compute the expected value and the standard deviation of the stress S.

13. In a simple model, there are only three cost items involved in Equation 7.73. The factors a_i and the initial cost C_I are taken as deterministic values. The total cost in $1,000 is written:

$$C = a_1C_1 + a_2C_2 + a_3C_3 + 300$$

in which $a_1 = a_2 = a_3 = 1$. Furthermore, the three costs C_1, C_2, and C_3 are all normal random variables with the following mean and standard deviation values:

C_1: Mean = 300; Standard deviation = 15
C_2: Mean = 550; Standard deviation = 27
C_3: Mean = 700; Standard deviation = 37

(a) Compute the mean and standard deviation of the total cost C
(b) Compute the probability that the total cost will be less than $1.5 million

14. The deflection Y of a cantilever beam of span L carrying a load P at its free end is:

$$Y = \frac{PL^3}{3EI}$$

in which E = modulus of elasticity and I = moment of inertia. If P, E, and I are independent random variables, derive expressions for the mean and COV of Y in terms of the mean values and coefficients of variations of P, E, and I.

15. In Exercise 14, the beam cross section is rectangular with a width equal to B and depth equal to D. The mean and standard deviation of B are 120 and 12 mm, respectively. The mean and standard deviation of D are 240 and 24 mm, respectively.

(a) Compute the mean and standard deviation of I, assuming D and B are independent
(b) Repeat (a), but assume B and D are correlated with a correlation coefficient equal to 0.7
(c) Discuss why B and D can be correlated

16. In Exercise 14, L = 1.8 m; and the following values are available for the means and coefficients of variation of the three random variables P, E, and I (Table P7.16).

Table P7.16

Random variable	Mean	Coefficient of variation
P	12 kN	0.15
E	2.0×10^4 MPa	0.10
I	1.4×10^8 mm^4	0.20

(a) Compute the mean and COV of the deflection Y
(b) If for any deflection more than 40 mm, the beam is considered failed, estimate the probability of failure of the beam, assuming Y is a lognormal distribution function

17. The duration of a construction project is treated as a random variable with a mean of 120 da and a standard deviation of 12 days. The target time for the completion of the project is uncertain. It is assumed to have a mean of 130 da with a standard deviation equal to 15 da.

    (a) Assume both the completion and target time follow normal distribution and compute the probability of delay in the project
    (b) Repeat (a), but use lognormal distributions

18. A study concerns the combined effect of duration of the yellow light (random variable X) and the traffic flow average speed (random variable Y) on accidents during rush hours in an intersection. Three possibilities for the duration of the yellow light and speed each are considered. Table P7.18 summarizes the possible values of X and Y and the corresponding probability of accidents $p(x,y)$ at the intersection.

    (a) Verify the validity of values given in Table P7.18 as joint probability mass function values
    (b) Compute the probability of accidents for durations of yellow light less than or equal to 7 sec and the speed less than or equal to 50 km/h

Table P7.18

	Yellow light duration (sec)		
Speed (Km/h)	$X = 4$	$X = 7$	$X = 10$
$Y = 40$	0.10	0.07	0.09
$Y = 50$	0.14	0.10	0.05
$Y = 55$	0.21	0.16	0.08

    (c) Compute and tabulate the values of the marginal probability functions $p_x(x)$ and $p_y(y)$
    (d) Compute and tabulate the values of the conditional probability mass function $p_{X|y}(x \mid y)$ and $p_{Y|x}(y \mid x)$

19. Figure P7.19 shows an activity network for a construction project. The four activity durations X_1, X_2, X_3, and X_4 are independent normal random variables with their respective mean and standard deviation values as indicated in the figure.

    (a) Compute the mean and standard deviation for each of the two paths ABCD and ABE
    (b) Compute the probability that each path will be completed within 65 da

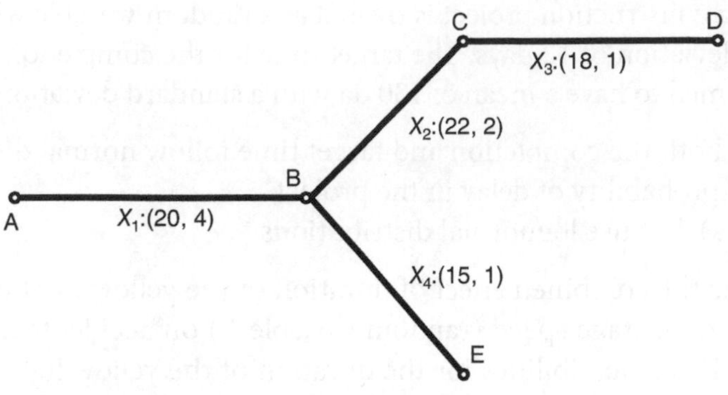

Figure P7.19 A small activity network

20. In Exercise 19, compute the correlation coefficient between the paths ABCD and ABE.

21. In Exercise 19, the target time for the completion of activity AB is 25 da. If the target time for the entire network (shown in Figure P7.19) is 65 da, compute the probability that the project will be completed on time.

Principles of Statistics: Part I—Estimation of Statistical Parameters and Testing Validity of Distribution Functions

8

8.1 INTRODUCTION

In Chapters 6 and 7 the basic theories of random variables, probability, and probability distribution functions were presented. In nearly all example problems cited in those chapters, the basic assumption was that the distribution model describing a random variable is known. However, in real-world problems, such information is not available. In fact, the available information comprises a set of data samples compiled through experiments, field investigations, or subjective judgment. Before one can apply theoretical distribution models to an engineering decision-making problem, a determination of the type of distribution model that can best describe the data must be made. Figure 8.1 shows the flow of activities in conducting an engineering design and decision making under uncertainties using the theory of probability. We observe that the link between engineering knowledge and the theoretical probability modeling is statistical analyses of data. This link is crucial in providing the necessary information as to what type of distribution model should be used and what parameters describe the distribution model. A wide variety of statistical analyses are available, many of which are specifically useful in engineering problems. A simple statistical data analysis consists of estimating such parameters as the mean and standard deviation and/or arranging the data in the form of a bar chart. More advanced analyses may include establishing confidence intervals for the mean, conducting a test for the validity of a distribution model to represent the data, testing hypotheses, estimating a correlation between two sets of data, and conducting regression analyses. In Chapters 8 and 9 we present several common statistical methods that are especially useful in the analysis of engineering data. Chapter 8 concentrates primarily on the estimation of statistical parameters, confidence intervals, and statistical tests for normality and for the validity of other distribution models. Chapter 9 discusses the analysis of variance, comparison between two or more populations, regression analysis, and correlation. Hypothesis testing is also discussed in Chapter 9.

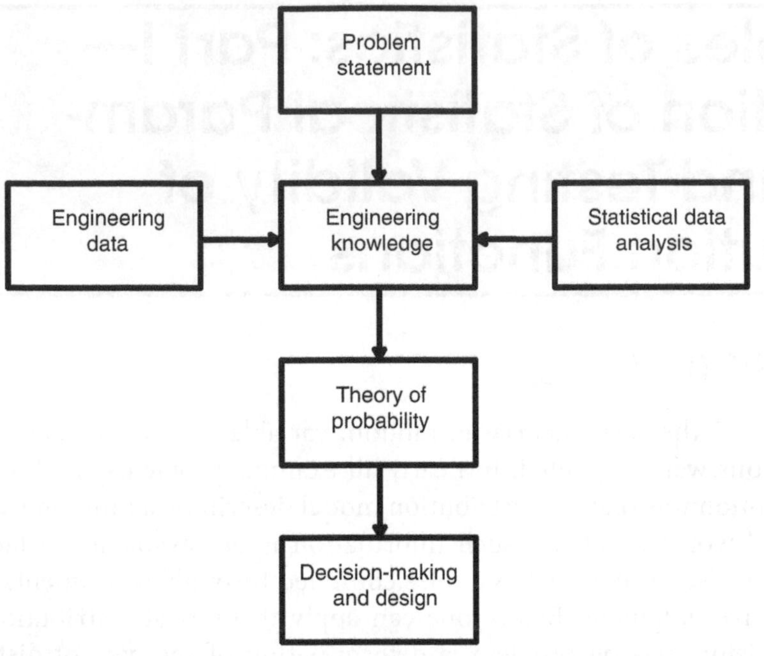

Figure 8.1 Flow of activities in an engineering decision-making and design problem under uncertainties

8.2 DATA COMPRESSION

One of the first steps in analyzing and processing data is *data compression*. Data compression simply means to filter, arrange, and summarize data in a form that can be used readily for the intended engineering decision making, analysis, and design. A simple data compression scheme is to arrange the data in the form of a frequency-of-occurrence diagram or a bar chart. Time-dependent data may be compressed into *exceedance* curves or into a frequency-domain graph. More advanced data compression methods may involve derivation of distribution models suitable to represent the data and the use of data to establish an empirical relationship between two or more variables. Depending on the type of data compiled, one or more of these methods may be used for data compression. Laboratory and field data often consist of a series of sample values for a quantity. Data compression for this type of data is generally composed of establishing frequency diagrams, generating exceedance rates, or a plot of maximum and minimum values. An exceedance rate simply presents the number of times the sample values compiled for the quantity was larger than (i.e., exceeded) certain predetermined thresholds. Subjective data is compiled through interviews or by conducting a survey among experts. Depending on the type of problem, the outcome of the subjective data collection process may be a set of numerical values obtained for a quantity. In such a case, the data is treated similar to the laboratory or field data. In certain cases the outcome of the data collection process may be in linguistic terms. For example, the outcome may be expressed in the form of a series of *yes* and *no* answers. In such a case, the data compression may simply involve the development of percentages describing the relative ratio of the yes responses to the no responses. In a more elaborate process, nonparametric statistical methods may be used to identify any correlation between different groups of data and to establish cause-and-effect relationships.

The objective of this section is to present several simple methods that can be helpful in the compression of most engineering data.

8.2.1 Processing Data into Occurrence Frequencies

This method is popular and results in a table or graph demonstrating the frequency of occurrence of data within various ranges. These ranges are selected *a priori* and cover all possible values obtained for the quantity for which the data has been compiled. The data compressed in this form is considered to be a first step in developing additional information on its trends and the estimates for key statistical parameters of its distribution and probability values. The frequency-of-occurrence graphs are usually used for cases where the sample values need not be presented in a time-dependent fashion. For example, in conducting laboratory investigations for the compressive strength of a batch of concrete, each sample value represents a compressive strength. Such data is arranged using frequency-of-occurrence diagrams without any specific reference to time because all samples tested for strength have the same age. In another example, when the change in the strength of concrete over, for example, a 28-day period is desired, the sample values are time-dependent. A simple frequency-of-occurrence diagram may not offer a suitable technique for data compression and processing in this case. Rather, the data is displayed in the form of a graph showing the strength as a percentage of the maximum strength versus curing time.

The procedure in developing frequency-of-occurrence diagrams involves the following steps:

1. Select a definite number of ranges within the compiled data
2. Count the number of times (i.e., the frequency of occurrence) that the sample values fall within these individual ranges
3. Plot the results in Step 2 versus the ranges; or alternatively, divide the frequencies by the total sample size and then plot the results versus the ranges

The following examples illustrate this method.

Example 8.1 Twenty measurements of the monthly maximum wind velocity at a station near the airport of a major city have been made. The sample data is provided using km/h as the unit.

22, 35, 48, 17, 42, 19, 52, 33, 8, 38, 24, 28, 40, 14, 9, 27, 33, 36, 18, 37

(a) Arrange the data in the form of occurrence frequencies. Plot the results.

(b) Using the relative frequencies (i.e., the frequencies divided by the sample size), estimate the probability that the maximum monthly wind velocity will be less than 30 km/h.

Continues

Example 8.1: *Continued*

Solution

(a) We select six ranges as 0–9, 10–19, 20–29, 30–39, 40–49, and 50–59. Notice that these ranges cover all possible values collected for the wind velocity. Within 0–9, there are only two values recorded; whereas within 10–19, there are four values recorded. For the entire data, see Table E8.1.

A plot of the compressed data is provided in Figure E8.1. Notice that the relative frequencies add up to 1.00. Also, the probability that maximum monthly wind velocity is within the limits of a specific range is equal to the relative frequency for that range. For example, the probability that the maximum monthly wind velocity is within 10 to 19 km/h is 0.20.

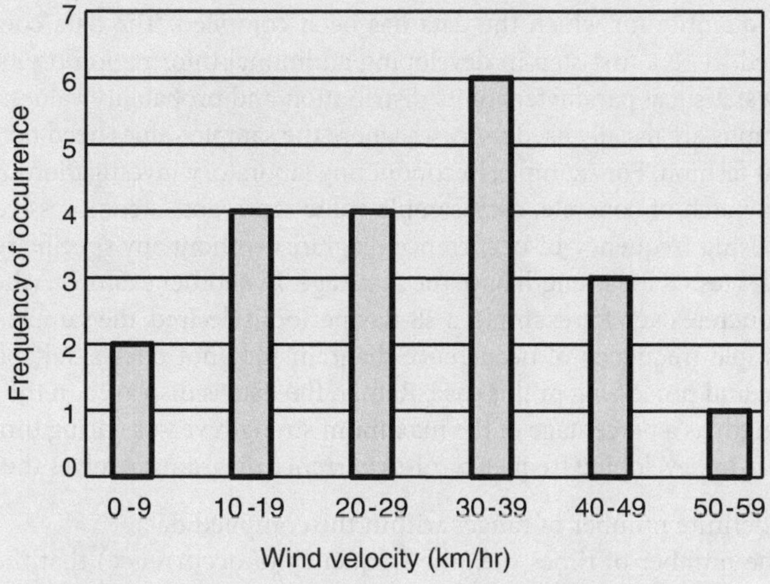

Figure E8.1 Frequency of occurrences of wind velocity in an area

Table E8.I

Range (km/h)	Frequency of occurrence	Relative frequency
0–9	2	2/20 = 0.10
10–19	4	4/20 = 0.20
20–29	4	4/20 = 0.20
30–39	6	6/20 = 0.30
40–49	3	3/20 = 0.15
50–59	1	1/20 = 0.05
Sum	20	1.00

(b) In this part, using X as the maximum monthly wind velocity, we will get $P(X < 30) = 0.10 + 0.20 + 0.20 = 0.50$.

Example 8.2 Twenty-five specimens taken from freshly-made batches of a type of concrete were tested for the 28-day compressive strength. The results are in MPa as listed:

22.75, 21.70, 23.80, 23.45, 21.74, 25.20, 22.61, 24.15, 24.92, 21.00, 25.00, 25.70, 20.75, 23.85, 23.95, 22.43, 27.56, 21.33, 22.19, 24.00, 23.75, 26.75, 20.40, 22.80, 23.42

(a) Compress the data into the occurrence frequency format.

(b) Plot the results.

Solution

(a) The selected ranges, the frequency of occurrence for each range and relative frequencies are summarized in Table E8.2:

Table E8.2

Range (MPa)	Frequency of occurrence	Relative frequency
20.00–20.99	2	2/25 = 0.08
21.00–21.99	4	4/25 = 0.16
22.00–22.99	5	5/25 = 0.20
23.00–23.99	6	6/25 = 0.24
24.00–24.99	3	3/25 = 0.12
25.00–25.99	3	3/25 = 0.12
26.00–26.99	1	1/25 = 0.04
27.00–27.99	1	1/25 = 0.04
Sum	25	1.00

(b) The plot is presented in Figure E8.2.

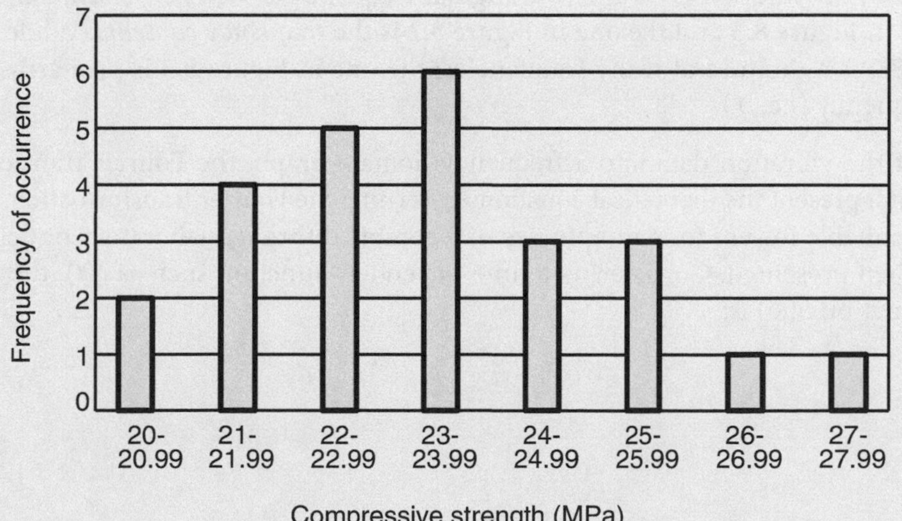

Figure E8.2 Frequency of occurrence of compressive strength in a sample of concrete

8.2.2 Processing Time-dependent Data

Time-dependent data is sometimes shown in the form of a time-history graph. It is simply a graph of a quantity versus time. Such graphs often contain many peaks and valleys and may not be readily used due to the large amount of information they display. Instead of the time-history graphs, data can be compressed into a more manageable form. Two common types of data processing for time-dependent quantities are:

- Conversion of data into a frequency-dependent function by conducting a Fourier transform.
- Conversion of data into exceedance graphs using a *cycle-counting* method.

These two methods are explained:

Frequency-dependent function: A typical sample of data for this type of data processing is the vibration of a mechanical or structural system. Ground vibration during an earthquake is another example of this kind of data. Such data is often compiled in the form of ground accelerations versus time. Figure 8.2 shows a record compiled in 1941 during the El Centro, California earthquake. The curve has been simplified for clarity. As seen in the record, the entire period of vibration is only about 30 sec. The sample data is compiled at short time intervals; as such, it may not immediately provide detailed information on the vibrational characteristics of the earthquake. Since the vibration is not steady, it is a mixture of many vibration frequencies. Figure 8.3 shows the vibrational characteristics similar to that generated by the operation of an engine. The accelerations generated are rather small in magnitude and are relatively constant for the entire vibration period. The period of vibration (T) is constant and the frequency of vibration is $f = 1/T$ (the angular frequency $\omega = 2\pi f$). The difference between the record in Figure 8.3 and the one in Figure 8.2 is the *frequency contents*. While the record in Figure 8.2 is a mixture of many frequencies; the one in Figure 8.3 is primarily composed of one frequency (i.e., f).

To convert the vibration data into a frequency-domain graph, the Fourier transformation is used. We first present the theoretical equations governing the Fourier transformation. Numerical equations suitable for use for a sample size of n compiled through laboratory or field observations are then presented. Considering a time-dependent function such as $g(t)$, the frequency-domain function $S(\omega)$ is:

$$S(\omega) = \frac{1}{2\pi}\int_{0}^{+\infty} g(t)e^{-i\omega t}\,dt \tag{8.1}$$

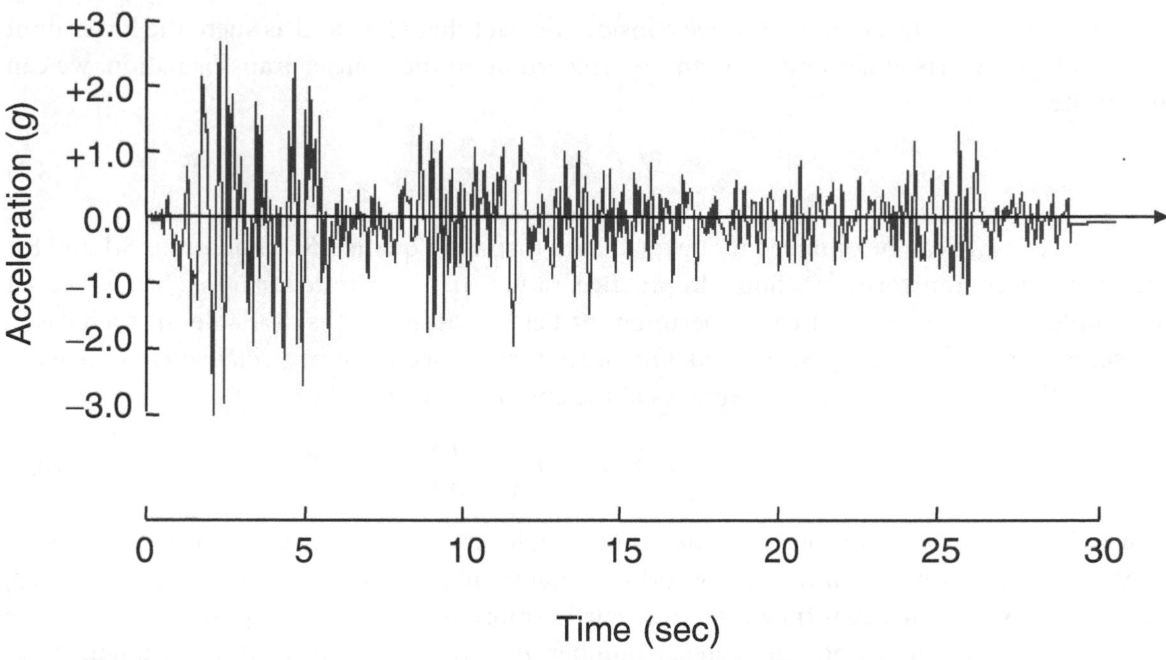

Figure 8.2 A sample of vibrations of ground from an earthquake

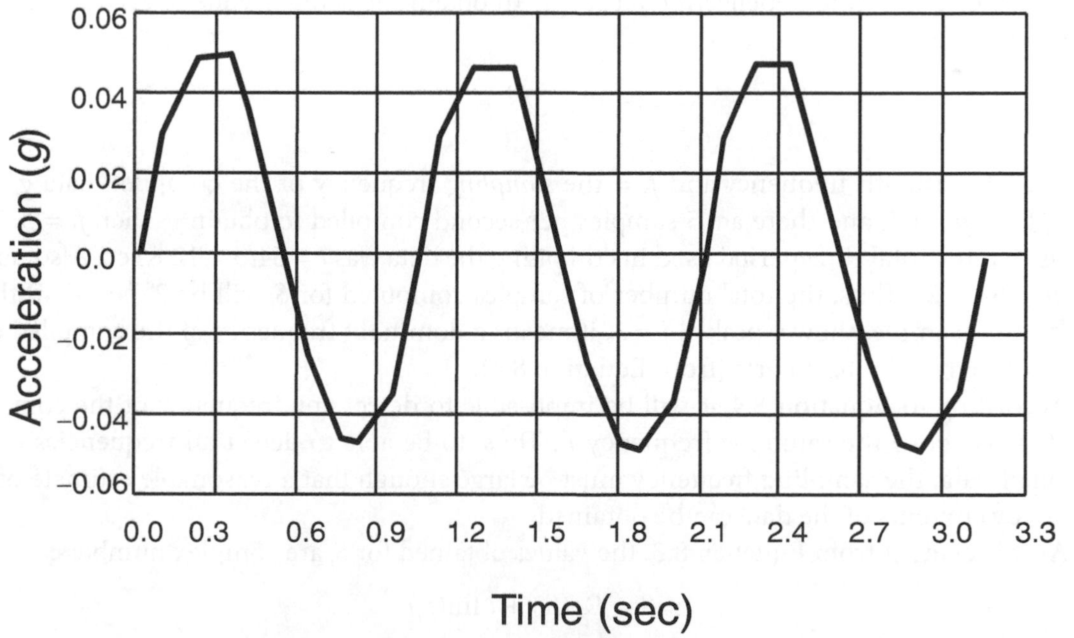

Figure 8.3 An example of vibrations generated by a machine

in which $i = \sqrt{-1}$. In Equation 8.1, we consider the fact that $t \geq 0$; and as such, the lower limit of the integral starts at zero rather than $-\infty$. According to the Fourier transformation, we can also write:

$$g(t) = \int_0^{+\infty} S(\omega) e^{i\omega t} d\omega \tag{8.2}$$

which means $g(t)$ can be obtained if $S(\omega)$ is known through Equation 8.2. Equations 8.1 and 8.2 are the Fourier transform functions. In practice, rather than a definite form for $g(t)$, only a set of sample values, compiled in an experiment or field investigation, is available. In such cases, a discrete form of Equation 8.1 is used. One form that is used in commercial software (e.g., in MathCAD as cited in the list of references at the end of this chapter) is:

$$S_j = \frac{1}{\sqrt{n}} \sum_k g_k \exp\left[i\left(2\pi \frac{j}{n}\right)k\right] \tag{8.3}$$

in which $g_k = k^{\text{th}}$ value compiled (in the set of sample values); S_j = the value of the frequency-domain function for the jth frequency and n = total number of sample values. In Equation 8.3, all g_k are compiled at equal time intervals. Furthermore, Equation 8.3 requires n to be in the form 2^m, in which m is a nonzero integer number. Also notice that instead of the angular frequency ω, the frequency term $j = \omega/(2\pi)$ is used. Using Equation 8.3, 2^{m-1} sample values for S_j at equal frequencies are obtained. Equation 8.3 does not give values in terms of the actual frequencies of the original time-dependent data set g_k. To obtain the actual frequencies, the following equation is used:

$$f_j = \frac{j}{n} f_S \tag{8.4}$$

in which f_j = the jth frequency and f_S = the *sampling* frequency of the compiled data g_k. For example, if $n = 64$, and there are 5 samples per second compiled to obtain g_k then $f_S = 5$. This means that the total time period used in compiling the data was $t = 64/5 = 12.8$ sec. Also notice that $n = 64 = 2^6$. Thus, the total number of samples computed for S_j will be $2^{6-1} = 32$. If the S_j results, for example, show a peak at $j = 3$, it means a dominant frequency of the compiled data was $f_3 = (3/64) \times 5 = 0.23$ hertz (from Equation 8.4).

According to Equation 8.4, it will be impossible to detect any frequency of the compiled data that is above the sampling frequency f_S. Thus, to be able to detect all frequencies of the compiled data, the sampling frequency must be large enough that a reasonable estimate of the frequency contents of the data can be obtained.

As it is evident from Equation 8.3, the values obtained for S_j are complex numbers:

$$S_j = \text{Re}(S_j) + i\,\text{Im}(S_j) \tag{8.5}$$

in which Re(S_j) = real part of S_j and Im(S_j) = imaginary part of S_j. Usually, to plot or present the frequency-domain values (i.e., S_j), we use $|S_j|$ which is:

$$|S_j| = \sqrt{[\mathrm{Re}(S_j)]^2 + [\mathrm{Im}(S_j)]^2} \tag{8.6}$$

Example 8.3 Table E8.3 summarizes a set of 32 sample values compiled for the accelerations data at the base of a shaking table subject to a sudden shock. The entire record was for 3.1 sec, and the samples were compiled at a rate of ten per second (sampling frequency = 10).

Table E8.3

Time (sec)	Acceleration (*g*)	Time (sec)	Acceleration (*g*)
0	0	1.6	0.221
0.1	0.471	1.7	0.170
0.2	0.709	1.8	0.080
0.3	0.754	1.9	0.015
0.4	0.374	2.0	−0.061
0.5	0.361	2.1	−0.091
0.6	0.073	2.2	−0.107
0.7	−0.175	2.3	−0.053
0.8	−0.313	2.4	−0.014
0.9	−0.324	2.5	0.036
1.0	−0.324	2.6	0.060
1.1	−0.180	2.7	0.074
1.2	−0.031	2.8	0.099
1.3	0.088	2.9	0.072
1.4	0.189	3.0	0.038
1.5	0.239	3.1	0.006

(a) Plot the data (g_k) in terms of time.

(b) Using Equation 8.3, compute S_j for $j = 1$ to 32 and plot S_j.

(c) Identify the dominant frequency content of the compiled data.

Solution

(a) Figure E8.3a shows the plot of g_k versus time. As is evident from this graph, there is a mix of several frequencies inherent in the record. In this form, it is difficult to identify the dominant frequency.

Continues

Example 8.3: *Continued*

(b) Using commercial software (e.g., MathCAD or Microsoft Excel) we compute S_j. Alternatively, Equation 8.3 can be solved numerically using a spreadsheet or a simple computer program for numerical integration. The plot of $|S_j|$ appears in Figure E8.3b.

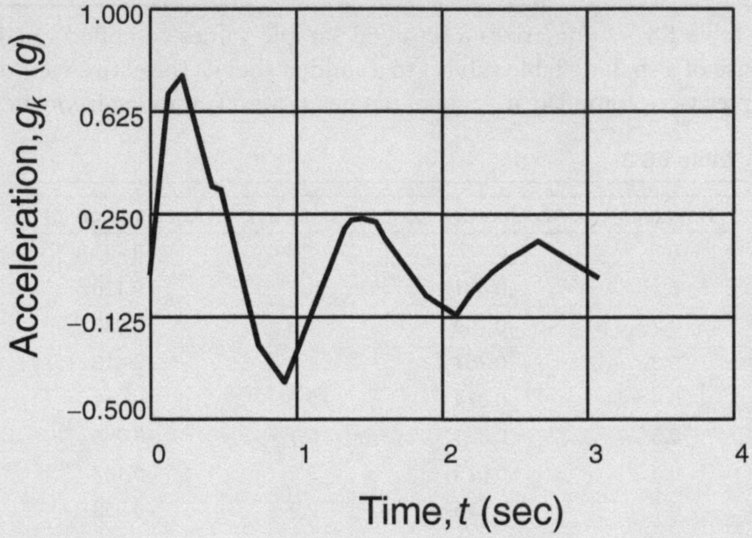

Figure E8.3a Time history plot of acceleration of a shaking table

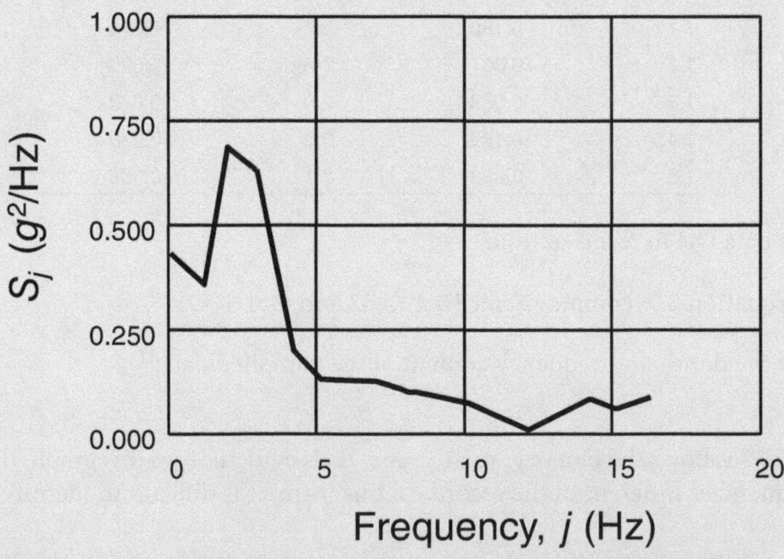

Figure E8.3b Frequency graph of the acceleration of a shaking table

Continues

Example 8.3: *Continued*

(c) Clearly, the dominant frequency is between $j = 2$ and $j = 3$. Since the sampling frequency $f_S = 10$, then from Equation 8.4:

$$f_2 = \left(\frac{2}{32}\right) \times 10 = 0.625 \text{ and } f_3 = \left(\frac{3}{32}\right) \times 10 = 0.938 \text{ hertz}$$

The average gives a frequency of 0.78 hertz.

Example 8.4 Apply a Fourier transformation to the vibration depicted in Figure 8.3.

Solution Clearly, there is only one frequency in this record. Using 32 samples from the data, $|S_j|$ is obtained and plotted in Figure E8.4. Since the dominant frequency is at $j = 3$, then the actual frequency with the sampling frequency of $f_S = 10$ is:

$$f_j = \frac{3}{32} \times 10 \approx 1.0 \text{ Hz}$$

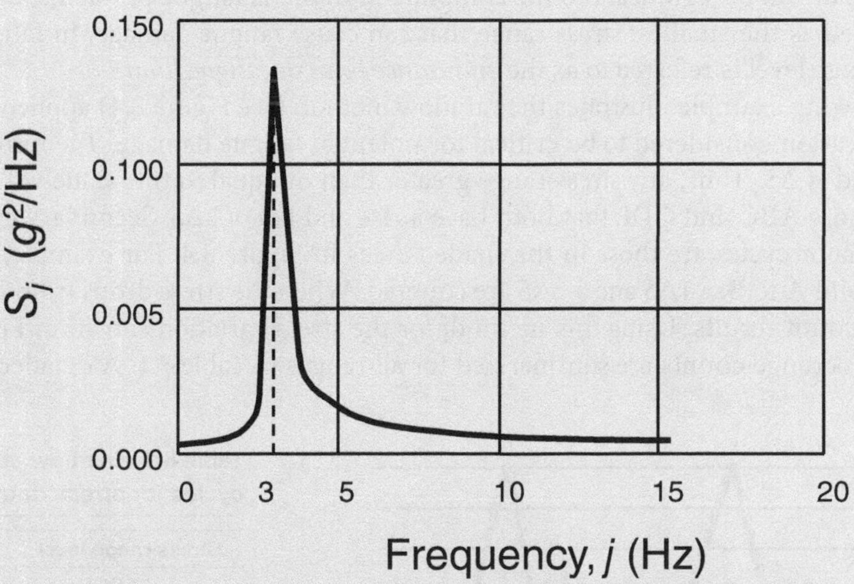

Figure E8.4 Fourier transformation of vibrations with a constant frequency

Cycle-counting process. In some engineering applications, the time-dependent data shows a cyclic fluctuation. In such cases, each cycle of the quantity can be described as a rise to a peak value followed by a fall (or inversely a fall to a valley followed by a rise). The change in the value of the parameter can be considered to be the complete rise from a valley to a peak or vice versa in one cycle of the data. The data compiled for the quantity may be compressed to show the number of times the change in the quantity is equal to a predetermined specific value. Various rules can be set up to count the number of cycles. For example, as a simple rule, any rise in a certain stress value above a specific stress (i.e., S_0) in a structural component can be counted as one cycle. Other rules may also be established for cycle counting. The most popular method, however, is called the *rainflow* method as described by the American Society for Testing and Materials (ASTM).

According to the rainflow method, a quantity (i.e., stress) starting from a reference level (i.e., zero) and rising to a peak should return back to the reference level before a cycle can be counted. A secondary fluctuation of the quantity within a major cycle will also be counted, if a rise followed by a fall (or inversely, a fall followed by a rise) equal to at least a predetermined minimum value of the quantity occurs. For example, if the quantity is the stress at a location in a mechanical or structural component, this minimum value will be taken equal to the smallest stress value that will be significant to the component. In metal fatigue problems, this minimum level is defined as the smallest stress range that can cause fatigue damage. In fatigue analysis, this stress range level is referred to as the *endurance limit* or *fatigue limit*.

The following example illustrates the rainflow method (see Figure 8.4) applied to the stress range at a location considered to be critical for potential fatigue damage. The minimum stress range selected is ΔS. Thus, any stress range greater than or equal to this value will be counted. Major cycles are ABC and CDE that both have a rise and fall of $7\Delta S$. Secondary cycles formed within the major cycles are those in the shaded areas in Figure 8.4. For example, as the stress rises from point A to B, a $1\Delta S$ and a $2\Delta S$ are counted. When the stress drops from D to E, a $1\Delta S$ stress range count results. Using this method, for the stress variation shown in Figure 8.4, the rainflow stress range counts are summarized for all ranges in Table 8.1. A detailed description

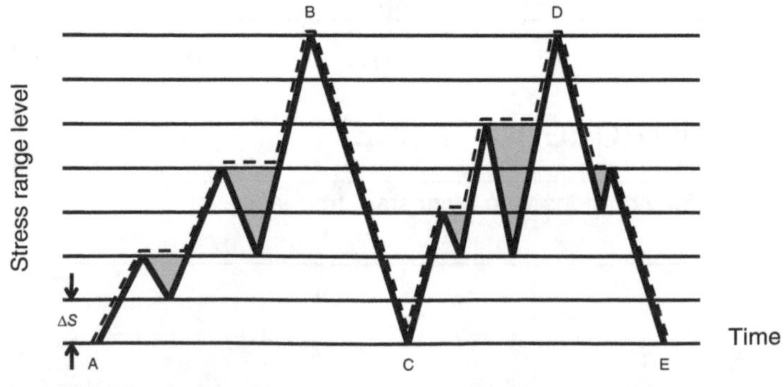

Figure 8.4 Cycle-counting using rainflow

Table 8.1 Rainflow stress range cycles for stress data in Figure 8.4

Stress range level	Cycle count
$7\Delta S$	2
$6\Delta S$	0
$5\Delta S$	0
$4\Delta S$	0
$3\Delta S$	1
$2\Delta S$	1
$1\Delta S$	3

of the method is provided by the ASTM. The cycle-counting method can easily be programmed into the data acquisition system used in the field or laboratory. As a result, compressed data in a form similar to those in Table 8.1 can be obtained as the data are being compiled.

8.3 ESTIMATION OF MEAN AND VARIANCE

As discussed in Chapter 6, the mean and variance are two important parameters describing a random variable. The exact values for these parameters require the knowledge of the probability mass function or the probability density function (depending on whether the random variable is discrete or continuous). Since any set of data essentially can be considered as a population belonging to a random variable, it will be possible to arrive at estimates for the mean and variance by making certain assumptions on the distribution model of the random variable. Methods that are used to estimate the mean and variance are either based on the *point estimation* or *interval estimation* techniques. In conducting the point estimation, one can assume that the distribution model is either uniform; or it follows a frequency-of-occurrence diagram (such as the one shown in Figure E8.1). In such cases, the point estimation is said to be based on *the method of moments*. In another point estimation process, one may use a distribution model along with the compiled data to come up with estimates for the parameters of the distribution function directly through maximization of the likelihood that the function will be able to represent the compiled data. The point estimation based on this technique is called *the method of maximum likelihood*. Of course, once the parameters of the distribution functions are estimated, the estimates can be used to compute the mean and variance. These methods are presented in this section. The interval estimation is discussed in Section 8.4.

Suppose n sample values for a quantity have been compiled as described by $x_1, x_2, \ldots, x_n$. If the sample size n is relatively small, we can assume x_i ($i = 1$ to n) belong to a discrete random variable X with a uniform probability mass function, that is:

$$P(X = x_i) = p(x_i) = c \text{ for all } x_i \tag{8.7}$$

in which c is a constant and $p(x)$ is the probability mass function of the random variable X. Assuming the n sample values make up the entire sample space, then:

$$p(x_1) + p(x_2) + \ldots + p(x_n) = 1 \tag{8.8}$$

or

$$np(x_i) = nc = 1 \tag{8.9}$$

This will result in $c = 1/n$. By applying Equations 6.20 and 6.22, the mean and variance of X can be obtained. However, since this will only be an estimate for the actual mean, we use m to define the mean:

$$m = \sum_{i=1}^{n} x_i p(x_i) = \frac{1}{n}\sum_{i=1}^{n} x_1 = \frac{1}{n}(x_1 + x_2 + \ldots + x_n) \tag{8.10}$$

This equation is the one most often used in computing the average of n values. As is evident, the estimate for the mean is based on the fact that all x_i have the same weight. Using Equation 6.22, the variance can also be estimated. We use s^2 to describe the estimate for the variance (i.e., s is the estimate for the standard deviation):

$$s^2 = \sum_{i=1}^{n} (x_i - m)^2 p(x_i) = \frac{1}{n} \sum_{i=1}^{n} (x_i - m)^2 \tag{8.11}$$

According to Freund (1962), this is only a biased estimate of the variance; an unbiased estimate can be obtained when $(n - 1)$ rather than n is used. In such a case:

$$s^2 = \frac{1}{n-1} \sum_{i=1}^{n} (x_i - m)^2 \tag{8.12}$$

If n is large, it may be more appropriate to compress the data in the form of the frequency-of-occurrence diagram (histogram); and then use the graph as an approximation for the probability mass function of X. By applying Equations 6.20 and 6.22, the effect of the weight of an individual sample value such as x_i in the population will then be considered. Usually, when the data is compressed in the form of a histogram, the frequency-of-occurrence is given for all different ranges of x_i. For example, a frequency-of-occurrence f_i represents the number of times the value of the compiled data was between x_i and x_{i+1}. If f_i is the relative frequency (i.e., the frequency divided by the sample size n), then we can write:

$$f_i \approx P(x_i < X \leq x_{i+1}) \tag{8.13}$$

We can simply assume that $f_i \approx p(x_i)$, if x_i, and x_{i+1} are very close to each other; or otherwise assume $f_i \approx p(x_{im})$ in which $x_{im} = (x_i + x_{i+1})/2$. Then the estimates for the mean and variance will be:

$$m = \sum_{i=1}^{n} x_{im} p(x_{im}) = \sum_{i=1}^{n} f_i x_{im} \tag{8.14}$$

$$s^2 = \sum_{i=1}^{n} (x_{im} - m)^2 p(x_{im}) = \sum_{i=1}^{n} f_i (x_{im} - m)^2 \tag{8.15}$$

Notice that the relative frequencies f_i satisfy the following equation:

$$f_1 + f_2 + \ldots + f_n = 1$$

Example 8.5 In Example 8.1:

(a) Compute the sample data mean and standard deviation, assuming all measured velocities have the same weight.

Continues

Example 8.5: *Continued*

(b) Recompute the sample data mean and standard deviation considering the relative frequencies obtained for the data.

Solution

(a) In this part, Equations 8.10 and 8.12 are used:

$$m = \frac{1}{20}(22 + 35 + 48 + \ldots) = 29 \text{ km/h}$$

$$s^2 = \frac{1}{20-1}[(22-29)^2 + (35-29)^2 + (48-29)^2 + \ldots] = 155.4 \text{ and } s = 12.46 \text{ km/h.}$$

(b) In Example 8.1, six ranges were used and relative frequencies were obtained. We will use Equations 8.14 and 8.15 to compute the mean and standard deviation. Table E8.5 summarizes the ranges, x_{im} for each range and the relative frequencies f_i:

Table E8.5

Range	x_{im}	Relative frequency
0–9	4.5	0.10
10–19	14.5	0.20
20–29	24.5	0.20
30–39	34.5	0.30
40–49	44.5	0.15
50–59	54.5	0.05

From Equation 8.14:

$$m = 0.10 \times 4.5 + 0.20 \times 14.5 + 0.20 \times 24.5 + 0.30 \times 3.45 + 0.15 \times 44.5 + 0.05 \times 54.5 = 28 \text{ km/h}$$

and from Equation 8.15:

$$s^2 = 0.1(4.5-28)^2 + 0.20(14.5-28)^2 + 0.30(24.5-28)^2 + \mathrm{K} + 0.05(54.5-28)^2 = 182.75$$

and $s = 13.52$ km/h.

As is evident from Example 8.5, the estimates for the mean and variance from Equations 8.14 and 8.15 will depend on the ranges selected to establish the frequency-of-occurrence diagram. The use of these equations over Equations 8.10 and 8.12 is preferred in cases where the sample size is large enough so that many narrow ranges can be selected. With a large sample size (i.e., $n \geq 50$), the frequency-of-occurrence diagram will better represent the distribution inherent in the data; and as such, Equations 8.14 and 8.15 may result in more accurate estimates for the mean and variance.

8.4 CONFIDENCE INTERVALS FOR STATISTICAL PARAMETERS

8.4.1 Confidence Intervals for the Mean

The mean and variance computed from the compiled data are considered to be estimates only. If additional data becomes available, the values obtained for m and s^2 will change. This simply means that the exact mean and variance are not known. However, it is possible to establish confidence intervals for the estimated mean, the standard deviation, and for other statistical parameters. For example, the confidence intervals for the mean simply indicate how good our estimate for it is. Usually, confidence intervals provide us with an upper and a lower bound value for the mean. The confidence level associated with these bounds is the probability that the actual mean will be within these bounds. This method of establishing bounds for the mean is called the *interval estimation* method.

Consider again the n sample values $x_1, x_2, \ldots, x_n$ compiled in a data acquisition session. We can assume that each sample value x_i belongs to an independent random variable X_i. The n independent random variables $X_1, X_2, \ldots, X_n$ will have a mean value M:

$$M = \frac{1}{n}(X_1 + X_2 + \ldots + X_n) \tag{8.16}$$

In light of Equation 8.16, we realize that the estimate for the sample mean (m) belongs to the random variable M. The expected value of the random variable M is:

$$E(M) = \frac{1}{n}[E(X_1) + E(X_2) + \ldots + E(X_n)] \tag{8.17}$$

Assuming that all X_i are identically distributed each with a mean value μ and standard deviation σ, Equation 8.17 becomes:

$$E(M) = \frac{n\mu}{n} = \mu \tag{8.18}$$

This indicates that the expected value of the random variable M is equal to the actual mean value of the sample data. Furthermore, the variance of M will be:

$$\mathrm{Var}(M) = \frac{1}{n^2}[\mathrm{Var}(X_1) + \mathrm{Var}(X_2) + \ldots + \mathrm{Var}(X_n)] \tag{8.19}$$

Since $\mathrm{Var}(X_1) = \mathrm{Var}(X_2) = \ldots = \mathrm{Var}(X_n) = \sigma^2$, then:

$$\mathrm{Var}(M) = \frac{n\sigma^2}{n^2} = \frac{\sigma^2}{n} \tag{8.20}$$

This implies that the random variable M has a mean value equal to μ and a standard deviation equal to $\sigma/\sqrt{n}$.

The term $\sigma/\sqrt{n}$ is also known as the *standard error of the mean*. Notice that when n approaches infinity, $\sigma/\sqrt{n}$ will approach zero; and as such, M will become a deterministic (non-random) value. In such a case, the estimate of the mean (i.e., m) will be the exact mean. In practice, $\sigma/\sqrt{n}$ will always be associated with the random variable M and indicate the error corresponding to the estimate of the mean (i.e., m).

When n is sufficiently large, based on the *central limit theorem* (Ang and Tang, 2007), we can assume that M will be a normal random variable with a mean equal to μ and a standard deviation $\sigma/\sqrt{n}$. We now introduce a random variable Z, as:

$$Z = \frac{M - \mu}{\sigma/\sqrt{n}}$$

This random variable follows the standard normal probability density function.

Considering now a specific probability value such as $(1 - \alpha)$, we can select two values $k_{\alpha/2}$ and $-k_{\alpha/2}$ such that the probability of Z being between $k_{\alpha/2}$ and $-k_{\alpha/2}$ will be equal to $(1 - \alpha)$. As seen in Figure 8.5:

$$P\left(-k_{\alpha/2} \leq \frac{M - \mu}{\sigma/\sqrt{n}} \leq k_{\alpha/2}\right) = 1 - \alpha \tag{8.21}$$

This probability is the area under the standard normal probability density function bounded by $k_{\alpha/2}$ and $-k_{\alpha/2}$. Notice that the area before $-k_{\alpha/2}$ and after $k_{\alpha/2}$ each is equal to $\alpha/2$. This means that:

$$k_{\alpha/2} = \Phi^{-1}(1 - \alpha/2) \tag{8.22}$$

This value is referred to as the *critical value* at α. Based on $(1 - \alpha)$ probability, and for a sample size equal to n, from Equation 8.21, the upper and lower bounds for the actual mean μ can be established as:

$$\left(m - \frac{k_{\alpha/2}\sigma}{\sqrt{n}}\right) \leq \mu \leq \left(m + \frac{k_{\alpha/2}\sigma}{\sqrt{n}}\right) \tag{8.23}$$

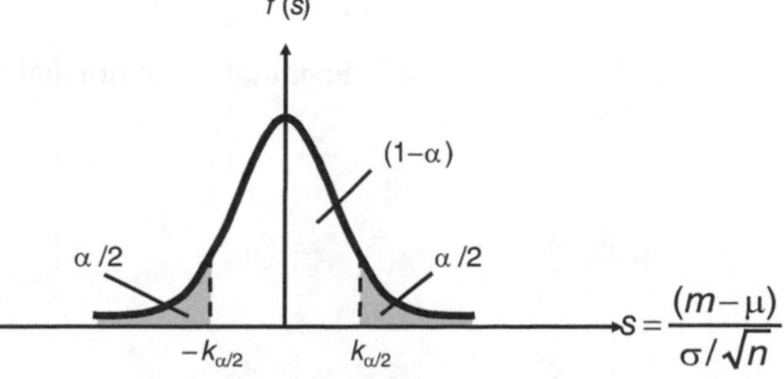

Figure 8.5 Confidence intervals

Equation 8.23 presents an interval estimation for the mean. In deriving Equation 8.23 from Equation 8.22, M was replaced by m, since m belongs to the random variable M. The interval is said to be based on a $(1 - \alpha)$ confidence level. Sometimes the estimated interval for the mean is said to be based on α. This means that the confidence level used to establish the interval is $(1 - \alpha)$.

The interval established for the actual mean is based on the assumption that the sample size is large and that the estimated standard deviation s is equal to the exact standard deviation σ. If the sample size is small, the t-distribution (as described in Chapter 6) is used. The t-distribution depends on the sample size n. As $n \rightarrow \infty$, the t-distribution will approach the normal distribution. Using the t-distribution, and considering the estimated standard deviation s, the interval for the mean μ will be:

$$\left(m - t_{\alpha/2}\frac{s}{\sqrt{n}}\right) \leq \mu \leq \left(m + t_{\alpha/2}\frac{s}{\sqrt{n}}\right) \tag{8.24}$$

in which:

$$t_{\alpha/2} = T_{\nu}^{-1}(1 - \alpha/2) \tag{8.25}$$

In Equation 8.25, T_ν plays the same role as function Φ and is the standard t-probability distribution function. This is to say that the random variable $\dfrac{M - \mu}{s/\sqrt{n}}$ follows the standard t-distribution function.

The parameter ν appearing as a subscript in T_ν is called the degree of freedom and is taken as $(n - 1)$. Values of $t_{\alpha/2}$ can be obtained from Table A.2 of Appendix A for various degrees of freedom.

Example 8.6 In Example 8.5, assume the standard deviation is known and is equal to the estimated value s.

(a) Using a confidence level equal to 99%, establish confidence intervals for the mean.

(b) Repeat (a) but use a 95% confidence level.

Solution

(a) Since $1 - \alpha = 0.99$, $\alpha = 0.01$ and $\alpha/2 = 0.005$. From tables of normal probability values $k_{\alpha/2} = \Phi^{-1}(1 - \alpha/2) = \Phi^{-1}(0.995) = 2.57$

and

$$m + \frac{k_{\alpha/2}\sigma}{\sqrt{n}} = 29 + \frac{2.57 \times 12.46}{\sqrt{20}} = 36.16$$

$$m - \frac{k_{\alpha/2}\sigma}{\sqrt{n}} = 29 - \frac{2.57 \times 12.46}{\sqrt{20}} = 21.48$$

Continues

Example 8.6: *Continued*

Thus, the confidence interval for the mean for 0.99 confidence (or at $\alpha = 0.01$) is $21.84 \le \mu \le 36.16$.

(b) In this part, since $1 - \alpha = 0.95$, $\alpha = 0.05$ and $\alpha/2 = 0.025$. From tables of normal probability values $k_{\alpha/2} = \Phi^{-1}(1 - \alpha/2) = \Phi^{-1}(0.975) = 1.96$

$$m + \frac{k_{\alpha/2}\sigma}{\sqrt{n}} = 29 + \frac{1.96 \times 12.46}{\sqrt{20}} = 34.46$$

$$m - \frac{k_{\alpha/2}\sigma}{\sqrt{n}} = 29 - \frac{1.96 \times 12.46}{\sqrt{20}} = 23.54$$

Thus, the confidence interval for the mean for 0.95 confidence (or at $\alpha = 0.05$) is $23.54 \le \mu \le 34.46$.

Notice that in Example 8.6, when the confidence level was reduced, the bounds for the mean became closer. This indicates that although the intervals estimated for the mean are narrower, we have less confidence in the intervals established.

Example 8.7 Repeat Example 8.6 but use *t*-distribution. This means we assume the standard deviation is not known and we only have an estimate for it.

Solution

(a) For $\alpha/2 = 0.005$, the *t*-distribution value from Table A.2 of Appendix A for a degree of freedom equal to $\nu = 20 - 1 = 19$ is:

$$t_{\alpha/2} = T_{\nu}^{-1}(1 - \alpha/2) = T_{19}^{-1}(0.995) = 2.861$$

Thus, lower and upper bounds for the mean are:

$$m + t_{\alpha/2} = \frac{s}{\sqrt{n}} = 29 + 2.861\frac{12.46}{\sqrt{20}} = 36.97$$

$$m - t_{\alpha/2} = \frac{s}{\sqrt{n}} = 29 - 2.861\frac{12.46}{\sqrt{20}} = 21.03$$

and the confidence interval for the mean for 0.99 confidence (or at $\alpha = 0.01$) is $21.03 \le \mu \le 36.97$.

Continues

Example 8.7: *Continued*

(b) For $\alpha/2 = 0.025$, the t-distribution value from Table A.2 of Appendix A for a degree of freedom equal to $\nu = 20 - 1 = 19$ is $t_{\alpha/2} = T_{19}^{-1}(0.975) = 2.093$, and the lower and upper bounds for the mean are:

$$m + t_{\alpha/2} = \frac{s}{\sqrt{n}} = 29 + 2.093\frac{12.46}{\sqrt{20}} = 34.83$$

$$m - t_{\alpha/2} = \frac{s}{\sqrt{n}} = 29 - 2.093\frac{12.46}{\sqrt{20}} = 23.17$$

The confidence interval for the mean for 0.95 confidence (or at $\alpha = 0.05$) is $21.03 \leq \mu \leq 36.97$.

As the degree of freedom increases, the critical values of the t-distribution become nearly equal to those of the normal distribution for the same α.

8.4.2 One-sided Confidence Limits

In many engineering applications, it may be desirable to establish one-sided confidence limits for the mean. Depending on the type of problem, either a lower or upper confidence limit can be established. Using the normal probability distribution function, with $(1 - \alpha)$ confidence level, the critical value k_α is established as shown in Figure 8.6. In this case:

$$k_\alpha = \Phi^{-1}(1 - \alpha) \tag{8.26}$$

and as shown in Figure 8.6a:

$$P\left(\frac{M - \mu}{\sigma\sqrt{n}} \leq k_\alpha\right) = 1 - \alpha \tag{8.27}$$

This results in the following inequality that describes a one-sided confidence level for the mean:

$$\mu \geq m - \frac{k_\alpha \sigma}{\sqrt{n}} \tag{8.28}$$

This is a lower confidence limit for the mean. Equation 8.28 implies that there is probability that the mean value will be at least equal to $(m - k_\alpha \sigma/\sqrt{n})$ based on a sample size n.

To establish an upper confidence limit, from Figure 8.6b, we observe that:

$$P\left(\frac{M - \mu}{\sigma\sqrt{n}} \geq k_\alpha\right) = 1 - \alpha \tag{8.29}$$

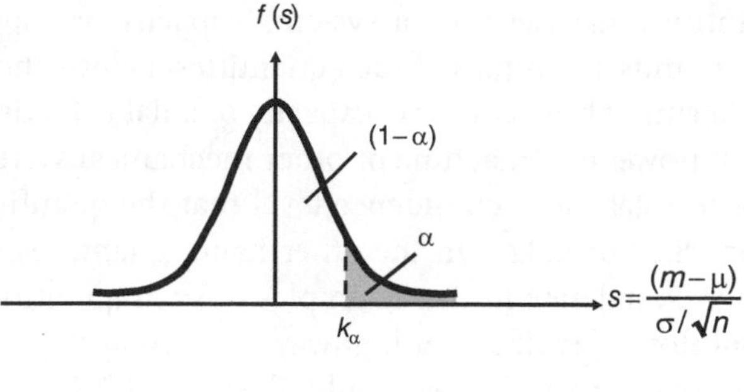

(a) Lower limit

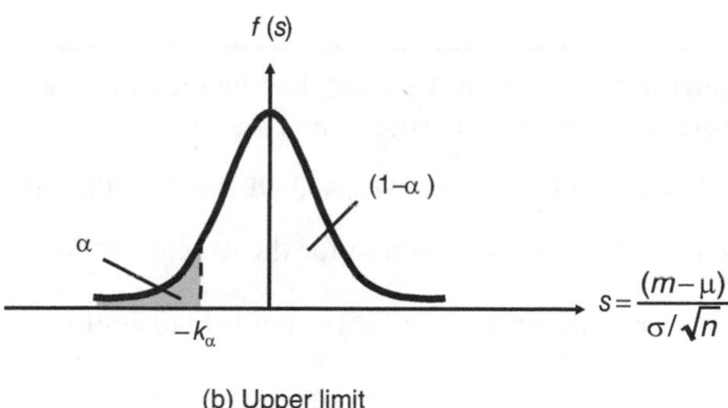

(b) Upper limit

Figure 8.6 One-sided confidence limits (a) lower limit and (b) upper limit)

This results in:

$$\mu \le m + \frac{k_\alpha \sigma}{\sqrt{n}} \tag{8.30}$$

In this case, there is $(1 - \alpha)$ probability that the mean will be at most equal to $(m + k_\alpha \sigma/\sqrt{n})$ based on a sample size n.

Again, for a small sample size and the case where only the estimate for the standard deviation is known, t-distribution will be used. In this case:

$$t_\alpha = T_v^{-1}(1 - \alpha) \tag{8 31 }$$

and

$$\mu \ge m - t_\alpha \frac{s}{\sqrt{n}} \tag{8.32}$$

for the lower confidence limit and

$$\mu \le m - t_\alpha \frac{s}{\sqrt{n}} \tag{8.33}$$

for the upper confidence limit.

In engineering, quantities that describe a system's capacity or *supply* are usually treated with the lower confidence limits. Examples of such quantities include the strength of a material, capacity of a highway in terms of hourly traffic, capacity of a data acquisition system in storing incoming data, the output power of a machine or other mechanical systems, and such. In these examples, we would like to establish a confidence level that the quantity will not be less than the limit set by Equations 8.28 or 8.32. On the other hand, quantities that represent *demand* are treated with the upper confidence limits. Examples of such quantities are the applied load on a structure, the actual flow of traffic in a highway, the actual flow of data entering a data acquisition system, etc. In these examples, we would like to establish a confidence level that the demand will not exceed the upper limit set by Equations 8.30 or 8.33.

Example 8.8 In measuring the maximum seasonal snow load in an area, the records from the past 15 yr are available. These records (using N/m^2) are:

833, 604, 432, 970, 1043, 777, 890, 345, 432, 991, 895, 1300, 1105, 704, 832

(a) Compute the mean and standard deviation for the maximum seasonal snow load.

(b) Establish a 95% upper confidence limit for the snow load assuming the standard deviation is known.

(c) Repeat (b) but assume that only an estimate for the standard deviation is known.

Solution

(a) The estimated mean and standard deviation using the point estimation method are:

$$m = 810.2 \text{ N/m}^2 \text{ and } s = 268.7 \text{ N/m}^2$$

(b) Assuming $\sigma = s = 268.7$, at 0.95 confidence, $\alpha = 0.05$ and $k_\alpha = \Phi^{-1}(0.95) = 1.64$ and

$$m + \frac{k_\alpha \sigma}{\sqrt{n}} = 810.2 + \frac{1.64 \times 268.7}{\sqrt{15}} = 924.0 \text{ N/m}^2$$

Thus, we can say with 95% confidence that the mean snow load will not exceed 924.0 N/m^2.

(c) In this part with $\nu = 15 - 1 = 14$ and with 0.95 confidence, from Table A.2, $t_\alpha = 1.761$ and

$$m + t_\alpha \frac{s}{\sqrt{n}} = 810.2 + \frac{1.761 \times 268.7}{\sqrt{15}} = 932.4 \text{ N/m}^2$$

Thus, the mean snow load will not exceed 932.4 N/m^2 at $\alpha = 0.05$ (which is at 95% confidence).

Example 8.9 In conducting a study of the traffic flow capacity of the northbound lanes in an urban highway, several observations during the peak traffic hours were made at various locations. In each observation a measure of traffic/hr causing traffic congestion was made. After 41 such measurements, the estimated mean value for the traffic/hr causing congestion was obtained as 8044 veh/hr. It was decided that this value can be used as the mean flow capacity. The corresponding standard deviation was 710. Establish a 95% lower confidence level for the mean value of the flow capacity using the *t*-distribution.

Solution

With a degree of freedom equal to $\nu = 41 - 1 = 40$ and $1 - \alpha = 0.95$ confidence, $t_\alpha = 1.684$ and

$$m - t_\alpha \frac{s}{\sqrt{n}} = 8.044 - \frac{1.684 \times 710}{\sqrt{41}} = 7{,}858$$

Thus, the traffic flow capacity of the northbound lanes of this highway will be at least equal to 7858 veh/hr based on $\alpha = 0.05$.

8.4.3 Determination of the Sample Size in an Experiment

An application of the interval estimation is in the determination of the sample size in a laboratory or field experiment. Considering the bounds established for the mean (Equation 8.23), the term $\pm\, k_{\alpha/2}\sigma/\sqrt{n}$ indicates an error term. Dividing this by the mean value μ, the error term will be:

$$\Delta = \frac{(\sigma/\mu)\, k_{\alpha/2}}{\sqrt{n}} = \frac{k_{\alpha/2}\,\delta}{\sqrt{n}} \tag{8.34}$$

in which δ is the coefficient of variation (COV). Solving Equation 8.34 for n:

$$n = \left(\delta \frac{k_{\alpha/2}}{\Delta}\right)^2 \tag{8.35}$$

Thus, by knowing an approximate value for δ, we can decide on the error Δ (above and below the mean) and a confidence level $(1 - \alpha)$ and then estimate the sample size n. If a confidence level equal to 99.75 percent is selected, then $\alpha = 0.0025$ and $k_{\alpha/2} = \Phi^{-1}(0.99875) \approx 3.0$. This results in:

$$n = 9\left(\frac{\delta}{\Delta}\right)^2$$

This equation has been commonly reported as a quick estimate for the sample size in the design of an experiment.

Example 8.10 In estimating the mean compressive strength of a concrete mix, an engineer is interested in knowing the desired sample size for testing. The COV is estimated to be 0.15.

(a) Compute the sample size based on 99.75% confidence and an accepted ±10% error for the mean compressive strength.

(b) If after 17 tests it is determined that the COV is 0.20, determine how many additional tests need to be performed.

Solution

(a) With $(1 - \alpha) = 99.75\%$, $k_{\alpha/2} \approx 3.0$ and with $\Delta = 0.10$, and $\delta = 0.15$, the estimate for the sample size is $n = 9 \times (0.15/0.10)^2 = 21$.

(b) Using $\delta = 0.20$, $n = 9 \times (0.21/0.10)^2 = 36$. Thus, $36 - 17 = 19$ additional tests must be conducted.

8.4.4 Confidence Intervals for Other Statistical Parameters

Using a method similar to that for the mean value, confidence intervals for the variance can also be established. The use of confidence intervals for the variance is limited and, for this reason, the discussion is not presented in this book. The reader is referred to Ang and Tang (2007) for a detailed description and equations for establishing confidence intervals for the variance.

One statistical parameter for which confidence intervals are often established is the *proportion*. We saw in Chapter 6 that for those engineering applications where the probability of occurrence of a sequence of events is required, the binomial distribution is used. The parameter p of the distribution represents the probability of occurrence for one trial. For example, considering the past performance of a construction company in winning jobs, p represents the probability of winning any job among a group of n jobs the company has been bidding on. Recall that this probability is constant for all trials (in this case winning individual jobs). The parameter p is estimated by proportion. The process of establishing confidence intervals for p is explained in this section.

The compiled data in this case constitutes a sample size of n. The n sample values are either 1 or 0, where 1 indicates success and 0 indicates failure. For example, the past record of a construction company shows that it bid for n jobs. If the company won the first job, then $x_1 = 1$; whereas if the company lost the bid on the second job, then $x_2 = 0$; and so on. This type of data is compiled by either reviewing historical records, past occurrences, or simply by observations. For example, we observe and record truck overload cases at a weigh station. If there is an overload, the sample value will be 1; otherwise 0. The parameter p can simply be obtained as the sum of the $x_1, x_2, \ldots, x_n$ divided by n (note that the x_i are either 1 or 0). Since the occurrence or nonoccurrence of the event of concern is independent of subsequent events in the sequence, we can assume that each sample value x_i belongs to an independent random variable X_i. A random variable P is then introduced as:

$$P = \frac{1}{n}\sum_{i=1}^{n} X_i \tag{8.36}$$

This represents the proportion of successful occurrences of the event of concern in n trials. Since P is a random variable, we can compute its expected value as:

$$E(P) = E\left(\frac{1}{n}\sum_{i=1}^{n} X_i\right) = \frac{1}{n}\sum_{i=1}^{n} E(X_i) \tag{8.37}$$

If the exact parameter of the binomial distribution is p, and since the possible values for X_i are either 1 or 0, we can write:

$$E(X_i) = 1(p) + 0(1 - p) = p \tag{8.38}$$

and

$$\text{Var}(X_i) = (1 - p)^2 p + (0 - p)^2 (1 - p) = p(1 - p) \tag{8.39}$$

Thus:

$$E(P) = \frac{np}{n} = p \tag{8.40}$$

Equation 8.40 indicates that the expected value of P is equal to the exact p. Furthermore, the variance of P is:

$$\text{Var}(P) = \frac{1}{n^2}\sum_{i=1}^{n} \text{Var}(X_i) = \frac{1}{n^2}\left[\sum_{i=1}^{n} E(X_i^2) - E^2(X_i)\right] \tag{8.41}$$

Since $\text{Var}(X_i) = p(1 - p)$, then:

$$\text{Var}(P) = \frac{1}{n^2}[n(p - p^2)] = \frac{p(1 - p)}{n} \tag{8.42}$$

This indicates that the random variable P has a variance that will approach zero at large n. For large n, P can be assumed to follow a normal probability distribution function, based on the central limit theorem. This assumption can be used to establish confidence intervals for the exact parameter of the binomial distribution p. If the estimate for p is available and is denoted as $\bar{p}$, then we can assume that the mean and standard deviation of P are approximately equal to $\bar{p}$ and $\sqrt{\bar{p}(1 - \bar{p})/n}$, respectively. Hence with $(1 - \alpha)$ confidence level:

$$P\left(-k_{\alpha/2} \le \frac{\bar{p} - p}{\sqrt{\bar{p}(1 - \bar{p})/n}} \le k_{\alpha/2}\right) = 1 - \alpha \tag{8.43}$$

From Equation 8.43, confidence intervals (at α) are established for the parameter p as:

$$\bar{p} - k_{\alpha/2}\sqrt{\frac{\bar{p}(1-\bar{p})}{n}} \leq p \leq \bar{p} + k_{\alpha/2}\sqrt{\frac{\bar{p}(1-\bar{p})}{n}} \tag{8.44}$$

Notice that for large n, $p \approx \bar{p}$.

Example 8.11 A small company manufactures prestressed, prefabricated floor joists. For quality control purposes, samples of manufactured joists are selected randomly and tested for the desired load capacity of 45 kN. The past data on 120 beams tested thus far indicated failures in 9 beams. Failure means a beam does not meet the desired capacity.

(a) Assuming the binomial distribution represents the failure among joists; estimate the parameter p of the distribution.

(b) Establish 95% confidence intervals for p.

Solution

(a) From the data provided:

$$\bar{p} = \frac{9}{120} = 0.075$$

(b) At 95% confidence, $\alpha = 0.05$ and $k_{\alpha/2} = 1.96$. Furthermore:

$$\sqrt{\bar{p}(1-\bar{p})/n} = \sqrt{\frac{0.075(1-0.075)}{120}} = 0.024$$

Thus:

$$\bar{p} - k_{\alpha/2}\sqrt{\frac{\bar{p}(1-\bar{p})}{n}} = 0.075 - 1.96 \times 0.024 = 0.028$$

and

$$\bar{p} + k_{\alpha/2}\sqrt{\frac{\bar{p}(1-\bar{p})}{n}} = 0.075 + 1.96 \times 0.024 = 0.122$$

Hence, $0.028 \leq p \leq 0.122$ based on $\alpha = 0.05$ (i.e., 95 percent confidence).

8.4.5 Determination of Sample Size for Estimating Proportions

In Section 8.4.3, we discussed a method for estimating the number of samples that need to be obtained in an experiment. In field observations and surveys where proportions are desired, we can select a limited sample size for estimating a proportion. However, the selected sample size must be capable of statistically representing the entire data population. As an example, consider a situation where the proportion of overloaded trucks on a highway is desired. We can keep track of all overloaded trucks on a continuous basis and update the proportion of overloaded trucks periodically. This method will be costly and may be unnecessary. Instead, we can randomly select a limited sample size and arrive at an estimate for the proportion within certain accepted marginal error and based on a desired confidence level. For this purpose we will utilize Equation 8.44 that was found for establishing bounds for the proportion p. Notice that in Equation 8.44 the term $k_{\alpha/2}\sqrt{\bar{p}(1-\bar{p})/n}$ is an indication of the difference between $\bar{p}$ and the actual p. Dividing this term by $\bar{p}$ and introducing the result by Δ we obtain the following:

$$\Delta = k_{\alpha/2}\sqrt{\frac{\bar{p}(1-\bar{p})}{\bar{p}^2 n}} \tag{8.45}$$

which means the ratio $p/\bar{p}$ differs by at most Δ. Solving Equation 8.45 for n results in:

$$n = \frac{1-\bar{p}}{\bar{p}}\left(\frac{k_{\alpha/2}}{\Delta}\right)^2 \tag{8.46}$$

As is evident, this equation depends on $\bar{p}$. Thus, it is applicable only if prior knowledge on the value of $\bar{p}$ is available. In many situations, $\bar{p}$ is not known. In these cases n can be determined by using certain assumptions. Recall that the term $\bar{p}(1-\bar{p})$ is the variance associated with the proportion. The maximum possible value for this term is 0.25 that corresponds to a $\bar{p}$ value equal to 0.5. Using this $\bar{p}$ value in Equation 8.46 will result in

$$n = \left(\frac{k_{\alpha/2}}{\Delta}\right)^2 \tag{8.47}$$

This equation is, of course, an approximation. If no prior knowledge on $\bar{p}$ is available, one may wish to use Equation 8.47 for estimating n. After compiling a reasonable amount of data, $\bar{p}$ can be computed and used along with Equation 8.46 for computing a revised value for n.

As indicated earlier, the term $\bar{p}(1-\bar{p})$ is the variance associated with the proportion. The standard error of the proportion $\bar{p}$ or the *sampling error* will then be:

$$\text{Sampling Error} = \sqrt{\bar{p}(1-\bar{p})/n} \tag{8.48}$$

This is similar to the standard error of the mean discussed in Section 8.4.1. As $n \to \infty$, the sampling error approaches zero and the estimate for the proportion becomes equal to the actual proportion.

Example 8.12 In selecting the sample size for the overloaded trucks along a highway, we wish to limit the error for the proportion p to ±10% (i.e., $\Delta = 0.10$). From previous data we know that the estimate for the proportion of the overloaded truck for this highway is 0.10 (i.e., $\bar{p} = 0.1$). Estimate the sample size n for a 99.75% confidence.

Continues

Example 8.12: *Continued*

Solution

For $(1 - \alpha) = 0.9975$, $\alpha/2 = 0.0025$ and $k_{\alpha/2} \approx 3$. Thus, from Equation 8.46:

$$n = \frac{1 - 0.10}{0.10}\left(\frac{3.0}{0.10}\right)^2 = 8100$$

Example 8.13 In conducting a survey of the proportion of commercial motor vehicles (CMVs) involved in traffic violations, a statistical sample size is desired.

(a) If no prior knowledge on $\bar{p}$ is available, estimate the sample size n for a 90% confidence and ±5% limit on the error for the actual proportion of CMVs involved in traffic violations.

(b) If after 400 samples selected at random, it is determined that $\bar{p} = 0.25$, compute the sampling error.

(c) If after 400 samples selected at random, it is determined that $\bar{p} = 0.25$, recompute it for the same requirements in (a).

Solution

(a) In this part, we use Equation 8.47 since no prior information on $\bar{p}$ is available. For a 90% confidence, $\alpha = 0.1$ and $k_{\alpha/2} = 1.65$. Thus:

$$n = (1.65/0.05)^2 = 1089$$

(b) With $n = 400$ and $\bar{p} = 0.25$, the sampling error is:

$$\text{Sampling Error} = \sqrt{\frac{0.25(1 - 0.25)}{400}} = 0.022$$

(c) In this part, since now a prior knowledge on $\bar{p}$ is available, we use Equation 8.46:

$$n = \frac{1 - 0.25}{0.25}\left(\frac{1.65}{0.05}\right)^2 = 3267$$

Thus, additional samples needed will be 3267 – 400 = 2867.

8.5 ESTIMATION OF STATISTICAL PARAMETERS USING THE METHOD OF MAXIMUM LIKELIHOOD

This method is used for estimating the parameters of a desired distribution model. The collected data are used to maximize the likelihood that the data will be represented by the distribution model. Once the parameters are estimated, Equations 6.20-6.23 may be used to estimate the mean and variance of the sample population.

Suppose $x_1, x_2, \ldots, x_n$ are n values compiled for a quantity. The data can be represented by a probability density function in the form of $f(x, \theta_1, \theta_2, \ldots, \theta_m)$ in which θ_i ($i = 1$ to m) are m parameters of the density function. A *likelihood* function L is then defined as:

$$L = f(x_1, \theta_1, \theta_2, \ldots, \theta_m)\, f(x_2, \theta_1, \theta_2, \ldots, \theta_m) \ldots f(x_n, \theta_1, \theta_2, \ldots, \theta_m) \tag{8.49}$$

The function L is maximized for parameters θ_i ($i = 1$ to m). This can be done by using the following set of equations:

$$\frac{\partial L}{\partial \theta_j} = 0 \quad (i = 1, 2, \ldots, m) \tag{8.50}$$

For certain functions it may be easier to use $\ell n(L)$ as the maximization function. Equation 8.49 assumes that x_i values belong to independent random variables taken from the probability density function f. The probability that $X_1 = x_1, X_2 = x_2, \ldots, X_n = x_n$ is proportional to the product L. Thus, Equation 8.50 will result in the optimization of this probability.

Example 8.14 The vehicle waiting time at a railroad crossing is being investigated. This is the time (in seconds) that vehicles have to wait before the crossing is clear and safe to pass. Twelve observations at several crossings are made and the waiting time is recorded. The results are:

25, 60, 73, 55, 62, 75, 33, 27, 58, 43, 62, 54.

(a) Using the method of maximum likelihood and assuming a lognormal probability density function, estimate the parameters and of the distribution model.

(b) Using the results in (a), compute the mean and standard deviation of the data and compare them with those from the point estimation method.

Solution

(a) In this problem, $m = 2$ (i.e., $\theta_1 = \lambda$, and $\theta_2 = \nu$). The function L will be:

$$L = \prod_{i=1}^{n} \frac{1}{\nu\, x_i \sqrt{2\pi}} \exp\left[-\frac{1}{2}\left(\frac{\ell n\, x_i - \lambda}{\nu}\right)\right]$$

$$= (\nu\sqrt{2n})^{-n} \frac{1}{x_1 x_2 \ldots x_n} \exp\left\{-\frac{1}{2}\left[\left(\frac{\ell n\, x_1 - \lambda}{\nu}\right)^2 + \left(\frac{\ell n\, x_2 - \lambda}{\nu}\right)^2 + \ldots\right]\right\}$$

Using $\ell n(L)$, we have:

$$\ell n(L) = \ell n(\nu\sqrt{2n})^{-n} + \ell n\left(\frac{1}{x_1 x_2 \ldots x_n}\right) - \frac{1}{2}\left[\left(\frac{\ell n\, x_1 - \lambda}{\nu}\right)^2 + \left(\frac{\ell n\, x_2 - \lambda}{\nu}\right)^2 + \ldots\right]$$

Continues

Example 8.14: *Continued*

Conducting the maximization:

$$\frac{\partial \ell n(L)}{\partial \lambda} = 0 \quad \text{and} \quad \frac{\partial \ell n(L)}{\partial v} = 0$$

These equations will result in the following equations for λ and v:

$$\lambda = \frac{1}{2}\sum_{i=1}^{n} \ell n(x_i) \quad \text{and} \quad v^2 = \frac{1}{2}\sum_{i=1}^{n} [\ell n(x_i) - \lambda]^2$$

In this problem, $n = 12$. Now using the 12 sample values:

$$\lambda = (\ell n25 + \ell n60 + \ell n73 + \ldots)/12 = 3.90$$

$$v^2 = \frac{1}{12}[(\ell n25 - 3.90)^2 + (\ell n60 - 3.90)^2 + (\ell n73 - 3.90)^2 + \ldots] = 0.127 \text{ and } v = 0.356$$

(b) The relationship between the mean (μ) and the COV (δ) and λ and v is:

$$\lambda = \ell n(\mu) - v^2/2 \quad \text{and} \quad v^2 = \ell n(1 + \delta^2)$$

Using these equations will result in $\mu = 52.9$, $\delta = 0.368$ and standard deviation $\sigma = 19.5$. Using the method of moments, the estimates for the mean and standard deviation are $m = 52.25$ and $s = 16.73$ (from Equations 8.10 and 8.12). Notice that the latter values correspond to assuming a uniform distribution for the sample data.

8.6 TESTING DATA FOR A DESIRED DISTRIBUTION MODEL

As discussed in Section 8.5, the knowledge of the type of distribution model that can best represent the data will be needed before one can estimate the parameters describing the model. Often we can use simple methods in arriving at a quick estimate as to what type of probability distribution functions may be suitable to represent the data. More elaborate methods may then be employed for testing these models to determine the one that best represents the data. As is evident, the process of estimating the parameters of a distribution model is done simultaneously with the process of determining the type of model. In this section, several methods for identifying and testing the validity of probability distribution models are presented. We begin by introducing simple methods and conclude the section with the description and application of more elaborate methods that are used for testing the validity of a desired distribution model.

8.6.1 Branch-and-leaves Method

The branch-and-leaves method offers a quick way at guessing what the shape of the distribution may be. No prior arrangement of the data will be necessary. For simplicity, all data values may be multiplied by a common factor. The data values can also be rounded up to the nearest whole number. To apply this method, a two-column table is formed. In the first column the *branches* are listed; the second column contains the *leaves* corresponding to the individual branches. The leaves are the digits to the right of the individual data values; the branches are the remaining digits. The leaves for a given branch are listed in the second column in a row (without any space or other separators between them). When all the data are processed, the cluster of leaves will form the shape of a distribution model. Example 8.15 illustrates this method.

Example 8.15 For the data in Example 8.2, use the branch-and-leaves method and provide a preliminary guess at the type of distribution model that may be suitable for the data.

Solution To apply this method, we first round up sample values to form the following data set:

22.8, 21.7, 23.8, 23.5, 21.7, 25.2, 22.6, 24.2, 24.9, 21.0, 25.0, 25.7, 20.8, 23.9, 24.0, 22.4, 27.6, 21.3, 22.2, 24.0, 23.8, 26.8, 20.4, 22.8, 23.4.

Since the smallest sample value is 20.4 and the largest 27.6, the branches are:

20, 21, 22, 23, 24, 25, 26 and 27.

Considering 20 as a branch, the only leaves are 4 and 8; considering 21, the leaves are 7, 7, 0, and 3, respectively. We then arrange all branches and leaves:

Branches	Leaves
20	48
21	7703
22	86428
23	85984
24	2900
25	207
26	8
27	6

The cluster of leaves shows a nonsymmetric distribution. A lognormal or Rayleigh distribution is perhaps suitable for this set of data. Notice the resemblance of the distribution to the graph shown in Figure E8.2. Further testing of a lognormal or Rayleigh distribution will be required to determine the validity of either distribution model to represent the data.

8.6.2 Determination of the Shape of Distribution from Frequency-of-occurrence Diagram

In this method, the data needs to be sorted and plotted in the form of a frequency-of-occurrence diagram. The frequency bars serve to point to the shape of a suitable diagram.

In Example 8.2 it is evident that a lognormal or Rayleigh distribution may be suitable for representing the data.

8.6.3 Probability-scaled Plotting

In this method, a cumulative distribution of the collected data is developed by sorting the data in ascending order. This cumulative distribution is then plotted against the cumulative values from a theoretical distribution model that is being considered. A straight line is an indication that the distribution model selected is suitable to represent the data. The cumulative values from the theoretical distribution can be used to actually construct a probability-scaled coordinate system as a tool for hand plotting the data. This type of plotting system is referred to as the probability paper. The plotting positions are obtained through the following model that is based on a stepwise cumulative probability function (note that the data values are first arranged in ascending order, i.e., $(x_1 < x_2 < x_3 < \ldots < x_n)$.

$$\begin{aligned} p_i &= 0 && \text{for} \quad x_i < x_1 \\ p_i &= \frac{i}{n+1} && \text{for} \quad x_1 \le x_i \le x_n \\ p_i &= 1 && \text{for} \quad x_i > x_n \end{aligned} \tag{8.51}$$

The x_i values are then plotted against the z_i values that are obtained from the following equation:

$$P(X \le z_i) = p_i \tag{8.52}$$

If the standard form of the theoretical probability distribution function that is being considered is $F_S(x)$, the z_i values are:

$$z_i = F_S^{-1}(p_i) \tag{8.53}$$

For example, if the normal probability distribution function is being considered, then:

$$z_i = \Phi^{-1}(p_i) \tag{8.54}$$

Example 8.16 Use the data in Example 8.15 and prepare a probability-scaled plot to determine whether the normal distribution will be suitable to represent the data.

Solution We must first rearrange the data in ascending order. This means $x_1 = 20.4$, $x_2 = 20.8$, $x_3 = 21.0$, ..., and $x_n = 27.6$. in which $n = 25$. From Equation 8.51, $p_1 = 1/26$, $p_2 = 2/26$, $p_3 = 3/26$, ..., and $p_n = 25/26$. Furthermore, from Equation 8.54:

$$\begin{aligned} z_1 &= \Phi^{-1}(1/26) = -1.77 \\ z_2 &= \Phi^{-1}(2/26) = -1.43 \\ &\ldots \\ z_n &= \Phi^{-1}(25/26) = -1.77 \end{aligned}$$

Continues

Example 8.16: *Continued*

A summary of the values for the sample data x_i, p_i, and z_i are provided in Table E8.16. Figure E8.16 presents a plot of x_i versus z_i values. As seen in the figure, a straight line can be used to approximately pass through the data points. This indicates that perhaps the normal probability density function may be used to represent the data.

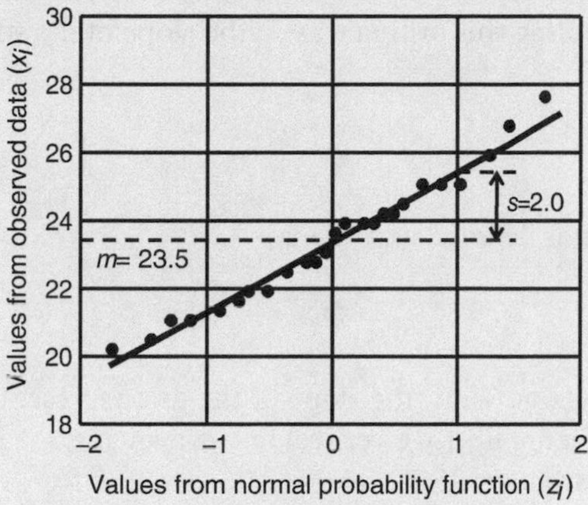

Figure E8.16 Probability-scaled plotting

Table E8.16

x_i	Rank	p_i	z_i	x_i	Rank	p_i	z_i
20.4	1	1/26 = 0.038	−1.77	23.8	14	0.538	0.10
20.8	2	2/26 = 0.077	−1.43	23.8	15	0.577	0.19
21.0	3	3/26 = 0.115	−1.20	23.9	16	0.615	0.29
21.3	4	0.153	−1.03	24.0	17	0.654	0.40
21.7	5	0.192	−0.87	24.0	18	0.692	0.50
21.7	6	0.231	−0.74	24.2	19	0.731	0.62
22.2	7	0.269	−0.62	24.9	20	0.769	0.74
22.4	8	0.308	−0.50	25.0	21	0.808	0.87
22.6	9	0.346	−0.40	25.2	22	0.846	1.03
22.8	10	0.385	−0.29	25.7	23	0.885	1.20
22.8	11	0.423	−0.19	26.8	24	0.923	1.43
23.4	12	0.462	−0.10	27.6	25	0.962	1.77
23.5	13	0.500	0				

Notice that for the normal probability density function, $P(X \leq \mu) = 0.5$. This probability value corresponds to $z_i = 0$. This indicates that the mean value of the sample data can be obtained by

Continues

Example 8.16: *Continued*

finding the value of x_i (i.e., from the ordinate of the probability-scaled plot) that corresponds to $z_i = 0$. In Figure E8.16, the estimated mean $m = 23.5$. If the plot is represented by a straight line, the mean value can be estimated by plotting a vertical line at $z_i = 0$ and reading the ordinate at the intersection of the vertical line and the straight line representing the $x_i - z_i$ relation. Furthermore, the ordinate corresponding to $z_i = 1$ can be used to estimate the standard deviation of the sample data. Denoting this ordinate as x_s, the slope of the straight line representing the plot will be equal to:

$$\text{Slope} = \frac{x_s - m}{1 - 0} = x_s - m$$

If the standard deviation is denoted as s, then:

$$P(X \leq x_s) = \Phi\left(\frac{x_s - m}{s}\right) = \Phi(1)$$

This indicates that $s = x_s - m$. Thus, the slope of the line representing the plot is equal to the standard deviation. In Figure E8.16, a vertical line drawn at $z_i = 1$ intersects the graph at 25.3. Thus, the standard deviation $s = 25.3 - 23.3 = 2.0$.

A similar technique can also be used to plot the data against the values from lognormal probability distribution function. In this case, rather than plotting x_i versus z_i, the logarithm of x_i is plotted against z_i. If the plot shows a straight line, the data can be represented with a lognormal probability density function. It can be shown that the ordinate of the straight-line plot at $z_i = 0$ corresponds to the sample data median, and the slope of the line corresponds to the parameter ν of the lognormal distribution. More discussion on this subject can be found in Ang and Tang (2007) under the subject of probability papers.

8.6.4 Chi-square Goodness-of-fit Test for Testing Validity of a Distribution Model

The techniques presented in Sections 8.6.1 to 8.6.3 are primarily intended as quick methods of arriving at a first estimate for the type of distribution model that may be suitable for a set of data. In this section, the chi-square goodness-of-fit test is explained. This is a statistical test that is conducted on the data against a desired theoretical distribution model. The differences between the *observed* frequencies (computed from the data) and the *theoretical* frequencies (computed from the theoretical distribution model) are calculated. These differences are then combined and represented as a single value. This value is then compared with a critical value to determine whether the test passes. The critical value is based on a predetermined *significance level*. To apply this method, the data is arranged in the form of occurrence frequencies as described in Section 8.2.1. For each range such as i, the theoretical distribution (that is being tested) is used to compute a theoretical frequency e_i as:

$$e_i = nP(x_{i-1} < X \leq x_i) \tag{8.55}$$

in which $(x_{i-1} < X \leq x_i)$ represents the limits of range i, and n is the sample size. The observed frequency for this range is n_i, and is simply the number of sample values of the data set that are within the $(x_{i-1} < X \leq x_i)$ limits. The differences between the theoretical frequencies and the observed frequencies are obtained for all the ranges and combined to form a value c according to the following equation:

$$c = \sum_{i=1}^{k} \frac{(n_i - e_i)^2}{e_i} \tag{8.56}$$

in which k is the total number of ranges. This value c belongs to a random variable C that follows a chi-square distribution. The chi-square probability density function is a one-tailed function that depends on the sample size. The density function is:

$$f(c) = \frac{1}{2^{\nu/2}\Gamma(\nu/2)} c^{(\nu/2-1)} e^{-c/2} \quad \text{for } c > 0 \tag{8.57}$$

in which ν is an integer describing the degrees of freedom and Γ is the gamma function. Using Equation 8.57, a critical value for a predetermined probability level (i.e., $1 - \alpha$) can be established. This critical value is said to be based on α significance. The degree of freedom used in Equation 8.57, when used for testing the validity of a distribution model, is taken as $\nu = k - 1$, if the parameters of the distribution (that is being tested) are known. However, if only the estimates for the parameters are available, $\nu = k - 1 - r$ in which r is the number of parameters in the distribution model (for example, $r = 1$ for the Poisson distribution and $r = 2$ for the normal and lognormal distribution). Notice that in computing ν, the number of ranges k is used (and not the sample size n). The ranges are selected in such a way that they cover all possible values of the random variable in the theoretical distribution model. For example, if the normal distribution is being tested against a set of data, ranges covered are between $\pm\infty$. The ranges do not need to be at equal intervals. The minimum number of ranges suggested is five.

The critical value at α significance is $C_{1-\alpha,\nu}$ and satisfies the following equation:

$$P(C \leq C_{1-\alpha,\nu}) = 1 - \alpha \tag{8.58}$$

Thus, if the computed c is less than $C_{1-\alpha,\nu}$, the value of c is too small to happen by chance; and as such, the test passes. This means that the theoretical distribution is suitable for representing the data. The chi-square critical values $C_{1-\alpha,\nu}$ are provided in Table A.3 in Appendix A.

Example 8.17 In an investigation of productivity in a construction project, the completion times for 12 identical projects were evaluated. The data given shows the completion times for the 12 projects in days:

141, 137, 124, 121, 132, 136, 139, 141, 144, 152, 161, 118

Conduct a chi-square test to determine if the normal distribution is suitable for the data at $\alpha = 0.05$ significance.

Solution Using the point estimation method, the estimates for the mean and standard deviation (i.e., the parameters of the normal distribution function) are $m = 137.2$ and $s = 12.4$ da, respectively. Since these are estimates, the actual parameters of the distribution are considered to be unknown. Next, we use five ranges ($k = 5$). These ranges, the corresponding observed and theoretical frequencies and the terms $(n_i - e_i)^2/e_i$, are listed in Table E8.17.

The critical value of chi-square at $\alpha = 0.05$ and $\nu = 5 - 1 - 2 = 2$ degrees of freedom is 5.99 (from Table A.3 in Appendix A). Since $c = 0.1027 < 5.99$, the test passes.

Table E8.17

Range	Observed frequency (n_i)	Theoretical frequency (e_i)*	$(n_i - e_i)^2/e_i$
$X \le 120$	1	0.99	0.0001
$120 < X \le 130$	2	2.38	0.0610
$130 < X \le 140$	4	3.72	0.0210
$140 < X \le 150$	3	3.09	0.0026
$X > 150$	2	1.82	0.0180
Sum	12	12.00	0.1027

* The theoretical frequencies are computed:
$e_1 = 12P(X \le 120) = 12\Phi[(120 - 137.2)/12.4] = 12 \times 0.0823 = 0.99$
$e_2 = 12P(120 < X \le 130) = 12\{\Phi[(130 - 137.2)/12.4] - \Phi[(120 - 137.2)/12.4]\} = 12 \times (0.2810 - 0.0823) = 2.38$ and so on.

8.6.5 Kolmogorov-Smirnov Test

The Kolmogorov-Smirnov (KS) test can also be used to test the validity of a distribution model. This test is preferred over the chi-square test because it is simpler and does not require the data be arranged in ranges. In this method, the cumulative distribution values $S_n(x_i)$ for the test data are computed using a simple stepwise model. Furthermore, using the theoretical distribution model $F(x_i)$ that is being tested, the cumulative probability values for all x_i are also computed.

For a given sample value such as x_i, the difference (D_i) between the cumulative values from the observed data and from the theoretical distribution model is computed as:

$$D_i = | S_n(x_i) - F(x_i) | \tag{8.59}$$

Considering all sample values, $x_1, x_2, \ldots, x_n$, the respective differences will be $D_1, D_2, \ldots, D_n$. The maximum among these n values D_{max} is a random variable that follows the KS distribution. The critical value from the KS distribution is obtained based on a significance level α and the sample size n. This critical value is denoted as $D_{\alpha,n}$. If D_{max} is less than this critical value, the KS test passes because D_{max} is too small to occur by chance. This implies that:

$$P(D_{max} \leq D_{\alpha,n}) = 1 - \alpha \tag{8.60}$$

which means there is (1 – α) probability that the maximum difference will be less than the critical value. If the test passes at α significance, we can decide that the theoretical distribution model is appropriate to represent the sample data. The values of $D_{\alpha,n}$ are provided in Table A.4 in Appendix A. To obtain the cumulative values for the observed data, first the sample data is arranged in ascending order such that $x_1 < x_2 < \ldots < x_n$. The cumulative values are then computed using the following stepwise function:

$$\begin{aligned} S_n(x) &= 0 \quad \text{for} \quad x < x_i \\ S_n(x) &= \frac{i}{n} \quad \text{for} \quad x_i \leq x \leq x_{i+1} \\ S_n(x) &= 1 \quad \text{for} \quad x > x_n \end{aligned} \tag{8.61}$$

Example 8.18 Seventeen samples of a new stainless steel wire were tested for tensile strength in kN. The following results were obtained:

153.0, 95.0, 123.0, 122.0, 108.4, 131.0, 138.1, 131.5, 107.0, 117.0, 126.0, 125.8, 112.3, 129.1, 124.8, 110.3, 140.0

(a) Estimate the mean and standard deviation of the tensile strength of the wire.

(b) Estimate the parameters λ and ν from the results in (a), if we assume a lognormal distribution will represent the data.

(c) Conduct a KS test to determine if the lognormal distribution will pass at α = 0.05 significance.

Solution

(a) Using the point estimation method, the mean (m) and the standard deviation (s) are, respectively:

m = 123.2 kN, and s = 14.1 kN.

Continues

Example 8.18: *Continued*

(b) The relationship between the mean (μ) and the COV (δ) and λ and ν is:

$$\lambda = \ell n(\mu) - \nu^2 \quad \text{and} \quad \nu^2 = \ell n(1 + \delta^2)$$

Using the estimates for the mean and standard deviation, $\delta = 14.1/123.2 = 0.114$. Thus:

$$\nu^2 = \ell n[1 + (0.114)^2] = 0.0129 \quad \text{and} \quad \nu = 0.114 \quad \text{and} \quad \lambda = \ell n(123.2) - (0.114)^2/2 = 4.807$$

(c) In this part we rearrange the data in ascending order; then we compute the cumulative values from the data using Equation 8.61 and from the lognormal distribution. For example, since $x_1 = 95.0$, then $S_n(x_1) = 1/17 = 0.0588$ and

$$F(x_1) = P(X \leq 96.0) = \Phi\left(\frac{\ell n95 - 4.807}{0.114}\right) = 1 - \Phi(2.22) = 1 - 0.9868 = 0.0132$$

Thus, $D_1 = |0.0588 - 0.0132| = 0.0456$. Table E8.18 summarizes the cumulative values and the computed differences for all sample values.

As seen in Table E8.18, the maximum difference $D_{max} = 0.0878$. The critical KS value at $\alpha = 0.05$ and $n = 17$ is taken from Appendix A. (i.e., $D_{0.05,17} = 0.32$). Since $D_{max} < D_{0.05,17} = 0.32$, we conclude that the test passes. This means that the lognormal distribution will be appropriate to represent the data.

Table E8.18

Sample data	$S_n(x)$	$F(x)$	$D_i = \lvert S_n(x) - F(x)\rvert$
$x_1 = 95.0$	1/17 = 0.0588	0.0132	0.0456
$x_2 = 107.0$	2/17 = 0.1176	0.1190	0.0014
$x_3 = 108.4$	3/17 = 0.1765	0.1446	0.0319
$x_4 = 110.3$	4/17 = 0.2353	0.1814	0.0539
$x_5 = 112.3$	5/17 = 0.2941	0.2266	0.0675
$x_6 = 117.0$	6/17 = 0.3529	0.3483	0.0046
$x_7 = 122.0$	7/17 = 0.4118	0.3974	0.0144
$x_8 = 123.0$	8/17 = 0.4706	0.5199	0.0493
$x_9 = 124.8$	9/17 = 0.5294	0.5675	0.0381
$x_{10} = 125.8$	10/17 = 0.5882	0.5948	0.0066
$x_{11} = 126.0$	11/17 = 0.6470	0.6026	0.0444
$x_{12} = 129.1$	12/17 = 0.7059	0.6808	0.0251
$x_{13} = 131.0$	13/17 = 0.7647	0.7257	0.0390
$x_{14} = 131.5$	14/17 = 0.8235	0.7357	0.0878
$x_{15} = 138.1$	15/17 = 0.8824	0.8554	0.0270
$x_{16} = 140.0$	16/17 = 0.9412	0.8810	0.0602
$x_{17} = 153.0$	17/17 = 1.0	0.9750	0.0250

SUMMARY

In this chapter several basic topics in statistics were discussed. The main objective of the chapter was to describe methods that are commonly used to analyze a set of data compiled through experimental investigations, field testing, or interviews with experts. The methods by which the data is compressed into more meaningful results were presented. The estimation of statistical parameters describing the sample data was also discussed. Point estimation and interval estimation methods for the mean and other statistical parameters were presented. The analyses presented in this chapter are especially important in decision making and design in engineering problems. Data compiled for various engineering parameters is essential and supplementary to the modeling inherent in engineering decision making and design. Through the compiled data and relevant statistical analyses, an engineer is able to provide the supporting elements needed to arrive at meaningful conclusions that bear the uncertainty inherent in the data. The materials presented in this chapter also provide the background needed for determining the type of probability distribution model and the estimate for key statistical parameters that make up the theories in Chapters 5, 6, and 7.

REFERENCES

Ang, A. H.-S. and W. H. Tang (2007). Chapters 5 and 6 in *Probability Concept in Engineering: Emphasis on Applications to Civil and Environmental Engineering*, Hoboken, NJ: John Wiley & Sons.

Fowler, J. R. and F. J. Couper (1993). Pages 26-31 of *Survey Research Methods*, 2nd ed., Applied Social Research Methods Series, vol I, Thousand Oaks, CA: Sage Publications.

Freund, J. E. (1962). *Mathematical Statistics*, Englewood Cliffs, NJ: Prentice-Hall.

Mathcad (2001). *User's Guide* with *Reference Manual.* Cambridge. MA: MathSoft.

Walpole. R. E., R. H. Myers, S. L. Myers, and K. Ye (2006). Chapters 8 and 9 in *Probability and Statistics for Engineers and Scientists*, 8th ed., Upper Saddle River, NJ: Prentice-Hall.

EXERCISES

1. A concrete manufacturing plant received an order to make concrete with a mean strength in excess of 42 MPa. For quality control purposes, 14 cylinders were prepared and tested. The following values were obtained for the strength:

   42.4, 56.7, 44.1, 35.7, 43.8, 63.4, 34.3, 51.1, 40.6, 44.8, 58.1, 45.9, 50.8, 52.5

   (a) Develop a frequency-of-occurrence graph. Select ranges such as 32.6–35.0, 35.1–37.5, 37.6–40.0, and so on. Plot the graph.

   (b) Based on results in (a), estimate the probability that the compressive strength of concrete will be at least equal to the desired 42.0 MPa.

2. In evaluating the quality of construction work conducted by a contractor, a group of 15 experts were invited to rank the quality on a scale of 1 to 7, where a rank of 1 means totally unsatisfied with the quality while 7 means total satisfaction with quality. The results of the survey are:

   $$5, 1, 6, 6, 5, 4, 3, 4, 4, 5, 2, 7, 5, 5, 3$$

   As is evident, the random variable describing quality, referred to as the *quality index*, is discrete. Thus, in this problem, instead of using ranges for the data simply use the numerical values 1, 2, 3.... 7 and construct a frequency-of-occurrence graph for the quality of construction work.

3. For the data in Exercise 1:

   (a) Using the method of moments, estimate the mean and standard deviation for the strength
   (b) Establish 95 percent confidence intervals for the mean assuming the standard deviation is known
   (c) Repeat (b) but assume that only an estimate for the standard deviation is known

4. For the data in Exercise 2:

   (a) Compute the estimate for the mean and standard deviation assuming all sample values have the same weight
   (b) Repeat (a), but use the frequency-of-occurrence results obtained in Exercise 2 as the probability mass function for the construction quality index

5. In a study of accidents involving CMVs on interstate highways, accident data for 15 4-lane rural highways in several states was compiled. The sample values are adjusted to present the annual number of accidents per kilometer of highway. The following results are obtained:

   0.05, 0.50, 0.25, 1.18, 0.37, 0.23, 0.65, 0.84, 0.33, 0.90, 0.25, 0.43, 0.55, 0.47, 0.71

   (a) Plot the data in the form of a frequency-of-occurrence graph using ranges such as 0–0.19, 0.2–0.39, and so on
   (b) Compute the mean and standard deviation using the results in (a)

6. In Exercise 5:

   (a) Using the method of maximum likelihood along with the lognormal distribution and estimate the distribution parameters λ and ν
   (b) Based on the results in (a), compute the mean and standard deviation

7. Repeat Exercise 6, but use the Rayleigh probability density function.

8. In compiling data for the failure rates among components in a piece of electronic measurement equipment, when exposed to harsh environments, 150 components were tested and only 4 failed.
   (a) Assuming failure among components follows a binomial distribution; estimate the parameter p of the distribution
   (b) Establish 95 percent confidence intervals for p
   (c) If the electronic equipment system is made up of 25 components, compute the probability of failure of the equipment when exposed to harsh environments
9. In Exercise 5, use the KS method to test the validity of the lognormal distribution to represent the data. Use the parameters you computed in Exercise 6. Use $\alpha = 0.05$.
10. Soil samples from nine locations at a site are taken to determine the bearing capacity in kN/m^2. The results are summarized:

    23.1, 42.4, 30.3, 31.0, 28.4, 39.3, 32.4, 27.4, 38.0

   (a) Estimate the mean, standard deviation, and COV of the data.
   (b) Compute the standard error of the mean.
   (c) Establish 95 percent confidence intervals for the mean.
   (d) Using the COV from (a), how many samples must be tested to have ±10 percent error for the mean at 95 percent confidence?
11. In Exercise 5:
   (a) Use the probability-scaled plotting and comment whether the normal distribution will be suitable to represent the data
   (b) Use the results in (a) and estimate the sample mean and standard deviation
12. In Exercise 5, conduct a chi-square test at $\alpha = 0.01$ to test the validity of the normal distribution in representing the data.
13. An engineer is hired to inspect the condition of terra cotta blocks used in the façade of a building. The engineer examined a random sample of 52 terra cotta blocks and found that 6 had loose connections. These were considered failures.
   (a) Based on the information found by the engineer and assuming the binomial distribution represents the failure among terra cotta blocks, estimate the parameter p.
   (b) Establish 95 percent confidence for p.
   (c) If the same bounds found for p in (b) must be maintained but at 99 percent confidence, how many additional terra cotta blocks must the engineer examine?
14. An engineer is planning to test the reliability of a small aluminum bracket that supports an automated control device in a brand of passenger cars. A mode of failure for the bracket is

fatigue that is initiated from steady vibrations. Fifty brackets were tested by simulating a designated level of vibration on a shaking table for a specific period of time. One bracket failed.

(a) If the conditions among brackets are assumed independent, estimate the failure probability for any given bracket
(b) Establish 95 percent confidence intervals for the failure probability in (a)

15. To study the misalignment in rotating blades of a turbine, vibration data at or around the hub was compiled. The sample of acceleration data, composed of 32 sample values, is given (the acceleration unit is g):

$$0.01, 0.08, 0.12, 0.07, 0, -0.08, -0.15. -0.04, 0.08, 0.09, 0.15, 0.07, \\ 0, -0.04, -0.16, -0.21, -0.14, -0.01, 0.04, 0.08, 0.12, 0.07, -0.04, -0.10, \\ -0.15, -0.08, 0.08, 0.10, 0.12, 0.18, 0.04, 0$$

These values were compiled at every 0.5 sec.

(a) Plot the acceleration data versus time
(b) Use commercial software and obtain the Fourier transform of the data
(c) Estimate the dominant frequency of the vibration data

16. A small manufacturing facility specializes in making prestressed, precast joists of various span lengths. With each production batch, a joist is selected at random and tested under the application of a load R applied at the midspan. The strength of a joist is taken equal to the load R (using kN unit) causing failure. The data provided in Table P8.16 shows the strength values for 103 samples of $L = 6$ m span length.

Table P8.16

Values of R	Number of samples
$0 < R \le 5$	0
$5 < R \le 10$	1
$10 < R \le 15$	2
$15 < R \le 20$	4
$20 < R \le 25$	8
$25 < R \le 30$	12
$30 < R \le 35$	18
$35 < R \le 40$	32
$40 < R \le 45$	17
$45 < R \le 50$	6
$50 < R \le 55$	2
$55 < R \le 60$	1
$60 < R$	0

(a) Estimate the mean and standard deviation of R.
(b) Establish a one-sided lower confidence level at $\alpha = 0.05$. Use the t-distribution.
(c) Using the results in (a), estimate the parameters of the lognormal distribution; then conduct a chi-square test for the lognormal distribution at $\alpha = 0.10$.

17. A new composite material was tested for strength evaluation purposes. Ten samples were tested and the following results were obtained for the compressive strength in MPa:

79, 81, 84, 86, 90, 91, 78, 69, 80

(a) Use the branch-and-leaves method and show whether the normal distribution is suitable to represent the data.
(b) Estimate the mean and standard deviation of the data.
(c) Establish a one-sided lower confidence level at $\alpha = 0.05$ for the compressive strength.
(d) How many more samples should be tested to obtain the same confidence level in (c) but at $\alpha = 0.01$? Use the same standard deviation as the one in (c).

18. Within the past 20 yr, a construction company has been bidding on 5 jobs every year. Based on the company's past record, the following data is compiled for the number of jobs won each year by this company (Table P8.18).

Table P8.18

Number of years	Jobs won out of 5 bids/yr
4	0
7	1
6	2
2	3
1	4
0	5

(a) Estimate the parameter p for use in the binomial distribution for modeling the number of wins in a total of 5 bids per year.
(b) Test whether the binomial distribution will be suitable for modeling the number of wins in a total of 5 bids per year. Use $\alpha = 0.05$.

19. The number of airplanes approaching a medium-sized airport in any hour is a random variable X. Seventy observations were made for X; and the results are summarized in Table P8.19.

Table P8.19

Number of aircraft/hr	Observed frequency
0	3
1	10
2	15
3	16
4	12
5	8
6	4
7	2
8 or more	0

(a) Assuming the Poisson distribution for X, estimate the parameter λ of the distribution

(b) Conduct a chi-square test to determine whether the Poisson distribution passes at $\alpha = 0.01$

20. A survey was conducted among a group of 18 experts for compiling data on the failure rate of fittings used in extension gas pipes attached to home gas appliances. Failure was defined as a leak initiated at a fitting. The experts were asked the following question: "In your opinion, how many fittings in extension pipes attached to home appliances will develop gas leaks in a 10-yr period? Please provide your estimates based on 1000 cases." The following responses were received:

0, 3, 1, 5, 2, 4, 7, 6, 4, 2, 8, 3, 5, 2, 4, 6, 3, 3

(a) Plot the data in the form of a frequency-of-occurrence diagram.

(b) Estimate the mean and standard deviation and the standard error of the mean using the method of moments.

(c) At $\alpha = 0.05$, what will be the one-sided upper confidence level for the failure rate?

21. In a survey of the number of vehicles that make a left turn at an intersection, 300 samples were selected at random. Fifteen of these vehicles made a left turn.

(a) Estimate the proportion of the vehicles making the left turn for this intersection.

(b) Compute the sampling error associated with the estimated proportion.

(c) Establish a 95 percent confidence interval for the actual proportion p.

(d) If the same interval in (c) is desired but with 99 percent confidence, how many more samples must be selected? Use the same estimated proportion as in (a).

22. An engineer is conducting a survey among medium-sized construction companies to determine the proportion of companies that use certain software for construction management (CM) purposes.

    (a) If no prior knowledge on the proportion of companies that use the software for CM purposes exists, determine the sample size for a ±10 percent error for the actual proportion (i.e., $\delta = 0.10$) and a 90 percent confidence level.
    (b) Before proceeding with the survey, the engineer comes across a research paper that indicates about 20 percent of medium size companies use the software for CM purposes. To comply with the requirements in (a) what sample size should the engineer select?

Principles of Statistics: Part II—Hypothesis Testing, Analysis of Variance, Regression, and Correlation Analysis

9

9.1 INTRODUCTION

In Chapter 8 we introduced several basic methods of statistics. Although these methods are adequate for treating engineering data, they may not be sufficient for providing a more comprehensive evaluation and analysis of data. In this chapter we introduce several additional topics in statistics that are useful in the design of an experiment or a field data collection session, comparing two or more sets of data compiled from similar sources, identifying correlations between two sets of data, and establishing a regression between two parameters for which data has been compiled. The methods presented in this chapter include *hypotheses testing* of the statistical parameters (e.g., the mean value) of a population, *analysis of variance* (ANOVA), *multiple range test* for comparing the mean values from two or more populations, and correlation and regression analyses.

Hypothesis testing for the mean or other parameters of a population is conducted to support or reject a notion that a desired parameter will have a predetermined limit. For example, in determining the applied snow load on a building, a design engineer believes that the maximum snow load will be less than 950N/m^2. The engineer compiles historical snow load data for the region where the building will be located. The data is then used to test whether the mean snow load will be less than 950N/m^2 based on a prescribed confidence level.

In certain applications we may wish to compare the mean values from two populations. As an example, consider a project dealing with measuring strain in a structural component in an airframe system. To measure the strain, two different methods may be used. One method is to employ a common type of strain gauge that is made up of an electric circuit of resistors. The gauge is mounted on the component. When the structural component experiences deformation and strain, changes in electric resistance will occur in the sensor. These changes are used to determine the strain induced in the component. A newer device for strain measurement is the optical fiber. The fiber is mounted on the component. A light source emits light that passes through the fiber from one end and is sensed on the other end. Any deformity or strain will cause a change in the amount of light received at the exiting end. This change is used to determine strain. Suppose we compile 100 sample values for the strain. However, 50 sample values are obtained using strain gauges, while the other 50 are obtained using optical fibers. We wish to test the hypothesis that the mean value of strain from measurements made by using optical fibers is the same as the mean value of strain from measurements made by strain gauges. Depending on whether the two populations have equal variances, different statistical methods are used for comparing the population means. These methods, which may involve either pooling or pairing data, are discussed in this chapter.

The ANOVA is an analytic procedure by which a population is divided into several subgroups. Each subgroup contains data obtained for a common parameter. However, different methods or systems are used in data compilation within each group. The variation in the measured parameter within each subgroup (also called a *treatment*) is then associated with an identified source and used to test a hypothesis on the equality of group means.

In connection with ANOVA, sometimes we wish to conduct further analyses to test whether the mean values computed for two or more sets of data are statistically different. We use a multiple range test for this purpose. To clarify when this method may be used, consider again the strain data for a component in the airframe system. However, we use the same type of gauge on several similar aircraft. Suppose the ANOVA results in the rejection of the hypothesis that the population means are identical. We would like to investigate the data sets further to identify where the difference between the means lies.

In conducting hypothesis and multiple range tests, critical values that determine whether or not a hypothesis is supported are obtained from the standard normal probability distribution function (when the sample size is large), or from the t-distribution (when sample size is small). Thus it is imperative that the data follows the normal or t-distribution functions. Accordingly, we often conduct a test of normality to determine whether the normal distribution is valid. If the data does not follow any distribution, then distribution-free methods must be used in testing hypotheses or when comparing two sample populations. These methods are also explained in this chapter.

The correlation analysis was discussed in theory in Chapter 7. In this chapter we present statistical methods that can be used to determine the correlation between two sets of data. The regression analysis is conducted to arrive at a suitable relationship that describes the expected value of a dependent variable in terms of one or more independent variables. Although the correlation and regression analyses are two different methods with different objectives, they are related to each other. When the compiled data for two random variables are ranked, the usual methods of correlation analysis may not be suitable. In this case we can use a *rank correlation analysis*. For data presented in linguistic terms, or in cases where one believes a correlation exists between two measured parameters yet the usual statistical methods do not result in an acceptable level of correlation, methods based on *fuzzy logic* may be used. Another method that can be used to determine correlation and establish the relation between a dependent variable and a series of independent variables is based on *artificial neural network* (ANN) systems. These systems require the use and programming of ANN software to recognize the relation between the dependent variable and the independent variables. These topics are not covered in detail in this chapter; however, pertinent references are provided at the end of this chapter for additional reading, (see Klir, et al., 1997; Nelson and Illingworth, 1991; and Yao, 1985).

9.2 HYPOTHESES TESTING

In Chapter 8 our discussion of data analysis focused on using data for computing estimates for parameters that describe a random variable. This was achieved by analyzing a set of data for estimating the mean, variance, or establishing confidence intervals for these parameters. However, the analysis was done without any prior information on the typical ranges or limits for these parameters. If there is a perceived notion that the value of an estimator, such as θ, will be within a limit, a statistical analysis can be conducted to determine whether this notion is supported. This type of analysis falls under the general subject of *hypothesis testing*. To further clarify this type of problem, consider an example involving the investigation of the rate of accidents by commercial motor vehicles (CMVs) operating along an interstate highway. Based on the engineer's previous experience, it is believed that on the average more than 20 percent of CMVs involved in accidents have a type of equipment safety problem. With sufficient data and a statistical analysis, it can be stated whether the engineer's notion can be supported or rejected. Upon collection of data, simple statistical analyses (such as those discussed in Chapter 8) can be used to arrive at an estimate for the mean value of the percentage of CMVs with safety problems that are involved in accidents. Notice that if the mean value is greater than 20 percent, we still need to provide statistical proof that this did not occur by chance. In this problem, the hypothesis is that the mean percentage of CMVs with safety violations that are involved in accidents is greater than 20 percent. Statistical tests can then be conducted to support or reject the hypothesis.

In another example, a production engineer, working in a concrete manufacturing plant, believes that a new batch of concrete produced in the plant will have a mean strength of at least 28 MPa. Suppose, upon collection of data on the strength of this batch of concrete, the mean value of the strength is found to be equal to 31 MPa. Although numerically this value is larger than 28 MPa, statistical tests can be conducted to prove that with a certain confidence level the mean strength value is too large to occur by chance. As a result, the hypothesis that the mean strength is at least equal to 28 MPa is supported.

In both these examples, a value has been assigned to a parameter that imposes a prior knowledge, belief, or simply a subjective estimate on the statistical parameter. This value (designated by θ_0) is referred to as the *null value*. The engineer's perceived notion on the statistical parameter is referred to as the *alternate* or *research hypothesis*, and is often shown with the symbol H_1. Whereas, the negation of this theory or hypothesis is called the *null hypothesis* and is described with H_0. Usually, inequality expressions describe H_0 and H_1. However, there may be applications where a specific value (presented by a statement of equality) is assigned to H_0. The objective of hypothesis testing is to support H_1 and to reject H_0. In any problem where hypothesis testing is desired, it is important to identify H_0 and H_1 and θ_0 before one can start compiling data to determine whether H_0 can be rejected in support of H_1. In identifying and formulating a hypothesis problem, it is noted that:

1. The H_0 and H_1 identities are complementary of each other and often described by two inequalities in terms of the null value θ_0. The statement of equality is included in H_0.
2. The objective of the analysis is to compile data and to determine whether H_0 can be rejected in support of H_1.
3. The data compiled for hypothesis testing follows the normal probability distribution function.

The following examples are intended to demonstrate H_0 and H_1 and the null value θ_0 in several hypothesis-testing problems.

Example 9.1 In the problem dealing with the accident rates of CMVs, the traffic engineer believes that the number with safety violations that are involved in accidents is greater than 20%. Identify the null value, the null hypothesis, and the alternate or research hypothesis.

Solution Let:

μ = Mean percentage of CMVs involved in accidents and with safety violation records. Then
μ_0 = Null value = 0.20

H_0: $\mu \leq \mu_0$ Null hypothesis
H_1: $\mu > \mu_0$ Alternate or research hypothesis

Notice that the statement of equality is included in H_0.

Example 9.2 In the example of the strength of the new batch of concrete, the engineer's perceived notion is that the mean strength is greater than 28 MPa. Identify the null value, the null hypothesis, and the alternate or research hypothesis.

Solution Let:

μ = Mean strength. Then
μ_0 = Null value = 28

H_0: $\mu \leq \mu_0$ Null hypothesis
H_1: $\mu > \mu_0$ Alternate or research hypothesis

Example 9.3 The design snow load in an area is not listed in the design code. An engineer believes that the mean snow load applied on the roof of a building is less than 950 N/m^2. Identify the null value, the null hypothesis, and the alternate or research hypothesis.

Solution Let:

μ = Mean snow load. Then
μ_0 = Null value = 950

H_0: $\mu \geq \mu_0$ Null hypothesis
H_1: $\mu < \mu_0$ Alternate or research hypothesis

To support H_1, sufficient data must be compiled and used in the analysis.

In conducting the hypothesis testing, the outcome will be one of the following:

1. The analysis shows that H_0 is rejected and H_1 is supported. This means that a correct decision was made when the null value and the hypotheses were established.
2. The analysis shows that we fail to reject H_0 and, thus, H_1 is not supported. In this case, the decision making based on the H_1 hypothesis results is an error. This is referred to as a Type I error. Thus Type I results in the rejection of H_0 when it is true.
3. The analysis shows that H_1 is supported and H_0 is rejected. Similar to 1, this will result in a correct decision.
4. The analysis shows that H_1 cannot be supported and, thus, H_0 cannot be rejected. The error associated with the decision-making process in this case is referred to as a Type II error. In other words, Type II results in the acceptance of H_0 when it is not true.

Table 9.1 presents a summary of the four outcomes.

9.2.1 Hypothesis and Significance Testing of a Population Mean

Engineering problems often involve the estimation of the mean of a sample population. Various design and decision-making strategies can be made by knowing the mean value. For this reason, the hypothesis and significance testing for the mean of a population constitutes an important part of an engineering decision-making process. Recall from Example 9.3 that the design

Table 9.1 Outcomes in hypotheses testing

Hypothesis	Situation	
	Rejected	**Fail to reject**
H_0	Correct decision	Type I error
H_1	Type II error	Correct decision

engineer believed the mean snow load was going to be less than 950 N/m^2. The design snow load value D is computed in terms of the mean μ and standard deviation σ as:

$$D = \mu + \beta\sigma \tag{9.1}$$

in which the parameter β is a positive value and is related to the probability that the load will be greater than the design value D (see Figure 9.1). As is evident from Equation 9.1, the design load is directly related to the mean load. After compiling data for the snow load, if the null hypothesis in this problem (i.e., H_0: $\mu \geq 950$) cannot be rejected, then the engineer's notion on the load cannot be supported; as such, a change in the design may have to be made.

In this section, the procedure for hypothesis and significance testing for the mean is presented along with several example problems. Depending on the type of problem, and the statement leading to the formation of H_0 and H_1, the hypothesis-testing problem can be classified as: (a) the right-tailed; (b) the left-tailed; and (c) the two-tailed significance problems.

Right-tailed significance problems: This type of problem is defined by the expressions:

$$H_0: \mu \leq \mu_0$$

$$H_1: \mu > \mu_0$$

Examples 9.1 and 9.2 represent a right-tailed significance problem. We recall from Chapter 8 that Equations 8.10 and 8.14 can only provide an estimate m for the mean. The mean from a sample population is a random variable with an expected value equal to the actual mean μ and a standard deviation equal to $\sigma/\sqrt{n}$ From the central limit theorem, the probability density function of the mean will be normal when $n \to \infty$. With a sufficiently large n, and assuming the sample standard deviation is known, the random variable $\dfrac{\mu - m}{\sigma/\sqrt{n}}$ will then follow the standard normal probability density function.

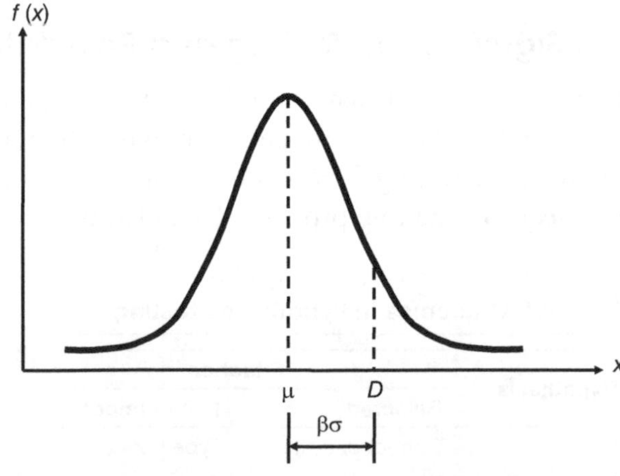

Figure 9.1 Design value *D* vs. mean value

In a right-tailed significance problem, a significance level (such as α) is selected and the corresponding critical value k_α is obtained from the tables of standard normal probability distribution function. As shown in Figure 9.2, the probability associated with H_1 can be formulated as:

$$P(\mu > \mu_0) = 1 - P(\mu \le \mu_0) = 1 - \Phi\left(\frac{\mu_0 - m}{\sigma/\sqrt{n}}\right) = \Phi\left(\frac{m - \mu_0}{\sigma/\sqrt{n}}\right)$$

Notice that when $\dfrac{m - \mu_0}{\sigma/\sqrt{n}} = k_\alpha$, then $P(\mu > \mu_0) = 1 - \alpha$. However, when $\dfrac{m - \mu_0}{\sigma/\sqrt{n}} > k_\alpha$, then $P(\mu > \mu_0) > (1 - \alpha)$. Or, in other words, $P(\mu > \mu_0)$ will be at least equal to the confidence level $(1 - \alpha)$. This means that in order to reject H_0 in favor of H_1 at α significance, we must have:

$$\frac{m - \mu_0}{\sigma/\sqrt{n}} > k_\alpha$$

The significance level α for the right-tailed test is also shown in Figure 9.3a.

If the standard deviation is not known, the random variable $\dfrac{\mu - m}{\sigma/\sqrt{n}}$ follows a standard t-distribution with $\nu = (n - 1)$ degrees of freedom. Again with α significance, the corresponding critical value of the t-distribution will be t_α. In this case, H_0 is rejected in favor of H_1 when:

$$\frac{m - \mu_0}{s/\sqrt{n}} > t_\alpha$$

in which s is the estimate for the population standard deviation.

Left-tailed significance problems: In this case, the null and alternate hypotheses are presented as:

$$H_0: \mu \ge \mu_0$$

$$H_1: \mu < \mu_0$$

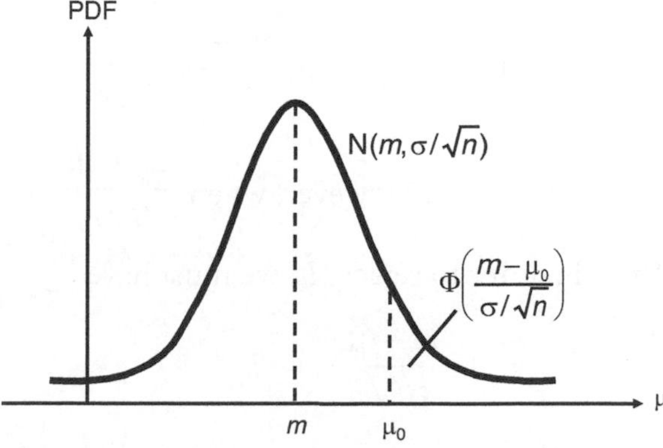

Figure 9.2 Probability density function of the mean μ

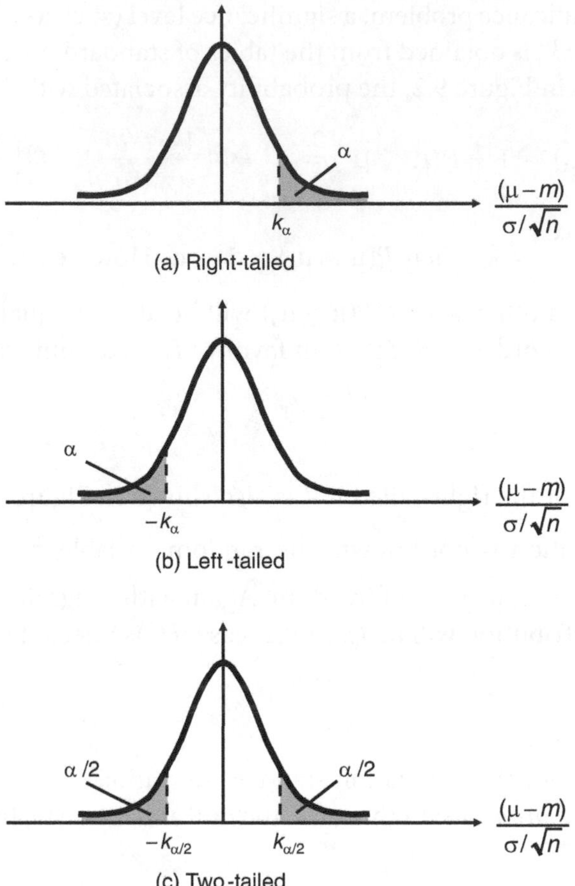

Figure 9.3 Types of significance problems

If the standard deviation is known, then:

$$P(\mu < \mu_0) = \Phi\left(\frac{\mu_0 - m}{\sigma/\sqrt{n}}\right)$$

If $\frac{\mu_0 - m}{\sigma\sqrt{n}} = k_\alpha$, then $P(\mu < \mu_0) \leq (1 - \alpha)$. However, when $\frac{\mu_0 - m}{\sigma\sqrt{n}} > k_\alpha$, then $P(\mu \leq \mu_0)$ will be at least equal to $(1 - \alpha)$. Thus, in order to reject H_0, we must have $\frac{\mu_0 - m}{\sigma\sqrt{n}} > k_\alpha$; or:

$$\frac{m - \mu_0}{\sigma/\sqrt{n}} < -k_\alpha$$

The significance level α for the left-tailed test is also shown in Figure 9.3b.

If σ is not known, a t-distribution with $\nu = (n - 1)$ degrees of freedom is used, and H_0 is rejected when:

$$\frac{m - \mu_0}{s/\sqrt{n}} < -t_\alpha$$

Two-tailed significance problems: In this case, the objective is to support the notion that μ will not be equal to μ_0. The null and alternate hypotheses are:

$$H_0: \mu = \mu_0$$

$$H_1: \mu \neq \mu_0$$

The two-sided significance involves the critical values $\pm k_{\alpha/2}$ as shown in Figure 9.3c. If μ is not equal to μ_0, then it must be either less than or greater than it. This implies that either:

$$\frac{\mu_0 - m}{\sigma/\sqrt{n}} < -k_{\alpha/2}$$

or

$$\frac{\mu_0 - m}{\sigma/\sqrt{n}} < -k_{\alpha/2}$$

Rearranging these inequalities, H_0 will be rejected if:

$$\frac{m - \mu_0}{\sigma/\sqrt{n}} > k_{\alpha/2}$$

or

$$\frac{m - \mu_0}{\sigma/\sqrt{n}} > -k_{\alpha/2}$$

Again, if σ is not known, the estimate for the standard deviation (i.e., s) is used along with a t-distribution with $\nu = (n - 1)$ degrees of freedom. The critical values of the t-distribution will be $\pm t_{\alpha/2}$; and the null hypothesis H_0 is rejected when:

$$\frac{m - \mu_0}{s/\sqrt{n}} > t_{\alpha/2}$$

or

$$\frac{m - \mu_0}{s/\sqrt{n}} < -t_{\alpha/2}$$

Example 9.4 In Example 9.1, suppose the traffic engineer conducts a survey of 17 highways and finds out that the fraction (p) of CMVs involved in accidents with safety violations has an estimated mean of 0.25 and standard deviation of $s = 0.08$, respectively. Determine whether the engineer's notion of $p > 0.2$ is supported at $\alpha = 0.05$ significance. If $p > 0.2$, then the engineer will propose a more stringent regulation for the safely inspection of CMVs.

Solution This is a right-tailed significance problem. Using the t-distribution, the critical value at $\alpha = 0.05$ (or at $1 - \alpha = 0.95$ probability) and $\nu = 17 - 1 = 16$ degrees of freedom is:

$$t_\alpha = 1.746$$

Based on $\mu_0 = 0.2$ and mean $m = 0.25$:

$$\frac{m - \mu_0}{s/\sqrt{n}} = \frac{0.25 - 0.20}{0.06/\sqrt{16}} = 2.5$$

This value exceeds $t_\alpha = 1.746$; as such the null hypothesis is rejected at $\alpha = 0.05$ in support of the engineer's notion that the mean fraction of the number of CMVs involved in accidents and with safety violations is greater than 0.20. In this problem, the engineer's decision to suggest a more stringent regulation for safety inspection of CMVs is a correct one. Notice that the standard deviation was considered to be unknown, and as such, the t-distribution was used.

Example 9.5 In Example 9.2, to support the belief that the concrete mean strength is greater than 28 MPa, the engineer conducts a laboratory test on 50 samples. The estimates for the mean and standard deviation are 29.75 and 2.8 MPa, respectively. Determine whether the engineer's notion is supported. Use $\alpha = 0.05$ significance.

Solution Again, this is a right-tailed problem. Since n is relatively large, we can use the normal distribution. The critical value of normal distribution at $\alpha = 0.05$ (i.e., 0.95 probability) is $k_\alpha = 1.65$. Notice that if we use the t-distribution, at $\alpha = 0.05$ significance and with $\nu = 50 - 1 = 49$ degrees of freedom, the critical value of the t-distribution is $t_\alpha = 1.67$, which is only marginally larger than k_α.

Based on $\mu_0 = 28$ and mean $m = 29.75$:

$$\frac{m - \mu_0}{\sigma/\sqrt{n}} = \frac{29.75 - 28.00}{2.8/\sqrt{50}} = 4.42$$

which is larger than the critical value. This means that the null hypothesis H_0: $\mu \le \mu_0$ is rejected in favor of the alternate hypothesis ($H_1 : \mu > \mu_0$). The decision by the engineer to use at least 28 MPa strength is correct.

Example 9.6 In Example 9.3, the design engineer plans to use a mean snow load of at most 950N/m^2. To support this, the records of the past 15 seasons are reviewed and used in the analysis. The mean maximum seasonal snow load is estimated to be equal to m = 1,140 N/m^2 and an estimated standard deviation of s = 380 N/m^2. Determine whether the engineer's decision is correct. Use $\alpha = 0.01$.

Solution This is a left-tailed significance problem since the alternate hypothesis is presented by $H_1 : \mu < \mu_0$ based on the engineer's belief (where $\mu_0 = 950$). At $\alpha = 0.01$ significance and using the t-distribution with a degree of freedom equal to 15 – 1 = 14, the critical value of the t-distribution is $-t_\alpha$ (where $t_\alpha = 2.62$). Based on $\mu_0 = 950$ and mean m = 1,140:

$$\frac{m - \mu_0}{s/\sqrt{n}} = \frac{1{,}140 - 950}{380\sqrt{15}} = 1.94$$

Certainly this is not smaller than the critical value which is –2.62. This means that the null hypothesis $H_1 : \mu \geq \mu_0$ cannot be rejected. Thus, the engineer has made an error in deciding on the mean snow load.

9.2.2 Hypothesis Testing of a Proportion

Hypothesis testing can also be applied to proportion p. Recall that p is the probability of occurrence for any one trial in the binomial distribution. The hypothesized value for p is p_0, which is usually defined as a prior knowledge or belief by an engineer. As in the case of the mean value, the three types of tests are:

$H_0 : p \leq p_0$	$H_0 : p \geq p_0$	$H_0 : p = p_0$
$H_1 : p > p_0$	$H_1 : p < p_0$	$H_1 : p \neq p_0$
Right-tailed test	Left-tailed test	Two-tailed test

The test statistic for p is based on the random variable selected for establishing the confidence intervals for p as discussed earlier. Denoting this test statistic as z, we can write:

$$z = \frac{p - p_0}{\sqrt{p_0(1 - p_0)/n}} \tag{9.2}$$

in which $p = x/n$ represents the proportion of the number of times (x) the event of concern occurred to the total number of trials (n) used in estimating the proportion. Using an α significance, the critical value from the normal distribution is k_α. In a right-tailed test, if z is less than this critical value, the null hypothesis cannot be rejected. In a left-tailed test, if z is less than $-k_\alpha$, then the null hypothesis is rejected. In a two-tailed test, the critical values of the normal distribution at α are $\pm\, k_{\alpha/2}$. In this case, the null hypothesis cannot be rejected if z is within the two limits.

Example 9.7 A small manufacturing plant produces prefabricated joists for use in the construction of floors in residential buildings. The joists are rated for a total load of 45 kN. On rare occasions the applied load may exceed this value and reach 90 kN. Under this heavy load the production engineer believes that the probability of failure will be less than 0.20. To further support this notion, the engineer tests $n = 15$ beams; 4 fail. At $\alpha = 0.10$, determine whether there is enough statistical evidence to support the engineer's belief.

Solution The null and alternate hypotheses follow those in the left-tailed test. We wish to test the alternate hypothesis $H_1 : p < p_0$. Based on the data, the estimate for the proportion p is computed as $4/15 = 0.267$. The test statistic is:

$$z = \frac{0.267 - 0.20}{\sqrt{0.20(1 - 0.20)/15}} = 0.65$$

The critical value for the normal distribution at $\alpha = 0.10$ is described as $-k_\alpha$, where:

$$k_\alpha = \Phi^{-1}(1 - \alpha) = \Phi^{-1}(0.90) = 1.28$$

Since z is larger than the critical value (i.e., -1.28), the null hypothesis cannot be rejected. Thus, the engineer's belief is not supported statistically.

9.3 COMPARING TWO POPULATION MEANS

In certain problems it is often necessary to make a judgment on whether the mean values from two populations (data sets) are statistically different from one another. As an example, consider data compiled on stress values at the midspan locations of certain girders from two highway bridges with similar structural conditions. The stress values are generated as a result of the passage of trucks on these bridges. A bridge engineer is interested in knowing whether there is enough statistical evidence to believe that the mean stress values from the two bridges are different. In another example, consider data compiled on the vertical accelerations of two identical aircraft. Upon compiling the data an engineer is interested in knowing whether the mean vertical accelerations from the two aircraft are statistically identical (i.e., no significant difference exists between the two mean values). In each of these examples the random samples from two populations can be assumed to be independent. If the variances of the two populations are identical, a *pooled* sample variance will be used in comparing the mean values. However, if the variances are not identical, pooling the variances cannot be used. In certain classes of problems, the sample populations are not independent. Consider the rate of accidents at several intersections before and after an improvement plan is implemented on the intersections. Sample Population I constitutes the number of accidents per year before implementing the plan. Sample Population II consists of the number of accidents per year after the improvement plan is implemented. The accident rates before and after the implementation of the plan may be related. In this case a *paired* sample population is used for comparing the mean values.

In this section, comparing means for three situations is described: (1) independent populations with equal variances; (2) independent populations with unequal variances; and (3) dependent populations. In cases where the sample populations are independent, we need to know whether the variances of the two populations are equal. The method for comparing two variances is explained.

9.3.1 Testing of Equality of Two Variances

To decide whether two variances are identical, we need to provide statistical evidence that the difference between the two variances is not significant. We cannot make a judgment on the equality of the two variances by simply computing the difference between the two and expressing an opinion based on the fact that the difference appears to be small. A random variable that can be used to establish a *test statistic* for testing a hypothesis on the equality of the variances must be defined. To establish the test statistic, the ratio of the two variances (rather than the difference between them) is used as a random variable for hypothesis testing. Since the ratio is used as a random variable here, the probability distribution used will be of those with one infinite end. The *F*-distribution is used for this purpose as described later.

The two population variances are σ_1^2 and σ_2^2. The idea is to provide enough statistical evidence that the ratio σ_1^2/σ_2^2 is close to 1. If the ratio is too small to reasonably occur by chance, then the observed σ_1^2/σ_2^2 is close to zero. On the other hand, if the ratio is too large to reasonably occur by chance, then it is much larger than 1. In both these cases, $\sigma_1^2 \neq \sigma_2^2$. In conducting the test of equality of the two variances, however, we use the estimates for the variances, since the exact variances are not known. The estimates are S_1^2 and S_2^2, and their ratio $\theta = S_1^2/S_2^2$. Note that we use capital letters (e.g., S) to define variances because they are taken as random variables. The hypothesis testing for the equality of the variances is a two-tailed one and is defined by the following null hypothesis H_0 and alternative hypothesis H_1:

$$H_0 : \theta = \theta_0$$

$$H_1 : \theta \neq \theta_0$$

As described earlier, the *F*-distribution is used for θ in testing the H_0 and H_0 hypotheses. The *F*-distribution is obtained as the ratio of two chi-square distributions. The numerator has $v_1 = n_1 - 1$ degrees of freedom describing S_1^2; whereas the denominator has $v_2 = n_2 - 1$ degrees of freedom describing S_2^2. Note that n_1 and n_2 are the sample sizes for the two distributions. The *F*-distribution is always described with the degrees of freedom v_1 and v_2. As shown in Figure 9.4 the critical values of the *F*-distribution based on $\alpha/2$ can be obtained such that:

$$p(f_{1-\alpha/2} < F \leq f_{\alpha/2}) = 1 - \alpha \tag{9.3}$$

The critical values of the *F*-distribution for various combinations of the degrees of freedom v_1 and v_2 are provided in Table A.5 of Appendix A. Note that as usual $\alpha/2$ represents the probability that the random variable F will be greater than $f_{\alpha/2}$; that is:

$$P(F > f_{\alpha/2}) = \alpha/2 \tag{9.4}$$

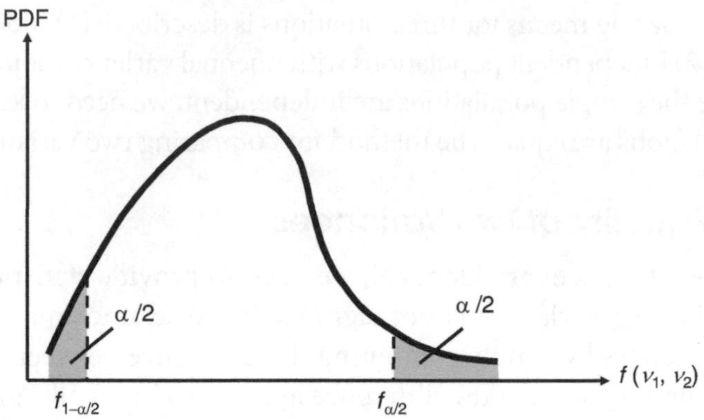

Figure 9.4 Probability density function for *F*-distribution

and

$$P(F \leq f_{\alpha/2}) = 1 - \alpha/2 \tag{9.5}$$

Furthermore, the following relationship holds between the critical values of the *F*-distribution:

$$f_{1-\alpha/2}(\nu_1, \nu_2) = \frac{1}{f_{\alpha/2}(\nu_2, \nu_1)} \tag{9.6}$$

For example, for $\alpha = 0.05$, $\nu_1 = 10$ and $\nu_2 = 15$.

$$f_{\alpha/2} = f_{0.025}(10, 15) = 3.06$$

This indicates that $P(F > 3.06) = 0.025$, and $P(F \leq 3.06) = 1 - 0.025 = 0.975$.

To obtain $f_{1-\alpha/2} = f_{0.975}$ for $\nu_1 = 10$ and $\nu_2 = 15$, we will need $f_{0.025}(15,10)$ which is 3.52. Thus, in light of Equation 9.6:

$$f_{0.975}(10, 15) = \frac{1}{f_{0.025}(15, 10)} = \frac{1}{3.52} = 0.28$$

We then use the critical values of the *F*-distribution to test whether the null hypothesis H_0 cannot be rejected. If the null hypothesis cannot be rejected, then we have enough evidence to support the equality of the two population variances.

Example 9.8 Samples of soils from two nearby sites with similar characteristics were taken and tested for bearing capacity. The following results provide the summary of the findings:

Site I	Site II
$n_1 = 10$ samples	$n_2 = 13$ samples
$m_1 = 182$ kN/m^2	$m_2 = 171$ kN/m^2
$S_1 = 15.2$ kN/m^2	$S_2 = 13.8$ kN/m^2

Continues

Example 9.8: *Continued*

Determine whether the variances from these populations are identical at $\alpha = 0.05$ significance.

Solution We wish to test the null hypothesis H_0. Since $S_1^2 = 231.04$ and $S_2^2 = 190.44$, then:

$$\theta = S_1^2/S_2^2 = 231.04/190.44 = 1.21$$

The critical values of the F-distribution at $v_1 = 10 - 1 = 9$ and $v_2 = 13 - 1 = 12$ are:

$$f_{\alpha/2} = f_{0.025}(9,12) = 3.44$$

$$f_{1-\alpha/2} = f_{0.975}(9,12) = \frac{1}{f_{0.025}(12,9)} = \frac{1}{3.87} = 0.26$$

Since $\theta = 1.21$ is within $f_{1-\alpha/2} = 0.26$ and $f_{\alpha/2} = 3.44$, the null hypothesis cannot be rejected; as such, the variances from the two populations are identical at $\alpha = 0.05$.

9.3.2 Comparing Mean Values from Independent Populations with Identical Variances (Pooled *t*)

When variances S_1^2 and S_2^2 are identical, the standard t random variable used in hypothesis is:

$$t = \frac{(M_1 - M_2) - (\mu_1 - \mu_2)}{\sqrt{\sigma^2(1/n_1 + 1/n_2)}} \tag{9.7a}$$

in which $(M_1 - M_2)$ is the point estimation of the difference in the two population means (a random variable); μ_1 and μ_2 are the population means, and σ^2 is the common variance of the two populations. Since this variance is not known, an unbiased estimate for σ^2 is computed based on S_1^2 and S_2^2:

$$S_p^2 = \frac{(n_1 - 1)S_1^2 + (n_2 - 1)S_2^2}{n_1 + n_2 - 2} \tag{9.8}$$

where S_p^2 is a *pooled* variance. The hypotheses to be tested are defined:

$H_0 : \mu_1 \le \mu_2$	$H_0 : \mu_1 \ge \mu_2$	$H_0 : \mu_1 = \mu_2$
$H_1 : \mu_1 > \mu_2$	$H_1 : \mu_1 < \mu_2$	$H_1 : \mu_1 \ne \mu_2$
Right-tailed test	Left-tailed test	Two-tailed test

Substituting S_p^2 for σ^2 in Equation 9.7a, the standard t variable can be written as:

$$t = \frac{(M_1 - M_2) - (\mu_1 - \mu_2)}{\sqrt{S_p^2(1/n_1 + 1/n_2)}} \tag{9.7b}$$

When the objective is to determine whether the difference between the two mean values is statistically insignificant (i.e., the means are statistically identical), the null and alternative hypothesis may be written as a two-tailed problem as:

$$H_0: D = 0$$

$$H_1: D \neq 0$$

in which the test statistic, D, is written as:

$$D = \frac{(m_1 - m_2) - d_0}{\sqrt{S_p^2(1/n_1 + 1/n_2)}} \tag{9.9}$$

and m_1 and m_2 are the estimates for the two population means. The parameter d_0 is the hypothesized difference in the two population means. In most applications, $d_0 = 0$. At α significance, the critical values of the t-distribution are $\pm t_{\alpha/2}$ with $v = (n_1 + n_2 - 2)$ degrees of freedom. If $-t_{\alpha/2} < D < +t_{\alpha/2}$, the null hypothesis H_0 cannot be rejected; as such, the difference between the means is not significant.

It is emphasized that the above procedure is valid when the two populations follow the normal probability distribution function. The test of normality for these populations will need to be conducted, using the methods described in Section 8.6.4 and 8.6.5, to determine the significance level at which the test passes.

Example 9.9 In Example 9.8, since the variances are statistically identical, determine whether the difference between the two population means is significant at $\alpha = 0.05$.

Solution The pooled variance S_p^2 is computed using Equation 9.8:

$$S_p^2 = \frac{9 \times 231.04 + 12 \times 190.44}{10 + 13 - 2} = 207.84$$

Using $d_0 = 0$, the test statistics D for the difference between the two means is:

$$D = \frac{(182 - 171) - 0}{\sqrt{(207.84)(1/10 + 1/13)}} = 1.81$$

The critical value of the t-distribution at $\alpha = 0.05$ and $\nu = (10 + 13 - 2) = 21$ degrees of freedom is:

$$t_{\alpha/2} = T_v^{-1}(1 - \alpha/2) = 2.08$$

Since $-2.08 < D < +2.08$, the difference between the two means is not significant at $\alpha = 0.05$.

Example 9.10 Flight data for two identical aircraft were compiled for several hours of test flight. The data compiled was the number of cases where the aircraft vertical acceleration exceeded 3.0*g*, where *g* is the gravity acceleration. Aircraft I had $n_1 = 11$ cases of vertical acceleration exceeding 3.0*g* with a mean vertical acceleration equal to 3.6*g* and a corresponding standard deviation equal to 0.40*g*. The data for Aircraft II showed $n_2 = 16$ cases of the vertical acceleration exceeding 3.0*g* with a mean value of 3.2*g* and corresponding standard deviation equal to 0.36*g*.

(a) Show that the variances corresponding to accelerations (exceeding 3.0*g*) from the two aircraft are identical at $\alpha = 0.10$ significance.

(b) Given the results in (a), determine whether the difference between the two means is statistically significant at $\alpha = 0.10$. Assume the two populations have been tested for and passed the normal probability distribution function.

Solution

(a) The variances for the two populations are:

$$S_1^2 = 0.16g^2 \quad \text{and} \quad S_2^2 = 0.13g^2, \quad \text{thus} \quad \theta = S_1^2/S_2^2 = 1.23$$

At $\alpha = 0.10$ and degrees of freedom $v_1 = 11 - 1 = 10$ and $v_2 = 16 - 1 = 15$:

$$f_{\alpha/2} = f_{0.05}(10, 15) = 2.54$$

and

$$f_{1-\alpha/2} = f_{0.95} = \frac{1}{f_{0.05}(15, 10)} = \frac{1}{2.85} = 0.35$$

Since the test statistic $\theta = 1.23$ is between the critical values 0.35 and 2.54, the two variances are statistically identical at $\alpha = 0.10$.

(b) The pooled variance S_p^2 is:

$$S_p^2 = \frac{10 \times 0.16g^2 + 15 \times 0.13g^2}{11 + 16 - 2} = 0.142g^2$$

Using the hypothesized difference $d_0 = 0$, the test statistic D is:

$$D = \frac{(3.6g - 3.2g) - 0}{\sqrt{(0.142g^2)(1/11 + 1/16)}} = 2.71$$

The critical value of the *t*-distribution at $\alpha = 0.10$ and $v = (11 + 16 - 2) = 25$ degrees of freedom is:

$$t_{\alpha/2} = T_v^{-1}(1 - \alpha/2) = 1.708$$

Since $D > 1.708$, the difference between the two population means is statistically significant.

9.3.3 Comparing Means from Independent Populations with Unequal Variances

In those cases where variances are statistically unequal, they cannot be combined by pooling. Instead, a combined variance in the form of:

$$\sigma^2 = \frac{\sigma_1^2}{n_1} + \frac{\sigma_2^2}{n_2} \tag{9.10}$$

must be used, where σ_1^2 and σ_2^2 are the variances of the two populations. The standard t variable describing the difference between the two means μ_1 and μ_2, will be:

$$t = \frac{(M_1 - M_2) - (\mu_1 - \mu_2)}{\sqrt{S_1^2/n_1 + S_2^2/n_2}} \tag{9.11}$$

in which S_1^2 and S_2^2 are the estimates for σ_1^2 and σ_2^2, respectively. The number of degrees of freedom is obtained from the data. Different methods are suggested for computing the degrees of freedom ν. The Smith-Satterthwaite procedure (Milton and Tsokos, 1983) is used in this book. The degree of freedom is:

$$\nu = \frac{(S_1^2/n_1 + S_2^2/n_2)^2}{\frac{(S_2^2/n_1)^2}{n_1 - 1} + \frac{(S_2^2/n_2)^2}{n_2 - 1}} \tag{9.12}$$

The value obtained for ν from Equation 9.12 is rounded down to the nearest integer for use in the t-distribution table. Again, we must first perform a normality test for the two populations before testing a hypothesis in this case. The three hypothesis problems are similar to those in Section 9.3.2. In particular, if the desired test is meant to investigate whether the difference between the two population means is significant, the hypothesis is written as a two-tailed problem in the following form:

$$H_0\text{: } D = 0$$

$$H_1\text{: } D \neq 0$$

in which the test statistic, D, is written as:

$$D = \frac{(m_1 - m_2) - d_0}{\sqrt{(S_1^2/n_1 + S_2^2/n_2)}} \tag{9.13}$$

Again the hypothesized difference d_0 is often taken as zero.

Example 9.11 To compare two brands of a piece of construction equipment, the data on the operation time until the first breakdown in each brand is studied. The operation time until

Continues

Example 9.11: *Continued*

the first breakdown is measured in hours. For the two brands, the following statistics are obtained:

Brand I	Brand II
$n_1 = 25$ samples	$n_2 = 16$ samples
$m_1 = 1253$ hr	$m_2 = 1433$ hr
$S_1 = 125$ hr	$S_2 = 208$ hr

Assume the operation time follows a normal probability distribution function.

(a) Determine whether the variances from the two populations are equal at $\alpha = 0.10$.

(b) Determine whether the two mean values are significantly different.

Solution

(a) To compare the variances, we establish $\theta = S_1^2/S_2^2 = 0.36$. At $\alpha = 0.10$ and degrees of freedom $v_1 = (25 - 1) = 24$ and $v_2 = (16 - 1) = 15$:

$$f_{\alpha/2} = f_{0.05}(24, 15) = 2.29$$

and

$$f_{1-\alpha/2} = f_{0.95} = \frac{1}{f_{0.05}(15, 24)} = \frac{1}{2.11} = 0.47$$

Since $\theta = 0.36 < 0.47$, the two variances are statistically unequal.

(b) Given the results in (a), to compare the means we can use the procedure in Section 9.3.3 for unequal variances. Using $d_0 = 0$ in Equation 9.13:

$$D = \frac{(1.253 - 1.433) - 0}{\sqrt{(125)^2/25 + (208)^2/16}} = -3.12$$

The degree of freedom v is:

$$v = \frac{[(125)^2/25 + (208)^2/16]^2}{\frac{[(125)^2/25]^2}{24} + \frac{[(208)^2/16]^2}{15}} = 22$$

The critical value of the t-distribution at $\alpha = 0.10$ and $v = 22$ degrees of freedom is:

$$t_{\alpha/2} = T_v^{-1}(1 - \alpha/2) = 1.717$$

Since $D = -3.12 < -1.717$, the difference between the two means is statistically significant.

9.3.4 Comparing Means from Dependent Populations (Paired *t*)

In certain problems the two random samples are not independent. This usually occurs when the sample data represents the same variable under two different conditions or environments. As an example, consider the rate of accidents in a series of intersections along a 10-km roadway before and after implementation of a traffic signal coordination. The traffic signal coordination is achieved by synchronizing the duration of red lights at intersections so that a motorist traveling at the posted maximum speed will encounter the smallest number of red lights. As another example, suppose a safety regulation is imposed on construction companies in an area to enhance the safety of construction workers. The two populations in this case may be (1) the number of yearly accidents at the sites before the implementation of the safety regulation and (2) the number of yearly accidents at the same sites after the implementation of the safety regulation. In these examples, the observed data sets are paired because of their dependence. If x_i and y_i are the *i*th observed values from the two populations X and Y, respectively, the difference between the two values x_i and y_i (i.e., d_i) belongs to a random variable $D = X - Y$. Estimates for the mean and standard deviation of D can be obtained using the point estimation method. Since the sample size for both populations is n, then:

$$m_D = \frac{1}{n}\sum_{i=1}^{n} d_i \tag{9.14}$$

$$S_D^2 = \frac{1}{n-1}\sum_{i=1}^{n} (d_i - m_D)^2 \tag{9.15}$$

The hypotheses for testing whether the difference between the two population means is significant in a two-tailed problem are as follows:

$$H_0 : \overline{D} = 0$$

$$H_1 : \overline{D} \neq 0$$

Where:

$$\overline{D} = \frac{m_D - 0}{S_D / \sqrt{n}} \tag{9.16}$$

The difference between the two means is significant when $\overline{D} < -t_{\alpha/2}$ or $\overline{D} > +t_{\alpha/2}$, in which $\pm t_{\alpha/2}$ are the critical values of the *t*-distribution at α significance and $\nu = (n - 1)$ degrees of freedom.

Example 9.12 In the problem of the traffic signal coordination, 15 intersections along a roadway were studied before and after implementation of a synchronized red light system. The results in Table E9.12 present the number of accidents in these intersections.

Continues

Example 9.12: *Continued*

Table E9.12

Intersection	Accidents/year before signal coordination (*X*)	Accidents/year after signal coordination (*Y*)	$D = X - Y$
1	3	2	1
2	9	8	1
3	3	5	−2
4	0	1	−1
5	0	2	−1
6	3	3	0
7	4	5	−1
8	0	0	0
9	2	1	1
10	7	8	−1
11	7	9	−2
12	2	2	0
13	0	1	−1
14	5	4	1
15	2	6	−4

Determine whether the mean values of the number of accidents before and after signal coordination are statistically significant.

Solution The values for the random variable *D* are listed in Table E9.12. The mean value and standard deviation of *D* are computed from Equations 9.14 and 9.15:

$$m_D = -0.67, \text{ and } S_D = 1.447$$

Clearly, the signal coordination has increased the mean number of accidents: but we need to conduct a test to determine whether the difference is significant. At $\alpha = 0.05$, the critical values of *t*-distribution are $\pm t_{\alpha/2}$, where $t_{\alpha/2} = 2.145$ for a degree of freedom $\nu = 15 - 1 = 14$. The test statistic $\overline{D}$ is:

$$\overline{D} = \frac{-0.67 - 0}{1.447/\sqrt{15}} = -1.79$$

Since $-2.145 < (\overline{D} = -1.79) < 2.145$, there is not enough evidence to prove the difference between the mean number of accidents before and after the implementation of the signal coordination is significant at $\alpha = 0.05$.

9.4 ANALYSIS OF VARIANCE

In Section 9.3, we discussed methods for comparing the means and variances from two populations. In this section, we extend these methods to include more than two populations. These methods are referred to as the analysis of variance (ANOVA). The purpose of ANOVA is to conduct hypothesis testing among the mean values of several data sets. To perform the testing, the total variation in the compiled data is divided into segments. Once the source of variations in each segment is known, the hypothesis testing on the differences between the population means can be performed. Consider as an example the vertical accelerations compiled for a fleet of five identical airplanes. Each data set contains measurements made for the accelerations of one of the five aircraft. No other variables are considered in the data sets. Upon conducting the normality test on the data sets, ANOVA can be used to test whether the differences between means are significant. This type of comparison is referred to as the *one-way classification, completely random design with fixed effect.*

9.4.1 One-way Classification, Completely Random Design with Fixed Effect

This method is called one-way because only one parameter is under study. For example, in the flight data acquisition problem, we are only interested in the vertical acceleration without any other difference between the aircraft (such as their age in service, history of flight, and such) being considered. The data sets are independent of each other; for this reason, the method is considered completely random. Fixed effect refers to the fact that the data is compiled from specific sources. In the flight data acquisition problem, the data is compiled from five specific aircraft rather than from a random group of aircraft.

In this method, we observe the following two general problems:

1. A total of N samples are going to be studied using k different treatments. The N samples are randomly divided into k groups of sizes $n_1, n_2, \ldots, n_k$. Each group will be subject to a different treatment. As an example of this type, assume three methods are suggested for enforcing the speed limits in N roadways. These methods are: Method A, using flashing lights along the roadway; Method B, using electronic boards that display the speed of an approaching motorist; and Method C, having a continuous patrol present along the roadway. We randomly assign n_1 roadways to receive Method A; n_2 roadways to receive Method B; and n_3 roadways to receive Method C. Note that $(n_1 + n_2 + n_3) = N$. We monitor the number of speed violations in these roadways for a year. Now we wish to compare the mean number of speed violations for the three groups of roadways that received the three methods (treatments).

2. There are k populations of sizes $n_1, n_2, \ldots, n_k$. All populations are subject to the same treatment. As an example, consider four groups of construction companies from different states. A new safety regulation is being implemented on the groups. We want to compare the mean

number of accidents per year for each group after the implementation of the safety regulation. In this case, only the differences in the results are attributed to the basic differences among the *k* populations, and not to the type of treatment.

The two problems cited are different. However, in both of these problems we wish to determine whether the mean values from the *k* populations are statistically identical. The hypotheses to be tested are:

$$H_0 : \mu_1 = \mu_2 = \ldots = \mu_k$$

$$H_1 : \mu_i \neq \mu_j \text{ for some } i \text{ and } j$$

The null hypothesis states that there is no difference between the mean values; the alternative hypothesis states that at least one mean value is different from the others.

Recall that group i has n_i sample values. We introduce X_{ij} as the jth sample value in group i (also known as the *factor level*), where $i = 1, 2, \ldots, k$ and $j = 1, 2, \ldots, n_i$. For example, the sample values in group 1 will be $X_{11}, X_{12}, X_{13}, \ldots, X_{1n_1}$, whereas, the sample values in group 2 will be $X_{21}, X_{22}, X_{23}, \ldots, X_{2n_2}$, and so on. We further introduce the following parameters:

$$T_i = \sum_{j=1}^{n} X_{ij} = \text{Sum of sample values in group } i \quad (9.17)$$

$$M_i = \frac{T_i}{n_i} = \text{Sample mean for group } i \quad (9.18)$$

$$T = \sum_{i=1}^{k} T_i = \sum_{i=1}^{k}\sum_{j=1}^{n_1} X_{ij} = \text{Sum of all sample values from all groups} \quad (9.19)$$

$$M = \frac{T}{N} = \text{Sample mean for all groups (grand mean)} \quad (9.20)$$

$$\sum_{i=1}^{k}\sum_{j=1}^{n_1} X_{ij}^2 = \text{Sum of squares of all sample values for all groups} \quad (9.21)$$

Notice that ($i = 1, 2, \ldots, k$) are the theoretical mean values for the *k* populations. If μ represents the theoretical mean for all data combined, we can then introduce a model for hypothesis testing. Furthermore, notice that μ ignores the group effect (factor level effect) on the outcome. If various groups have no effect on the outcome, then all μ_i will be the same and equal to μ. However, the factor levels will usually affect the outcome; and as such $(\mu_i - \mu)$ will represent the effect of the group (or factor level) *i*. As it is expected due to the variations in the data, within each group there will also be certain levels of variability associated with the population means. The difference between a specific sample value X_{ij} and the group mean μ, (i.e., $X_{ij} - \mu_i$) is referred to as the *random error*. The term $(\mu_i - \mu)$ on the other hand represents the error in μ_i due to the differences between methods (or treatments) applied to various groups. A model for hypothesis testing can be developed as follows:

$$X_{ij} = \mu + (\mu_i - \mu) + (X_{ij} - \mu_i) \quad (9.22)$$

For:

$$i = 1, 2, \ldots, k$$

$$j = 1, 2, \ldots, n_i$$

The model represented by Equation 9.22 simply indicates that the outcome for the sample value X_{ij} is written as the overall mean plus two error terms; one from the fact that group i received a different treatment; and the other due to the variations within the group. If groups are not receiving different treatments, most of the errors will be due to the variations within the group. It is emphasized that the basic assumptions underlying the model of Equation 9.22 are that: (a) the k populations are independent; (b) each population follows a normal probability distribution function; and (c) each sample population has the same variance σ^2. Furthermore, since μ_i and μ are not known, their estimates are used in the model of Equation 9.22. Thus:

$$X_{ij} = M + (M_i - M) + (X_{ij} - M_i) \tag{9.23}$$

or

$$(X_{ij} - M) = (M_i - M) + (X_{ij} - M_i) \tag{9.24}$$

where M_i and M are estimates for μ_i and μ, respectively. In Equations 9.23 and 9.24, capital M is used to define means as random variables. If each side of Equation 9.24 is squared and summed over all possible values of i and j, the following equation is obtained. This is called the *sum of square identity.*

$$\sum_{i=1}^{k}\sum_{j=1}^{n_i}(X_{ij} - M)^2 = \sum_{i=1}^{k} n_i(M_i - M)^2 + \sum_{i=1}^{k}\sum_{j=1}^{n_i}(X_{ij} - M_i)^2 \tag{9.25}$$

In this identity, let:

$$SS_T = \sum_{i=1}^{k}\sum_{j=1}^{n_i}(X_{ij} - M)^2 = \sum_{i=1}^{k}\sum_{j=1}^{n_i}X_{ij}^2 - \frac{T^2}{N} \tag{9.26}$$

$$SS_M = \sum_{i=1}^{k} n_i(M_i - M)^2 = \sum_{i=1}^{k}\frac{T_i^2}{n_i} - \frac{T^2}{N} \tag{9.27}$$

$$SS_E = \sum_{i=1}^{k}\sum_{j=1}^{n_i}(X_{ij} - M_i)^2 \tag{9.28}$$

Where:

SS_T = The sum of squares of deviations of sample values from the grand mean. This is a measure of total variability in the data.

SS_M = The weighed sum of squares of deviations of group means from the grand mean. This is a measure of variability in different groups due to different methods or treatment applied to them.

SS_E = The sum of squares of deviations of sample values from the group mean. This is a measure of variability within each group and is called *error* (or *residual*) sum of squares.

In light of the identity represented by Equations 9.25-9.28:

$$SS_T = SS_M + SS_E \tag{9.29}$$

To test the null hypothesis on the equality of the group means, we define two statistics as functions of SS_M and SS_E. These are referred to as the *treatment mean square* (MS_M) which is found by dividing SS_M by $(k - 1)$; and the *error mean square* (MS_E) which is found by dividing SS_E by the value $(N - k)$. Thus:

$$MS_M = \frac{SS_M}{k-1} \tag{9.30}$$

$$MS_E = \frac{SS_E}{N-k} \tag{9.31}$$

Both MS_M and MS_E are random variables that follow some type of a distribution model. The expected values of MS_M and MS_E (presented here without proofs) are:

$$E(MS_M) = \sigma^2 + \sum_{i=1}^{k} \frac{n_i(\mu_i - \mu)^2}{k-1} \tag{9.32}$$

$$E(MS_E) = \sigma^2 \tag{9.33}$$

in which σ^2 is the population common variance. To test the null hypothesis (H_0), we need to provide statistical evidence that $\mu_1 = \mu_2 = \ldots = \mu_k = \mu$. This means that the second term in Equation 9.32 must be zero. This also implies that if H_0 is true, then MS_M and MS_E will be close in value and the ratio MS_M/MS_E will be close to unity. This ratio can be used as a test statistic. The ratio follows an F-distribution with $\nu_1 = (k - 1)$ and $\nu_2 = (N - k)$ degrees of freedom. The null hypothesis is rejected when the ratio is larger than the critical value of the F-distribution at α significance and with ν_1 and ν_2 degrees of freedom. The test is always a right-tailed one. In practice, the ANOVA process is usually shown in tabular form as in Table 9.2. Furthermore, SS_T can easily be computed. Having computed SS_M, the value of SS_E can be computed as $SS_T - SS_M$.

Table 9.2 ANOVA: One-way classification, completely random design with fixed effect

Sources of variation	Degrees of freedom	*SS*	*MS*	*E(MS)*	*F*
Treatment	$k - 1$	SS_M	$\frac{SS_M}{k-1}$	$\sigma^2 + \sum_{i=1}^{k} \frac{n_i(\mu_i - \mu)^2}{k-1}$	$\frac{MS_M}{MS_E}$
Error	$N - k$	SS_E	$\frac{SS_E}{N-k}$	σ^2	
Total	$N - 1$	SS_T			

Example 9.13 In the example of speed control methods for roadways, suppose we wish to determine whether there is enough evidence to show that the three methods (identified as Methods A, B, and C) result in an identical number of speed violations. Upon implementation of the three methods, the data on speed violations for the three groups of roadways is obtained and summarized in Table E9.13a.

Test the hypothesis that the mean values of the number of speed violations/mo for the three methods are identical at $\alpha = 0.05$. Assume the data in each group follows the normal probability distribution function.

Table E9.13a Number of speed violations/mo

Method A (flashing lights)	Method B (displaying speed)	Method C (continuous patrol)
6	3	8
3	4	9
7	3	2
6	8	4
10	2	3
11	10	3
4	9	3
0	4	6
3	6	1
7	8	3
2	6	
	3	
	0	
	1	
$n_1 = 11$	$n_2 = 14$	$n_3 = 10$
Total = 11 + 14 + 10 = 35		

Solution Based on the data presented in Table E9.13a:

$M_1 = (6 + 3 + 7 + \ldots)/11 = 5.36$
$M_2 = (3 + 4 + 3 + \ldots)/14 = 4.79$
$M_3 = (8 + 9 + 2 + \ldots)/10 = 4.20$

Furthermore:

$$T_1 = 59,\ T_2 = 67,\ T_3 = 42,\quad \text{and} \quad T = 168.$$

The grand mean M is:

$$M = (6 + 3 + 7 + \ldots + 3 + 4 + 3 + \ldots + 8 + 9 + 2 + \ldots)/35 = 4.80$$

Continues

Example 9.13: *Continued*

and

$$SS_T = \sum_{i=1}^{3}\sum_{j=1}^{n_i} X_{ij}^2 - \frac{T^2}{N} = 1,112 - \frac{(168)^2}{35} = 305.60$$

$$SS_M = \sum_{i=1}^{3}\frac{T_i^2}{n_i} - \frac{T^2}{N} = \frac{(59)^2}{11} + \frac{(67)^2}{14} + \frac{(42)^2}{10} - \frac{(168)^2}{35} = 7.10$$

$$SS_E = SS_T - SS_M = 305.60 - 7.10 = 298.50$$

$$MS_M = \frac{SS_M}{k-1} = \frac{7.10}{3-1} = 3.55$$

$$MS_E = \frac{SS_E}{(N-k)} = \frac{298.50}{35-3} = 9.33$$

The test statistics $F_{2,32} = MS_M/MS_E = 3.55/9.33 = 0.38$.

With the degrees of freedom (DF) $\nu_1 = 3 - 1 = 2$ and $\nu_2 = 35 - 3 = 32$, the critical value of the F-distribution at $\alpha = 0.05$ is $f_\alpha = f_{0.05}(2,32) = 3.30$ (using interpolation). Since $F_{2,32} = 0.38 < f_\alpha = 3.30$, the null hypothesis H_0 cannot be rejected. Thus, the three methods result in a statistically identical number of speed violations. The results are summarized in Table E9.13b.

Table E9.13b ANOVA summary

Variation source	DF	*SS*	*MS*	*F*	f_α
Method (treatment)	2	7.10	3.55	0.38	3.30
Error	32	298.50	9.33		
Total	34	305.60			

We emphasize that for ANOVA, the populations must follow the normal distribution. Thus, the test of normality must be conducted before ANOVA. Furthermore, we need to test whether the population variances are statistically identical. The reader is referred to Bartlett's or Cochran's test for this purpose (see Guttman, et al., 1982; Milton and Tsokos, 1983).

9.4.2 Multiple Range Test

In conducting ANOVA for one-way classification, completely random design with fixed effect, we often need to conduct further analyses for the population means. This will be the case where the null hypothesis is rejected and there is evidence that the means are statistically different. Occasionally, the ANOVA results in a very small significance (small α) for the equality of the population means. For example, we find the test statistic F to be smaller than $f_{0.025}$ but larger than $f_{0.05}$. This indicates that the probability that the critical value becomes larger than F is small. In this situation, we may wish to further investigate the data to analyze where the differences between the two populations lie. A method used for this purpose (when H_0 in ANOVA is rejected) is called the *multiple range test*. The test was originally designed by D. B. Duncan for equal sample sizes, and later extended by C. Y. Kramer for different sample sizes.

Multiple range test for equal sample sizes: In this method, the population means are arranged in ascending order; thus, $\mu_1 < \mu_2 < \ldots < \mu_k$. Considering any two population means such as μ_i and μ_j where $j > i$, the corresponding difference $(\mu_j - \mu_i)$ is significant if it is larger than a value called the *shortest significant range*, R_p. The value of R_p is computed from the following equation:

$$R_p = r_p \sqrt{\frac{MS_E}{n}} \tag{9.34}$$

The subscript p identifies the sample means that are being compared. For example, if μ_i and μ_j are compared, $p = (j - i + 1)$. Furthermore, n = common sample size for each population, and MS_E is the error mean square from ANOVA. The values of r_p for a degree of freedom $\nu = (N - k)$, which is the degree of freedom associated with MS_E, are provided in Table A.6 of Appendix A.

Example 9.14 In conducting an inspection of conditions of single span steel girder bridges, bridge engineers use a rating scale ranging from 0 to 9. A rating of 0 indicates the bridge is in a critical situation requiring immediate attention; a rating of 9 means the bridge is in perfect condition. To investigate the differences in subjective ratings, four bridge engineers were invited to independently rate twelve bridges. The results are summarized in Table E9.14a. Assume appropriate tests have been conducted and passed for the normality of the populations and the equality of variances.

(a) Conduct a one-way classification completely random design ANOVA with fixed effect to test the equality of the population means at $\alpha = 0.05$.

(b) Given the results in (a), conduct a multiple range test at $\alpha = 0.05$ to determine whether differences among means are statistically significant.

Table E9.14a Bridge ratings

Bridge	Engineer 1	Engineer 2	Engineer 3	Engineer 4
1	5	6	5	3
2	3	6	4	3
3	7	6	8	5
4	4	6	6	3
5	6	8	6	5
6	6	7	8	5
7	6	3	7	4
8	3	4	7	4
9	8	7	8	6
10	3	6	4	4
11	7	6	6	5
12	4	5	6	3
Total	12	12	12	12

Continues

Example 9.14: *Continued*

Solution Based on the data given in Table E9.14a:

$$T_1 = 62, T_2 = 70, T_3 = 75, \text{ and } T_4 = 50, \text{ and } T = 257.$$

Also:

$$\sum_{i=1}^{4}\sum_{j=1}^{12} X_{ij}^2 = 1493$$

(a) The population means are: 62/12 = 5.17, 70/12 = 5.83, 75/12 = 6.25, and 50/12 = 4.17. Furthermore:

$$SS_M = \sum_{i=1}^{4}\frac{T_i^2}{n_i} - \frac{T^2}{N} = \frac{(62)^2}{12} + \frac{(70)^2}{12} + \frac{(75)^2}{12} + \frac{(50)^2}{12} - \frac{(257)^2}{48} = 29.73$$

$$SS_T = \sum_{i=1}^{4}\sum_{j=1}^{12} X_{ij}^2 - \frac{T^2}{N} = 1{,}493 - \frac{(257)^2}{48} = 116.98$$

$$SS_E = SS_T - SS_M = 116.98 - 29.73 = 87.25$$

With $\nu_1 = 4 - 1 = 3$, and $\nu_2 = 48 - 4 = 44$:

$$MS_M = \frac{SS_M}{3} = \frac{29.73}{3} = 9.91$$

$$MS_E = \frac{SS_E}{44} = \frac{87.25}{44} = 1.98$$

$$F_{3,44} = MS_M/MS_E = 9.91/1.98 = 5.01 \text{ and at } \alpha = 0.05, f_\alpha = 2.38.$$

As is evident, since $F_{3,44} > f_\alpha$ the null hypothesis is rejected. This means there is not enough evidence to prove that the means are identical. The results of ANOVA are provided in Table E9.14b.

Table E9.14b ANOVA summary

Variation source	DF	*SS*	*MS*	*F*	f_α
Method (treatment)	3	29.73	9.91	5.01	2.83
Error	44	87.25	1.98		
Total	47	116.98			

(b) To further investigate the source of variation, we must conduct a multiple range test. We first rearrange means in ascending order. Thus:

$m_1 = 4.17$	$m_2 = 5.17$	$m_3 = 5.83$	$m_4 = 6.25$
(Engineer 4)	(Engineer 1)	(Engineer 2)	(Engineer 3)

Continues

Example 9.14: *Continued*

Comparing m_2 and m_1, $(m_2 - m_1) = 5.17 - 4.17 = 1.0$. At $\alpha = 0.05$, $r_p = r_2 \cong 2.85$ (by interpolation) for $v = 44$. Also $r_3 = 3.00$ and $r_4 = 3.096$ (for $\nu = 44$). The corresponding R_p values are:

$$R_p = r_p \sqrt{\frac{MS_E}{n}}$$

$$R_2 = 2.85\sqrt{\frac{1.98}{12}} = 1.16$$

$$R_3 = 3.00\sqrt{\frac{1.98}{12}} = 1.22$$

$$R_4 = 3.096\sqrt{\frac{1.98}{12}} = 1.26$$

Comparing m_1 and m_2, $(m_2 - m_1) = 1.0 < 1.16$. Thus, the difference is not significant.
Comparing m_1 and m_3, $(m_3 - m_1) = 1.66 > 1.22$. Thus, the difference is significant.
Comparing m_2 and m_3 $(m_3 - m_2) = 0.66 < 1.16$. Thus, the difference is not significant.
Comparing m_2 and m_4, $(m_4 - m_2) = 1.08 < 1.22$. Thus, the difference is not significant.
Comparing m_3 and m_4, $(m_4 - m_3) = 0.42 < 1.16$. Thus, the difference is not significant.

The results are summarized in Table E9.14c. As shown in the table, if differences are not significant, an underline is used to connect the means.

Table E9.14c Multiple range test summary

Test group	Difference (d)	p	R_p	Is $d > R_p$?	Summary
I	$(m_2 - m_1) = 1.0$	2	1.16	No	
	$(m_3 - m_1) = 1.66$	3	1.22	Yes*	$m_1\ m_2\ m_3\ m_4$
II	$(m_3 - m_2) = 0.66$	2	1.16	No	
	$(m_4 - m_2) = 1.08$	3	1.22	No	$m_2\ m_3\ m_4$
III	$(m_4 - m_3) = 0.42$	2	1.16	No	$m_3\ m_4$

*Difference is significant

Note that in Group I, no comparison between m_1 and m_4 is made. This is because m_3 and m_1 showed a significant difference; and since $m_3 < m_4$, the difference between m_4 and m_1 will be significant. As shown in Table E9.14c, the difference between m_1 and m_4 and between m_1 and m_3 are significant. This means $m_1 \neq m_4$, and $m_1 \neq m_3$. No other differences have been detected. The significance level ($\alpha = 0.05$) indicates there is 0.05 probability that at least one of these conclusions is incorrect.

Multiple range test for unequal sample sizes: Duncan's method was extended by C. Y. Kramer in 1956 for cases with unequal sample sizes. In this modified method, the shortest significant range is adjusted as:

$$R'_p = r_p\sqrt{MS_E} \tag{9.35}$$

The test statistic is also modified as:

$$d' = (m_j - m_i)\sqrt{\frac{2n_i n_j}{n_i + n_j}} \tag{9.36}$$

in which n_i and n_j are the sample size for the data sets i and j, respectively. Again, the means are arranged in ascending order; thus, $m_j > m_i$. The significant studentized range r_p is obtained as before for significance α and a degree of freedom ν, where:

$$\nu = \sum_{i=1}^{k}(n_i - 1) \tag{9.37}$$

in which k is the number of populations (data sets). If $d' > R'_p$, the difference between m_j and m_i is significant.

Example 9.15 Fly ash is an additive used in manufacturing high-strength concrete. Suppose three different types of fly ash are tested on several samples of concrete in three batches. Upon testing the materials, the compressive strengths are obtained in a laboratory. All other parameters (i.e., water cement ratio, aggregate type) are identical for all samples. Assume normality and equality of variance tests for the three populations have been conducted and passed. The results obtained for the three batches using ANOVA are:

$$MS_E = 4.10$$

For Batch I, number of cylinders tested, $n_1 = 12$, the mean strength $m_1 = 49.5$ MPa.
For Batch II, number of cylinders tested, $n_2 = 13$, the mean strength $m_2 = 64.1$ MPa.
For Batch III, number of cylinders tested, $n_3 = 11$, the mean strength $m_3 = 68.6$ MPa.

Conduct a multiple range test to further investigate whether the differences among the means are significant at $\alpha = 0.05$.

Solution The degree of freedom $\nu = (12 - 1) + (13 - 1) + (11 - 1) = 33$.

At $\alpha = 0.05$ and $\nu = 33$, $r_2 = 2.879$. The corresponding R'_p value is:

$$R'_p = 2.879 \times \sqrt{4.10} = 5.83$$

Comparing m_1 and m_2:

$$d' = (64.1 - 49.5)\sqrt{\frac{2 \times 12 \times 13}{12 + 13}} = 51.6 > R'_p = 5.83 \text{ Difference is significant.}$$

Continues

Example 9.15: *Continued*

Comparing m_2 and m_3:

$$d' = (68.6 - 61.4)\sqrt{\frac{2 \times 13 \times 11}{13 + 11}} = 15.5 > R'_p = 5.83 \text{ Difference is significant.}$$

The conclusion is that $m_1 \neq m_2 \neq m_3$. The results are shown as:

$$m_1 \; m_2 \; m_3$$

9.4.3 Random Effect

In certain cases, the k data groups are selected from a larger set of populations. Thus, the procedure will not be one with fixed effect. Rather, it is one with random effect. A somewhat different method is then used to investigate whether some variability exists among the data groups within the larger set of population. As an example, consider the statistics on the number of accidents per year in medium-sized construction sites in eight Midwestern states. We randomly select three states for conducting a test on the variability in the number of accidents per year in construction companies in Midwestern states. The procedure to conduct this test can be found in Milton and Tsokos (1983) and Walpole, et al., (2006).

9.5 DISTRIBUTION-FREE METHODS

So far, a basic requirement for statistical procedures, which we have discussed, is that the sample data must follow a normal probability distribution function. In certain problems, the normality test does not pass; as such, we face the question of whether we can still use these methods. This is especially the case when the sample size is small (say $n < 10$). In general, if the normality assumption is not valid, the use of these statistical methods will lead to results that may not be reliable. And as such, distribution-free methods may offer a better solution. These methods generally require ranking the data and deriving the test statistics using simple counting methods. In this section, several such methods are presented for: (a) hypothesis testing; (b) comparing two populations using paired data; and (c) comparing two independent populations. These are parallel to the statistical methods discussed in Section 9.3.

9.5.1 Test of Location for a Given Sample Data

Recall that when normality assumption is valid, the population mean is the measure of the center of location of the distribution of the random variable. In this case, the mean and median are theoretically at the same location along the x-axis of the distribution function. In distribution-free methods, the center of location is usually the median. In general, the median represents the *50-percentile* value. Denoting M as the median of the population X, $P(X \leq M) = 0.5$ for a continuous X. A more general description of the median is defined by the two following expressions:

$$P(X < M) \leq 0.5 \quad \text{and} \quad P(X \leq M) \geq 0.5$$

The idea of a test location is to test the median of a population against a hypothesized value such as M_0. The hypothesis testing will involve the following:

$H_0 : M \leq M_0$	$H_0 : M \geq M_0$	$H_0 : M = M_0$
$H_1 : M > M_0$	$H_1 : M < M_0$	$H_1 : M \neq M_0$
Right-tailed test	Left-tailed test	Two-tailed test

Several methods are available to test the median hypotheses; one is called the *sign test*, which is based on the binomial distribution for the number of sample values that are smaller than M_0; the other is the *Wilcoxon signed-rank test*. The latter is discussed here.

9.5.2 Wilcoxon Signed-rank Test

The null hypothesis in this test is that the distribution of the data is symmetric about the hypothesized value for the median (M_0). In a set of data consisting of n sample values $X_1, X_2, \ldots, X_n$, the differences from the median are $(X_1 - M_0), (X_2 - M_0), \ldots, (X_n - M_0)$. For the null hypothesis to be true, these differences must be from a random variable that is symmetric about zero. In conducting the test, the absolute values of these differences are ranked in ascending order. The smallest absolute value of the difference receives a rank of 1, the largest n. If in conducting the ranking a tie occurs, the average of the ranks will be used for the tied scores. Furthermore, if $(X_i - M_0)$ is negative, the corresponding rank will be assigned a negative sign. As a result, if we designate the rank for $(X_i - M_0)$ as r_i, some r_i will be negative. All positive ranks are added together to form a value W_+. The absolute value of the sum of negative ranks is also computed and denoted as W_-, that is:

$$W_+ = \sum r_i \text{ for all positive ranks} \tag{9.38}$$

$$W_- = \left|\sum r_i\right| \text{ for all negative ranks} \tag{9.39}$$

If the null hypothesis is true, then W_+ and W_- should be about the same. The test statistic is W which is the smaller of W_+ and W_-, that is:

$$W = \min\,(W_+, W_-) \tag{9.40}$$

This value is then compared against a critical value from the table of values for the Wilcoxon signed-rank test (see Table A.7 in Appendix A). If W is equal to or smaller than the critical Wilcoxon value, then the null hypothesis is rejected.

Example 9.16 In conducting subjective ratings for a steel girder in single span, two-lane bridges, the median rating is believed to be 6.7 (a scale of 0 to 9 is used for bridge rating). Eight such bridges are rated for their steel girders. The results are:

4, 8, 8, 3, 5, 6, 7.5, and 7

Conduct a Wilcoxon test for the null hypothesis at $\alpha = 0.05$.

Solution In this problem $M_0 = 6.7$. Table E9.16 summarizes the data given, X_i, $(X_i - M_0)$ and the ranks. Note that instead of Ranks 4 and 5, the average (4.5) is listed because of the two tied scores for $(X_i - M_0) = 1.3$. From Equations 9.38 and 9.39:

$$W_+ = 4.5 + 4.5 + 3 + 1 = 13$$

$$W_- = |-7 - 8 - 6 - 2| = 23$$

The smaller of the two is $W = 13$. At $\alpha = 0.05$ (i.e., 0.05 significance for one-sided test in Table A.7 of Appendix A) and for $n = 8$, the critical value for the test will be 6. Since $W = 13 > 6$, the null hypothesis cannot be rejected and there is sufficient evidence that the median is equal to 6.7.

Table E9.16

X_i	$(X_i - M_0)$	Rank (r_i)	Sign
4	−2.7	7	−
8	1.3	4.5*	+
8	1.3	4.5*	+
3	−3.7	8	−
5	−1.7	6	−
6	−0.7	2	−
7.5	0.8	3	+
7	0.3	1	+

*Tied scores

9.5.3 Wilcoxon Signed-rank Test for Paired Data

In this case there are two populations from two continuous random variables with paired sample values of $(X_1, Y_1), (X_2, Y_2), \ldots, (X_n, Y_n)$. The difference between the paired values (i.e., $X_i - Y_i$ for $i = 1, 2, \ldots, n$) is also from a continuous random variable. The null hypothesis in this case is that the differences are symmetric about zero. To conduct this test, the absolute values of the differences are arranged in ascending order. They are ranked from 1 to n. A negative sign is then assigned to the ranks (r_i) that correspond to the differences that are negative. Using Equations 9.38 and 9.39, W_+ and W_- are computed. The test statistic W is the smaller of W_+ and W_-. If W is smaller than the critical values from the Wilcoxon table, then the null hypothesis is rejected.

In assigning the ranks, r_i, tied differences will receive the average rank for the tied values. If a difference becomes zero, its rank is usually considered to be negative. This will increase the size of W_- and, thus, make it harder to reject the null hypothesis.

Example 9.17 In the evaluation of construction productivity, an engineer believes that if daily work shift hours are shortened by 0.5 hr, there will be an increase in productivity. Productivity in this example is measured as the ratio of the volume of work completed to the volume of work scheduled to be completed in a given period of time. In a survey, eight construction companies agree to change their daily work shift hours by 0.5 hr. The engineer conducts a research and compiles the data presented in Table E9.17a.

Determine if there is enough evidence to prove that the modification of the daily work shift will improve productivity using the Wilcoxon signed-rank test by pairing data.

Table E9.17a Construction productivity (%) before and after daily work shift hour modification

Case	Productivity before modification (X_i)	Productivity after modification (Y_i)	Difference ($X_i - Y_i$)
1	84	73	11
2	80	86	−6
3	78	90	−12
4	85	85	0
5	90	88	2
6	76	71	5
7	85	90	−5
8	78	81	−3

Solution Table E9.17b summarizes the differences in construction productivity, the ranking and sign used for the ranks.

Notice that for $(X_i - Y_i) = 0$, a negative sign is assigned to the corresponding rank. Furthermore, ranks 4 and 5 are tied; thus, the average (4.5) is used for both, although one received a negative sign and one a positive sign. From Table E9.17b

Table E9.17b

Case	Difference ($X_i - Y_i$)	Rank (r_i)	Rank sign
1	11	7	+
2	−6	6	−
3	−12	8	−
4	0	1	−
5	2	2	+
6	5	4.5	+
7	−5	4.5	−
8	−3	3	−

W_+ = Sum of all positive ranks = 7 + 2 + 4.5 = 13.5

W_- = Absolute value of sum of all negative ranks = $|-6 - 8 - 1 - 4.5 - 3| = 22.5$. Thus, $W = 13.5$. Using $\alpha = 0.05$ (i.e., 0.05 significance) and $n = 8$, the Wilcoxon critical value from Table A.7 of Appendix A is 6. Since $W = 13.5 > 6$, the null hypothesis cannot be rejected. This means that the reduction in daily work shift hours will not increase the productivity significantly.

9.5.4 Wilcoxon Rank-sum Test for Unmatched Data

In this case, $X_1, X_2, \ldots, X_m$, and $Y_1, Y_2, \ldots, Y_n$ represent data from two populations with different sample sizes m and n where $m < n$. The null hypothesis is that these populations are identical. If the two populations are different in location, then the null hypothesis is likely to be rejected. The procedure in this case is to pool the populations to form $m + n$ observations. Then the pooled data is arranged in ascending order and ranked from 1 to $m + n$. The test statistic W is the sum of ranks for X_i populations (the smaller-sized population). This decision can well be explained considering the case where X_i is located above or below the location for the Y_i population. If X_i is below, then the X_i values will have smaller ranks. Thus, a small value for W will be obtained. On the other hand, if X_i is above Y_i then the X_i values will have larger ranks. This will result in a large W. The null hypothesis is either too small or too large compared to the critical values for this test. Thus, it cannot be rejected if W is within the critical Wilcoxon values based on a two-tailed test. However, if we suspect that the X_i values are above the Y_i values, then the critical Wilcoxon value will be based on a one-sided (right-tailed) test. In this case if W is less than this critical value, then the null hypothesis cannot be rejected. On the other hand, if we suspect that the X_i values are below the Y_i values, then the critical Wilcoxon value will be based on a one-sided (left-tailed) test, and the null hypothesis cannot be rejected if W is larger than this critical value. The critical Wilcoxon values for the rank-sum test are summarized in Table A.8 of Appendix A. Notice that two critical values are listed for each α (significance level). For example, if $\alpha = 0.05$, $m = 7$ and $n = 10$, the critical right-tailed value is 80; whereas the critical left-tailed value is 46. However, at $\alpha = 0.05$, $m = 7$ and $n = 10$, the two-tailed critical values are 43 and 83.

Example 9.18 In comparing the soil bearing capacities for two adjacent sites, 7 samples from Site 1 and 9 samples from Site 2 are taken and tested. The corresponding capacities in kN/m^2 are summarized in Table E9.18a.

Conduct a Wilcoxon test by pooling the data to test whether the two populations are identical at $\alpha = 0.05$.

Table E9.18a Soil bearing capacity (kN/m^2)

Site 1	Site 2
190	185
174	147
184	193
136	110
200	90
188	117
153	178
	159
	165
$m = 7$	$n = 9$

Continues

Example 9.16: *Continued*

Solution After pooling the data, we arranged them in ascending order and ranked them. The results are summarized in Table E9.18b.

Table E9.18b Ranking for Site 1 and Site 2 pooled data

Sample value	Group	Rank
90	2	1
110	2	2
117	2	3
136	1	4
147	2	5
153	1	6
159	2	7
165	2	8
174	1	9
178	2	10
184	1	11
185	2	12
188	1	13
190	1	14
193	2	15
200	1	16

Adding the ranks for Site (Group) 1:

$$W = 4 + 6 + 9 + 11 + 13 + 14 + 16 = 73$$

From Table A.8 of Appendix A for $m = 7$ and $n = 9$ and at $\alpha = 0.05$, the critical values of the Wilcoxon two-tailed test are 41 and 78. Since $W = 73$ is bounded by 41 and 78, the null hypothesis cannot be rejected. This means the two populations are statistically identical. If we suspect that Site 1 bearing capacity value tends to be larger than those in Site 2, then we use a right-tailed test. In this case for $\alpha = 0.05$, $m = 7$ and $n = 9$, the critical Wilcoxon value is 76. Since $W = 73$ is less than this critical value, then the null hypothesis cannot be rejected; as such, the two populations are identical.

It is noted that the Wilcoxon test is usually applicable when we know populations are symmetric. If we are not sure of this, then other tests of locations, for example the sign test (Guttman, et al., 1982; Milton and Tsokos, 1983; and Walpole, et al., 2006), may be used instead of the Wilcoxon test.

9.5.5 Tests for Several Populations

There are other tests available that can be used for comparing k populations. One such test is the Kruskal-Wallis test used for k populations for both matched and unmatched data. The details of these tests can be found in Milton and Tsokos (1983).

9.6 REGRESSION AND CORRELATION ANALYSIS

In Chapter 7 we discussed the dependence of a random variable on another. A correlation coefficient was introduced and described the degree of linear correlation between a dependent variable and independent variable. In many engineering problems, we may have to go beyond simply finding the correlation. In fact, the interest may be in investigating whether a relationship exists that would describe the dependent variable Y in terms of the independent variable X. It is noted that the basis for establishing such a relationship is a set of data compiled each for the dependent and independent variables. A *regression analysis* refers to the analysis of data to arrive at this relationship. The correlation and regression analyses are conceptually different. However, they are related. Any established relation between Y and X is within the range of data compiled. The validity of such relation is evaluated through the correlation between Y and X.

In problems where only one independent variable is involved, the regression analysis is called simple regression. When several independent variables X_i ($i = 1$ to q) are involved, the regression is said to be *multiple*. Furthermore, recall that the correlation coefficient (ρ) described in Chapter 7 is a measure of linear correlation between X and Y. As a result, if a nonlinear relation between Y and X exists, the correlation coefficient will be substantially less than 1 (recall that $\rho = 1$ is an indication of a perfect linear correlation between X and Y). In most engineering applications, the regression analysis is primarily conducted to arrive at an empirical equation for the dependent variable. The equation is then adopted for use in design and decision-making problems when the exact relation between Y and X is not known or cannot easily be found through theoretical formulations.

Regression analysis problems can be classified into one of the following cases:

- Linear regression of the dependent variable Y on the independent variable X
- Linear regression of the dependent variable Y on q independent variables X_i ($i = 1, 2, \ldots, q$)
- Nonlinear regression of Y on X or Y on several independent variables X_i

In addition, we may be interested in investigating the correlation between Y and X (or between Y and X_i) in these cases. When the data set involves ranks or the sample size is small enough that it can be easily ranked, a measure of linear correlation can also be obtained using the ranks. In any correlation and regression analyses, it is important to note that the statistical samples play a crucial role. A regression equation is only valid within the range of the data collected. As such, one should not extrapolate the regression relationship beyond the boundaries of the compiled data. Finally, it is noted that the correlation coefficient represents a statistical parameter. Therefore, confidence levels may also be established for this parameter, similar to other statistical parameters discussed in Chapter 8.

We begin the discussion with simple linear regression analysis. This is followed by the nonlinear analysis and a comprehensive discussion on correlation analysis.

9.6.1 Simple Linear Regression Analysis

This problem involves regression analysis of a dependent variable Y on an independent variable X. In theory, we will be interested in finding $E(Y|x)$ that yields the expected value of Y given a specific outcome for X such as x. We can write:

$$E(Y|x) = a + bx \tag{9.41}$$

in which a and b are constants. The estimates for a and b are found using the *least square analysis*. Suppose in a data acquisition session, we compile n pairs of sample values for Y and X. These pairs are (x_1, y_1), (x_2, y_2), . . ., (x_n, y_n). Each x_i corresponds to a specific value for Y such as y_i. The pair of values for X and Y are obtained in an experiment, through field data collection, by surveys and so on. A plot of y_i versus x_i often appears in the form of a cluster of points. If a strong linear correlation between X and Y exists, the data will cluster close to each other and show a linear trend. Very weak correlation between X and Y will result in a large scatter in the plotted points and no specific trend between the two. Figure 9.5 shows these two situations. Assuming the exact values for the constants a and b are known, Equation 9.41 can also be plotted along with the scatter plot of y_i versus x_i. As seen in Figure 9.6, for any given value for X such as x_i, there are two values for Y. These are y_i, which is obtained from the data, and y_i^*, which is from Equation 9.41. The difference between the two values $\Delta_i = (y_i^* - y_i)$ is called the *residue*. In

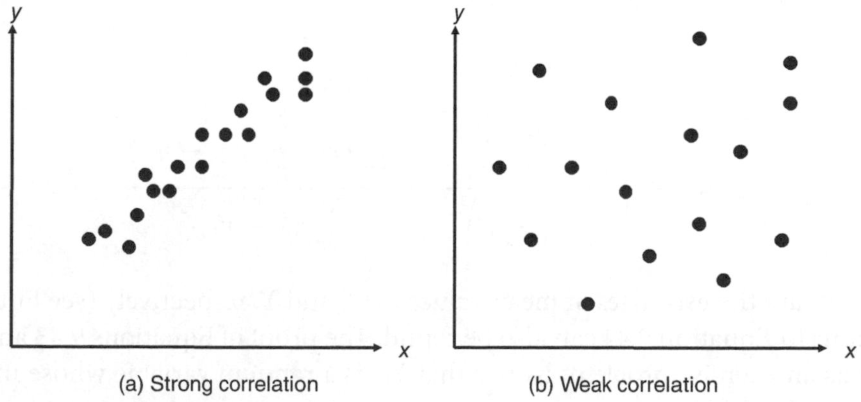

Figure 9.5 Strong and weak correlation cases

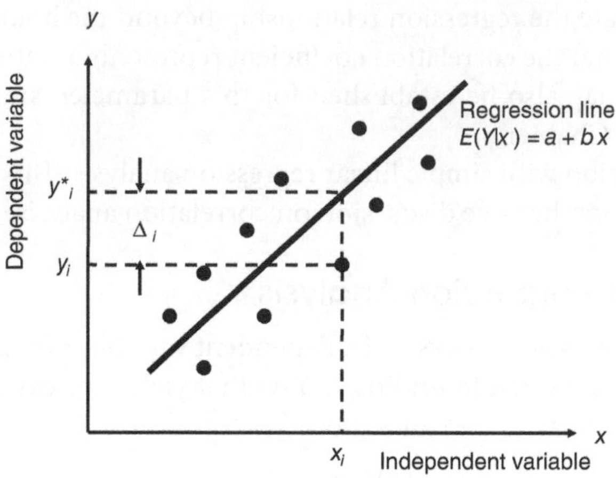

Figure 9.6 Regression line

conducting the least square method, the sum of square values of the residues is minimized to obtain estimates for a and b. These estimates will represent the *best-fit* equation for the data. Denoting the sum of squares of the residues as SS_E:

$$SS_E = \sum_{i=1}^{n} (y_i^* - y_i)^2 = \sum_{i=1}^{n} (a + bx_i - y)^2 \tag{9.42}$$

The minimization involves the following operations:

$$\frac{\partial SS_E}{\partial a} = 0 \quad \text{and} \quad \frac{\partial SS_E}{\partial b} = 0$$

These will result in the following two equations for a and b. Note that because the equations provide only the estimates for a and b, the symbols $\hat{a}$ and $\hat{b}$ are used to indicate that these are only the estimates:

$$\hat{a} = m_Y - \hat{b} m_X \tag{9.43}$$

$$\hat{b} = \frac{n\sum_{i=1}^{n} x_i y_i - \sum_{i=1}^{n} x_i \sum_{i=1}^{n} y_i}{n\sum_{i=1}^{n} x_i^2 - \left(\sum_{i=1}^{n} x_i\right)^2} = \frac{\sum_{i=1}^{n} (x_i - m_X)(y_i - m_Y)}{\sum_{i=1}^{n} (x_i - m_X)^2} \tag{9.44}$$

Here, m_X and m_Y are the estimates of mean values for X and Y, respectively (see Equation 8.10). Other variations to Equation 9.44 can also be found. The proof of Equations 9.43 and 9.44 is left to the reader as an exercise problem. Notice that $Y|x$ is a random variable whose mean value is

$E(Y|x)$. An unbiased estimate for the variance of this random variable is denoted by $S^2_{Y|x}$, and is computed as:

$$\mathrm{Var}(Y \mid x) = S^2_{Y|x} = \frac{1}{n-2}\sum_{i=1}^{n}(y_i^* - y_i)^2 = \frac{1}{n-2}\sum_{i=1}^{n}(a + bx_i - y_i)^2 = \frac{SS_E}{n-2} \quad (9.45)$$

The standard deviation of $Y|x$ (i.e., $S_{Y|x}$) is sometimes shown with two parallel lines along with the regression line on the (Y, X) graph. We must emphasize that this standard deviation is different from s_Y since s_Y is the standard deviation of Y without any reference to X. As it is explained in the next section the two standard deviations (i.e. s_Y, and $S_{Y|x}$) are related to each other through the correlation coefficient. Furthermore, it is noted that the regression of Y on x, that is, $E(Y|x)$, is different from that of X on y, which is $E(Y|y)$. If the respective regression equations are plotted on the same coordinate system, they appear as two different lines. However, it can be shown that the coordinates of the intersection point of these lines are (m_X, m_Y).

9.6.2 Correlation Coefficient

Recall from Section 7.3.2 that the linear correlation coefficient is computed from Equation 7.36. In theory, Equation 7.36 requires information on the joint probability density (or mass) function of X and Y. An estimate for ρ can be found using the data compiled for X and Y. Assuming a uniform joint probability distribution for X and Y, the covariance of X and Y can be written:

$$\mathrm{Cov}(X, Y) = E[(X - \mu_X)(Y - \mu_Y)] = \frac{1}{n}\sum_{i=1}^{n}(x_i - m_X)(y_i - m_Y) \quad (9.46a)$$

An unbiased estimate for the covariance can be found when $n - 1$ instead of n is used:

$$\mathrm{Cov}(X, Y) = \frac{1}{n-1}\sum_{i=1}^{n}(x_i - m_X)(y_i - m_Y) \quad (9.46b)$$

Using s_X and s_Y, that is, the estimates for the standard deviation of X and Y, and denoting the correlation coefficient as r as an estimate of ρ, in light of Equation 7.36:

$$r = \frac{\sum_{i=1}^{n}(x_i - m_X)(y_i - m_Y)}{\sqrt{\sum_{i=1}^{n}(x_i - m_X)^2 \sum_{i=1}^{n}(y_i - m_Y)^2}} \quad (9.47a)$$

$$r = \frac{1}{(n-1)} \times \frac{\sum_{i=1}^{n} x_i y_i - n(m_X m_Y)}{s_X s_Y} \quad (9.47b)$$

From Equations 9.44 and 9.47, it can be shown that:

$$r = \frac{\sum_{i=1}^{n}(x_i - m_X)(y_i - m_Y)}{\sum_{i=1}^{n}(x_i - m_X)^2} \times \frac{s_X}{s_Y} = \hat{b}\frac{s_X}{s_Y} \tag{9.47c}$$

Considering the unbiased estimate for $S_{Y|x}$, (Equation 9.45), and in light of Equation 9.47c, it can be shown that:

$$S^2_{Y|x} = \frac{n-1}{n-2}s^2_Y(-r^2) \tag{9.48}$$

Solving this equation for r^2:

$$r^2 = 1 - \frac{n-2}{n-1} \times \frac{S^2_{Y|x}}{s^2_Y} \tag{9.47d}$$

For large values of n, the ratio $(n - 2)/(n - 1) \approx 1$, and:

$$r^2 = 1 - \left(\frac{S_{Y|x}}{s_Y}\right)^2 \tag{9.47e}$$

For a perfect correlation, $r = 1$ and $S_{Y|x} = 0$. This means that there will be no scatter about the regression line. However, it is emphasized that this does not mean s_Y is zero because s_Y is an estimate of the standard deviation of Y without any reference to its relationship to X.

Example 9.19 In studying the productivity in construction sites discussed earlier, we want to establish a linear regression equation between the productivity (dependent variable) and the work shift hr/da (independent variable). The productivity is measured as the percentage of work completed as opposed to the scheduled work to be completed within a specific time period. In a survey, nine observations were made and the results in Table E9.19a were obtained.

Table E9.19a Productivity versus work shift hr/da

Productivity (%), *Y*	Work shift hr/da, *X*
90	7
85	6.5
75	9.5
83	7
88	7
78	10
90	7.5
87	7
69	11

Continues

Example 9.19: *Continued*

(a) Establish the regression equation $E(Y|x) = a + bx$.

(b) Estimate the standard deviation of $Y|x$.

(c) Estimate the correlation coefficient.

Solution

(a) Equations 9.43 and 9.44 are used for estimating a and b. Table E9.19b summarizes the statistical values needed for a and b computation.

Table E9.19b

x_i	y_i	x_i^2	y_i^2	$x_i \times y_i$
7	90	49	8,100	630
6.5	85	42.25	7,225	552.5
9.5	75	90.25	5,625	712.5
7	83	49	6,889	581
7	88	49	7,744	616
10	78	100	6.084	780
7.5	90	56.25	8,100	675
7	87	49	7,569	609
11	69	121	4,761	759
$\sum x_i = 72.5$	$\sum y_i = 745$	$\sum x_i^2 = 605.75$	$\sum y_i^2 = 62,097$	$\sum x_i \times y_i = 5,915$

$$\hat{b} = \frac{9 \times 5,915 - 72.5 \times 745}{9 \times 605.75 - (72.5)^2} = -3.977$$

$$m_X = \frac{72.9}{9} = 8.056 \quad \text{and} \quad m_Y = \frac{745}{9} = 82.778$$

$$\hat{a} = 82.778 + 3.977 \times 8.056 = 114.817$$

(b) In this part we utilize Equation 9.45. By expanding this equation we obtain:

$$\mathrm{Var}(Y \mid x) = \left(n\hat{a}^2 + \hat{b}^2 \sum x_i^2 + \sum y_i^2 + 2\hat{a}\hat{b} \sum x_i - 2\hat{a} \sum y_i - 2\hat{b} \sum x_i y_i\right)/(n-2)$$

$$\mathrm{Var}(Y|x) = [9(114.817)^2 + (-3.977)^2(605.75) + (62,097) + 2(114.817)(-3.977)(72.5) - 2 \times (114.817)(745) - 2(-3.977)(5,915)]/(9-2) = 84.0/7 = 12.00$$

The standard deviation $S_{Y|x} = 3.46$.

Continues

Example 9.19: *Continued*

(c) In this part we can either use Equation 9.47a or 9.47b. We first must compute s_X and s_Y:

$$s_X^2 = \frac{\sum x_i^2 - n \times m_X^2}{n-1} = \frac{605.75 - 9 \times (8.056)^2}{9-1} = 2.707 \text{ and } s_X = 1.745$$

Similarly:

$$s_Y^2 = \frac{\sum y_i^2 - n \times m_Y^2}{n-1} = \frac{62{,}097 - 9 \times (82.778)^2}{9-1} = 53.403 \text{ and } s_Y = 7.308$$

Thus:

$$r = \frac{1}{9-1} \times \frac{(5{,}915 - 9 \times 8{,}056 \times 82.778)}{1.645 \times 7.308} = -0.90$$

Figure E9.19 presents the data and the regression line. Notice that $S_{Y|x} = 3.46$ is shown as two parallel lines above and below the regression line.

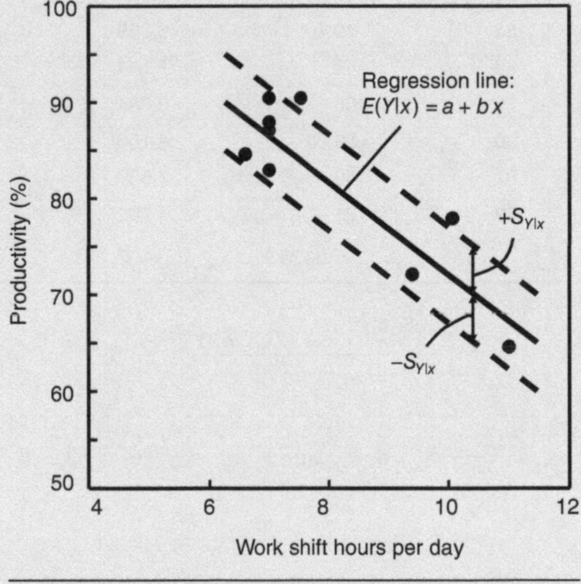

Figure E9.19 Productivity vs. work shift hours per day

9.6.3 Strength of Linear Correlation

The least square method used in establishing a straight-line fit to the X, Y data is merely a mathematical tool that always results in a linear equation between $E(Y|x)$ and x. However, one must use this equation with caution. There may be a case for which no linear relation between Y and X exists; or at best the linear relation may be weak. However, using the procedure explained in Section 9.6.1, we will still come up with an expression for the regression line. It is therefore necessary to investigate how well the straight line obtained through the least square analysis fits the data. Of course, the correlation coefficient can be used as a measure to determine how well the linear equation fits the data. More elaborate methods are conducted by employing a statistic called the *coefficient of determination* and by using ANOVA. Both of these methods are explained in this section.

In Section 9.6.2, we established a direct relationship between the estimates of b (i.e., the slope of the regression line) and r (see Equation 9.47c). Equation 9.47c indicates that b and r always have the same algebraic sign. The positive values of r will always be accompanied by a rise in the regression line (positive slope), while the negative r will be accompanied by a negative slope. We further derived an expression between the correlation coefficient r and the random variation in Y about x (as denoted by $S^2_{Y|x}$) and the variation in Y alone (as denoted by s^2_Y). The following notations are introduced:

$$S_{XY} = \sum_{i=1}^{n} (x_i - m_X)(y_i - m_Y)$$

and

$$S_{XX} = \sum_{i=1}^{n} (x_i - m_X)^2$$

and

$$S_{YY} = \sum_{i=1}^{n} (y_i - m_Y)^2$$

and it is known that $SS_E \backsim S^2_{Y|x}$ and $S_{YY} \backsim s^2_Y$, thus, Equation 9.47e can also be written as:

$$r^2 = 1 - \frac{SS_E}{S_{YY}} = \frac{S_{YY} - SS_E}{S_{YY}} \tag{9.47f}$$

The numerator in this equation is a measure of variability in Y that concerns the linear relationship with X. This is to say that r^2 is the ratio of the measure of variation in linearity of Y to the measure of total variation in Y (without any reference to X). This ratio is called the *coefficient of determination*. The maximum value for this ratio is 1. If r^2 is large (closer to 1), it can be concluded that there is a strong linear relation between X and Y. Smaller r^2 values (i.e., below 0.4) indicate a somewhat weaker linear relation.

The ANOVA can also be used as a means to test whether the straight-line equation obtained for $Y|x$ is associated with statistically significant variation in Y. To conduct ANOVA, we need to identify components, such that the sources of the variability can be recognized through these components. Recall from Section 9.6.1 that the sum of squares of the difference between the observed values of Y and those from the linear equation was shown as SS_E where:

$$SS_E = \sum_{i=1}^{n} [y_i - (a + bx_i)]^2$$

It can be shown that:

$$SS_E = S_{YY} - bS_{XY}$$

or

$$S_{YY} = bS_{XY} + SS_E$$

Thus, the total variation in Y (i.e., S_{YY}) is now written in terms of two components. These are: (1) the variation that is attributed to the linear regression between Y and X (i.e. bS_{XY}); and (2) the *error*, or *residual, sum of squares* (unexplained error). Denoting bS_{XY} as SS_R, we notice that if the relationship between X and Y is close to linear, then most of the variability in Y will be in SS_R. The relative size of SS_R to SS_E can then be used as a basis for ANOVA.

To develop a model to conduct ANOVA, we assume that k specific values of the random variable X are selected. These are $x_1, x_2, \ldots, x_k$. Considering the regression of Y on X, there will be k random variables $Y|x_i$ where $i = 1, 2, \ldots, k$. At this point we assume that these random variables are independent normal with a common variance σ^2. If the regression of Y on X is linear, then the mean values of these k random variables must be all located on the line represented by the regression equation. The regression line representing the mean values can be expressed with an expression $\mu_{Y|x} = \alpha + \beta x$ as shown in Figure 9.7. We further assume that a random sample of size n_i is selected from each normal distribution describing the random variable $Y|x_i$. If y_{ij} is denoted as the jth element of the random sample i, then it has a mean value equal to $\alpha + \beta x_i$ and a variance equal to σ^2. There will also be an observed value associated with y_{ij}; however, this observed value will differ from the mean of the population of y_{ij} by a random amount Δ_{ij}. Denoting this observed value as y_{ij}^* we can write:

$$y_{ij}^* = \alpha + \beta x_i + \Delta_{ij} \tag{9.49}$$

This equation is valid for $i = 1, 2, \ldots, k$ and $j = 1, 2, \ldots, n_i$. Note that if $\beta = 0$, then the variation in Y will not be due to the linear model since all variation in Y will be random. On the other hand, if $\beta \neq 0$, Equation 9.49 will depend on x_i. This means that a portion of variation in Y will be due to the regression line. The null and alternate hypotheses can then be written:

$$H_0 : \beta = 0$$

$$H_1 : \beta \neq 0$$

If the null hypothesis cannot be rejected, it indicates a lack of linear regression at the significance level tested. However, if the null hypothesis is rejected, then there is enough statistical evidence to suggest that the linear regression is valid.

To establish the test statistics for ANOVA, we utilize SS_R and SS_E as described earlier, to define: (1) MS_R, which is called the *regression mean square,* and (2) MS_E, which is referred to as

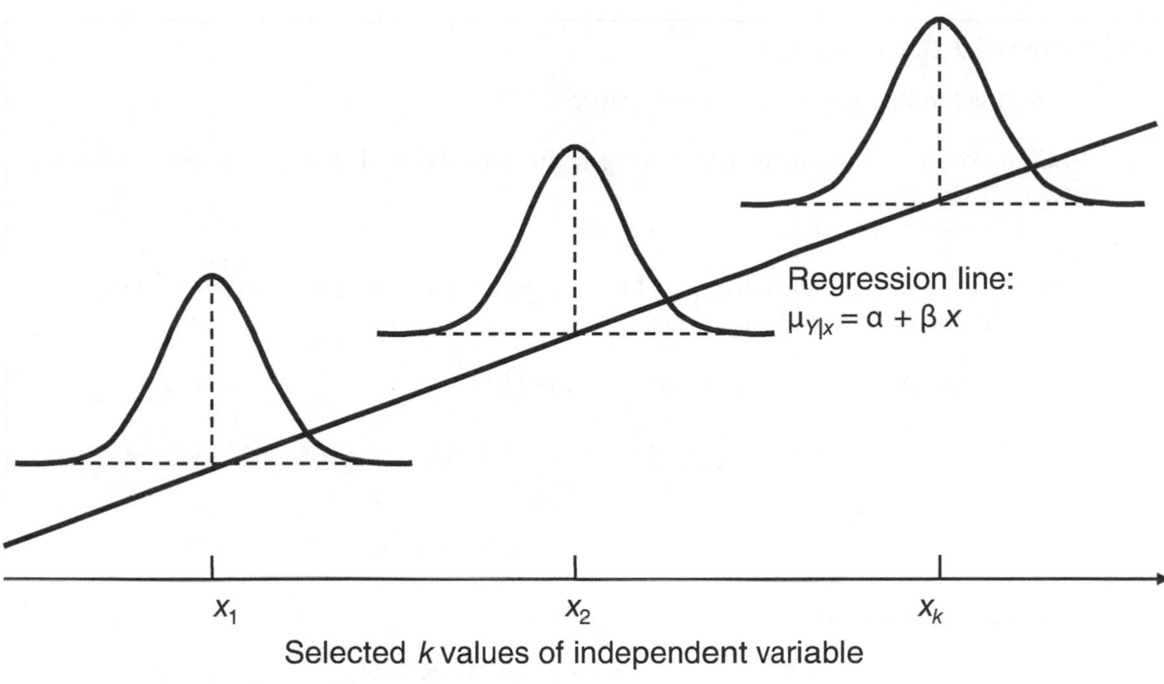

Figure 9.7 Regression line representing the mean value

the *error mean square*. The degrees of freedom associated with these statistics are 1 and $n - 2$, respectively, where $n = n_1 + n_2 + \ldots + n_k$. We can thus write:

$$MS_R = \frac{SS_R}{1} \quad \text{and} \quad MS_E = \frac{SS_E}{(n-2)} \tag{9.50}$$

in which $SS_R = bS_{XY}$ and $SS_E = S_{YY} - SS_R$.

Conceivably, if the null hypothesis is true, then the observed value of the ratio MS_R/MS_E will be close to 1. If the null hypothesis is rejected, the ratio is said to be inflated and, as such, the regression equation is linear. This ratio will follow an F-distribution function with degrees of freedom equal to 1 and $(n - 2)$. In practice, the necessary calculations for ANOVA are arranged in tabular form as shown in Table 9.3.

Table 9.3 ANOVA for linear regression

Variation source	Degrees of freedom	Sum of squares (*SS*)	Mean square (*MS*)	*F*
Regression	1	$SS_R = bS_{XY}$	$MS_R = SS_R/1$	MS_R/MS_E
Error	$n - 2$	$SS_E = S_{YY} - bS_{XY}$	$MS_E = SS_E/(n - 2)$	
Total	$n - 1$	S_{YY}		

Example 9.20 In Example 9.19:

(a) Discuss the strength of the linear regression.

(b) Conduct an ANOVA to further investigate the statistical evidence of the linear relationship between Y and X.

Solution

(a) In this part, we compute r^2 to equal 0.81. This ratio is large enough to suggest that the linear relationship between X and Y is strong.

(b) To conduct ANOVA we must compute S_{XY} and S_{YY}:

$$S_{XY} = \sum_{i=1}^{n}(x_i - m_X)(y_i - m_y) = \sum_{i=1}^{n}x_i y_i - n(m_X m_Y) = 5{,}915 - 9 \times 8.056 \times 82.778 = -86.74$$

$$S_{YY} = \sum_{i=1}^{n}(y_i - m_Y)^2 = \sum_{i=1}^{n}y_i^2 - n(m_Y^2) = 62{,}097 - 9 \times (82.778)^2 = 427.22$$

Based on these, we have:

$$SS_R = bS_{XY} = -3.977 \times (-86.74) = 344.96$$

and

$$SS_E = S_{YY} - SS_R = 427.22 - 344.96 = 82.26$$

Thus:

$$MS_R = SS_R/1 = 344.96, \text{ and } MS_E = SS_E/(n-2) = 82.26/7 = 11.75$$

The test statistic is:

$$F = MS_R/MS_E = 344.96/11.75 = 29.35$$

The critical value of the F-distribution for $\nu_1 = 1$ and $\nu_2 = 9 - 2 = 7$ degrees of freedom is $f_{0.05}(1,7) = 5.59$ at $\alpha = 0.05$ significance. Since $F = 29.35$ is larger than the critical value, the ratio F is *inflated*; and as such the null hypothesis is rejected. This means there is enough statistical evidence to have a linear regression of Y on X. The ANOVA results are summarized in Table E9.20.

Table E9.20 ANOVA results

Variation source	Degrees of freedom	Sum of squares (SS)	Mean square (MS)	F	$f_{0.05}(1,7)$
Regression	1	$SS_R = 344.96$	$MS_R = 344.96$	29.35	5.59
Error	7	$SS_E = 82.26$	$MS_E = 11.75$		
Total	8	$S_{YY} = 427.22$			

9.6.4 Multiple Linear Regression Analysis

Occasionally, in certain engineering problems, it may be necessary to establish the linear regression between the dependent variable Y and q independent variables $X_1, X_2, \ldots, X_q$. The regression equation in this case is written:

$$E(Y \mid x_1, x_2, \ldots, x_q) = a + b_1 x_1 + \ldots + b_q x_q \tag{9.51}$$

The procedure for estimating the constants a and b_j (where $j = 1, 2, \ldots, q$) is the same as that in single linear regression. First, the equation for SS_E (i.e., the sum of the square of the difference between the observed values of Y and the theoretical values from Equation 9.51) is established, then, it is minimized. This will result in $q + 1$ simultaneous equations that are then solved for a and b_j ($j = 1, 2, \ldots, q$). It can be shown that:

$$a = m_Y - \sum_{j=1}^{q} b_j m_j \tag{9.52}$$

and that the constants b_j can be obtained from the following q simultaneous equations:

$$\begin{aligned}
&b_1 \sum (x_{1i} - m_1)(x_{1i} - m_1) + b_2 \sum (x_{1i} - m_1)(x_{2i} - m_2) + \ldots \\
&\qquad + b_q \sum (x_{qi} - m_1)(x_{qi} - m_q) = \sum (x_{1i} - m_1)(y_i - m_Y) \\
&b_1 \sum (x_{2i} - m_2)(x_{1i} - m_1) + b_2 \sum (x_{2i} - m_2)(x_{2i} - m_2) + \ldots \\
&\qquad + b_q \sum (x_{2i} - m_2)(x_{qi} - m_q) = \sum (x_{2i} - m_2)(y_i - m_Y) \\
&\qquad \vdots \\
&b_1 \sum (x_{qi} - m_q)(x_{1i} - m_1) + b_2 \sum (x_{qi} - m_q)(x_{2i} - m_2) + \ldots \\
&\qquad + b_q \sum (x_{qi} - m_q)(x_{qi} - m_q) = \sum (x_{qi} - m_q)(y_i - m_Y)
\end{aligned} \tag{9.53}$$

in which m_j is the mean value of the variable X_j ($j = 1, 2, \ldots, q$); and m_Y is the mean value of Y. Furthermore, in Equation 9.53 the summations are from $i = 1$ to n (note that n is the size of each of the variables Y and X_j); and a term such as x_{ji} means the ith element of the variable X_j. For simplicity, in Equations 9.52 and 9.53 we have dropped the symbol ^ that is often used to describe the estimates for a and b_j. The variance of Y in the regression equation can also be obtained as in the case of the single linear correlation, that is:

$$\mathrm{Var}(Y \mid x_1, x_2, \mathrm{K}, x_q) = \frac{SS_E}{(n - q - 1)} \tag{9.54}$$

in which:

$$SS_E = \sum (y_i - a - b_i x_{1i} - b_2 x_{2i} - \ldots - b_q x_{qi})^2 \tag{9.55}$$

9.6.5 Nonlinear Regression

In many engineering problems, the data set may show a nonlinear trend between Y and X. In these problems, it is desirable to establish a nonlinear regression between Y and X. Usually the scatter plot of Y data versus X sample values will suggest the type of function that may be suitable for the relation between X and Y. If it is determined that the relation follows a specific type of function, such as $g(x)$, then the regression equation can simply be established by linearization. However, if the type of function cannot be identified, a polynomial may be attempted to establish the regression equation. In this case, one may wish to closely examine the scatter plot of Y versus X to decide on the order of the polynomial that would best fit the data. In this section these two methods are explained.

Regression by using a predetermined function: Suppose the scatter plot of Y versus X suggests that the relationship is in the form of a function $g(x)$. For example, this function is exponential or logarithmic (i.e., $g(x) = e^2$ or $g(x) = \ell n\, x$). The regression of Y on X can then be written:

$$E(Y|x) = a + bg(x) \tag{9.56a}$$

The linearization is achieved by substituting an auxiliary variable z for $g(x)$. Since $z = g(x)$, then Equation 9.56a becomes:

$$E(Y|z) = a + bz \tag{9.56b}$$

Constants a and b can now be estimated from Equations 9.43 and 9.44. However, in these equations all x_i will be substituted by $z_i = g(x_i)$.

Nonlinear regression using a polynomial function. In this case, the regression equation is written as a polynomial of the mth order:

$$E(Y|x) = a_0 + a_1x + a_2x^2 + \ldots + a_mx^m \tag{9.57}$$

Once m is selected, a procedure similar to the linear regression method is used to minimize the sum of squares of the differences between the observed values of Y (i.e., y_i) and the theoretical value from Equation 9.57. This results in $m + 1$ simultaneous equations for computing the estimates for the constants $a_0, a_1 \ldots \ldots a_m$ in the form of:

$$\frac{\partial SS_E}{\partial a_j} = 0 \text{ for } j = 0 \text{ to } m$$

These $m + 1$ equations can be written in matrix as:

$$[C]\{a\} = \{R\} \tag{9.58}$$

in which it can be shown that:

$$[C] = \begin{bmatrix} n & \sum x_i & \sum x_i^2 & \cdots \sum x_i^m \\ \sum x_i & \sum x_i^2 & \sum x_i^3 & \cdots \sum x_i^{m+1} \\ \cdots & \cdots & \cdots & \cdots\ \cdots \\ \sum x_i^m & \sum x_i^{m+1} & \sum x_i^{m+2} & \cdots \sum x_i^{2m} \end{bmatrix}$$

and

$$\{R\} = \begin{Bmatrix} \sum y_i \\ \sum y_i x_i \\ \sum y_i x_i^2 \\ \dots \\ \sum y_i x_i^m \end{Bmatrix}$$

and

$$\{a\} = \begin{Bmatrix} a_0 \\ a_1 \\ a_2 \\ \dots \\ a_m \end{Bmatrix}$$

For more information on regression analysis in general, see Myers (1990).

Example 9.21 Bus fares in suburban communities of a major metropolitan area are a function of rides per bus line per day. In a survey of eight communities, the data in Table E9.21a was gathered on the bus fare in dollars (dependent variable, *Y*) and number of rides per line per day (independent variable, *X*).

A regression equation in the form of $E(Y|x = a + b\ell n(x)$ has been suggested for the data. Compute the estimates for *a* and *b* and discuss the correlation of *Y* on $\ell n(X)$.

Table E9.21a

Rides/line/da	Fare ($)
1,050	0.60
1,100	0.75
1,255	0.70
1,270	0.65
1,420	0.80
1,530	0.75
1,625	0.85
1,750	0.85

Solution We introduce $Z = \ell n(X)$ and establish a linear regression between *Y* and *Z*. Table E9.21b summarizes the results:

$$\hat{b} = \frac{8 \times 43.004 - 57.693 \times 5.95}{8 \times 416.30 - (57.698)^2} = 0.396$$

Continues

Example 9.21: *Continued*

$$m_Z = \frac{57.693}{8} = 7.212 \quad \text{and} \quad m_Y = \frac{5.95}{8} = 0.744$$

$$\hat{a} = 0.744 - 0.396 \times 7.212 = -2.112$$

$$E(Y|z) = -2.112 + 0.396z \quad \text{and} \quad E(Y|x) = -2.112 + 0.396\ell n(x)$$

To compute the correlation coefficient between Y and Z, we can either use Equation 9.47a or 9.47b. We first must compute s_Y and s_Z:

$$s_Z^2 = \frac{\sum z_i^2 - n(m_Z)^2}{n-1} = \frac{416.30 - 8 \times (7.212)^2}{8-1} = 0.0281 \quad \text{and} \quad x_Z = 0.168$$

Similarly:

$$s_Y^2 = \frac{\sum y_i^2 - n(m_Y)^2}{n-1} = \frac{4{,}485 - 8 \times (0.744)^2}{8-1} = 0.0081 \quad \text{and} \quad x_Y = 0.090.$$

Thus, using Equation 9.47b:

$$r = \frac{1}{8-1} \times \frac{43.004 - 8 \times 7.212 \times 0.744}{0.168 \times 0.090} = 0.74$$

The r^2 factor is about 0.6. This indicates a reasonable strength for the relationship.

Table 9.21b

x_i	$z_i = \ell n(x_i)$	y_i	z_i^2	y_i^2	$z_i\,y_i$
1,050	6.957	0.60	48.40	0.360	4.174
1,100	7.003	0.75	49.04	0.563	5.252
1,255	7.135	0.70	50.91	0.490	4.995
1,270	7.147	0.65	51.08	0.423	4.646
1,420	7.258	0.80	52.68	0.640	5.806
1,530	7.333	0.75	53.77	0.563	5.500
1,625	7.393	0.85	54.66	0.723	6.284
1,750	7.467	0.85	57.76	0.723	6.347
Sum	57.693	5.95	416.30	4.485	43.004

9.6.6 Spearsman's Rank Correlation Coefficient

In estimating the correlation coefficient between X and Y, we may find out that the data consists of ranks or that the data set is small and it can be ranked readily. In these cases, the Spearsman's rank method is used for measuring the correlation coefficient. Suppose we have compiled n sets of values for X and Y denoted by (x_i, y_i) where $i = 1$ to n. We rank both the X and Y data in ascending order. The ranks for the X values are denoted by r_{xi} and those for Y by r_{yi}. Thus, we now have n

pairs of ranks such as (r_{xi}, r_{yi}). For tied scores, again we use the average of the ranks for those scores that are tied. Spearsman's rank correlation coefficient Y, is obtained from the equation:

$$r_S = \frac{n\sum r_{xi}r_{yi} - \sum r_{xi}\sum r_{yi}}{\sqrt{[n\sum r_{xi}^2 - (\sum r_{xi})^2][n\sum r_{yi}^2 - (\sum r_{yi})^2]}} \tag{9.59}$$

The summations in Equation 9.59 each is from $i = 1$ to n. The results from this equation are usually slightly different from those from Equation 9.47. For large n, the two results are in close agreement with each other. If there are no ties in ranks, the Spearsman's coefficient is given by the simple equation:

$$r_S = 1 - \frac{6\sum d_i^2}{n(n^2 - 1)} \tag{9.60}$$

in which $d_i = r_{xi} - r_{yi}$ and again the summation is from $i = 1$ to n. As discussed previously, the values of r_S close to ± 1 indicate a strong correlation between X and Y.

Example 9.22 In Example 9.19, compute the correlation coefficient using Spearsman's equation.

Solution We need to rank the values of X and Y in ascending order from 1 to 9. Table E9.22 summarizes the ranks for X (i.e., r_{xi}) and ranks for Y (i.e., r_{yi}), r_{xi}^2, r_{yi}^2, and $r_{xi}r_{yi}$. Note that tied scores receive the average of the ranks in the tied-score group. Using Equation 9.59:

$$r_S = \frac{9 \times 192.25 - 45 \times 25}{\sqrt{[9 \times 280.00 - (45)^{22}][9 \times 284.50 - (45)^2]}} = -0.57$$

Table E9.22 Ranked data

Work shift hr/da, X	Ranks r_{xi}	Productivity %, Y	Ranks r_{yi}	r_{xi}^2	r_{yi}^2	$r_{xi}r_{yi}$
7	3.5*	90	8.5**	12.25	72.25	29.75
6.5	1	85	5	1.00	25.00	5.00
9.5	7	75	2	49.00	4.00	14.00
7	3.5*	83	4	12.25	16.00	14.00
7	3.5*	88	7	12.25	49.00	24.50
10	8	78	3	64.00	9.00	24.00
7.5	6	90	8.5**	36.00	72.25	51.00
7	3.5*	87	6	12.25	36.00	21.00
11	9	69	1	81.00	1.00	9.00
Sum	45		45	280.00	284.50	192.25

*This is the average of Ranks 2, 3, and 4 that are tied.
**This is the average of Ranks 8 and 9 that are tied.

SUMMARY

In this chapter, we reviewed several useful statistical methods for the analysis of engineering data. The focus of the discussion was on hypothesis testing, comparing means from two or more populations, the ANOVA, and correlation and regression analysis. The discussion on these methods is rather comprehensive. We presented only those methods that are commonly used in statistical analysis of data. The applications of these methods in engineering are numerous. One may wish to apply these methods as statistical evidence that the data sets compiled in two groups are either identical or different. Furthermore, an engineer may wish to apply these methods in connection with providing the necessary evidence as to whether a presumed notion is valid. In general, we recommend the use of the methods presented in this chapter for those engineers who are involved in research that require field or laboratory data compilation.

REFERENCES

Guttman, I., S. S. Wilks, and J. S. Hunter (1982). *Introductory Engineering Statistics*, 3rd ed., New York, NY: John Wiley & Sons.

Klir, G. J., U. H. St. Clair, and B. Yuan (1997). *Fuzzy Set Theory, Foundations and Applications*, Upper Saddle River, NJ: Prentice-Hall.

Milton, J. S. and J. O. Tsokos (1983). *Statistical Methods in the Biological and Health Sciences*, New York, NY: McGraw-Hill Publishing.

Myers, R. H. (1990). *Classical and Modern Regression with Applications*, 2nd ed., Boston, MA: Duxbury Press.

Nelson, M. and W. Illingworth (1990). *A Practical Guide to Neural Nets*, Reading, MA, USA, Addison-Wesley Publishing.

Walpole, R. E., R. H. Myers, S. L. Myers, and K. Ye (2006). *Probability and Statistics for Engineers and Scientists*, 8th ed., Upper Saddle River, NJ: Prentice-Hall.

Yao, J. T. P. (1985). *Safety and Reliability of Existing Structures*, Marshield, MA: Pittman Publishing.

EXERCISES

1. In designing the traffic light for an intersection, a traffic engineer believes that the duration of the yellow light is 3.0 sec on the average. The engineer collects sample data from 10 similar intersections. The results are:

   2.5, 2.5, 4.0, 3.0, 1.5, 2.0, 1.5, 3.0, 2.0, 3.5

   (a) Estimate the mean and standard deviation of the compiled data
   (b) Conduct a hypothesis test at $\alpha = 0.05$ to determine whether the engineer's belief on the mean duration of yellow light is supported

   Hint: Use a two-tailed hypothesis testing.

2. Use the least square analysis to derive Equations 9.43 and 9.44.

3. In conducting an inspection of the condition of main steel girders in several railroad bridges, an engineer believes that the mean effective cross section area is at least equal to 90 percent of the original cross-sectional area as indicated on the design drawings. In a survey of 12 different girders, the cross-sectional area was measured as a fraction of the "design cross-sectional area" at the midspan of each girder. The results were:

   0.76, 0.95, 0.73, 0.82, 0.85, 0.85, 0.89, 0.91, 0.78, 0.81, 0.86, 0.87.

   (a) Compute the mean and standard deviation of the data given
   (b) Test the hypothesis that the mean ratio of the cross section area to the design cross-sectional area is larger than 0.90 per engineer's notion

4. Two groups of structural engineers were invited to rate the condition of several single family unreinforced masonry houses following a moderate earthquake in a seismic area. One group (Group I) consisted of $n_1 = 16$ surveys; the other (Group II) consisted of $n_2 = 13$ surveys. The engineers rated the houses on a scale system of 1-5. These scales are interpreted as follows:

   1 = Severe damage to the house; house needs to be demolished
   2 = Major damage including large cracks, some foundation displacements; major repair is needed
   3 = Moderate damage including some cracks; moderate levels of repair will be needed
   4 = Minor damage to masonry and plaster; some repair will be needed
   5 = No damage occurred to the house

   Upon the completion of the survey, the scales from each group were analyzed. The mean (m) and standard deviation (s) for each group were then computed as given:

Group I	Group II
$n_1 = 16$	$n_2 = 13$
$m_1 = 3.6$	$m_2 = 4.0$
$s_1 = 0.70$	$s_2 = 0.95$

   (a) Determine whether the variances from the two groups are identical at $\alpha = 0.05$
   (b) Given the result in (a), compare the two mean values at $\alpha = 0.05$ and determine whether the difference between them is significant

5. In a traffic engineering research, the efficiency of gates at toll booths is being investigated. Two different devices are considered for automatic opening of gates. One uses coins; the other uses a prepaid card. The average time to collect the toll for each car and open the gate for the car to pass through is estimated in a random sample of $n_1 = 26$ cars for the coin

collector system; and $n_2 = 31$ for the card system. The results for the mean (m) and standard deviation (s) are summarized:

Coin system	Card system
$n_1 = 26$	$n_2 = 31$
$m_1 = 6.2$ sec	$m_2 = 5.0$ sec
$s_1 = 0.70$	$s_2 = 0.95$

(a) Determine whether the variances from the two populations are identical at $\alpha = 0.05$

(b) Given the result in (a), compare the two means at $\alpha = 0.05$ and determine whether the difference between them is significant

6. A concrete manufacturing plant is testing the effectiveness of two types of agents that delay the curing of concrete without compromising the compressive strength and concrete workability. In a series of tests, concrete samples were prepared using the two agents (designated as Agents I and II). The delay time was measured for these samples. The results are summarized:

Agent I	Agent II
$n_1 = 9$	$n_2 = 9$
$m_1 = 95$ min	$m_2 = 106$ min
$s_1 = 15$	$s_2 = 12$

(a) Determine whether the variances from the two populations are identical at $\alpha = 0.10$

(b) Based on the result in (a), compare the two means at $\alpha = 0.10$ and determine whether the difference between them is significant

7. In Exercise 6, the compressive strength was also computed. The results were:

Agent I	Agent II
$n_1 = 9$	$n_2 = 9$
$m_1 = 53.0$ MPa	$m_2 = 49.0$ MPa
$s_1 = 5.0$ MPa	$s_2 = 3.9$ MPa

(a) Determine whether the two variances from these populations are identical at $\alpha = 0.10$

(b) Based on the finding in (a), compare the two means at $\alpha = 0.10$ and determine whether the difference between them is significant

8. For quality assurance purposes, an engineer wishes to compare three brands of high-strength bolts for their shear capacity in kN. Samples of the three brands were obtained and

tested for shear capacity. All test samples had 19 mm diameters. The results for the three brands appear in Table P9.8.

Table P9.8 Bolt shear capacity in kN

Brand I	Brand II	Brand III
81	76	82
75	92	80
83	83	76
86	79	77
78	88	89
85	82	85
90	88	84
	77	80
	90	79
		88
		86
$n_1 = 7$	$n_2 = 9$	$n_3 = 11$

Assume the distribution for each sample is normal.

(a) Compute the mean shear strength in each group and the grand mean for all groups

(b) Conduct a one-way classification completely random design ANOVA with fixed effect to test the hypothesis that the means are identical at $\alpha = 0.10$ and $\alpha = 0.05$

9. In Exercise 8, conduct a multiple range test to further compare the means at $\alpha = 0.05$.

10. In Exercise 1, assume the engineer notion is that the median for the yellow light duration is 3 sec. Furthermore assume that the test of normality does not pass; and as such a distribution-free method is more appropriate. Conduct a Wilcoxon signed-rank test for the null hypothesis that the median yellow light duration is equal to 3 sec. Note that there are several groups of tied scores in the data.

11. In Exercise 8, assume the normality test does not pass.

    (a) Compare Brands I and II using the Wilcoxon signed-rank test (unmatched data) to determine whether the two populations are identical at $\alpha = 0.10$. (Hint: you need to pool the data.)

    (b) Repeat (a) for comparing Brands I and III.

    (c) Repeat (a) for comparing Brands II and III.

12. An engineer managing a construction project is developing a scheduling and activity network for use in the critical path method. There is an uncertainty associated with the

estimated duration of certain activity in the network. The engineer believes that the median duration of this activity is 12 da. To further support this notion, she compiles actual data for the same activity in 9 previous projects. The results in days are:

10, 8, 8.5, 14, 11, 7, 9.5, 9.5, 7.5

Conduct a Wilcoxon signed-rank test for the null hypothesis that the median for the duration of the activity is equal to 12 da at $\alpha = 0.05$.

13. In Exercise 12, assume the test of normality passes. Furthermore assume the mean duration for the activity is believed to be equal to 12 da. Use the data given in Exercise 12 and test the null hypothesis that the mean is equal to 12 da at $\alpha = 0.10$.

14. The stress value at a location on the flange of an aluminum front spar in the right wing of a test aircraft is affected by the aircraft's vertical accelerations that exceed $2.0g$, where g is the gravity acceleration. In a flight survey, the incidences of acceleration exceeding $2.0g$ were measured during 300 flight hr. At each occurrence of acceleration exceeding $2.0g$, the stress at the aforementioned location was also measured. In total there were 19 such occurrences. The results appear in Table P9.14.

Table P9.14

Acceleration in *g* (Variable *X*)	Stress in MPa (Variable *Y*)	Acceleration in *g* (Variable *X*)	Stress in MPa (Variable *Y*)
3.2	72	2.1	38
3.4	80	2.7	60
2.8	60	3.6	87
3.0	43	2.2	49
3.3	56	2.3	57
3.0	64	2.4	63
3.5	71	2.9	75
2.2	47	3.1	79
2.7	54	2.5	49
3.0	76		

(a) Plot the dependent variable Y versus X.
(b) Compute the constants a and b in the regression equation $E(Y|x) = a + bx$.
(c) Compute Var $(Y|x)$.
(d) Compute the correlation coefficient between X and Y.
(e) What will be the estimated stress at an acceleration equal to $3.25g$?

15. In Exercise 14

(a) Discuss the strength of the linear regression between X and Y
(b) Conduct ANOVA to further investigate the linear regression of Y on X

16. The annual growth of the volume of CMV for a highway depends on the average daily traffic (ADT) volume along the highway. Assume in a survey of eleven roadways, the average percentage of annual CMV growth over a 5-year period versus the ADT volumes was obtained as summarized in Table P9.16.

Table P9.16

ADT volume (Variable X)	CMV yearly growth in % (Variable Y)
16,350	3.2
17,690	3.0
18,390	3.6
20,430	3.5
21,740	3.6
22,400	4.0
23,390	3.9
25,630	4.0
26,600	4.1
27,950	4.3
28,840	4.3

(a) Establish a linear regression of Y on X.
(b) Compute the correlation coefficient between X and Y.
(c) Given an ADT volume of 24,500 vehicles, what will be the estimate for the CMV growth per year?

17. In Exercise 16

(a) Discuss the strength of linear regression
(b) Conduct an ANOVA to further discuss the linearity of the regression of Y on X

18. In Exercise 16, a nonlinear regression equation in the form of logarithm of X is suggested. Compute a and b, and estimate the annual CMV growth for an ADT volume of 24,500 vehicles.

19. Two groups of experts were invited to rank the quality of construction conducted by 10 contractors of public works. One group consisted of engineers from the public sector; the other from private consulting companies. The ranking was done on a scale of 1 to 10; in which a ranking of 10 means an excellent quality of work and 1 means an unacceptable quality. Scales 2-9 are intermediate measures of quality between the two extreme scales. Table P9.19 summarizes the rankings from the two groups of experts. Compute the correlation coefficient using Spearsman's equation for the opinions expressed by the two groups.

Table P9.19

Contractor	Ranking by Group I experts	Ranking by Group II experts
A	1	1
B	3	2
C	10	10
D	2	4
E	6	7
F	5	5
G	7	6
H	8	3
I	9	8
J	4	9

20. A company is marketing an additive for gasoline to improve car mileage per gallon of gasoline consumed. To conduct test runs, 10 cars were selected. The highway miles per gallon (MPG) for each car before and after using the additive were measured. The results are provided in Table P9.20.

Table P9.20

Case	MPG without additive	MPG with additive
1	25.5	25.0
2	19.0	20.1
3	18.2	19.2
4	17.5	18.2
5	24.0	24.0
6	21.0	21.5
7	27.5	27.4
8	18.5	20.0
9	19.4	20.1
10	26.0	26.2

(a) Compute the mean MPG before and after using the additive.
(b) Determine whether there is enough statistical evidence that the MPG is improved upon using the additive. Use $\alpha = 0.05$. (Hint: Compare means using paired t by assuming the data in each group follows the normal distribution.)

21. In Exercise 20, use a distribution-free method (Wilcoxon) to compare the two populations.

Basic Hard Systems Engineering: Part I 10

10.1 INTRODUCTION: HARD SYSTEMS ANALYSIS

In Chapter 1 we dealt briefly with the rudiments of the systems approach to problem solving. In this chapter, we begin by elaborating the meaning of some specific terms associated with the systems approach. This approach offers systemic (holistic rather than piecemeal) and/or systematic (step-by-step rather than intuitive) guidelines to problem solving. Both systemic and systematic methodologies and techniques are used by engineers. Techniques, in general, are precise, specific programs of action that will produce a standard result. Methodology, on the other hand, lacks the precision of a technique, but is a more definitive guide to action when compared to a philosophy.

Nearly all important real-world problems that we face on a day-to-day basis are systemic problems, such as sustainability and environmental problems, homelessness and poverty problems, social and economic problems, and so forth. These complex problems are truly systemic in the sense that they cannot be attacked on a piecemeal basis, partly because of their interconnectedness. From the early 1950s, systems analysis (which is the economic appraisal of different means of meeting a defined end) and systems engineering (which involves the design of complex, technical systems to insure that all components operate in an integrated, efficient way) have been widely used in problem solving all over the world. When systems analysis and systems engineering are put to use for solving problems concerning natural and physical systems, we describe this approach as the *hard* systems approach. Essentially, the hard systems approach defines the objectives to be achieved and then *engineers* the system to achieve these objectives. However, when dealing with problems involving human activity systems, one notices that the problem is usually ill-defined. Such cases are defined as *soft* systems. In contrast to hard systems engineering, soft systems methodology does not seek to mechanically design a solution as much as it orchestrates a process of learning (Checkland, 1981; Khisty, 1993).

Operations research (OR) and management science can also be classified as a hard systems methodology comprising a range of techniques that are typical of the means-end approach. It is well known that OR emerged as a means of tackling the vast logistical problems that were encountered during World War II. Later, a variety of formal quantitative techniques based on the principles of OR were developed for use in every conceivable area, including manufacturing, production, transportation, and construction management (CM).

This chapter deals with several techniques that form the basis of hard systems methodology. First, we deal with methods based on calculus. Next, three of the best known methods of network analysis (Critical Path Method [CPM], Program Evaluation and Review Technique [PERT], and Line-of-Balance [LOB]) used extensively in CM are presented, followed by three other methods of network analysis: shortest path, minimal spanning tree and maximal flow. Lastly, the basic ideas of linear programming (LP) are described. LP is a quantitative method of analysis used extensively in business and engineering.

10.2 METHODS BASED ON CALCULUS

The classical methods of calculus provide elegant and powerful solutions to a relatively large number of problems encountered in engineering and economics. At the same time, one of the principal assumptions on which it rests is that the variables that describe a problem must be continuous along all points. This assumption limits its use for practical problem solving, that is, network systems in transportation or choosing between discrete projects in CM. Many of the tools described in this chapter are predominantly linear and involve the solution of sets of linear equations. A large number of managerial problems consist of one or more nonlinear relationships, where traditional linear solution methods are not applicable. Fortunately, you have already been exposed to the basic principles of microeconomics in Chapter 4, and some of the examples worked out with respect to demand, supply, and elasticity are revisited in this section.

The optimization of a nonlinear objective function may be constrained or unconstrained. The former may be solved by the method of substitution or by the use of Lagrange multipliers. The best way to get familiar with these techniques is to work through the examples given in this chapter.

10.2.1 Production Function Characteristics

A production function is a basic representation for the conversion of resources to products. A production function could be represented as:

$$Z = k(x_1, x_2, \ldots x_n)$$

For example, Z could be the maximum number of houses provided by a city, where x_1 represented the land provided, x_2 the labor supplied, and so on. The shape of the production function has important implications regarding where to search for an optimum solution. The following example illustrates the use of calculus.

Example 10.1 If the total cost (TC) of providing labor for the repair of motors is given by:

$$\text{TC} = 40 + 24X - 5.5X^2 + \frac{1}{3}X^3$$

where X = number of labor involved in repairs and TC = total cost, find the relative minimum and maximum labor force required for this cost function. What is your recommendation?

Solution

$$\text{TC} = 40 + 24X - 5.5X^2 + \frac{1}{3}X^3$$

The necessary and sufficient condition for a maximum or minimum are:

$$\frac{d(\text{TC})}{dX} = 24 - 11X + X^2 = 0$$

$$\therefore X = 8 \quad \text{or} \quad X = 3$$

Taking the second derivative:

$$\frac{d^2(\text{TC})}{dX^2} = -11 + 2X$$

At $X = 8$, $\quad -11 + (2)(8) = 5 > 0$
At $X = 3$, $\quad -11 + (2)(3) = -5 < 0$

Thus, at $X = 8$, TC = $40 + 24(8) - 5.5(8)^2 + (8)^3$ = \$50.67 (minimum) and at $X = 3$, TC = $40 + 24(3) - 5.5(3^2) + (3)^3/3$ = \$71.50 (maximum).

It is recommended that the labor force be kept at 8 laborers to minimize the cost function.

10.2.2 Relationship among Total, Marginal, and Average Cost Concepts and Elasticity

You were introduced to the price elasticity and cost functions in Chapter 4. We make use of these concepts in this section. Remember that the price elasticity (e) of demand is:

$$e = \frac{dq}{dp} \times \frac{p}{q}$$

which is frequently expressed as:

$$e = \frac{dq/dp}{q/p} = \frac{\text{marginal cost}}{\text{average cost}}$$

The next example uses this relationship.

Example 10.2 If the demand for bus travel tickets between two cities is $q = 800 - 5p - p^2$, where p, the price of the ticket, is \$10, and q is the number of tickets sold, what is the price elasticity of demand?

Solution

$$q = 800 - 5p - p^2$$

$$\frac{dq}{dp} = -5 - 2p$$

Substituting the value of $p = 10$ in the equation:

$$\frac{dq}{dp} = -5 - 2 \times 10 = -25$$

Next, find the number of tickets sold when $p = \$10$:

$$q = 800 - 5(10) - (10)^2 = 650$$

Substituting these values:

$$e = \frac{dq}{dp} \times \frac{p}{q} = (-25)\left(\frac{10}{650}\right) = -0.3846$$

Hence, inelastic.

Example 10.3 If a company's demand function for machines is $p = 45 - 0.5q$ and its average cost function is:

$$\text{AC} = q^2 - 8q + 57 + \frac{2}{q}$$

find the level of output which (a) maximizes total revenue, (b) minimizes marginal cost (MC), and (c) maximizes profits.

Solution

(a) Demand function is $p = 45 - 0.5q$

Total revenue (TR) is $(p)(q) = (45 - 0.5q)q = 45q - 0.5q^2$

To maximize q, $\frac{d(\text{TR})}{dq} = 45 - (0.5)(2)q$ and equating this to zero:

$$q = 45$$

Testing the second-order condition:

$$\frac{d^2(\text{TR})}{dq^2} = -1 < 0$$

Continues

Example 10.3: *Continued*

Thus, at $q = 45$, TR is a maximum.

(b) From the average cost function $AC = q^2 - 8q + 57 + 2/q$

$$\text{Total cost, TC} = (\text{AC})(q) = \left(q^2 - 8q + 57 + \frac{2}{q}\right)q = q^3 - 8q^2 + 57q + 2$$

Marginal cost:

$$\text{MC} = \frac{d(\text{TR})}{dq} = 3q^2 - 16q + 57$$

MC is minimized when:

$$\frac{d(\text{MC})}{dq} = 6q - 16 = 0, \quad \text{and} \quad q = 2\frac{2}{3}$$

Testing the second-order condition $\frac{d^2(\text{MC})}{dq^2} = 6 > 0$

$$\therefore \text{at } q = 2\frac{2}{3} \text{ MC is at a minimum}$$

(c) Profit = TR − TC:

$$= (45q - 0.5q^2) - (q^3 - 8q^2 + 57q + 2) = -q^3 + 7.5q^2 - 12q - 2$$

for maximizing profit (Pr):

$$\frac{d(\text{Pr})}{dq} = -3q^2 + 15q - 12 = 0$$

$$\therefore q = 1 \text{ or } q = 4$$

testing the second-order conditions:

$$\frac{d^2(\text{Pr})}{dq^2} = -6q + 15$$

At $q = 1$; this results in $9 > 0$
At $q = 4$; this results in $-9 < 0$
$\therefore$ Profits are maximized at $q = 4$

and Profit $= -(4)^3 + 7.5(4)^2 - 12(4) - 2 = 6$

10.2.3 The Method of Lagrange Multipliers

The method of Lagrange multipliers can be used for solving constrained optimization problems consisting of a nonlinear objective function and one or more linear or nonlinear constraint equations. The constraints, as multiples of a Lagrange multiplier, λ, are subtracted from the objective function resulting from a unit change in the quantity value of the constraint equation. This characteristic will be obvious when the following example is worked through.

Example 10.4 The cost function of a firm selling two products A and B is $C = 8A^2 - AB + 12B^2$. However, the firm is required by contract to produce a minimum quantity of A and B totaling 42. What are the values of A and B, and what is your interpretation of the value of λ?

Solution Set the constraint to 0, multiply it by λ and form the Lagrange function:

$$C = 8A^2 - AB + 12B^2 + \lambda(A + B - 42)$$

Take the first-order partials:

$$C_A = 16A - B + \lambda = 0$$

$$C_B = -A + 24B + \lambda = 0$$

$$C_\lambda = A + B - 42 = 0$$

$$(A = 25;\ B = 17;\ \text{and}\ \lambda = -383$$

which means that a one-unit increase in the production quota will lead to an increase in cost by approximately $383.

10.3 CRITICAL PATH METHOD

One of the popular uses of network analysis is for the planning and monitoring of projects before and during execution. Such analysis is vital in order to finish a project within the budget allotment and prescribed time limit. CPM and PERT are the two most popular network analysis techniques used for project planning. Developed in the late 1950s to aid in the planning and scheduling of large projects, today, CPM and PERT are used worldwide.

Both techniques have many characteristics in common, although CPM is deterministic whereas PERT is probabilistic. Both involve the identification and proper sequencing of specific tasks or activities to complete projects in time. Also, the relationship between specific tasks and the logic of precedence is important, as is their duration and quantification. Coupled with these qualities is the classification and quantity of labor, along with their periods of time and wages. The planning of cash flows and financial assistance is also a crucial part of CPM and PERT.

Project planning involves the identification and sequencing of specific tasks, their duration, and their relationships. This process is represented by a network, not necessarily drawn to scale. Two types of networks are currently in use; an Activity-on-Arrow (AOA) and an Activity-on-Node (AON). We describe only the AOA network. The AOA network consists of arrows (branches) and nodes. The arrows represent activities (or tasks) while the nodes represent the

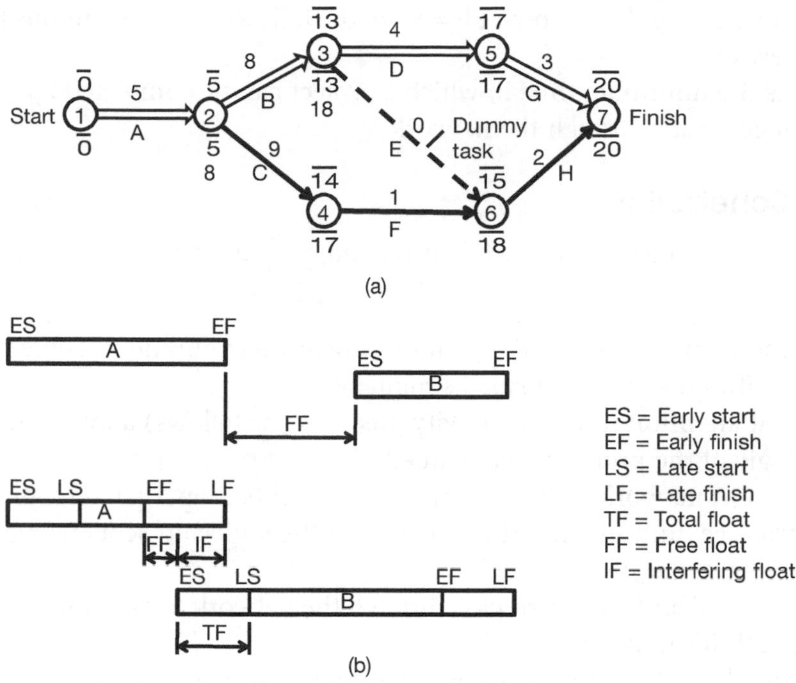

Figure 10.1 (a) CPM network showing nodes and arrows and (b) total, free, and interfering float

beginning and end of activities referred to as events. Since a number of terms are used in CPM, it is best to begin describing them with the help of a typical diagram (see Figure 10.1).

10.3.1 Key Concepts

1. CPM is a linear graph consisting of nodes and arrows as shown in Figure 10.1.
2. Two methods of diagramming can be used: AON or AOA. We use the AOA method.
3. Dummy activities have zero duration. For example, Activity (3-6) is a dummy activity which means that Activity (6-7) cannot start before Activity (2-3) is completed.
4. The forward pass gives the early start (ES) and the early finish (EF) time of an activity. The forward pass establishes the earliest time for each event.
5. The backward pass gives the late start (LS) and the late finish (LF) time of an activity. The backward pass is simply a reversal of that for calculating the earliest event time.
6. When ES = LS for an activity, it lies on the critical path.
7. The critical path (CP) is the set of activities that cannot be delayed if the project is to be completed on time.
8. Total float (TF) is the amount of time that an activity may be delayed without delaying the completion of the project. TF = LF – EF = LS – ES; also, TF = FF + IF. Free float (FF) is the time that the finish of an activity can be delayed without delaying the ES

time of any activity that follows. FF = ES of the following activity minus EF of the activity in question.

9. The CP is the minimum time in which a project can be completed and is the duration of the longest path through the network.

10.3.2 CPM Scheduling

To keep control over a CPM network while it is being prepared and worked out, the following steps are useful:

(a) List all the activities sequentially and estimate their duration. There may be two or more activities that are performed simultaneously.
(b) Pay special attention to which activity precedes (or follows) another activity, so that a proper logic of the project is maintained.
(c) Draw an AOA network with the activities and events properly interconnected. If necessary, introduce dummy activities to maintain the logic and sequencing (in time) of all the activities in question.
(d) Make a forward and backward pass through the network to establish ES, LS, EF, and LF times for all the activities.
(e) Determine the CP and the corresponding critical activities.
(f) Prepare a table with all the details as shown in Table 10.1.

Note that the first activity starts at zero and we add the duration to its ES to obtain its EF time. In this manner, you can progress through the network calculating ES and EF times for all activities, always choosing the preceding EF with largest time, at that node. Next, we can work backward from right to left, which is called the backward pass. On the last activity the EF time becomes the LF time, in order to finish the project as soon as possible. The LF time of the last activity is its LF time minus its duration. Working backward, the LF and LS times for preceding activities can be determined, noting always that the smaller value has to be taken into account. The CP is the longest interconnected path through the network. All activities on this path have the same ES and LS times (and similarly they have the same EF and LF times). Note that these activities have no float to their durations. Finally, all values of ES, LS, and LF times are put in a table (see Table 10.1) and the TF and FF are calculated as per definitions given before.

10.3.3 The Time-grid Diagram and Bar Charts

CPM networks are not generally drawn to scale and, therefore, the lengths of the arrows do not represent the duration of tasks. However, the arrows in time-grid diagrams are drawn to scale in the horizontal direction (but not in the vertical scale). FTs are represented by broken horizontal lines whose lengths indicate time. (See Figure E10.5.) Project network activities can also be represented by bar charts (or Gantt charts) as shown in Figure E10.5.

Table 10.1 Activities, times, and floats

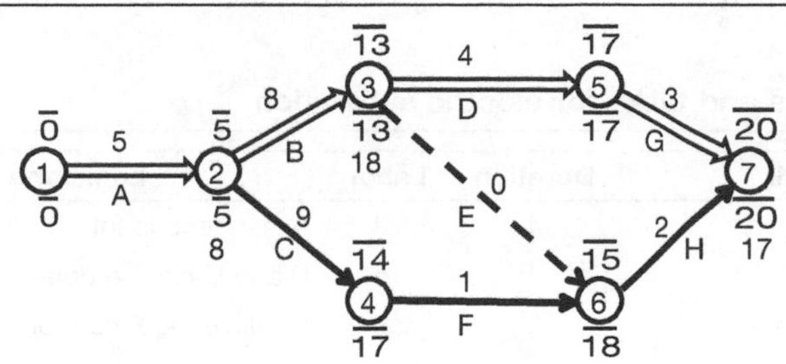

Activity		Duration	ES	EF	LS	LF	TF	FF	IF
*A	1–2	5	0	5	0	5	0	0	0
*B	2–3	8	5	13	5	13	0	0	0
C	2–4	9	5	14	8	17	3	0	0
*D	3–5	4	13	17	13	17	0	0	0
E	3–6	0	—	—	—	—	—	—	—
F	4–6	1	14	15	17	18	1	0	0
*G	5–7	3	17	20	17	20	0	0	0
H	6–7	2	15	17	18	20	2	0	0

*Activities on the critical path

10.3.4 Resource Scheduling

Project managers generally want to proceed with projects under the condition that they will be executed efficiently at scheduled rates. They would like to verify that the resources, in the shape of man power, cash flows, machinery, and requirements are available to them. It is worthwhile for managers to draw up a resource allocation diagram, say with an ES timing, to see whether it could be improved. One can, of course, only tinker with those activities that are not on the critical path. This is exactly what has been done through the resource leveling procedure shown in Figure E10.5. Notice how the fluctuations of labor have been leveled off.

Example 10.5 An electric substation is to be installed and the following basic tasks are identified (Table E10.5). The site will be cleared and leveled and the necessary materials for preparing the foundations and fencing will be procured and stored at site. Next, the foundation excavation and pouring of concrete will be done, a fence will be constructed, and the site will be cleared up and handed over to the authorities. Draw the activities network, perform the forward and backward passes, complete the activities, times, and floats table, draw the time-grid and Gantt chart, and finally sketch resource allocation diagrams, first with ES times followed by the procedure of leveling the labor resource.

Continues

Example 10.5: *Continued*

Table E10.5a Activities and tasks for electric substation

Activity	Task	Duration	Labor	Logic and sequence
A	Select site	4	3	A is independent
B	Clear site	4	5	B and C can be done simultaneously
C	Procure materials	5	6	C follows A; F can follow C and D
D	Excavate foundation	3	10	
E	Fix fence	6	4	E can be done only after A and B are finished
F	Pour concrete and cure	8	8	F can be done only after D is finished
G	Fix gate to fence	5	2	After E and F are completed
H	Dummy	0	0	Clean up cannot start till gate is fixed
I	Clean up and hand over	2	7	

Solution Table E10.5b presents activities and their corresponding events.

Table E10.5b Activities, times, and floats for electric substation

Activity	Event	Duration	ES	EF	LS	LF	TF	FF	IF
A*	1-2	4	0	4	0	4	0	0	0
B*	2-3	4	4	8	4	8	0	0	0
C	2-4	5	4	9	6	11	2	2	0
D*	3-4	3	8	11	8	11	0	0	0
E	3-6	6	8	14	13	19	5	5	0
F*	4-6	8	11	19	11	19	0	0	0
G	3-5	5	8	13	14	19	6	6	0
H	5-6	0	19	19	19	19	0	0	0
I*	6-7	2	19	21	19	21	0	0	0

*On the critical path

Figures E10.5 shows (a) the plan of the substation, (b) the activities network, (c) the forward and backward passes, (d) a typical way of showing ES, EF, LS, and LF, (e) the time-grid diagram, (f) the Gantt chart, (g) resource allocation based on ES, and (h) resource leveling.

Continues

Example 10.5: *Continued*

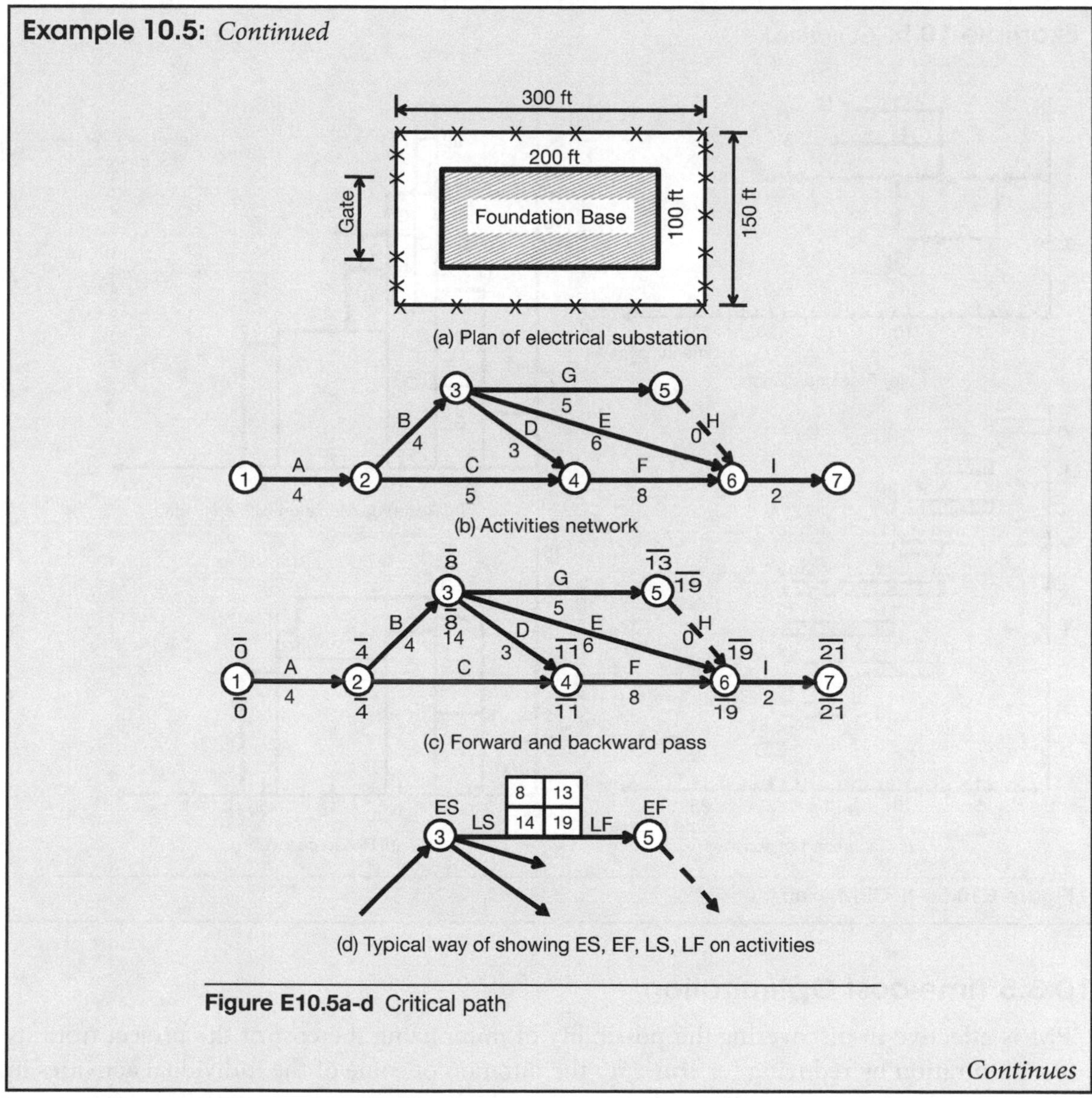

Figure E10.5a-d Critical path

Continues

Example 10.5: *Continued*

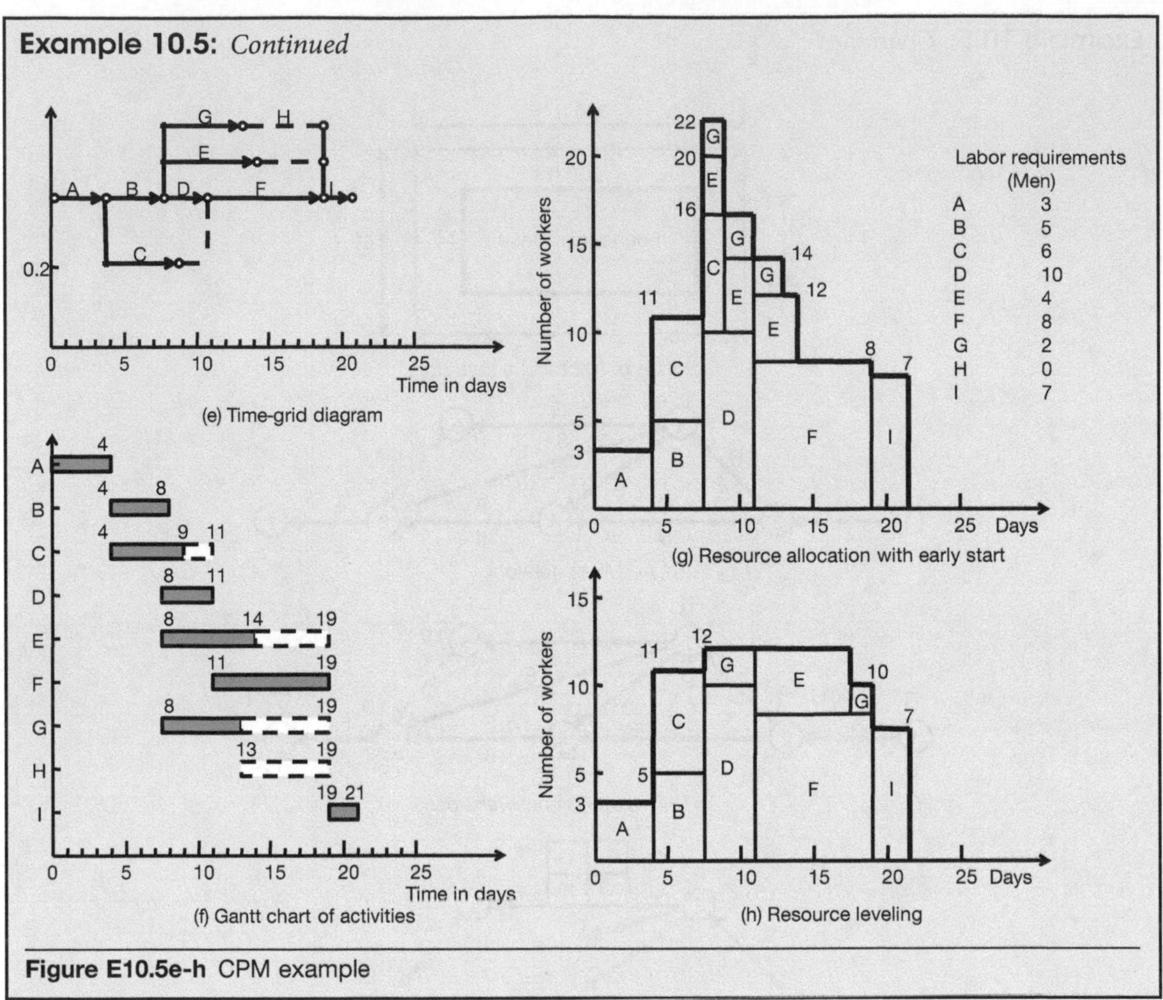

Figure E10.5e-h CPM example

10.3.5 Time-cost Optimization

CPM is effective in discovering the possibility of minimizing the cost of the project from its normal duration by reducing (or *crashing*) the duration of some of the individual activities in the network that lie on the critical path, by paying a higher cost (or *crash* cost). This extra cost is often offset by savings gained through lower overall indirect cost (or overhead cost). This technique of compressing certain tasks and thereby optimizing a project cost is often called *project crashing*. Typically, the crash cost per unit of time for an activity can be found as follows:

$$\text{Crash cost per unit time} = \frac{\text{Crash cost} - \text{Normal Cost}}{\text{Normal time} - \text{Crash time}}$$

The first step in crashing a project is to identify the critical activity with the minimum crash cost per unit of time and crash it with the allowable limit, taking into consideration the amount of float available with respect to multiple critical paths. The second step is to revise the network to identify new or multiple critical paths until all of the activities available for crashing have been utilized. Finally, all the crashing steps together with the corresponding cost increases are compared with the savings derived from overhead costs. See the following example for clarification.

Example 10.6 Determine the optimal completion time for a project whose logic is shown in Figure E10.6. Indirect cost is $150 per da. All other details are given in Table E10.6a.

Table E10.6a Activities and times

Activity	Normal duration T_N (days)	Total cost (Normal) C_N	Crash duration T_C (days)	Total cost (Crash) C_C
A	4	$600	2	$1000
B	6	$800	3	$1400
C	8	$500	3	$1200
D	7	$600	2	$1200
E	2	$500	2	$500
F	1	$100	1	$100

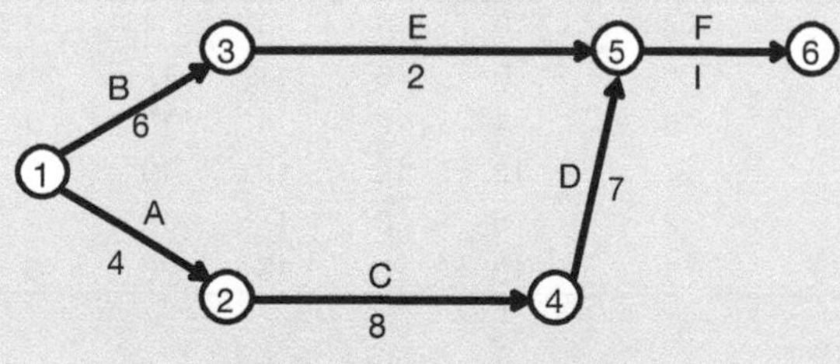

(a) Activities diagram

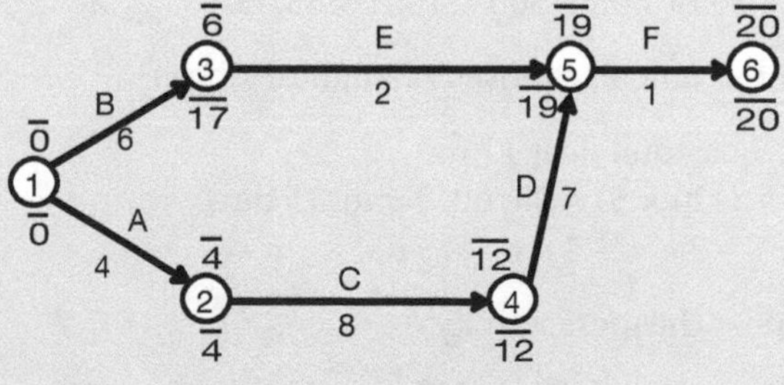

(b) Forward and backward passes

Figure E10.6a-b Activities and activity slopes

Continues

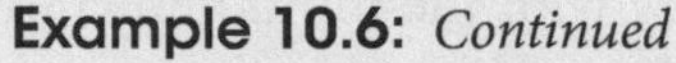

Example 10.6: *Continued*

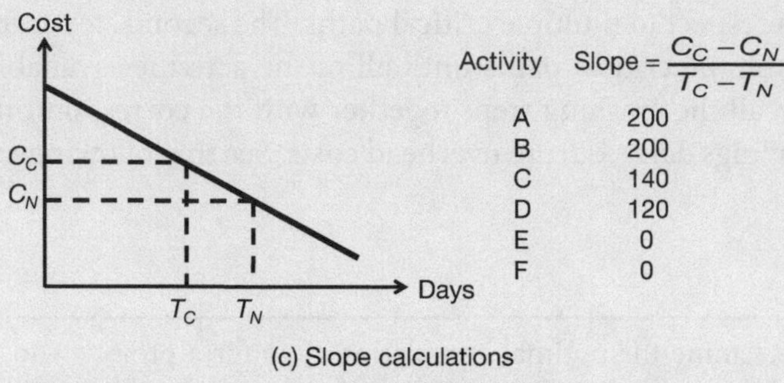

(c) Slope calculations

Figure E10.6c Activities and activity slopes

Solution The first task is to complete the activities table shown, after performing the forward and backward passes, as shown in Figure E10.6. Table E10.6b is completed followed by working out the time schedules to find the one that is the cheapest. These schedules are graphically shown in Figure E10.6a. The crash cost per unit time is the slope of the cost line and is computed for various activities as shown in Figure E10.6.

Table E10.6b ES, EF, LS, LF and TF

Activity	Duration	ES	EF	LS	LF	TF
A	4	0	4	0	4	0
B	6	0	6	11	17	11
C	8	4	12	4	12	0
D	7	12	19	12	19	0
E	2	6	8	17	19	11
F	1	19	20	19	20	0

Schedule 1: Normal Schedule; Critical Path 1 – 2 – 4 – 5 – 6 = 20 da

Cost = (600 + 800 + 500 + 600 + 500 + 100) = $3100

Schedule 2: Select activities on CP; choose D (smallest slope)

Compress D (limit 5 da; total float 11 da)
Direct cost = 3100 + (120 × 5) = $3700; Time: 15 da
TF: 11 – 5 = 6, CP: 1 – 2 – 4 – 5 – 6 = 15 da

Schedule 3: Compress next cheapest activity

Compress C; TF = 6 Crash 5 da @$140/da
Direct cost = 3700 + (140 × 5) = $4400
Time 9 da; TF: 6 – 5 = 1 CP: 1 – 2 – 4 – 5 – 6 = 10 da

Continues

Example 10.6: *Continued*

Schedule 4: Compress next cheapest activity

Compress A; TF = 1 Crash = 2
Compress A for only 1 da @$200/da
Direct cost = $4400 + (200 × l) = $4600
Time 9 da TF: 1 – 1 = 0 CP: 1 – 2 – 4 – 5 – 6 = 9 da and CP: 1 – 3 – 5 – 6 = 9 da

Schedule 5: Any other compression will affect both CPs.

Also, A can be compressed only 1 more day.
Compress A and B, one day each = $400
Direct cost = $4600 + $400 = $5000
Cost $5000; Time 8 da.

The time-cost optimization is shown in Figure E10.6d; and a summary of costs for schedules is given in Table E10.6c. As indicated in the table, Schedule 3 is the cheapest at $5900.

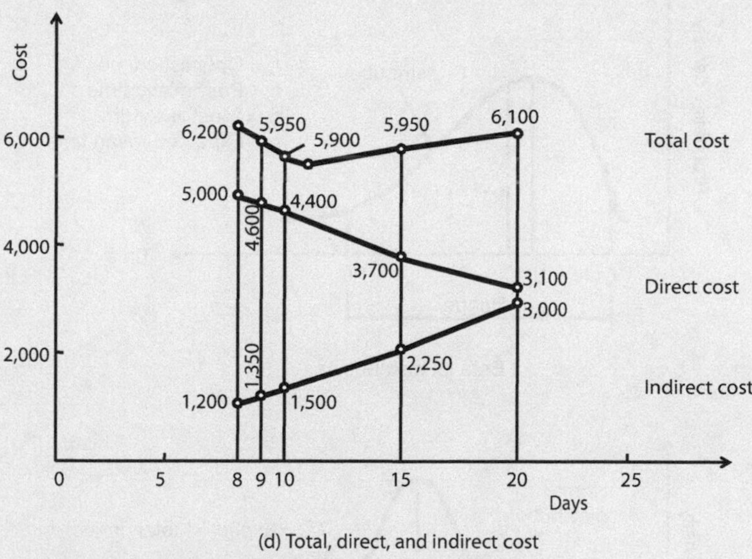

Figure E10.6d Total, direct, and indirect cost

Table E10.6c Schedule costs

Schedule	5	4	3	2	1
Days	8	9	10	15	20
Direct cost ($)	5000	4600	4400	3700	3100
Indirect cost ($)	1200	1350	1500	2250	3000
Total cost	6200	5950	5900	5950	6100

10.4 PROGRAM EVALUATION AND REVIEW TECHNIQUE AND THE LINE-OF-BALANCE

The PERT and the LOB techniques are both closely associated with CPM. PERT was developed to analyze projects in an environment of uncertainty, particularly with projects where the specific duration of activities could not be estimated with reliability. PERT uses two probability density functions: (1) the beta (β) distribution for each activity and (2) the normal distribution for estimating the completion time of the entire project (see Figure 10.2). Other than the use of these probability functions, PERT is similar to CPM. PERT is described first in Section 10.4.1. The LOB technique was developed by the U.S. Navy for controlling and managing production processes. It has since been used in the construction industry. LOB is described in Section 10.4.2.

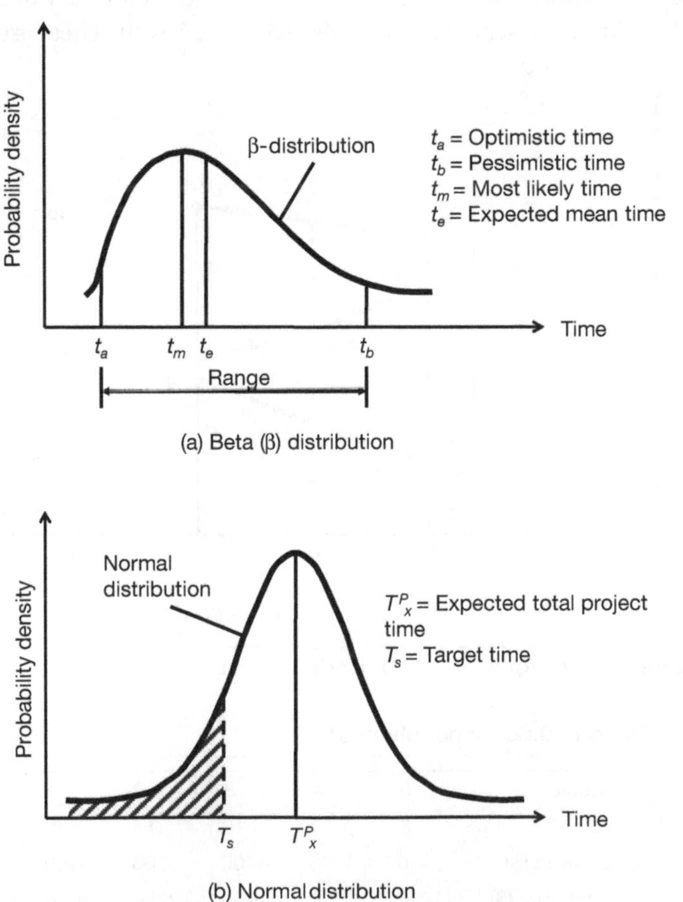

Figure 10.2 Key concepts (PERT): (a) beta distribution and (b) normal distribution

10.4.1 Key Concepts of PERT

- PERT introduces the concepts of uncertainty into time estimates as opposed to CPM, which is essentially deterministic.
- PERT uses expected mean time (t_e) with standard deviation, σ_{te} or variance v_{te}.
- The expected mean time (t_e) of an individual task is an estimate having an approximate chance of 50 percent success.
- The value of t_e is calculated from: t_a the optimistic time of completion of an individual task; t_m, the most likely time; and t_b, the pessimistic time of completion of this task. This forces the planner to take an overall view of each task's duration.
- The beta (β) distribution uses t_a, t_b, and t_m to estimate the expected mean time. The expected mean time t_e is:

$$t_e = \frac{t_a + 4t_m + t_b}{6}$$

with a standard deviation,

$$\sigma_{te} = \frac{t_b - t_a}{6}$$

and a variance,

$$v_{te} - (\sigma_{te})^2 - \left(\frac{t_b - t_a}{6}\right)^2$$

- t_a and t_b have small probabilities, around 5 to 10 percent.
- Once t_e and v_{te} are found for each activity, the critical path is found in the same fashion as in CPM.
- Project duration = Expected mean duration T_X^P which is the expected mean time along the CP.
- Once the expected mean time for an event (T_x) and its standard deviation σ_{T_X} are determined, calculate the event schedule time (T_s). This has a normal probability distribution with mean T_x and σ_{T_X}.
- The effect of adding a series of independent β-distribution gives a normal distribution.
- To determine the probability that the expected total project time T_x will exceed some target time T_s first calculate the value of Z, a dimensionless parameter expressing the horizontal axis of the standardized normal distribution function, where:

$$Z = \frac{(T_S - T_X)}{\sigma_{T_X}}$$

- Refer to Table 10.2 to find the corresponding probability associated with the value of Z.

Table 10.2 Values of Z and probability

Z	Probability	Probability	Z
−2.0	0.02	0.98	+2.0
−1.5	0.07	0.93	+1.5
−1.3	0.10	0.90	+1.3
−1.0	0.16	0.84	+1.0
−0.90	0.18	0.82	+0.90
−0.8	0.21	0.79	+0.8
−0.7	0.24	0.76	+0.7
−0.6	0.27	0.73	+0.6
−0.5	0.31	0.69	+0.5
−0.4	0.34	0.66	+0.4
−0.3	0.38	0.62	+0.3
−0.2	0.42	0.58	+0.2
−0.1	0.46	0.54	+0.1
0	0.50	0.5	0

Example 10.7 An activity network for a small house is shown in Figure E10.7. Table E10.7a shows the optimistic, most likely, and pessimistic times for various activities under Columns 2, 3, 4 of the table. What is the probability of finishing this project in 110, 115, 117, 119, 124 da?

Solution The expected mean time of all the activities are calculated using the formula $t_e = (t_a + 4t_m + t_b)/6$ and entered in Column 5 of Table E10.7. These values are also shown against each activity on the network. One can now do a forward and backward pass to determine the CP (in a manner similar to what was explained in the section on CPM). Now that the CP is known, one can calculate the standard deviation σ_{te} and variance v_{te} of the critical activities and enter these in Columns 6 and 7 respectively. The critical activities are marked with an asterisk and the total project duration works out to be 117 da with a corresponding variance of 74.41 da. Taking the square root of 74.41 da gives a standard deviation of 8.63 da. Notice that we cannot add values under Column 6 to get the standard deviation. We add the variance of critical activities and then take the square root of this total to get the standard deviation.

While we have found that the expected mean duration of this project is 117 da with a standard deviation of 8.63 da, the probability of finishing the project at a target time of 110, 115, 119, or 124 da is needed. For finding these probabilities we must calculate the corresponding values of Z, as shown in Table E10.7b. Notice that the probability of completing the project in 117 da is merely 50%.

Continues

Example 10.7: *Continued*

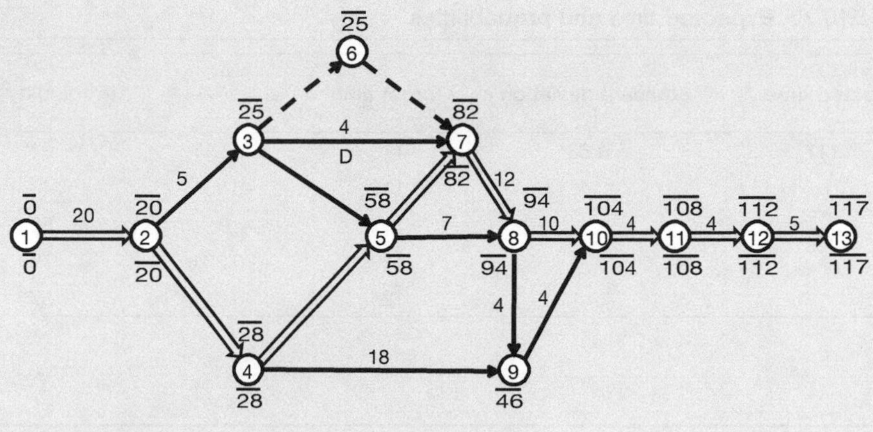

(i)⟹(j) *i j* indicates the critical path

Figure E10.7 PERT network

Table E10.7a Activities and their statistics

1	2	3	4	5	6	7
Activity	t_a	t_m	t_b	t_e	σ_{te}	v_{te}
*1–2	18	20	22	20	0.67	0.44
2–3	4	5	6	5		
*2–4	6	7	14	8	1.33	1.77
3–5	8	10	12	10		
3–7	14	16	18	16		
*4–5	20	25	60	30	6.67	44.44
4–9	14	18	22	18		
*5–7	18	20	46	24	4.67	21.77
5–8	6	7	8	7		
*7–8	11	12	13	12	0.33	0.11
8–9	4	10	16	10		
9–10	4	4	4	4		
* 8–10	8	8	20	10	2.00	4.00
*10–11	2	3	10	4	1.33	1.77
*11–12	3	4	5	4	0.33	0.11
*12–13	5	5	5	5	0	0
				117	8.63	74.41

*On the critical path. Note that $\sqrt{74.41} = 8.63$

Continues

Example 10.7: *Continued*

Table E10.7b Expected time and probabilities

Expected time T_{Xe}	Standard deviation σ	Target time	$Z = \frac{T_s - T_x}{\sigma}$	Probability (%)
117	8.63	110	−0.811	21
		115	−0.231	43
		117	0	50
		119	+0.231	79
		124	+0.811	57

10.4.2 The LOB Technique

The LOB technique is a management-oriented tool for collecting, measuring, and presenting information relating to the time and accomplishment of repetitive tasks during production. One of the major problems facing managers is obtaining information on the status of various operations soon enough to take effective action. LOB is particularly useful in repetitive construction work such as multi-house projects, road pavement construction, and the manufacturing of hundreds of identical units, such as small septic tanks, pylons, and beams.

The LOB technique consists of four elements: (1) the objective chart, (2) the program chart, (3) the progress chart, and (4) the comparison. The objective chart is a graph showing the cumulative end product to be manufactured over a period of time, while the program chart is a flow process diagram showing sequenced tasks and their interrelationship with lead times. Lead time is the number of time periods by which each activity must precede the end event to meet the objective. The progress chart consists of vertical bars representing the cumulative progress of each monitoring point based on site visits to the production area indicating the actual performance. The comparison activity is derived from the objective, program, and progress charts to draw the LOB. When one draws the LOB on the program chart, it represents the number of completed units that should have passed through each control point at the time of the study in order to deliver the completed units according to the contract schedule. Many of these terms will become clear in working through the illustrative examples that follow.

10.4.3 Progress Charts and Buffers

Suppose a company has been awarded a contract to erect ten steel pylons. The sequential operations involved are A = excavate; B = pour concrete; and C = erect pylon, as shown in Figure 10.3a.

This sequence needs to be repeated ten times to complete the work. However, to provide for a margin of error in the time taken to complete each operation, a time buffer is provided between two operations, as shown in Figure 10.3b. If the work order to begin work is given on Day 1, and one pylon has to be handed over at the end of every fifth working day, then the first pylon

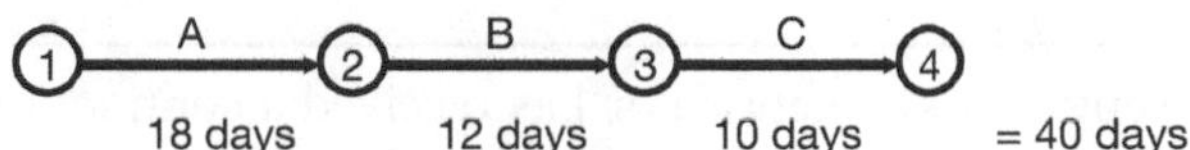

(a) Activities A, B and C without buffers

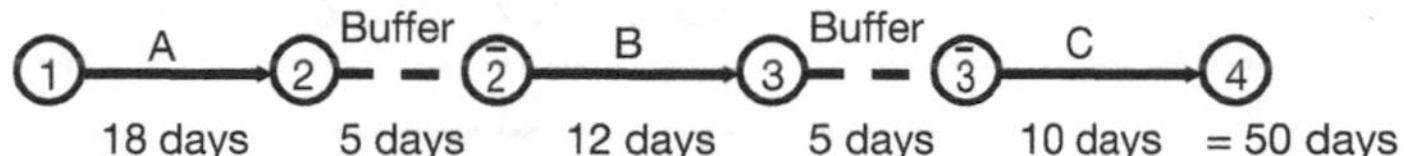

(b) Activities A, B and C with buffers

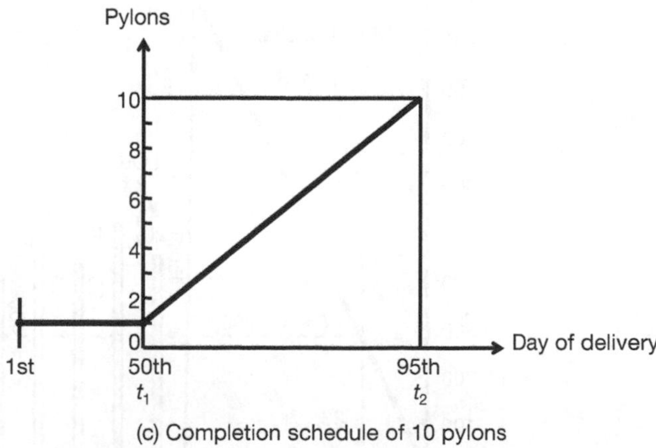

(c) Completion schedule of 10 pylons

Figure 10.3 Line of balance: (a) activities A, B, and C without buffer; (b) activities A, B, and C with buffers; and (c) completion schedule of 10 pylons

will have to be handed over at the end of the fiftieth day, and subsequent pylons will be handed over every fifth day. Refer to Figure 10.3c, if we use the straight-line equation $Q = mt + c$:

$$Q_1 = 1;\ t_1 = 50;\ Q_2 = 10;\ m = 1/5;\ c = 1, \text{ and } t_2 = ?$$

$$\text{then } t_2 = \left(\frac{10 - 1}{1/5}\right) + t_1 = (9 \times 5) + 50 = 95 \text{ da}$$

At the end of the ninety-fifth day, the tenth pylon will be completed. Notice that when we graph this problem, we must be mindful that on the fiftieth day we are ready to complete the first pylon, with nine more to go. Also, notice that we must begin work on the first pylon on the first day to get it ready by the fiftieth day.

We will now consider another example to illustrate a more realistic case of LOB.

Example 10.8 A septic tank prefabricator has received a contract to supply 1000 septic tanks, and to deliver 40 units each mo beginning on the first of the twentieth month. The major control points of the production scheme are shown in Figure E10.8a.

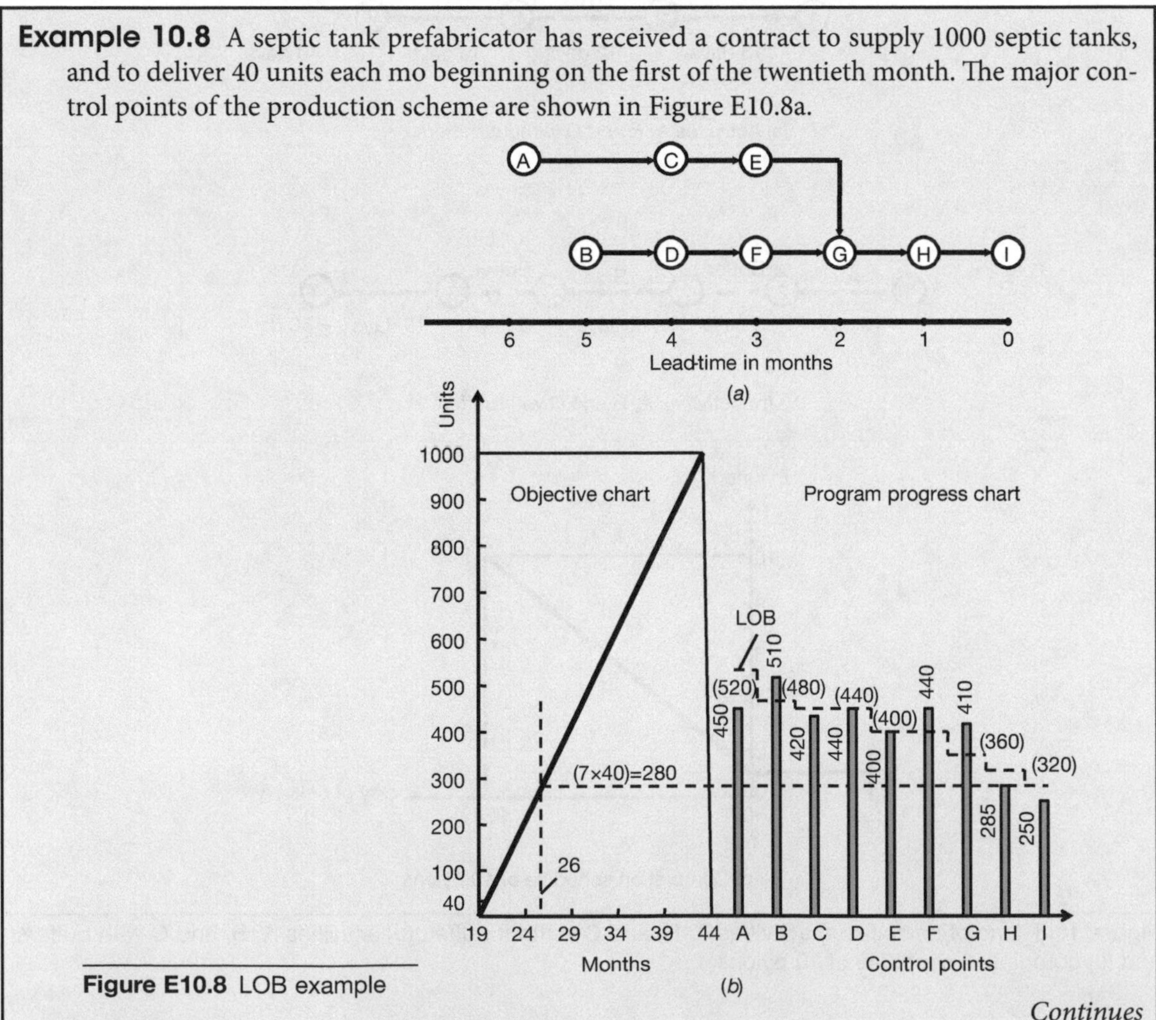

Figure E10.8 LOB example

Continues

Example 10.8: *Continued*

An LOB study performed on the first of the twenty-sixth month revealed that the number of completed units that actually passed through each control point is as follows (Table E10.8a):

Table E10.8a Control points and units

Control point	Units
A	450
B	520
C	420
D	440
E	400
F	440
G	410
H	285
I	250

Draw the objective chart, the program progress chart, and the LOB chart. Determine the deviation of units.

Solution Table E10.8b presents the results of calculations. The table indicates that the performance of control Points A, C, E, H, and I are behind schedule and that corrective action is needed.

Table E10.8b Control point performance results

Control point	Cumulative units to be delivered	Units actually completed	Deviation
A	520	450	−70
B	480	510	+30
C	440	420	−20
D	440	440	0
E	400	380	−20
F	400	440	+40
G	360	410	+50
H	320	285	−35
I	280	250	−30

10.4.4 Resource and LOB Schedule

Estimation of resources including labor requirements are an important feature of LOB schedules. In general such estimation is best done by: preparing a logic diagram of all the activities and tasks including sequenced and parallel (or simultaneous) activities; estimating the man-hours required to complete each task; choosing realistic buffer times that reflect the risk entailed in not completing sequenced activities; calculating the required output target to meet a given project completion date; and finally putting all the information in the form of a convenient table as shown in Table 10.3.

Table 10.3 Calculation sheet for LOB schedule

1	2	3	4	5	6	7	8
Activity	Man power per activity	Men per activity	Theoretical gang size	Actual gang size	Actual rate of output	Time in days for 1 activity	Time in days

Explanation of columns:

1: Major activities or tasks
2: Estimate of man-hours needed for each activity
3: The optimum number of laborers needed for each task (which is labor in each team)
4: The theoretical gang size needed to maintain the output rate (R) given by:

$$\frac{R \times (\text{Column 2})}{\text{Number of hr/wk}}$$

5: The actual gang size is chosen as a number that is a multiple of men required for one team

$$\text{6: Actual rate of output} = \frac{(\text{Actual gang size}) \times (\text{Target rate})}{\text{Theoretical gang size}}$$

7: Time taken for one activity =

$$\frac{\text{Man-hours for activity}}{(\text{Number of men in one team}) \times (\text{Number of hours in a working day})}$$

8: the time in days from the start of the first section to the start of the last section is:

$$\frac{(\text{Number of sections} - 1) \times (\text{Number of working days per week})}{\text{Actual rate of build}}$$

Table 10.3 helps us to draw the various activities which show the sections completed against time. Also, the slope of the activities is a function of several factors, such as the total units of time required to complete Q repetitive units; F, the resource unit factor, is the number of units

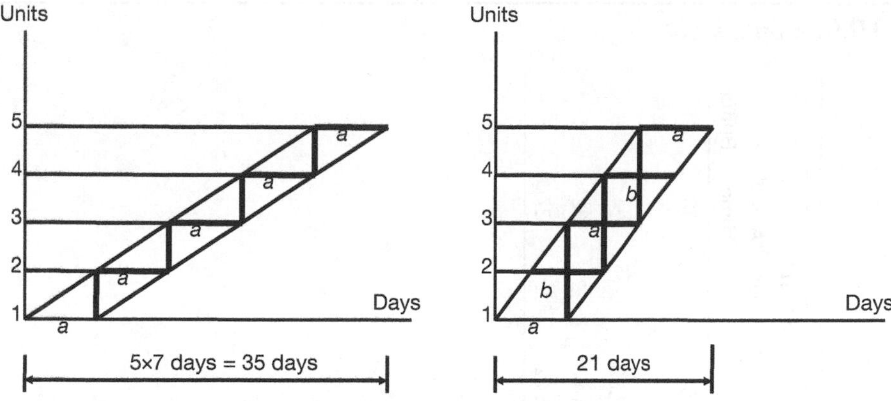

Figure 10.4 Logic diagrams for teams

of the resource that are required to achieve the rate of working necessary to meet the handover program; *d*, the activity duration; and *m*, the rate of handover. The buffer time allowed plays an important role in the entire completion time. Lastly, the actual gang size in relation to the men per activity is important as it determines the slope of the activity as shown in Figure 10.4. If a task takes 7 da, with one team shifting from one unit to the next, it would take 35 da for the completion of 5 units. If the same task is arranged with two teams (*a* and *b*), the 5 units would be completed in 21 da. These alternatives are shown in Figure 10.4.

The following example explains the procedure of LOB.

Example 10.9 A contract has been awarded to erect 15 steel pylons at the rate of three per wk, (assuming 5 da of 8 hr each) as per details given in Table E10.9a.

Table E10.9a Tasks and man-hours

	Task	Man-hours	Optimum number of men per operation
A	Excavate pit	180	6
B	Concrete pit	320	4
C	Fix pylon	200	4
D	Adjust/finish	60	2

A minimum buffer of 5 da is provided between tasks to take care of delays. Prepare a LOB schedule showing gang sizes and rates of build. Finally, draw a sketch showing the LOB schedule with buffers for all 15 pylons.

Solution The calculation of gang sizes and rates of build, with a target rate of three per wk is shown in Tables E10.9b and c; and the LOB schedule for the four tasks is sketched in Figure E10.9.

Continues

Example 10.9: *Continued*

Table E10.9b Details of calculations for Example 10.9

Activity	Man power per activity	Man per activity	Theoretical gang size	Actual gang size	Actual rate of output	Time in days for 1 activity	Actual time	Time from start on first section to start on last section	Actual time	Buffer
1	**2**	**3**	**4**	**5**	**6**	**7**	**8**	**9**	**10**	**11**
A	180	6	13.50	12	2.67	3.75	4	26.22	27	5
B	320	4	24.00	24	3.00	10.00	10	23.33	24	5
C	200	4	15.00	16	3.20	6.25	7	21.88	22	5
D	60	2	4.50	4	2.67	3.75	4	26.22	27	5

Note: Numbers in Column 3 must be multiples of those in Column 5.

TABLE E10.9c Details of calculations for Example 10.9 (continued)

	First unit		Last unit		
	Begin	**End**	**Begin**	**End**	**Buffer**
A	0	4	27	31	5
B	12	22	36	46	5
C	29	36	51	58	5
D	41	45	68	72	—

Continues

Example 10.9: *Continued*

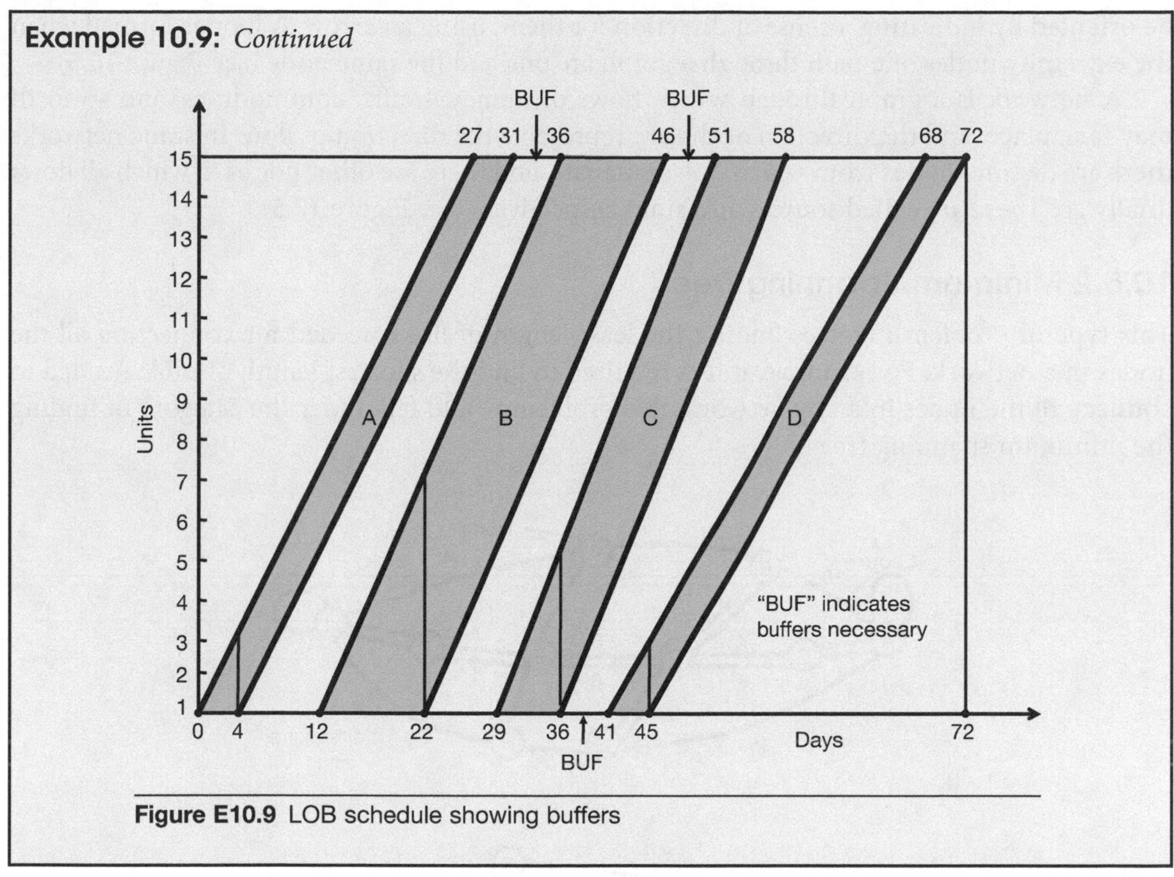

Figure E10.9 LOB schedule showing buffers

10.5 NETWORK FLOW ANALYSIS

Everybody is familiar with highway, telephone, and cable networks. They are all arrangements of paths and crossings for facilitating the flow of goods, people, and communication. In this section we deal with three types of network flow problems: (1) the minimum spanning tree, (2) the maximum flow, and (3) the shortest-path techniques.

10.5.1 Key Concepts

Graph theory and network theory are branches of mathematics that have grown in the last 75 years. The building of large physical systems such as highways, railroads, and pipelines has created wide interest in network theory. We have already seen many problems of sequencing and scheduling such as CPM and PERT that can also be looked upon as problems in graph and network theory.

A graph is formed when a number of points, nodes, or vertices are joined together by one or more lines, arcs, links, branches, or edges. Nodes may be joined by more than one arc and may be oriented by indicating a sense of direction for them, using an arrow. A loop is formed when the extremity nodes of a path through a graph are one and the same node (see Figure 10.5).

A network is a graph through which flows of money, traffic, commodities, and so forth may take place, and the direction of the arc represents the direction of flow. In some networks there are distinct nodes from which flows emanate, and there are other nodes to which all flows finally go. These are called sources and sinks respectively (see Figure 10.5).

10.5.2 Minimum Spanning Tree

This type of problem involves finding the least length of links needed for connecting all the nodes in a network. For example, if it is required to find the shortest length of cable needed to connect all the nodes in a city network, this problem would fall under the category of finding the minimum spanning tree.

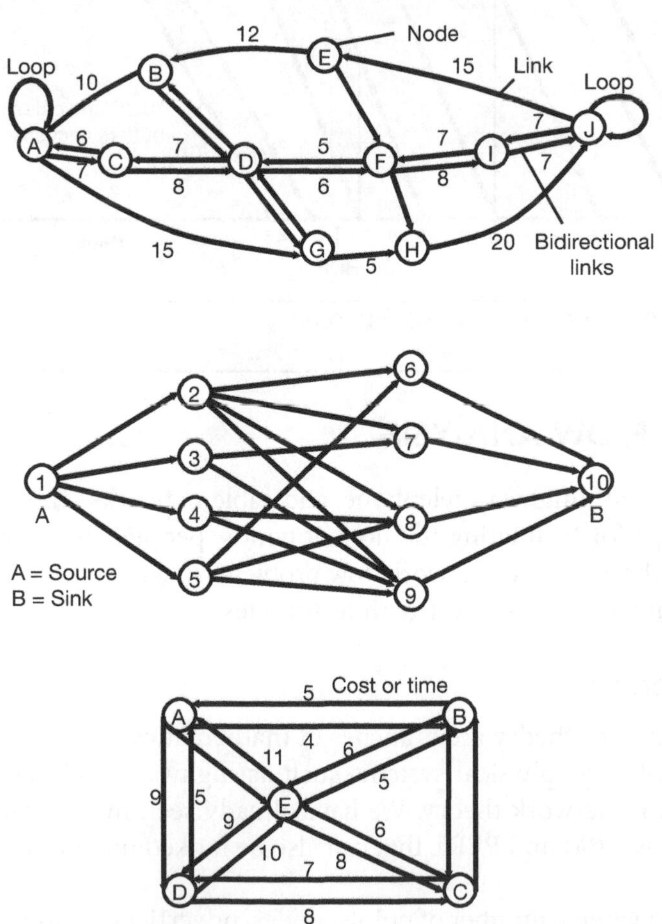

Figure 10.5 Network details

The procedure for finding the minimum spanning tree of a network is:

1. Select the shortest link in the network.
2. Identify an unconnected node with the shortest distance to the node in Step 1.
3. Connect these two nodes. In case of a tie, select one arbitrarily.
4. Continue connecting one node after another till all nodes are connected.

Example 10.10 A small village needs to be connected by cable to seven main nodes. What is the minimum length of cable needed to connect all the nodes? Distances are marked on the nodes.

Solution See Figure E10.10

1. The shortest link in the network is EG = 11.
2. The next shortest link connected to either E or G is GF = 13. Join GF.
3. The next shortest link, either E or F, is ED = 15. Join ED.
4. The next link at D, F or G is DC = 26. Join DC and so on.

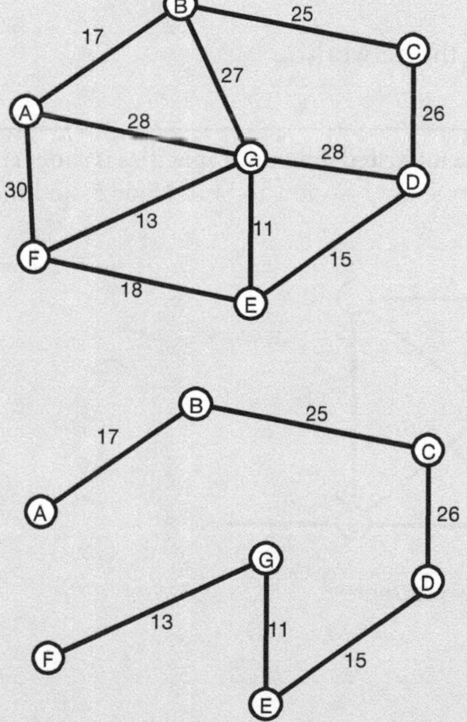

Total length of minimum spanning tree = 107 units

Figure E10.10 Minimum spanning tree

Continues

Example 10.10: *Continued*

The sequence of linking is EG, GF, ED, DC, CB, and BA, totaling 107 units, which is the minimum length of cable needed to join all the nodes in the network.

10.5.3 The Maximal Flow Problem

There are countless instances in which one would like to know the maximum number of trucks or wagons that can flow on a railroad or highway network from a source (or origin) to a sink (or destination). For example, in a highway network you could have traffic flow on a one-way street or a two-way street, whereas in a pipeline network, oil could flow in both directions. The procedure for determining the maximum flow in a network is:

1. Identify the source node and the sink node of the network
2. Determine all the feasible paths from source to sink that would be able to handle the flow
3. Determine whether there are possibilities of reverse flows, depending on the information supplied
4. Sum up all the flows through each link of the network

Example 10.11 A small railroad network with the indicated link flow capacities is shown in Figure E10.11. Determine the maximum flow from Source Node 1 to Sink Node 5, and indicate the flow on each link.

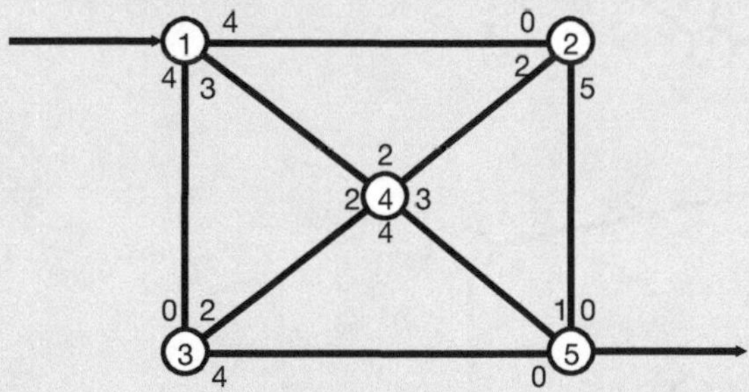

Figure E10.11 Maximal flow problem

Solution

Source node = 1 Sink node = 5

Continues

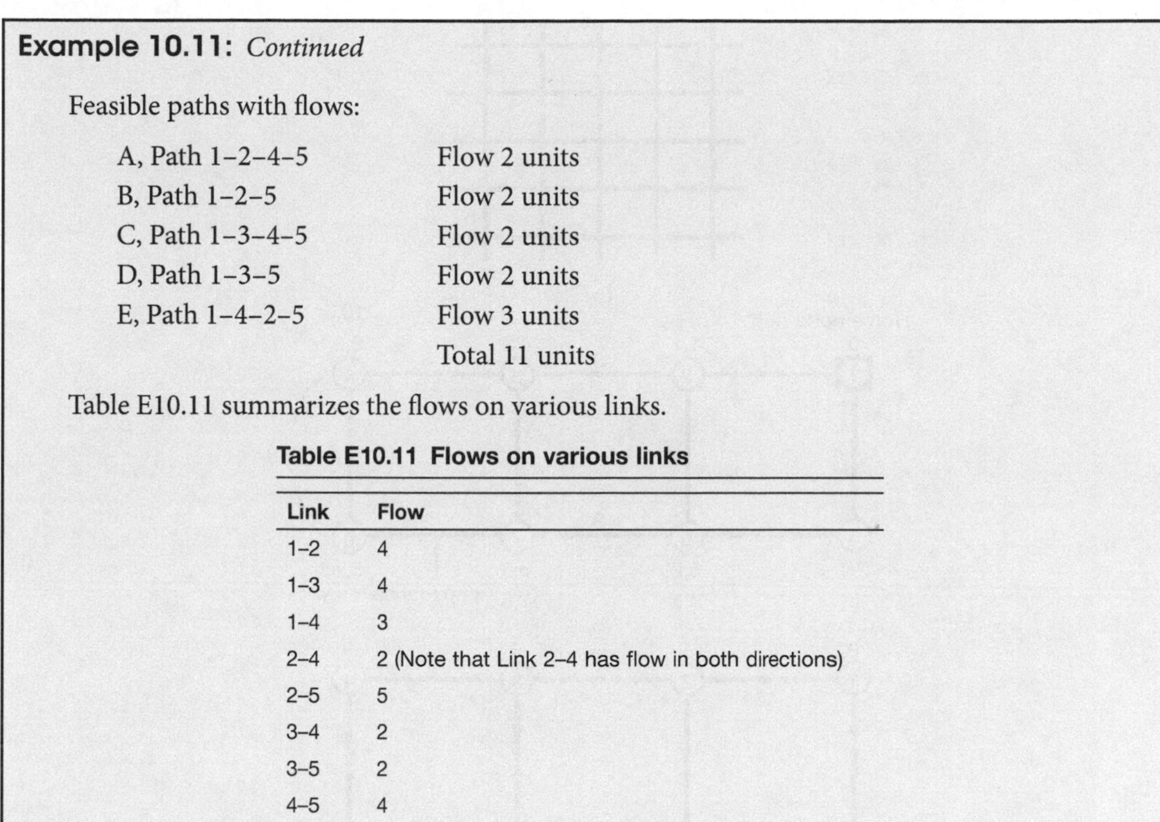

Example 10.11: *Continued*

Feasible paths with flows:

A, Path 1–2–4–5	Flow 2 units
B, Path 1–2–5	Flow 2 units
C, Path 1–3–4–5	Flow 2 units
D, Path 1–3–5	Flow 2 units
E, Path 1–4–2–5	Flow 3 units
	Total 11 units

Table E10.11 summarizes the flows on various links.

Table E10.11 Flows on various links

Link	Flow
1–2	4
1–3	4
1–4	3
2–4	2 (Note that Link 2–4 has flow in both directions)
2–5	5
3–4	2
3–5	2
4–5	4
4–2	3

10.5.4 Shortest-path or Minimum-path Technique

The shortest-path or minimum-path technique is based on the assumption that travelers want to use the minimum impedance route between two points. Efficient methods of determining minimum paths were developed, because manual determinations would be nearly impossible. In Figure 10.6, for example, 40 different paths must be tested to determine the minimum between A and B. You can imagine the problem of finding the shortest path in a network, such as a city, with thousands of links and nodes.

Many years ago the work that was undertaken to determine the minimum paths for long-distance telephone calls provided the help that planners needed. Rather than simply testing each path, algorithms allowed planners to find minimum paths for complete networks. The algorithm used most commonly is Moore's algorithm.

Using Moore's algorithm, minimum paths are developed by fanning out from their origin to all other nodes. Determining the minimum path from Node 1 to each of the other nodes results in a *skimtree* from Node 1 to all other nodes, as shown in Figure 10.6.

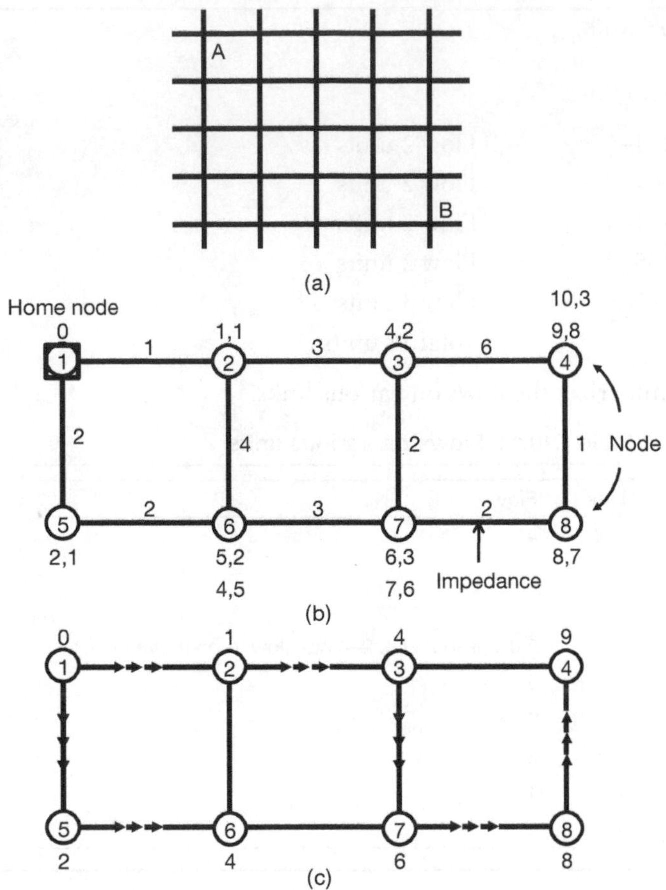

Figure 10.6 Minimum path technique: (a) small network—24 links, 16 nodes; (b) minimum path through a network; and (c) skim tree from Node 1 to all other nodes

Moore's algorithm is now applied to Figure 10.6b:

1. Start at Node 1 (the origin) and determine the shortest time (or distance) to get to a directly connected node. The two nodes directly connected to Node 1 are Nodes 2 and 5, with the shortest time of 1 and 2 units, respectively.
2. Since there is no doubt the shortest times to reach Nodes 2 and 5 are 1 and 2, respectively, we refer to these nodes as comprising the *permanent set*.
3. Repeat the foregoing steps by determining all the nodes directly connected to the nodes in the permanent set (Nodes 1, 2, and 5). Now, Nodes 3 and 6 are directly connected to Nodes 2 and 5.
4. Two paths can be identified connected to Node 6 (2-6 and 5-6), with times of 4 and 2 units respectively. Also, the path along 1-2-6 takes 5 units of time, while Path 1-5-6

takes 4 units. Therefore, the shortest path to Node 6 from Node 1 is 1-5-6. And, this path is indicated in Figure 10.6c.

5. Nodes 1, 2, 5, and 6 are now part of the permanent set, and one can proceed to other nodes repeating the process described in Steps 1 through 4.
6. The shortest path from the starting (or home) node to all other nodes can be tabulated as follows (see Table 10.4):

Once the minimum paths are found, the trips between the zones are loaded onto the links making up the minimum path. This technique of assigning trips to the network is sometimes referred to as all-or-nothing, because all trips between a given origin and destination are loaded on links comprising the minimum path, and nothing is loaded on the other links. After all possible interchanges are considered, the result is an estimate of the volume on each link in the network. Moore's algorithm can be stated as follows:

1. Label the start node (or home node) as zero.
2. Calculate working values for each node that is directly connected to the node labeled zero, using working values indicated on the links. Select the minimum of the labeled values plus the distance from the labeled node. This establishes a permanent set.
3. Select an unlabeled node with the lowest working value and label it with that value.
4. Repeat Steps 2 and 3 until all nodes have been labeled.
5. Mark all links lying on the shortest path; this is the skim tree rooted in the start (or home) node.
6. Repeat Steps 1 through 5, selecting successive nodes as start nodes.

Table 10.4 Shortest paths

From	To	Path	Duration
1	2	1–2	1
1	3	1–2–3	4
1	4	1–2–3–7–8–4	9
1	5	1–5	2
1	6	1–5–6	4
1	7	1–2–3–7	6
1	8	1–2–3–7–8	8

Example 10.12 A highway network consisting of 4 nodes and 10 links is shown in Figure E10.12. A trip table showing the number of vehicles wanting to use the network per hour from one node to another is also provided. Assign the trips to the network.

Continues

Example 10.12: *Continued*

Solution Refer to Figure E10.12. The origin-to-destination flows corresponding to each node of the trip table are assigned with the links that make up the minimum. The aggregate flow on each link is then shown.

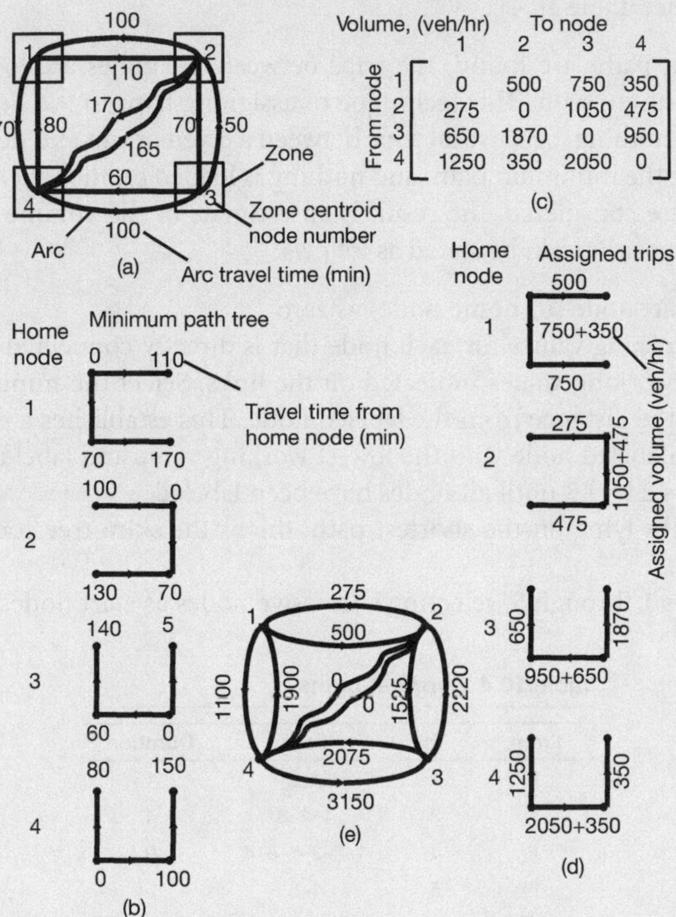

Figure E10.12 Example of all-or-nothing traffic assignment: (a) network; (b) minimum-path trees; (c) origin-destination trip table; (d) assignment of trips to minimum-path trees; and (e) assigned traffic volume

10.6 LINEAR PROGRAMMING

The objective of linear programing (LP) is to determine the optimal allocation of scarce resources among competing products or outputs. In most engineering and economic problems, one is frequently called upon to optimize a function that is subject to several constraints. When a single constraint is involved, the Lagrange method is used because of its simplicity. However, when more than one constraint is involved, LP is usually adapted. On the other hand, if the constraints are limited to only two variables (or at most three in some special cases), the graphical approach is the easiest to use. When a problem involves more than two variables, the best way of dealing with it is by using the *simplex* algorithm. We first deal with the graphical method and then follow up with the simplex.

10.6.1 The Graphical Method

This method is generally used for solving maximization or minimization problems involving two variables. It is best described by means of an example.

Example 10.13 A steel firm produces two products, small beams (X_1) and small poles (X_2). Each beam requires 2.5 hr for cutting and welding, 3 hr for finishing, and 1 hr for checking and testing. Each pole requires 1 hr for cutting and welding, 3 hr for finishing, and 2 hr for checking and testing. The firm is limited to no more than 20 hr for cutting and welding, 30 hr for finishing, and 16 hr for checking and testing. The firm makes a profit of \$3 per beam and \$4 per pole. How many beams and poles should the firm produce to maximize profit?

Solution The firm's objective function is to maximize profit, and the total profit is the sum of the individual profit gained from each of the two products. First, express the information given in the form of equations or inequalities. The objective function can be written:

$$Z = 3\,X_1 + 4\,X_2, \text{ subject to constraints}$$

$$2.5\,X_1 + X_2 \le 20$$

$$3\,X_1 + 3\,X_2 \le 30$$

$$X_1 + 2\,X_2 \le 16$$

$$X_1, X_2 \ge 0$$

The first three inequalities are technical constraints dictated by the availability of time, while the last constraint is imposed on all such problems to avoid negative values from the solution.

The three inequality constraints are treated as:

(1) $X_2 = 20 - 2.5\,X_1$
(2) $X_2 = 10 - X_1$
(3) $X_2 = 8 - 0.5X_1$

The graph of these three equations is shown in Figure E10.13. The feasibility area satisfying the equations above and the inequalities originally derived are shown by the area included in OABCD.

Continues

Example 10.13: *Continued*

To find the optimal solution within the feasible area, graph the objective function:

$$Z = 3X_1 + 4X_2 \quad \text{and} \quad \therefore X_2 = (Z/4) - (3/4)X_1$$

The objective function has a slope of (– 3/4). Raising the objective function from its initial position at (0,0) when the profit is zero, and testing it at the four extreme corners of the feasible area (A, B, C, D), one finds that the maximum profit is derived at Point B (4, 6). Substituting the values of $X_1 = 4$, and $X_2 = 6$ in the objective function equation gives:

$$Z = 3(4) + 4(6) = 36.$$

As a check, the amount of profit at the other extreme points can also be determined.

At	A(0,8) :	3(0) + 4(8) = 32
	B(4,6) :	3(4) + 4(6) = 36 (maximum profit)
	C(7,3) :	3(7) + 4(3) = 33
	D(8,0) :	3(8) + 4(0) = 24

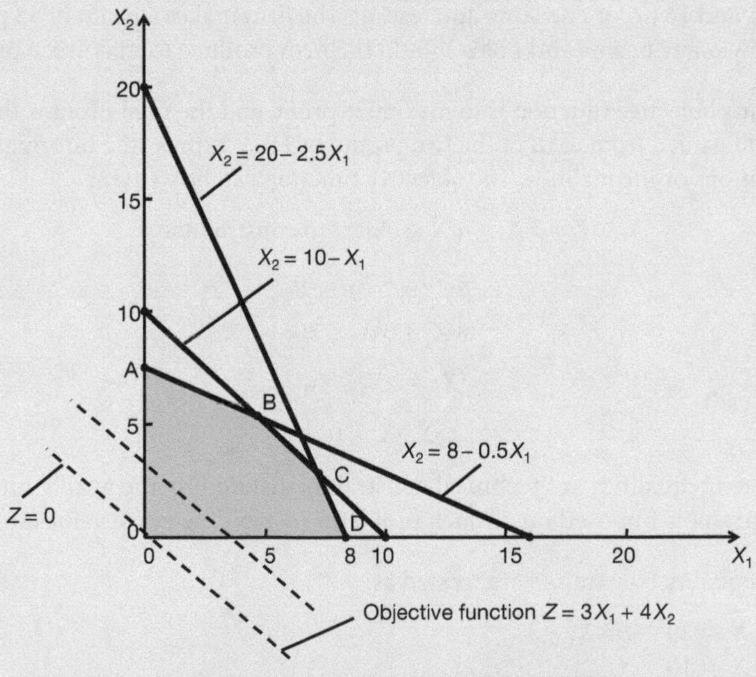

Figure E10.13 Graphical method of LP (maximization)

Example 10.14 A nutritionist wants patients in a hospital to receive the minimum requirements of three vitamins in two kinds of diets, as given in Table E10.14. Diet X_1 costs \$1.00/lb, while Diet X_2 costs \$2.00/lb. What combination of A and B will produce an adequate diet at a minimum cost?

Table E10.14 Diets and vitamin types

	Units/lb of Diet X_1	Units/lb of Diet X_2	Minimum requirement units
Vitamin A	1	3	90
Vitamin B	5	1	100
Vitamin C	3	2	120

Solution Note that this is a minimization problem and therefore one will have to be careful in deciding what the feasible region will be. The constraints are:

$$\text{Vitamin A:} \quad X_1 + 3\,X_2 \geq 90$$
$$\text{Vitamin B:} \quad 5\,X_1 + X_2 \geq 100$$
$$\text{Vitamin C:} \quad 3\,X_1 + 2\,X_2 \geq 120$$

The objective function $Z = 1.0\,X_1 + 2.0\,X_2$

Therefore, $X_2 = (Z/2) - (1/2)\,X_1$

Slope of $Z = -1/2$

The feasible area is the shaded area shown to the northeast of the Points OPQR in Figure E10.14. When Z is moved up from the origin, the first point it hits is Point Q, whose coordinates are (25.8, 21.4). This point gives the minimum cost for the combination of the two diets containing the specified requirements of Vitamins A, B, and C.

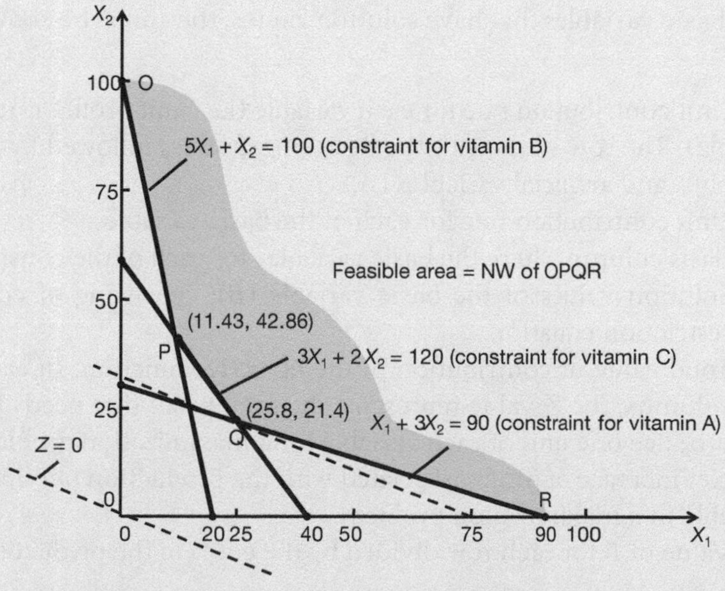

Figure E10.14 Graphical method of LP (minimization)

Continues

Example 10.14: *Continued*

The cost of such a diet is $Z = 1(25.8) + 2(21.4) = \$68.60$

Notice that the amount of Vitamin B = 5(25.8) + 1(21.4) = 150.4 units, which is more than the 100 units prescribed by the nutritionist.

10.6.2 Simplex Algorithm

An algorithm is a set of rules or procedures for finding a solution to a problem. The simplex algorithm first developed by George Dantzig in 1947 is a computational procedure for determining basic feasible solutions to a system of equations and testing the solution for optimality (see Tables 10.5 and 10.6). The algorithm moves from one basic feasible solution to another, always improving upon the previous one, until the optimal solution is reached. The best way to understand this algorithm is through a simple example as described in Example 10.15. The explanation of the simplex tableau is given:

Table 10.5 Simplex tableau example

C_j / C_b	Basis	Solution "B"	C_1 X_1	C_2 X_2	... ...	... S_1	... S_2	... ...	... A_1	... ...	θ
	Z_j										
	$C_j - Z_j$										

To determine the basic variables that have solution values, they must be converted into equalities:

$C_j =$	Unit contribution rate for each variable (i.e., unit profit or unit cost, for example). The row starts with the decision variables followed by the slack (s_i), surplus, and artificial variables (A_i).
$C_b =$	Unit contribution rate for each of the basic variables.
Basis =	Basis column where the basic variables for each of the constraints are listed.
Solution =	Solution values of the basic variable (B); consisting of constants from the restriction equation.
$Z_j =$	Total value or contribution of the objective function. In each of the variable columns, the Z_j value represents the total profit that needs to be sacrificed to produce one unit of each variable for a maximization problem.
$C_j - Z_j =$	Net increase of profit associated with the production of one unit of each variable in a maximization problem.
$\theta =$	Value of B for each row divided by the entry in the pivot column for that row.

To determine the basic variables that have solution values, they must be converted into equalities as follows:

Table 10.6 Constraint types and adjustments

Constraint type	Adjustment required
Less than or equal to, ≤	Add a slack variable, S_i
Exactly equal to, =	Add an artificial variable, A_i
Greater than or equal to, ≥	Subtract a surplus variable and add an artificial variable, $(-S_i + A_i)$

Example 10.15 A manufacturer produces two types of machines from parts X_1 and X_2. The resources needed for producing machines X_1 and X_2 and the corresponding profits are (Table E10.15):

Table E10.15 Resources needed for machines

Machine	Labor hrs/unit	Capital ($/unit)	Profit ($)
X_1	10	40	40
X_2	20	30	50

There are 400 hr of labor and $1200 worth of capital available per day for assembling the machines. How many machines, X_1 and X_2, should be produced per day to maximize profit?

Solution This problem is first worked out graphically as shown in Figure E10.15.

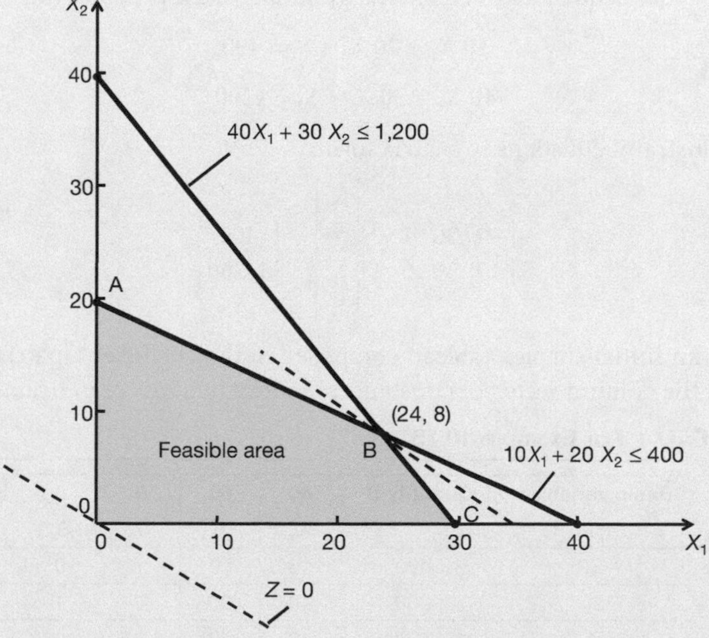

Figure E10.15 Graphical solution to manufacturer producing two types of machines

Continues

Example 10.15: *Continued*

$$Z = 40\,X_1 + 50\,X_2$$

Subject to:

$$10\,X_1 + 20\,X_2 \le 400$$

$$40\,X_1 + 30\,X_2 \le 1200$$

$$X_1, X_2 \ge 0$$

$$X_2 = (Z/50) - (40/50)X_1 \text{ with a slope of } -4/5$$

The feasible area is OABC. Point B (24, 8) gives the maximum profit. Therefore, 24 units of Machine X_1 and 8 units of Machine X_2 should be produced to maximize profit. The profit is worked out as:

$$Z = (40)(24) + (50)(8) = \$1{,}360$$

We will now use the simplex algorithm to maximize profits represented by $Z = 40\,X_1 + 50\,X_2$, which is the objective function, subject to:

$$10\,X_1 + 20\,X_2 \le 400$$

$$40\,X_1 + 30\,X_2 \le 1200$$

Step 1: Convert the inequalities to equations by adding slack variables (S):

$$10\,X_1 + 20\,X_2 + S_1 = 400$$

$$40\,X_1 + 30\,X_2 + S_2 = 1200$$

Express the constraint equations in matrix form:

$$\begin{bmatrix} 10 & 20 & 1 & 0 \\ 40 & 30 & 0 & 1 \end{bmatrix} \begin{bmatrix} X_1 \\ X_2 \\ S_1 \\ S_2 \end{bmatrix} = \begin{bmatrix} 400 \\ 1{,}200 \end{bmatrix}$$

Step 2: Set up an initial simplex tableau composed of the coefficient matrix of the constraint equations and the column vector of constants set above in a row of indicators, as shown:

TABLEAU 1 For Example 10.15

C_i / C_j	Basic variables	Quantity *B*	40	50	0	0	
			X_1	X_2	S_1	S_2	θ
0	S_1	400	10	20	1	0	20
0	S_2	1,200	40	30	0	1	40
	Z_j	0	0	0	0	0	
	$C_j - Z_j$		40	50	0	0	

Continues

Example 10.15: *Continued*

The meaning of the last column (θ) will be explained later. The values of the Z_j row are computed by multiplying each C_j column value (on the left side) by each column value under quantities X_1, X_2, S_1, and S_2, and then aggregating each of these sets of values.

Step 3: Start with a feasible initial solution (at the origin), $X_1 = 0$, $X_2 = 0$. The first column C_j gives their C values, which are zero.

Step 4: Compute the row marked $Z_j = \sum a_{ij} C_i$ where a_{ij} = any cell value.

$Z_j = 400(0) + 1200(0) = 0$ under Column B
$Z_j = 10(0) + 40(0) = 0$ under Column X_1 and so on.

Step 5: The $C_j - Z_j$ row shows the profit to be derived from adding a unit of that variable. The best variable to bring in is X_2, because every unit brought in brings $50 to the profit function (i.e., the column with the highest $[C_j - Z_j]$ value).

Step 6: Determine the row to be removed from the solution (i.e., the row with the lowest contribution). Calculate the θ column: $\theta_1 = 400/20 = 20$; $\theta_2 = 1200/30 = 40$; θ_1 is the lowest value and is therefore the pivot row.

Step 7: Calculate the rows and columns of the second tableau.

(a) New X_2 row replaces S_1 row. Divide each element of the old S_1 row by the pivot element 20. Thus, in the new row the old pivot element is replaced by 1, and the new row values are:

X_2 20 1/2 1 1/20 0

(b) A new S_2 row is obtained from the old one by deducting from it, element by element, the new S_2 row, multiplied by the pivot element of the old row, as shown:

New	20	1/2	1	1/20	0
Old S_2	1,200	40	30	0	1
−30 (new X_2)	−600	−15	−30	−3/2	1
New S_2	600	−25	0	−3/2	1

Step 8: Compute the Z_j and $C_j - Z_j$ row values:

	$Z_j = (50)(20) + (0)(60)$	$= 1000$
X_1:	$Z_1 = (50)(1/2) + (0)(25)$	$= 25$
X_2:	$Z_2 = (50)(1) + (0)(0)$	$= 50$
S_1:	$Z_3 = (50)(1/20) + (0)(-3/2)$	$= -5/2$
S_2:	$Z_4 = (50)(0) + (0)(1)$	$= 0$

The largest value of $C_j - Z_j$ is X_1, which is 15 and therefore the pivot column is X_1. Now, calculate the θ column values: $\theta_1 = 20/(1/2) = 40$, $\theta_2 = 600/25 = 24$. The lowest value = 24, therefore, row S_2 goes out.

Continues

Example 10.15: *Continued*

TABLEAU 2 For Example 10.15

C_j	Basic variables	Quantity *B*	X_1	X_2	S_1	S_2	θ
50	X_2	20	1/2	1	1/20	0	40
0	S_2	600	25	0	–3/2	1	24
	Z_j	1,000	25	50	–5/2	0	
	$C_j - Z_j$		15	0	–5/2	0	

Step 9: Repeat Step 7 for the third tableau.

(a) New X_1 replaces old S_2 row by dividing old S_2 by pivot cell (25)

X_1 24 1 0 –3/50 1/25

(b) New X_2 row is obtained from old one by deducting from it, element by element, the new X_1 row multiplied by the pivot element of the old row:

Old X_2	20	1/2	1	1/20	0
–1/2 (new X_1)	–12	–1/2	0	3/100	–1/50
New X_2	8	0	1	8/100	–1/50

TABLEAU 3 For Example 10.15

C_j	Basic variables	*B*	X_1	X_2	S_1	S_2
50	X_2	8	0	1	8/100	–1/50
40	X_1	24	1	0	–3/50	1/25
	Z_j		40	50	8/5	3/5
	$C_j - Z_j$		0	0	–8/5	–3/5

Step 10: Compute the Z_j and $C_j - Z_j$ row values:

	$Z_j = (50)(8) + (40)(24)$	= 1360	$C_j - Z_j$
X_1:	$Z_1 = (50)(0) + (40)(1)$	= 40	$40 - 40 = 0$
X_2:	$Z_2 = (50)(1) + (40)(0)$	= 50	$50 - 50 = 0$
S_1:	$Z_3 = (50)(8/100) + (40)(-3/50)$	= 8/5	$0 - 8/5 = -8/5$
S_2:	$Z_4 = (50)(-1/50) + (40)(1/25)$	=3/5	$0 - 3/5 = -3/5$

all $C_j - Z_j$ values are either zero or negative and therefore the solution is optimal.

$$X_1 = 24;\ X_2 = 8$$

Step 11: Verify the results:

$$Z = 40\ X_1 + 50\ X_2$$

$$Z = (40)(24) + (50)(8) = \$1{,}360 \text{ per da}$$

This result matches the one derived by the graphical method.

Example 10.16 An oil company produces two products, X_1 and X_2, that bring in profits of \$12 and \$36 for each gallon respectively. Both products require processing by Machines A and B. X_1 requires 8 hr in A and 16 hr in B while X_2 requires 12 hr in A and 8 hr in B. Machines A and B have 24 and 32 hr of capacity respectively. (a) How many gallons of each product should be produced to get optimal profit? (b) How would you interpret the optimal solution? (c) Determine the effect of changes in the constraints.

Solution

(a) The objective function and constraints are:

$$\text{Maximize } Z = 12\,X_1 + 36\,X_2$$

Subject to:

$C_1 \quad 8\,X_1 + 12\,X_2 \le 24$
$C_2 \quad 16\,X_1 + 8\,X_2 \le 32$
and $X_1, X_2 \ge 0$

TABLEAU 1 For Example 10.16

C_b	Basis	B_i	X_1	X_2	S_1	S_2	θ
0	S_1	24	8	12	1	0	2
0	S_2	32	16	8	0	1	4
	Z_j	0	0	0	0	0	
	$C_j - Z_j$		12	36	0	0	

TABLEAU 2 For Example 10.16

C_b	Basis	B_i	X_1	X_2	S_1	S_2	θ
36	X_2	2	0.667	1	0.083	0	
0	S_2	16	10.667	0	−0.667	1	
	Z_j	72	24	36	3	0	
	$C_j - Z_j$		−12	0	−3	0	

Final optimal solution: $Z = \$72$; X_1 is not in solution; $X_2 = 2$ gallons.

Before we proceed with the rest of this example, let us obtain the solution graphically, as this will help us to visualize the interpretation through Figure E10.16.

(b) Interpretation of the final tableau:

Variables in solution have ones and zeros. X_2 is in solution with a value of 2 units. S_2 is in solution with a value of 16 units. Remember that each unit of X_2 requires 8 hr of Machine B, so 2 units of X_2 require 16 hr of the 32 hr available. This leaves 16 hr of Machine B slack, as indicated under the third column for S_2, which is also in the solution.

Continues

Example 10.16: *Continued*

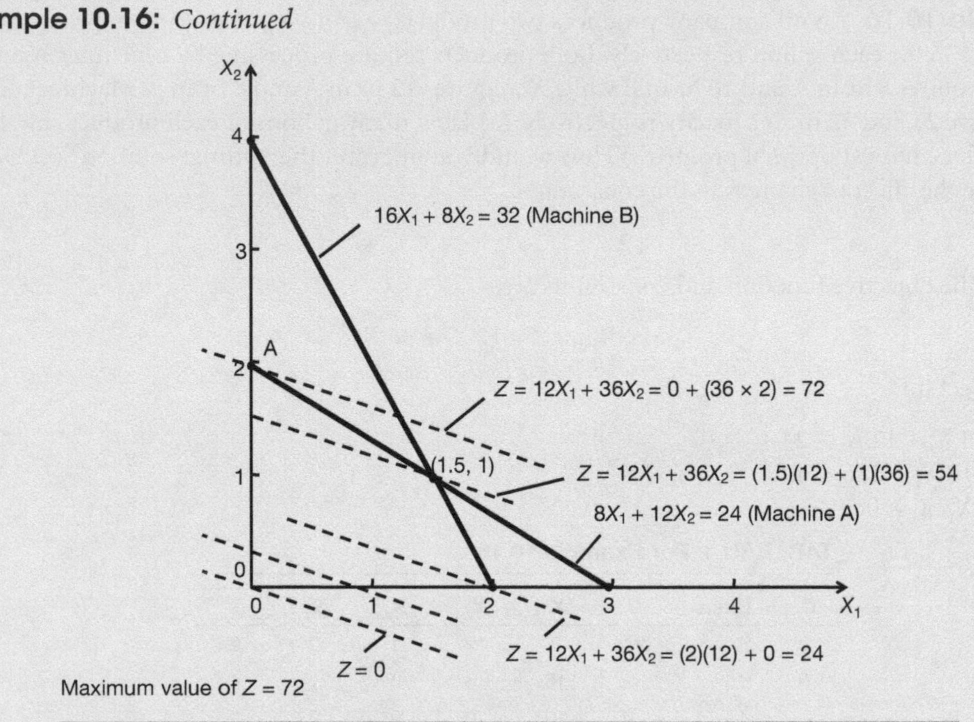

Figure E10.16 Graphical solution to interpreting the simplex algorithm results

The objective function $Z = \$72$ represents the profit that comes from producing two units of X_2 at \$36 each, plus 16 units of slack at \$0 each. The figures in the $(C_j - Z_j)$ row of the tableau mean the following:

(i) The (−12) to produce one gallon of X_1 would reduce profits by \$12 because it would take time away from Machine A from the production of X_2.

The \$12 amount is explained by Column X_1. Introducing one unit out of variable X_1 would reduce X_2 by 0.667 units at \$36 per unit = \$24 reduction; and reduce S_2 by 10.667 units at \$0 = \$0 reduction, for a total amount of \$24 − \$0 = \$24 cost that is offset by \$12 profit from each unit of X_1. The result is a net loss contribution of $C_j - Z_j = \$12 - \$24 = -\$12$.

(ii) The zero under the X_2 column indicates that X_2 is in solution (which means it is being produced).

(iii) The two values (−3.0) and (0) under Columns S_1 and S_2 respectively, are referred to as shadow prices. Shadow prices go with constraints and show the amount of change in the objective function that would result from each unit of change in the constraint. They indicate the net effect of increasing (or decreasing) the slack or idle time of Machines A and B by one unit.

Example 10.16: *Continued*

(iv) Because Machine A is fully utilized, to take one hour out of production and acquire one hour of idle time would reduce profit by $3 (Note that the profit from X_2 is $36 for each 12 hr of work on A, that is, a rate of $3 per hour). Conversely, if another hour could be made available, say, by shifting a current job from A, the time on A could be profitably utilized at a profit of $3 per hour.

(v) The zero corresponding to the constraint of Machine B signifies that Machine B already has slack time. Increasing B's available time (or decreasing it) by one unit would have no effect on profits.

(c) Determine the effect of changes in the constraints.

Machine A is the only active constraint. The sensitivity ratios for this constraint are:

For X_2:	(Column B_j)/($-S_1$)	=	2/(−0.083)	=	−24
For S_2:	(Column B_j)/($-S_1$)	=	16/(0.667)	=	24

The smallest positive ratio is 24 associated with S_2 which suggests that Constraint A may be relaxed by 24 hr before Machine B constraint begins to limit the solution.

If we examine the graphic solution, it is obvious that as the constraint for Machine A is relaxed (i.e., more hours are added), the constraint for Machine B takes effect at $X_2 = 4$. At that point, the profit would be $Z = 12\ X_1 + 36\ X_2 = \$(12)(0) + \$(36)(4) = \$144$. Also, at $X_2 = 4$, both Machines A and B would be fully utilized as shown in Table E10.16.

Table E10.16 Profits with revised limits

Old constraints	Revised limit	At $X_1 = 0$; $X_2 = 4$
Machine A: $8\ X_1 + 12\ X_2 \leq 24$	$8\ X_1 + 12\ X_2 \leq 48$	(8)(0) + (12)(4) = 48
Machine B: $16\ X_1 + 8\ X_2 \leq 32$	No change	(16)(0) + (8)(4) = 32

10.6.3 Marginal Value or Shadow Pricing

The value of the indicator under each slack variable in the final tableau expresses the marginal values or *shadow prices* of the input associated with the variable, that is, how much the objective function would change as a result of a one unit increase in the output. Thus, the simplex method allows one to examine how much one would be paying for additional quantities of any resource.

Example 10.17 Consider the following problem. A manufacturer makes two types of chemicals X_1 and X_2 that require three processes A, B, and C to manufacture each chemical. The number of hours needed by the machines for each chemical X_1 and X_2 and the constraint are given in Table E 10.17. The profit from each chemical per pound is also indicated. How much of each chemical should be produced to maximize profit? Also, provide an analysis of how profits would change as a result of a one-unit increase in the output.

Table E10.17 Hours needed and constraints

Process	X_1	X_2	Constraint
A	6	2	36
B	5	5	40
C	2	4	28
Profit	$50/unit	$30/unit	

Solution The objective function $Z = 50\, X_1 + 30\, X_2$.
Subject to:

$$6\, X_1 + 2\, X_2 \le 36$$

$$5\, X_1 + 5\, X_2 \le 40$$

$$2\, X_1 + 4\, X_2 \le 28$$

$$X_1, X_2 \ge 0$$

The final tableau is shown:

TABLEAU For Example 10.17

C_j / C_b	Basis	B_i	5 X_1	3 X_2	0 S_1	0 S_2	0 S_3
50	X_1	5	1	0	0.25	−0.10	0
30	X_2	3	0	1	−0.25	0.30	0
0	S_3	6	0	0	0.50	−1	1
	Z_j	340	50	30	5	4	0
	$C_j - Z_j$		0	0	−5	−4	0

The value under each slack variable in the final tableau expresses the marginal value or shadow price of the input associated with the variable. In other words, how much would the objective function change, as a result of a one-unit increase in the input? For example, profit would increase by $5 for a one-unit change in the constant value of constraint 1; by $4 for a one-unit increase in the constant value of constraint 2; and by $0 for a one-unit increase in value in constraint 3. Since constraint 3 has a positive slack variable, it is not fully utilized in the optimal

Continues

Example 10.17: *Continued*

solution and its marginal value is zero, which means that an addition of yet another unit would add nothing to the profit function, the optimum value of the objective function will always equal the sum of the marginal value of each input times the amount available of each input. If A, B, and C symbolize the constants in constraints 1, 2, and 3 respectively, then:

$$\text{Profit} = 50\,X_1 + 30X_2 = (50 \times 5) + (30 \times 3) = \$340$$

$$\text{A: } 6\,X_1 + 2\,X_2 + S_1 = (6 \times 5) + (2 \times 3) + 0 = 36$$

$$\text{B: } 5\,X_1 + 5\,X_2 + S_2 = (5 \times 5) + (5 \times 3) + 0 = 40$$

$$\text{C: } 2\,X_1 + 4\,X_2 + S_3 = (2 \times 5) + (4 \times 3) + 6 = 28$$

$$\begin{aligned}\text{Profit} &= MP_A(A) + MP_B(B) + MP_C(C)\\ &= 5(36) + 4(40) + 0(28)\\ &= 180 + 160 + 0\\ &= \$340\end{aligned}$$

10.6.4 Primal and Dual Problem Formulation Characteristics and Interpretation

Every maximization problem in LP has a corresponding minimization problem, and vice versa. If the original problem is called the primal, the corresponding problem is called the dual. The relationship between the two can best be expressed through the use of the parameters they share in common. An example will help to show the connection.

Example 10.18 Consider a small toy company making two kinds of toys X_1 and X_2 on an hourly basis. Toy X_1 produces a profit of \$5 while X_2 produces \$3 in profit. The resources for producing the toys are as follows (Table E10.18):

TABLE E10.18 Resources for producing toys

Resources	X_1	X_2	Total available/hr
Labor (hr)	6	2	36
Parts	5	5	40
Packing material	2	4	28
Profit	\$5	\$3	

The company wants to know the number of toys of Type X_1 and X_2 that should be produced per hour to maximize profit. The model can be formulated:

Continues

Example 10.18: *Continued*

Solution

Maximize $Z = 5X_1 + 3X_2$

Subject to

$$6X_1 + 2X_2 \leq 36$$

$$5X_1 + 5X_2 \leq 40$$

$$2X_1 + 4X_2 \leq 28$$

$$X_1, X_2 \geq 0$$

This is the primal form representing a maximization model. The dual form is the minimization model as shown:

Minimize $Z = 36Y_1 + 40Y_2 + 28Y_3$

Subject to:

$$6Y_1 + 5Y_2 + 2Y_3 \geq 5$$

$$2Y_1 + 5Y_2 + 4Y_3 \geq 3$$

$$Y_1, Y_2, Y_3 \geq 0$$

(a) We have used the coefficients along the vertical lines and interchanged ≥ for ≤.

(b) The dual variables Y_1, Y_2, and Y_3 correspond to the model constraints in the primal.

(c) The quantity values 36, 40, and 28 in the primal form the objective function in the dual.

$$Z = 36Y_1 + 40Y_2 + 28Y_3$$

Let us interpret the primal simplex solution.

TABLEAU For Example 10.18

		Quantity	**5**	**3**	**0**	**0**	**0**
C_j	**Basis**	B_i	X_1	X_2	S_1	S_2	S_3
5	X_1	5	1	0	0.25	−0.10	0
3	X_2	3	0	1	−0.25	0.30	0
0	S_3	6	0	0	0.50	−1.0	1.0
	Z_j	34	5	3	0.50	0.40	0
	$C_j - Z_j$		0	0	−0.50	−0.40	0

Toy $X_1 = 5$; Toy $X_2 = 3$; $S_3 = 6$ (Packing); $Z = \$34.00$. This primal simplex tableau also contains information about the dual. In the $C_j - Z_j$ row, the negative values of −0.5 and −0.4 under the

Continues

Example 10.18: *Continued*

S_1 and S_2 columns indicate that if one unit of either S_1 or S_2 were entered into the solution, the profit would decrease by 50¢ and 40¢, respectively. Also note that S_1 represents unused labor and S_2 represents unused parts. In this example they are not basic variables and are both equal to zero, which means that they are fully utilized (or that there is no slack). The values 50¢ and 40¢ are the *marginal* values of labor (S_1) and parts (S_2), respectively. These values are also referred to as *shadow prices*, since they reflect the maximum *price* one would be willing to pay to obtain one more unit of the resource. However, the $C_j - Z_j$ value for S_3 representing packaging is zero, which indicates that the marginal value of S_3 is zero, or that we would be unwilling to pay anything for one unit of packaging. In fact, there is a surplus of 6 units of packaging, left unused after the two types of toys were produced, as shown in the row marked S_3.

Example 10.19 A farmer has three resources to produce two types of wheat *A* and *B* as shown (Table E10.19).

Table E10.19 Resources for wheat types

Resources	Requirement for Wheat *A*	Requirement for Wheat *B*	Total available
Water/unit	10	15	300
Fertilizer/unit	5	9	100
Labor/unit	3	6	120
Profit	$110	$180	

(a) Write the primal linear program model.

(b) Write the dual model.

(c) Explain the units of the dual model.

Solution

(a) The primal model is:

Maximize $Z = 110\,A + 180\,B$

Subject to:

$$10\,A + 15\,B \leq 300$$

$$5\,A + 6\,B \leq 100$$

$$3\,A + 6\,B \leq 120$$

$$A, B \geq 0$$

Continues

Example 10.19: *Continued*

(b) In the primal model there are 2 primal variables and 3 constraints; while the dual model has 3 variables and 2 constraints, as follows:

Minimize $Z = 300\,C + 100\,D + 120\,E$

Subject to: $10\,C + 5\,D + 3\,E \geq 110$

$15\,C + 9\,D + 6\,E \geq 180$

$C, D, E \geq 0$

(c) The units of the dual model are:

C = The dollar value of a unit of water
D = The dollar value of a unit of fertilizer
E = The dollar value of a unit of labor
Z = The cost function that is being minimized

10.6.5 Solving Minimization Problems with Simplex

When minimization problems need to be solved using simplex, several adjustments must be made. First, constraints in the form of ≥ are converted to equalities by subtracting the surplus or slack amounts. Second, to get an initial solution to the problem, an artificial variable is added to the constraint, because merely introducing a slack variable is insufficient. Third, the cost of a surplus variable is $0, while the cost of an artificial variable is *M*, a large positive number that will prevent the artificial variable from having a nonzero value. Fourth, because the objective is to minimize, we will be computing $Z_j - C_j$ in the bottom row of the simplex tableau, instead of $C_j - Z_j$.

Example 10.20 A farmer feeds his cattle with nutrients X_1 and X_2 made up of two basic ingredients, carbohydrates (C) and protein (P). The details of the nutrients and their costs per pound are given in Table E10.20.

Table E10.20 Details of nutrient types

Ingredients	Nutrient X_1	Nutrient X_2	Minimum requirements (units)
Carbohydrates (C)	1	2	80
Protein (P)	3	1	75
Profit	4	6	

What combination of C and P should be provided as an adequate diet at minimum cost?

Continues

Example 10.20: *Continued*

Solution

Minimize $Z = 4X_1 + 6X_2$

Subject to: $X_1 + 2X_2 \geq 80$

$3X_1 + X_2 \geq 75$

$X_1, X_2 \geq 0$

Convert the inequalities to equations by subtracting slack variables (S) and adding artificial variables (A). Thus:

$$X_1 + 2X_2 - S_1 + A_1 = 80$$

$$3X_1 + X_2 - S_2 + A_2 = 75$$

where A_1 and A_2 must not appear in the final solution and therefore must be of high value $= M$.

SIMPLEX TABLEAU 1 For Example 10.20

C_b \ C_j	Basis	B_i	4 X_1	6 X_2	0 S_1	0 S_2	M A_1	M A_2	θ
M	A_1	80	1	2	–1	0	1	0	80
M	A_2	75	3	1	0	–1	0	1	25 (OUT)
	Z_j	155M	4M	3M	–M	–M	M	M	
	$C_j - Z_j$		4–4M (IN)	6–3M	M	M	0	0	

SIMPLEX TABLEAU 2 For Example 10.20

C_b \ C_j	Basis	B_i	4 X_1	6 X_2	0 S_1	0 S_2	M A_1	M A_2	θ
M	A_1	55	0	1.667	–1	0.333	1	–0.333	33 (OUT)
4	X_1	25	1	0.333	0	–0.333	0	0.333	75
	Z_j	10099	4	M	–M	–M	M	–M	
	$C_j - Z_j$		0	M (IN)	M	M	0	–M	

Continues

Example 10.20: *Continued*

SIMPLEX TABLEAU 3 For Example 10.20

C_b \ C_j	Basis	B_i	4 X_1	6 X_2	0 S_1	0 S_2	M A_1	M A_2
6	X_2	33	0	1	−0.6	0.2	0.6	−0.2
4	X_1	14	1	0	0.2	−0.4	−0.2	0.4
	Z_j	254	4	6	−2.8	−0.4	2.797	0.406
	$C_j - Z_j$		0	0	−2.8	−0.4	$-M$	$-M$

Final optimal solution = 254 (see Figure E10.20).

An alternative method of solving the minimization problem concerning the farmer's feed for his cattle can be done by converting this primal problem into its dual, which will be a maximization problem. Recall that we had the following equations representing the primal:

Minimize $Z = 4\,X_1 + 6\,X_2$

Subject to: $X_1 + 2\,X_2 \geq 80$ ———————— Y_1

$3\,X_1 + X_2 \geq 75$ ———————— Y_2

$X_1, X_2 \geq 0$

Rewriting, the dual (maximization) form:

$Y_1 + 3\,Y_2 \leq 4$

$2\,Y_1 + Y_2 \leq 6$

Maximize $Z = 80\,Y_1 + 75\,Y_2$

Notice that the dual variables Y_1 and Y_2 correspond to the model constraints in the primal, and the quantity values 80 and 75 in the primal form, are the coefficients in the objective function of the dual.

Example 10.21 Rework the previous examples using the simplex method using the dual form.

Solution

Maximize $Z = 80\,Y_1 + 75\,Y_2$

Subject to:

$Y_1 + 3\,Y_2 \leq 4$

$2\,Y_1 + Y_2 \leq 6$

Example 10.21: *Continued*

SIMPLEX TABLEAU 1 For Example 10.21

C_b \ C_j	Basis	B_i	80 Y_1	75 Y_2	0 S_1	0 S_2
0	S_1	4	1	3	1	0
0	S_2	6	2	1	0	1
	Z_j	0	0	0	0	0
	$C_j - Z_j$		80	75	0	0

SIMPLEX TABLEAU 2 For Example 10.21

C_b \ C_j	Basis	B_i	80 Y_1	75 Y_2	0 S_1	0 S_2
0	S_1	0	1	2.5	1	–0.5
80	Y_1	3	1	0.5	0	0.5
	Z_j	240	80	40	0	40
	$C_j - Z_j$		0	35	0	–40

SIMPLEX TABLEAU 3 For Example 10.21

C_b \ C_j	Basis	B_i	80 Y_1	75 Y_2	0 S_1	0 S_2
75	Y_2	0.4	0	1	0.4	–0.2
80	Y_1	2.8	1	0	–0.2	0.6
	Z_j	254	80	75	14	33
	$C_j - Z_j$		0	0	–14	-33

Final optimal solution $Z = 254$ (see Figure E10.21).

Continues

Example 10.21: *Continued*

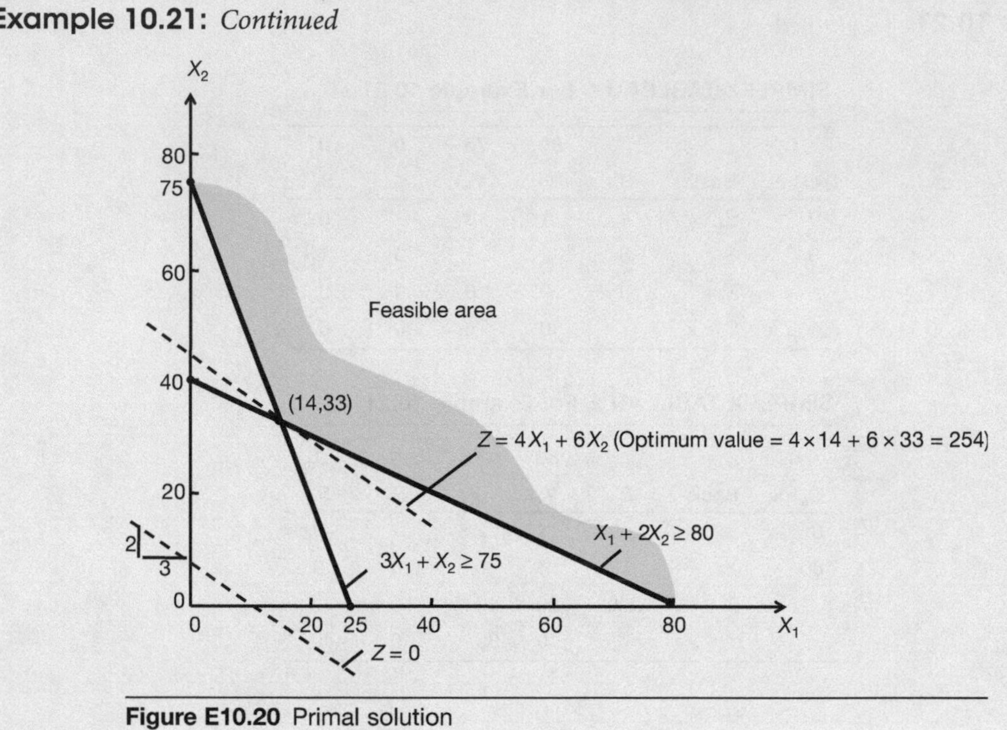

Figure E10.20 Primal solution

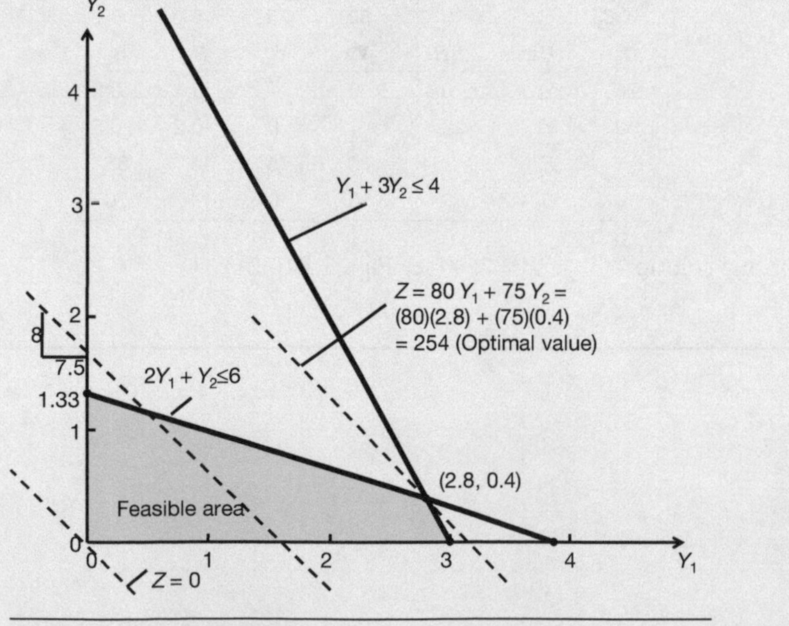

Figure E10.21 Dual solution

10.6.6 Interpretation of the Primal and Dual Models

Let us interpret the final tableau of the primal model given in Example 10.20. We observe that for minimizing the total cost, nutrient $X_1 = 14$ units and nutrient $X_2 = 33$ units, which works out to \$254. This primal tableau also contains information under the S_1 and S_2 columns indicating that if one unit of either S_1 or S_2 were entered into the solution, the cost would increase by \$2.8 and \$0.4 respectively, remembering that S_1 represents unused carbohydrates (C) and S_2 represents unused protein (P). This means that all of the C and P are being used for producing X_1 and X_2 and that there is no slack left over. In fact, the negative $C_j - Z_j$ row values of 2.8 and 0.4 units are the marginal values of C and P respectively. These dual values are also referred to as shadow prices. Note that the minimum cost is \$254. Now, let us look at the final tableau of the dual model (the maximization model). Here, we must understand the meaning of Y_1 and Y_2 as well as S_1 and S_2. The interpretation is:

Y_1 = marginal value of one unit of carbohydrate (C)
Y_2 = marginal value of one unit of protein (P)
S_1 = value of a unit of nutrient X_1
S_2 = value of a unit of nutrient X_2

The cost of this optimum combination is \$254.

The graphical solutions to the primal and dual problems are shown in Figures E10.20 and E10.21. The economic significance of Y_1 is that it is worth it to the producer to provide one unit of carbohydrate C; Y_2 is the worth per unit of protein P. The objective function $Z = 80\ Y_1 + 75\ Y_2$ represents the worth to the producer of meeting the needs. Its optimal value must be identical to the original (least-cost) optimum.

The simplex solution to the dual is shown in the first tableau of example 10.21. The solution is for optimal values of Y_1 and Y_2 but actually the optimal values of X_1 and X_2 appear as the ($C_j - Z_j$) entries for S_1 and S_2 respectively. The converse is also true; $Y_1 = 2.8$ and $Y_2 = 0.4$ appear as $C_j - Z_j$ entries for S_1 and S_2 respectively. Notice that $Z = \$254$ is the same in both cases.

Check:

$$\text{Optimal } Z = 80\ Y_1 + 75\ Y_2$$
$$= (80)(2.8) + (75)(0.4)$$
$$= 224 + 30 = \$254$$

or,

$$Y_1 + 3\ Y_2 + S_1 = 2.8 + 3(0.4) = 4 \text{ ---------- (A)}$$
$$2\ Y_1 + Y_2 + S_2 = 2(2.8) + (0.4) = 6 \text{ ---------- (B)}$$

Therefore:

$$\text{Optimal Z} = \text{mP(A)} + \text{mP(B)}$$
$$= (4)(14) + (6)(33) = \$254.$$

EXERCISES

1. A metal box company is designing hundreds of steel container boxes per day with the following specifications: the box must have a square base with an open top, and a volume of 64 cubic feet internal, and must use a minimum of sheet metal. What should be the optimum dimensions of the box?
2. Calculate the output of computer parts Q per day that should be produced in order to earn the maximum possible profit, remembering that for profit maximizing the marginal revenue should be equal to MC. The cost and revenue functions are:

   Total cost = $0.016Q^3 - 6Q^2 + 800Q + 60$
   Total revenue = $50Q$, where Q represents the number of parts.
3. A shopping center expert has observed that the number of customers visiting the center depends on the number of parking spaces provided. The estimate is given by the equation $Q = Kx^a$, where Q is the number of customers visiting the center and K and a are constants. What should be the optimal number of parking spaces provided?
4. If the total revenue of a company is TR: (a) derive an expression for the marginal revenue in terms of the elasticity of demand and the price; and (b) show that an increase in quantity sold results in an increase in revenue if the elasticity is greater than 1? (c) What would you advise a bicycle manufacturer, if the elasticity of demand is 0.9, and he wants to increase the unit price of bicycles? Set up a sample case to prove that your advice is correct.
5. A cloth manufacturer produces two types of cloth, A and B. His profit function is:

   $$\text{Profit} = 128A - 4A^2 + 8AB + 64B - 28 - 8B^2$$

   What should be the level of output of A and B to maximize profit?
6. The cloth manufacturer in Exercise 5 is faced with a constraint. He can only manufacture a total of 50 units of cloth A and B together. What should be his level of production to maximize profit?
7. A company has the following production function:

   $Q = 20L^{0.5}K^{0.5}$, where Q = quantity of output produced per hour, L = amount of labor, and K = capital expended. The company's total cost function is TC = $10L + 40K$. To meet customer demand, the firm needs to produce 100 units of the product per day at the minimum cost. How many units of labor and capital are needed to meet these conditions?
8. Solve Exercise 7 using Lagrange multiplier.
9. The network shown in Figure P10.9 indicates the activity times in weeks for completing a small culvert construction. Complete a table of ES, EF, LS, LF, and TF times and indicate the critical path and project duration.

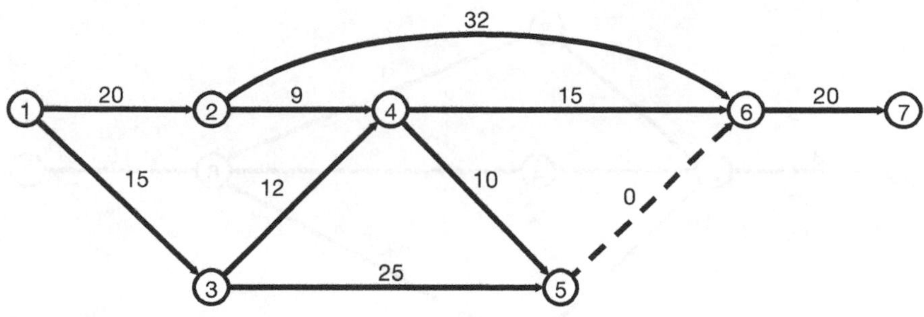

Figure P10.9

10. The network in Figure P10.10 represents the activities for constructing a culvert. Set up the ES, EF, LS, LF, TF, FF, and interfering float (IF) table and indicate the critical path. Draw a time-grid diagram and a man power resource schedule (see Table P10.10).

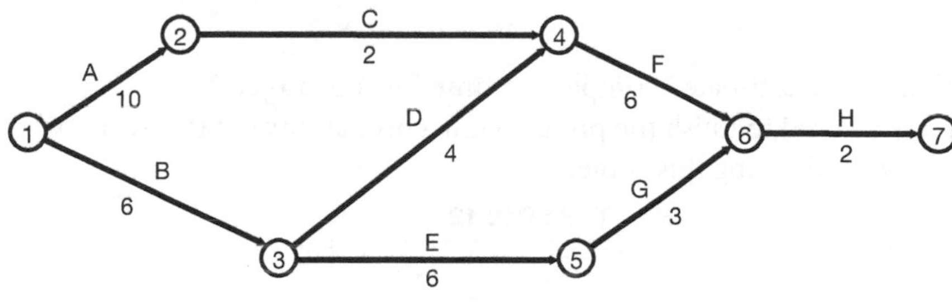

Figure P10.10

Table P10.10

A	5 men
B	7 men
C	8 men
D	9 men
E	6 men
F	10 men
G	4 men
H	8 men

11. An engineer draws up an activity network of 8 activities, assigning the most likely time in weeks for finishing each task, as shown in Figure P10.11:

   (a) Compute the ES, EF, LS, LF, and TF times for the various tasks.
   (b) What is the critical path and the most likely time the project will be completed?

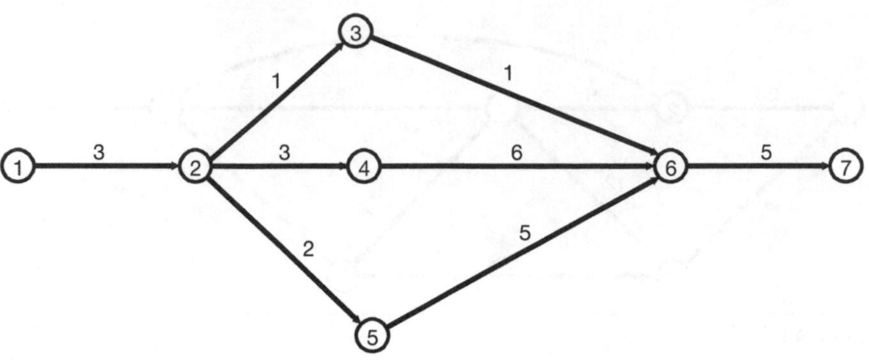

Figure P10.11

12. In Exercise 11, the engineer solicits the ideas of an architect to estimate the time required for each task. Table P10.12 shows the optimistic, most likely and pessimistic times for each task.

    (a) What is the estimated completion time for this project?
    (b) If it is desired to finish the project with a probability of 90 percent, what is the probability of finishing this project?

Table P10.12

Task	a–m–b
1–2	2–3–8
2–3	1–1–3
2–4	2–3–5
2–5	1–2–4
3–6	1–1–1
4–6	5–6–8
5–6	2–5–7
6–7	4–5–6

13. An engineer has estimated the duration of each activity under normal and crash conditions for constructing an office extension. The cost of crashing each task in $/da is also given (Table P10.13).

    (a) What would be the normal time of completion of this project?
    (b) By how many days can the project be compressed?
    (c) If the overhead charges are $50/da, what crash schedule would be optimal?

Table P10.13

Activity	Time (days) Normal	Crash	Cost of crashing in $/da
1–2	9	6	20
1–3	8	5	25
1–4	15	10	30
2–4	5	3	10
3–4	10	6	15
4–5	2	1	40

14. A shelter is planned for construction consisting of five major activities as shown in the Figure P10.14.

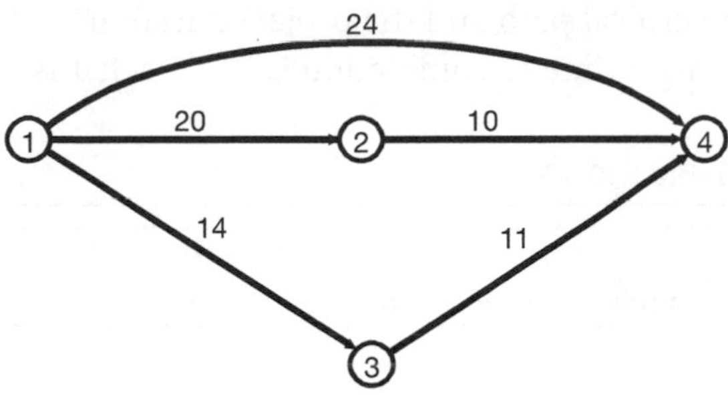

Figure P10.14

Details of the normal and crash times in weeks along with the corresponding costs in hundreds of dollars are indicated in Table P10.14.

Table P10.14

		Time in weeks		Cost in hundreds of $/wk	
Activity	***i* – *j***	**Normal**	**Crash**	**Normal**	**Crash**
1	1 – 2	20	8	10.00	14.80
2	1 – 4	24	20	12.00	14.00
3	1 – 3	14	7	7.00	11.90
4	2 – 4	10	6	5.00	8.20
5	3 – 4	11	5	5.50	7.30

(a) How much can the project time be compressed?
(b) Compute the normal and crash costs of the project.
(c) If the overhead cost of construction is $10,000 per wk, what would be an optimum crashing schedule?

15. A garage is planned to be an extension to a house with the following activities shown (Figure P10.15) with data given in Table P10.15:

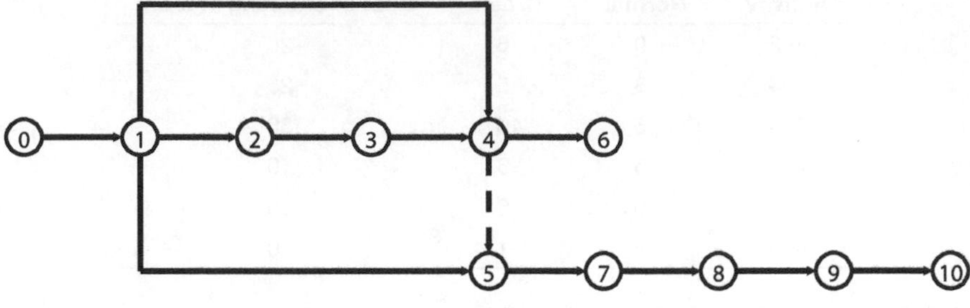

Figure P10.15

(a) Determine the critical path and the project duration.
(b) If the project target time of completion is 25 da, what is the probability of completion?

Table P10.15

		Time			
Activity	***i – j***	**a**	**m**	**b**	
0	0 –1	4	4	4	
1	1 – 2	1	3	5	
2	1 – 4	2	3	5	
3	1 – 5	2	4	7	
4	2 – 3	1	3	4	
5	3 – 4	2	5	8	
	4 – 5	0	0	0	Dummy
6	4 – 6	1	2	2	
7	5 – 7	3	6	8	
8	6 – 8	2	4	5	
9	7 – 8	3	4	6	
10	8 – 9	1	2	4	
11	9 – 10	1	1	1	

16. An architect lists 15 tasks for constructing a house. The logic and duration of the activities and the labor needed for completing the house are shown in Figure P10.16. Determine the ES, EF, LS, LF, TF and IF times for the activities given in Table P10.16. How would you allocate the labor so that there would be more or less equal man power requirements throughout the duration of the project? (Draw a time-grid diagram to help you out.) All times are in weeks.

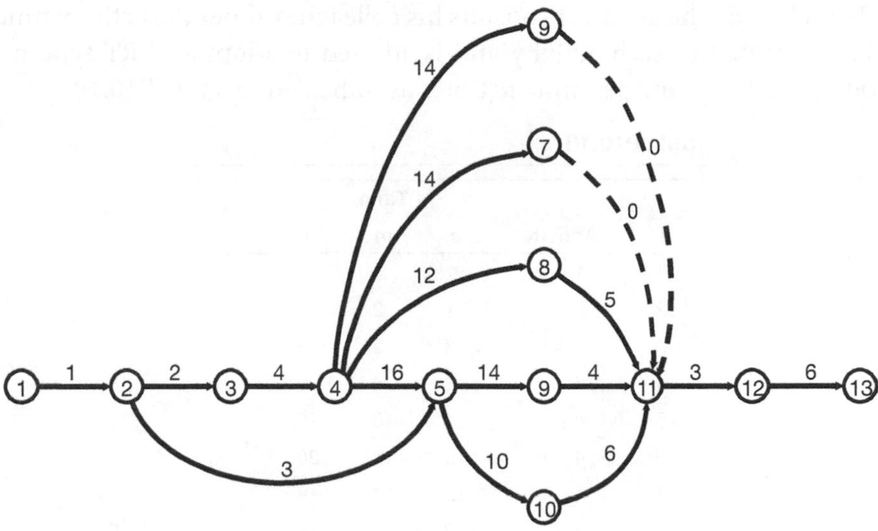

Figure P10.16

Table P10.16

#	Task	Predecessor activity	Time	Labor
1	1 – 2	–	1	1
2	2 – 3	1 – 2	2	1
3	2 – 5	1 – 2	3	1
4	3 – 4	2 – 3	4	1
5	4 – 5	3 – 4	16	2
6	4 – 6	3 – 4	14	1
7	4 – 7	3 – 4	14	2
8	5 – 8	(4 – 5), (2 – 5)	12	2
9	5 – 9	(4 – 5), (2 – 5)	14	3
10	5 – 10	(4 – 5), (2 – 5)	10	1
11	6 – 11	(4 – 6), (Dummy)	0	0
12	7 – 11	(4 – 7), (Dummy)	0	0
13	8 – 11	5 – 8	5	1
14	9 – 11	5 – 9	4	2
15	10 – 11	5 – 10	6	1
16	11 – 12	(6 – 11), (7 – 11), (8 – 11), (9 – 11), (10 – 11)	3	2
17	12 – 13	11 – 12	6	1

17. Refer to Exercise 16. The architect feels that the time of the tasks on the critical path could be easily revised as follows: (2-3) 1 wk, (3-4) 3 wk, (4-5) 15 wk, (5-9) 13 wk, (9-11) 3 wk, (11-12) 2 wk, and (12-13) 5 wk. What is the new critical path and what is the duration of the project?

18. Refer to Exercise 16. The architect consults his colleagues about the activity time (in weeks) that he has allocated to each activity and is advised to adopt a PERT-type network with optimistic, most likely, and pessimistic times as indicated in Table P10.18.

Table P10.18

			Time		
#	**Activity**	**a**	**m**	**b**	
1	1 – 2	1	1	3	
2	2 – 3	1	2	3	
3	2 – 5	2	3	5	
4	3 – 4	2	4	6	
5	4 – 5	14	16	25	
6	4 – 6	13	14	20	
7	4 – 7	12	14	19	
8	4 – 8	10	13	18	
9	5 – 9	12	14	15	
10	5 – 10	8	10	14	
11	6 – 11	0	0	0	Dummy
12	7 – 11	0	0	0	Dummy
13	8 – 11	4	5	6	
14	9 – 11	2	4	5	
15	10 – 11	5	6	8	
16	11 – 12	3	3	5	
17	12 – 13	5	6	7	

(a) What is the expected time of completion of the project?
(b) What is the critical path?
(c) If the architect wants to be assured that the project is completed with 90 percent confidence, what is the timeline?

19. A manager notes that one of his electricians takes anywhere between 30 min to 1 hr to do repairs to electric motors. However, 40 min is the most frequent duration. If this electrician's task were on a PERT project:

(a) What would be the expected duration of a repair job?
(b) What is the variance and standard deviation?
(c) If he performs 8 such repairs per day with a 10 minute break between two repairs, what would be the expected time he spends on repairs per day?
(d) What is the variance and standard deviation of his working day?
(e) What is the probability he spends (1) 6.5 hr (2) 6.75 hr (3) 7 hr at work per day (including breaks)?

20. The project network for a repair job is shown in Figure P10.20; and data is given in Table P.10.20.

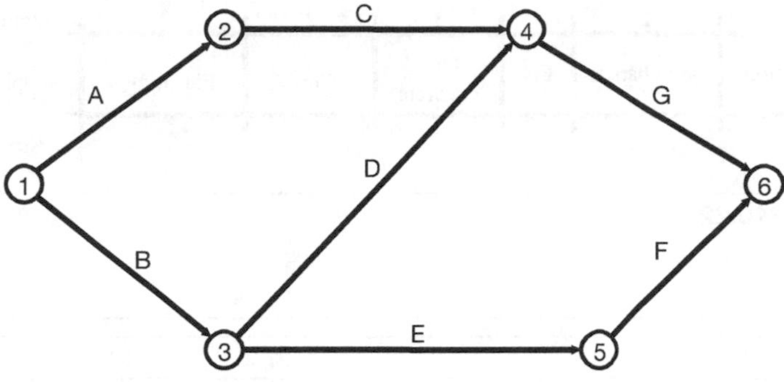

Figure P10.20

Table P10.20

Activity	Link	t_a	Time t_m	t_b
A	1 – 2	4	5	6
B	1 – 3	9	12	15
C	2 – 4	2	2	2
D	3 – 4	8	10	18
E	3 – 5	2	4	12
F	5 – 6	1	2	3
G	4 – 6	4	6	8

(a) Find the expected duration and variation of each activity.
(b) What are the expected project length and its variance and standard deviation?
(c) What is the probability that the project will be completed (1) at least 2 da earlier than expected or (2) no more than 2 da later than expected?
(d) If the project due date is 30 da, what is the probability of lateness?
(e) What due date has about 98 percent chance of being met?

21. Refer to Example 10.8, worked out in this chapter. The prefabricator has now agreed to deliver 50 units of septic tanks each month beginning with the thirtieth month. All other conditions remain the same. If an LOB study is to be conducted in the fortieth month, what would an inspector expect to find with respect to each control point?

22. A precast concrete beam fabricator is awarded a contract to produce 10,000 beams to be delivered as per schedule shown. The production schedule consists of 6 distinct stages and one supply stage as shown in Figure P10.22. No buffers are provided. An LOB study was done on the twelfth day after production started, and the report indicated that the actual

cumulative number in hundreds of beams in stages 1 through 7 were as shown in Table P.10.22.

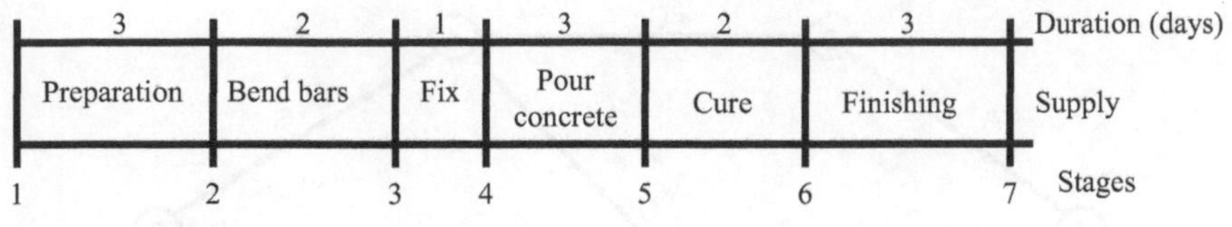

Figure P10.22

Table P10.22

Stage				1	2	3	4	5	6	7
Cumulative (100s) found in 12th day				88	85	83	81	76	70	30

Delivery schedule:										
Day	1	2	3	4	5	6	7	8	9	10
Quantity (100s)	2	1	2	2	1	2	2	2	2	2
Day	11	12	13	14	15	16	17	18	19	20
Quantity (100s)	2	7	8	9	8	9	7	9	8	2
Day	21	22	23	24	25	26	27	28		
Quantity (100s)	1	2	1	1	2	2	2	2		

Draw the LOB chart and indicate the deviations between the ideal and actual production of beams at various stages.

23. A concrete road is being planned for repetitive stretches of 30 sections each. The sequence of activities is: (a) prepare foundation, (b) pour concrete, and (c) grade and finish. The man hours and teams sizes are shown (Table P10.23):

Table P10.23

Task	**A**	**B**	**C**
Man hr/section	240	580	300
Men/team	6	12	8

(a) Assuming there are no buffers between operations, and that there are 5 da of work per week at 8 hr per da, prepare a LOB for 4 units per wk

(b) If a buffer of 5 da is provided between operations, and everything else remains the same, prepare a revised LOB for 4 units/wk

24. A small city wants to develop a bikeway plan connecting nine major attractions. The existing street plan shows distance units (see Figure P10.24). What is the minimal spanning tree?

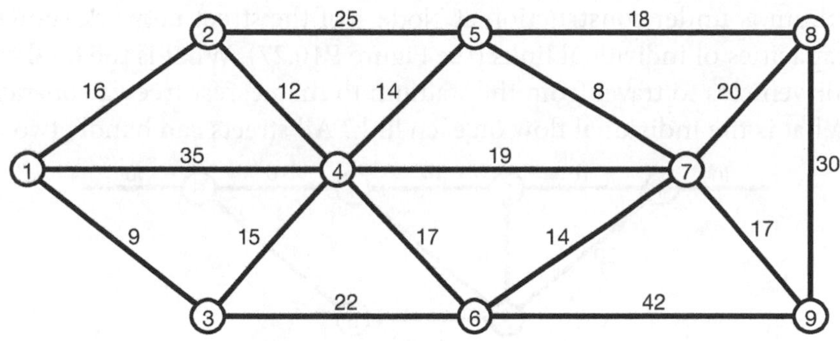

Figure P10.24

25. A street plan of a historic city is illustrated in Figure P10.25, showing 16 nodes that have ancient artifacts. A bikeway plan connecting all 16 nodes is planned. What is the minimal spanning tree for this city?

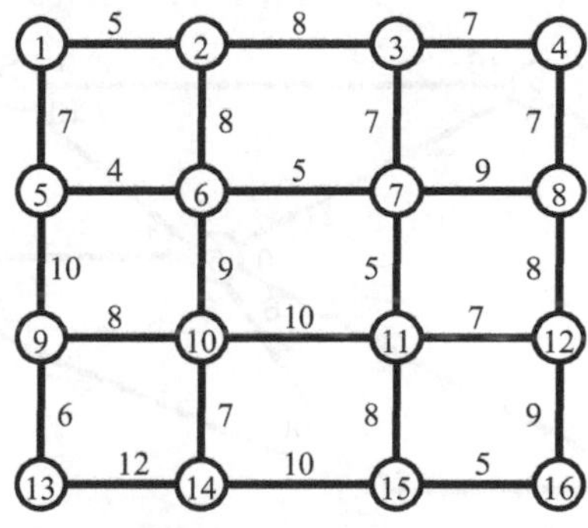

Figure P10.25

26. An amusement park with 10 attractions needs to be connected with the minimum length of paths (see Figure P 10.26). How would you plan it?

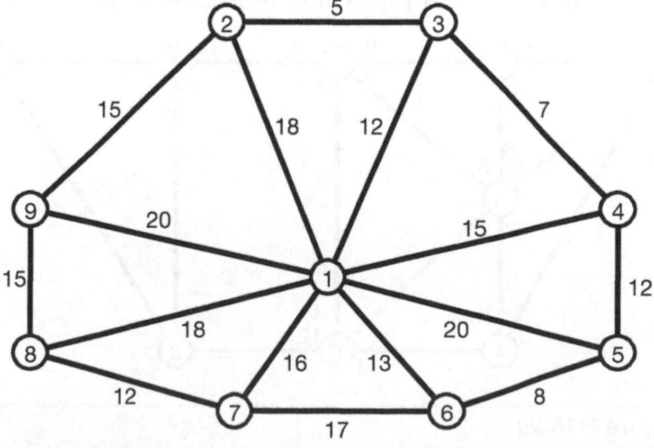

Figure P10.26

27. A new stadium is under construction at Node 1 of the street network shown, along with one-way capacities of individual links (see Figure P10.27). What is the total capacity of the network for vehicles to travel from the stadium to the nearest freeway on-ramp located at Node 6? What is the individual flow on each link? All streets can handle two-way flow.

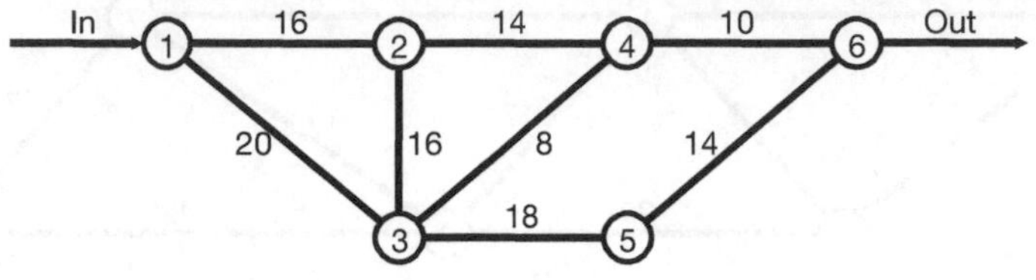

Figure P10.27

28. The network shown (Figure P10.28) carries trucks at a construction site that has link capacities in the hundreds as shown. What is the possible flow of trucks from Node 1 to Node 6? How many trucks do individual links carry? [Link capacities: 1 – 2 = 32; 1 – 3 = 24; 2 – 4 = 16; 2 – 5 = 28; 3 – 5 = 12; 3 – 6 = 8; 4 – 5 = 10; 4 – 7 = 20; 5 – 6 = 6; 5 – 7 = 12; 6 – 7 = 14.]

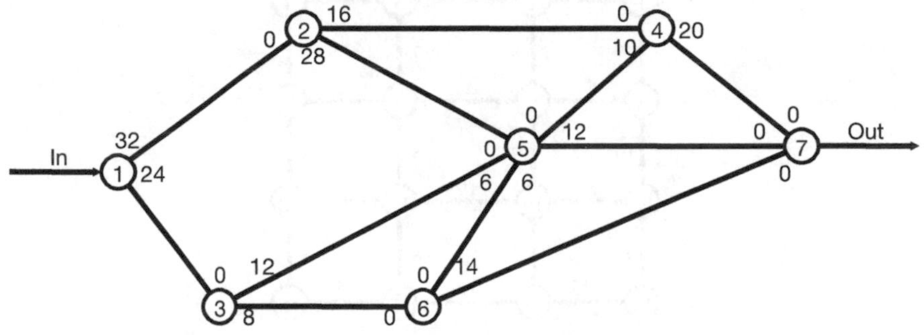

Figure P10.28

29. An oil pipeline system carries oil from Node 1 to Node 9 through individual pipes as shown in Figure P10.29 with data given in Table P.10.29. What is the capacity of the system and how much does each link carry? All pipes can handle flow in both directions.

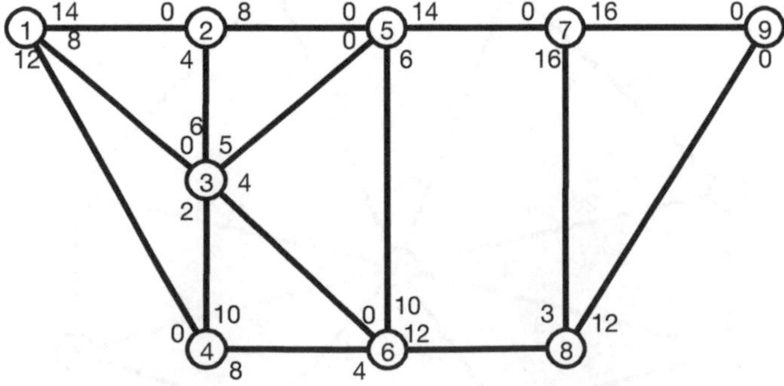

Figure P10.29

Table P10.29

	Flow	Inverse flow
Link	1 – 2 = 14	0
Link	1 – 3 = 8	0
Link	1 – 4 = 12	0
Link	2 – 3 = 4	6
Link	2 – 5 = 8	0
Link	3 – 4 = 2	10
Link	3 – 5 = 5	0
Link	3 – 6 = 4	0
Link	4 – 6 = 8	4
Link	5 – 6 = 6	10
Link	5 – 7 = 14	0
Link	6 – 8 = 12	0
Link	7 – 8 = 16	0
Link	7 – 9 = 16	0
Link	8 – 9 = 12	0

30. A highway network consisting of five nodes and eight links is shown in Figure P10.30. The cost of transportation is also shown. A trip table (Table P10.30) shows the numbers of vehicles per hour wanting to go from one node to another. Assign the trips to the network.

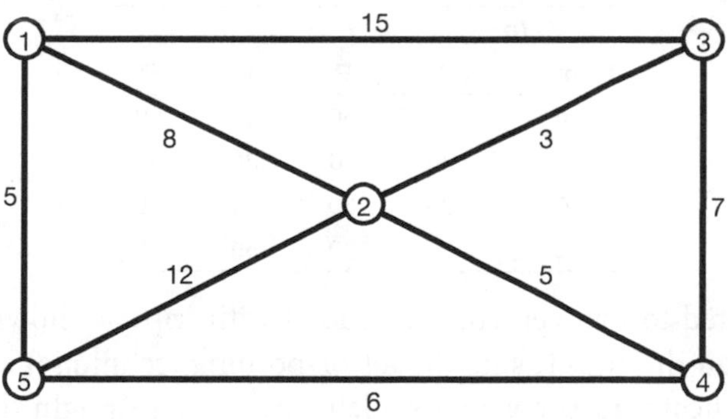

Figure P10.30

Table P10.30

From	To 1	2	3	4	5	Total
1	0	50	60	70	30	210
2	40	0	30	60	80	210
3	90	40	0	20	50	200
4	80	70	90	0	30	270
5	30	40	50	60	0	180
Total	240	200	230	210	190	1070

31. A simple network (shown in Figure P10.31) has two-way links. The time cost is also shown. Find the shortest path from Nodes A, B. C, and D to all other nodes and intersections.

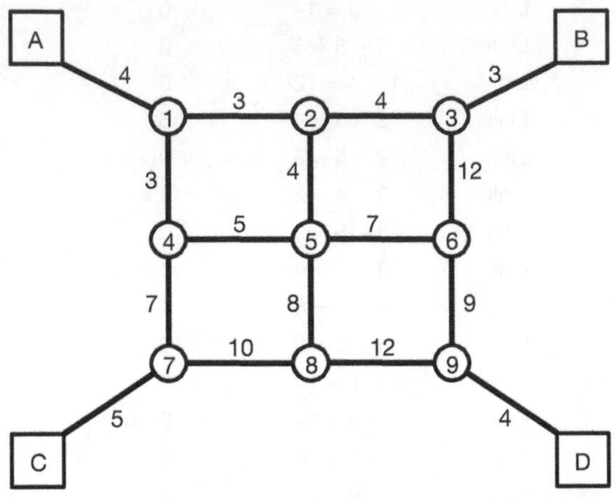

Figure P10.31

32. A trip table (veh/hr) needs to be loaded on the network shown for Exercise 31 (see Table P.10.32). Find the total volume on each link, assuming an all-or-nothing assignment.

Table P10.32

	To			
From	**A**	**B**	**C**	**D**
A	0	50	40	20
B	30	0	80	10
C	90	80	0	20
D	60	70	50	0

33. A network connected to four centroids is loaded with trips as shown in Figure P10.33 and Table P10.33. Assign the trips (using the all-or-nothing technique) assuming the following: figures on links indicate travel cost; plus a left turn, going straight through an intersection, and turning right carry penalties of 3, 2, and 1 units, respectively; and, all links are two-way.

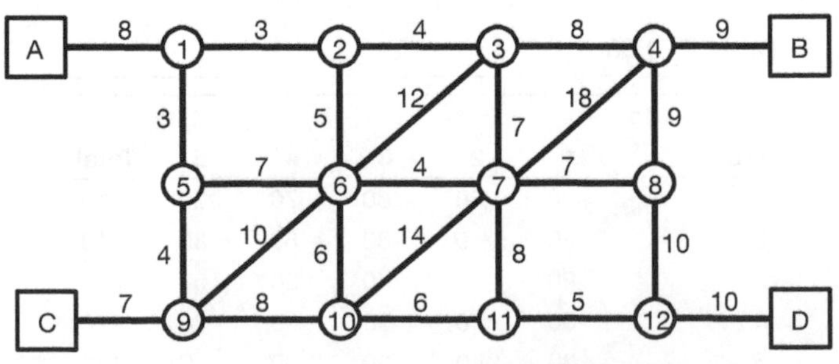

Figure P10.33

Table P10.33

	To			
From	**A**	**B**	**C**	**D**
A	0	900	400	700
B	200	0	700	300
C	600	800	0	400
D	100	200	500	0

34. A manufacturer produces two types of Machines X_1 and X_2. The resources needed for producing these machines and the corresponding profits are given in Table P10.34:

Table P10.34

Machine	Labor (hrs/unit)	Capital ($/unit)	Profit ($)
X_1	10	40	40
X_2	20	30	50

There are 400 hr of labor and $1200 worth of capital available per day. How many machines of Type X_1 and X_2 should be produced per day to maximize profits?

35. A company produces two kinds of septic tanks: a basic Type A and a special Type B. Each Type B tank takes twice as long to produce as Type A, and the company has the time to make a maximum of 2000 per wk, if it produced only Type A. The supply of materials per day is sufficient to produce a total of 1500 septic tanks/wk of both A and B types. The Type B septic tank requires special fixtures of which there are 600 per wk available. If the company makes a profit of $300 for an A type and $500 for a B type tank, how many of each should it produce per week to maximize profits? Solve this problem using the simplex method and then check your answer graphically.

36. A manufacturer produces two kinds of boxes, A and B. Their requirements are shown in Table P10.36:

Table P10.36

Product	A	B	Constraint
Steel sheets	8 units	10 units	80,000
Hinges	2.5 units	1 unit	20,000
Welding and paint	2 units	4 units	30,000

Both types of boxes can be produced at the same time and 12 boxes can be put together per hour; 750 per hr are available per week. There is a restriction that at least 3000 boxes of Type A have to be produced per week. How many boxes of each type should be produced?

37. A concrete beam manufacturer produces two products. Type A requires 2 hr of preparation, 4 hr of casting, and 10 hr of finishing; while Type B requires 5, 1, and 5 hr, respectively. There are time constraints in hours for the two products as shown.

$$2A + 5B \leq 40$$

$$4A + B \leq 20$$

$$10A + 5B \leq 60$$

The profit on Type A is 24 and 8 for Type B. How many of each type should be produced for maximizing profits?

38. A country suffering from famine wants to be assured that the people get the minimum daily requirements of three basic nutrients A, B, and C, which they can get from products X_1 and X_2. Daily requirements are 14 units of A, 12 of B, and 18 of C. Product X_1 has 2 units of A and 1 unit each of B and C; Product X_2 has one unit of A and B, and 3 units of C. Products X_1 and X_2 cost \$2 and \$4, respectively. What is the least cost combination of X_1 and X_2?

39. Refer to Example 10.13. Because of improved manufacturing techniques, the firm is able to make a profit of \$10 on beams and \$15 on poles. How many beams and poles should be produced to maximize profits? Solve this problem using the graphical method.

    (a) Rework the problem using the simplex method.
    (b) How would you interpret the final tableau and the optimal solution?

40. Refer to Example 10.15. A new management has taken over this firm and changed the production schedule and the quality of the machines such that X_1 and X_2 reap profits of \$50 and \$30 each respectively.

    (a) What is the revised production for optimizing profits (solve the problem graphically and using simplex)?
    (b) Interpret the final tableau and the optimal solution.

41. Refer to Example 10.16. The management of the company estimates that it can bring in profits of \$30 and \$40 per gallon of the two products X_1 and X_2, respectively.

    (a) How many gallons of each product should be produced to gain the maximum profit?
    (b) How would you interpret the final tableau and the optimal solution?

42. Refer to Example 10.20. The farmer has revised his minimum requirements of carbohydrates and protein to 100 and 85 units respectively. Rework this problem graphically as well as using simplex to find the optimum combination of carbohydrates and proteins at a minimum cost.

Basic Hard Systems Engineering: Part II

11

11.1 INTRODUCTION

In Chapter 10 we examined a variety of methods of optimization ranging from the use of calculus, application of the critical path method (CPM), the program evaluation and review technique (PERT), and network flow analysis to linear programming. Some problems can best be solved graphically while others are best tackled through matrix applications, such as the simplex method. This chapter may be considered a continuation of Chapter 10. It deals with such topics as forecasting and decision analysis that are techniques employed by planners and engineers in their day-to-day work.

11.2 FORECASTING

The ability to forecast or predict future events, such as the population of a city in the year 2030, or the traffic flow on Interstate 75 through Cincinnati over the Ohio River in 2020, are examples of some of the work entrusted to engineers and planners. They use judgment, past experiences, and mathematical models coupled with intuition to predict the future. Although there are semantic differences in the use of such terms as projecting, predicting, and forecasting, we use these terms interchangeably.

The science of forecasting encompasses an enormous field. We cover just the basics to enable you to get a feel for the topic. As engineers and planners we are concerned with two types of forecasts: long- and short-range. Long-range forecasts are related to problems connected with, for example, building new facilities that need billions of dollars of capital in the next 25 years. For instance, if we are planning the water supply facilities connected with the expansion of a city, we are looking at projections of the population and water demand for a 50-year period. On the other hand, short-range forecasts could encompass periods from 1 year to 5 years. For example, a small construction company that has been in business for 15 years may like to project its profits for the next 5 years based on past trends.

Forecasting methods can also be categorized in two groups: qualitative and quantitative. The former are usually nonstatistical techniques using the Delphi method and brainstorming sessions. Quantitative or statistical methods require appropriate historical data for their application. We concern ourselves with some pertinent quantitative methods: the time-series method and methods based on regression. In making decisions regarding forecasting, consider the following: What is to be forecast, why, and in what detail? How far into the future do you need to forecast and with what accuracy? What methods are you prepared to use? What prior information and data are needed to conduct a forecast and are these available and reliable?

We deal with some simple methods of forecasting commonly used in engineering and planning. It is a good idea to review available historical data pertaining to your problem situation to learn whether it has any predominant characteristics such as a linear, cyclical, or exponential trend. In addition, you may want to note seasonal trends as well as irregular or erratic variations connected with, such things as the weather or local floods.

11.2.1 Regularity

It is good practice to examine data presented to you so that you can observe some degree of regularity in the numbers. A few common forms of regularity are illustrated through the following example.

Example 11.1

(a) The series 27, 27, 27, 28, 27, 27, 27 has no trend. The next observation will most likely be 27.

(b) The series 27, 29, 31, 33, 35 is equally predictable. The next observation is most likely to be 37, because the increase is linear.

(c) The series 9, 27, 81, 243, 729 is tripling each period, thus, the most likely value of the next period is 2187. This is a case of exponential growth. By the same token, the series 64, 16, 4, 1, 0.25, 0.0625 appears to be an exponential *decay* series and the next value is likely to be 0.015625.

(d) The series 39, 24, 50, 29, 40, 25, 51, 30 corresponding to the sale of a product in four quarters of a year for a 2-yr period appears to be cyclical for four periods at a time. It is likely that the next four periods may have values 41, 26, 52, and 31.

(e) There are possibilities where a linear and cyclical series may result from a combination of a linear series 32, 33, 34, 35, 36, 37, 38, 39 and a cyclical series 40, 45, 48, 50, 40, 45, 48, 50, resulting in 72, 78, 82, 85, 76, 82, 86, 89. The next value would likely be 40 + 40 = 80.

11.2.2 Use of Time Series

There are several cases in engineering and financial forecasting dealing with continuous though fluctuating demand. Such cases can be dealt with through the use of time series. A *time series* is a set of observations of a variable over a period of time. An example of a time series forecast could consist of considering the yearly consumption of gasoline by a family over the past 5 years in order to predict what the consumption is likely to be in the next 2 years. This problem could be solved easily by plotting consumption per year as a graph and then drawing a freehand curve through the historical data and projecting it to the desired year. This result would undoubtedly be subjective, but may be sufficiently accurate for the family's requirements. However, for more accurate results we describe three basic methods that are commonly used in practice: (1) moving average (MA) forecast, (2) trend projection models (linear, exponential, and modified exponential), and (3) simple linear regression.

When historic data is available, one can detect a trend in the shape of a gradual long-term directional movement (either a growth or decline). Cyclical swings are long-term swings around a trend line, generally associated with business cycles. Seasonal effects and variations occurring during certain periods of the year can also be seen in the data presented.

Simple moving average (SMA): Let us start with the SMA method. It is calculated by adding up the last n observations and dividing the sum by n. When the next observation is available, the oldest observation in the earlier calculation is dropped, the new one is added, and a new average is calculated. It is best to describe this procedure by solving a practical problem.

Example 11.2 A sales manager wants to forecast the demand for personal computers for the month of August based on a 7-mo record, January through July: 180, 215, 225, 220, 210, 205, 220. He wants to dampen the fluctuations in sales by using MAs of 3 mo at a time. What will be his sale in the month of August?

Solution The MA is calculated for the 3-mo period, which in this case is for the month of August, by considering the demand for the last 3 mo in the sequence:

$$\text{MA} = (210 + 205 + 220)/3 = 635/3 = 211.667$$

The 3-mo MAs for April through July are also shown in Table E11.2. This allows us to compare the forecasts with the actual demands. We could have solved this problem using a 4- or 5-mo MA that would have yielded smoother results.

Forecast accuracy: When one performs a forecasting exercise, the question that is always asked is, "How accurate is your forecast?" The difference between the forecast and the actual is called the *forecast error*.

- The mean absolute deviation (MAD) is one of the more common measures of a forecast error. It is the difference between the forecast and the actual demand as calculated by the expression:

$$\text{MAD} = \frac{\sum \text{abs}(D_t - F_t)}{n}$$

Continues

Example 11.2: *Continued*

Where:

t = The period number
D_t = Demand in period t
F_t = Forecast for period t
n = Total number of periods
abs = Absolute value (sometimes indicated by two vertical lines ||)

For our example, the computation is based on Periods 4 through 7 (220, 210, 205, 220) and the forecast error for individual months is shown. If the absolute value is taken, it amounts to 44.999 and dividing it by $n = 4$, we get 11.25.

- The mean square error (MSE) is derived by squaring each forecast error and dividing by n. Thus, $(13.333)^2 + (10)^2 + (13.333)^2 + (8.333)^2 = 524.98$. Dividing this by $n = 4$, we get 131.25.
- The mean forecast error (MFE) is the running sum of forecast error (taking positive and negative signs into account, divided by n). Thus:

$$-13.333 + 10 + 13.333 - 8.333 = 1.6667/4 = 0.4167$$

Table E11.2 Forecast values

Period	Data	Forecast	Forecast error
1	180.000		
2	215.000		
3	225.000		
4	220.000	206.667	−13.333
5	210.000	220.000	+10.000
6	205.000	218.333	+13.333
7	220.000	211.667	−8.333
8		211.667	

Absolute value: 45

MAD: 11.25
MSE: 131.25
MFE: 0.4167

Weighted moving average (WMA): The SMA method can be adjusted to reflect fluctuations in the data provided by assigning weights to the most recent data. We will demonstrate the WMA method by amending the previous problem by assigning weights.

Example 11.3 With the same data but assigning weights of 0.5, 0.3, and 0.2 for Periods 7, 6, and 5, respectively, predict the sales in August (Period 8). Apply all three tests and compare the results with those obtained before (for SMA).

Solution Here we apply the weights to the last three entries:

$$\text{WMA} = (210 \times 0.2) + (205 \times 0.3) + (220 \times 0.5) = 213.50$$

The other forecasts are also shown.

Table E11.3a Forecast values (Example 11.3)

Period	Data	Forecast	Forecast error
1	180.000		
2	215.000		
3	225.000		
4	220.000	213.000	−7.000
5	210.000	220.000	10.500
6	205.000	216.000	11.000
7	220.000	209.500	−10.500
8		213.500	

The accuracy tests are:

(a) MAD for Periods 4 through 7 is 39/4 = 9.75
(b) MSE = 97.625
(c) MFE = 4/4 = 1.0

Table 11.3b Summary of SMA and WMA

Comparison	SMA	WMA
Forecast for August	211.667	213.500
MAD	11.25	9.75
MSE	131.240	97.625
MFE	0.4167	1.00

In this case, the WMA forecast is superior.

Exponential smoothing: This method of forecasting is really an averaging technique that weights the more recent past data point more strongly than the more distant data point. The method can be summarized as follows:

New forecast = old forecast + α (latest observation − old forecast), where α is a fraction between 0 and 1. Naturally, if α = 0, the new forecast is always the same as the old forecast, regardless of all the latest observations. If α = 1, then the new forecast = latest observation. We will apply this technique to the data given in Example 11.1 and use α = 0.2. The simple exponential smoothing forecast is computed using the formula:

$$F_{(t+1)} = \alpha D_t + (1 - \alpha) F_t$$

Where:

$F_{(t+1}$ = Forecast for the next period
D_t = Actual demand in the present period
F_t = Previously determined forecast for the present period
α = Weighting factor known as the *smoothing constant*

Example 11.4 Apply the exponential smoothing method to the data given in Example 11.2, with α = 0.2, and apply all the tests of accuracy.

Solution Since *a* = 0.2, it means that our forecast for the next period is based on 20 percent of recent demand (D_1) and 80 percent of past demand.

Forecast for Period 2: $F_2 = (0.2)(180) + (0.80)(180) = 180$
Forecast for Period 3: $F_3 = (0.2)(215) + (0.80)(180) = 187$
Forecast for Period 4: $F_4 = (0.2)(225) + (0.80)(187) = 194.6$

And so on, until:

Forecast for Period 8: $F_8 = (0.2)(220) + (0.80)(202.395) = 205.916$

Table E11.4 Forecast values (Example 11.4)

Period	Data	Forecast	Forecast error
1	180.000	180.000	
2	215.000	180.000	−35.000
3	225.000	187.000	−38.000
4	220.000	194.600	−25.400
5	210.000	199.680	−10.320
6	205.000	201.744	−3.256
7	220.000	202.395	−17.605
8		205.916	

The accuracy tests are:

(a) MAD for the Periods 2 through 7 is (129.5808)/6 = 21.5968

Continues

Example 11.4: *Continued*

(b) MSE = 623.5322
(c) MFE is (129.5808)/6 = 21.5968

Least-square method: Regression methods have already been dealt with in Chapter 9, but as a comparison we will demonstrate how regression methods can be applied in contrast to time series techniques.

Example 11.5 Refer to Example 11.2 and the data provided for seven time periods. What is the projection for Period 8 (August)?

Solution Refer to Chapter 9 to help you solve this problem. The equation is $Y = 198.5714 + 3.0357X$

Coefficient of determination: 0.1882
Correlation coefficient: 0.4338
Standard error: 14.9224

Table E11.5 Observed and predicted values

Observation	Observed value	Predicted value	Residual
1	180.000	201.607	−21.607
2	215.000	204.643	10.357
3	225.000	207.679	17.321
4	220.000	210.714	9.286
5	210.000	213.750	−3.750
6	205.000	216.786	−11.786
7	220.000	219.821	0.179

MAD: 12.3810

Therefore, $D_8 = (3.0357 \times 8) + 198.57 = 222.86$ units

11.3 TRANSPORTATION AND ASSIGNMENT PROBLEM

11.3.1 Introduction

Some management problems are concerned with transporting goods from a number of sources to several destinations across the country, or even across a city. The overall objective is to minimize total transportation costs. Some basic assumptions are:

- Transportation costs are a linear function of the number of units shipped
- All supply and demand are expressed in homogeneous units

- Shipping costs per unit do not vary with the quantity shipped
- Total supply must equal total demand, but if demand is larger than the supply, create a dummy supply and assign a zero transportation cost to it; if supply is larger than demand create a dummy demand

There are at least three methods for obtaining an initial solution: (1) the northwest (NW) corner method; (2) the minimum cost method; and (3) the penalty or Vogel's method. In all three cases the distribution uses a matrix, showing the demand requirements and supply availabilities. The costs associated with the route between the supply and demand points are indicated in the upper corner of the cells. The supply is allocated to meet the demand by writing down the number of units transported by a particular route from a supply source to a demand destination.

In all three methods the solution procedure is iterative, beginning with an initial solution and improving it until a satisfactory feasible distribution has been achieved. This feasible solution may not be an optimal one. Since applying any one or all of the three methods does not necessarily guarantee an optimal solution giving the lowest total cost (TC) of shipment, there are at least two methods to test the solution for optimality using the initial basic solution as a starting point. Let us begin by describing the northwest corner method first by working out an example.

11.3.2 Northwest Corner Method

Example 11.6 Steel factories A, B, and C need to supply 560 tn of steel products to their warehouses X, Y, and Z during the course of a day, demanding the same amount as shown in the matrix. Assign the product from the plants to the warehouses and find the TC using the NW method.

Table E11.6 Data for Example 11.6

	Warehouses			
Steel plants	**X**	**Y**	**Z**	**Supply**
A	24	15	18	240
B	45	30	36	160
C	9	27	30	160
Demand	300	140	120	560

Solution The NW method is a quick and systematic way of assigning the supply to the demand points, but is by no means scientific. However, it is a good start for an initial feasible solution. Simply put, you start from the northwest corner of the tableau, allocate as many units as possible into each cell, and work your way toward the southeast corner of the tableau. More specifically:

Continues

Example 11.6: *Continued*

TABLEAU for Example 11.6

Steel plants	Warehouses			Supply
	X	Y	Z	
A	24 **240 (1)**	15	18	240, 0
B	45 **60 (2)**	30 **100 (3)**	36	160, 100, 0
C	9	27 **40 (4)**	30 **120 (5)**	160, 120, 0
Demand	300, 60, 0	140, 40, 0	120, 0	560

(a) Assign as many units as possible to the NW corner Cell AX from the total available in Row A. Given the 240 units available in Row A and the 300 unit demand in Column X, the maximum number that can be assigned to Cell AX is 240, and this is shown as 240 in Cell AX and the demand column under Row X is reduced by 240 to (300 – 240) = 60.

(b) Assign additional units of supply from Row B, until the demand in Column X is satisfied. This requires 60 units in Cell BX, leaving 100 units of B yet to be assigned. The demand at Column X has now been satisfied. It is a good plan to keep an arithmetical account of what is happening at each allocation, including the sequencing of the allocation, as indicated on the matrix. The number in parenthesis indicates the sequence of allocation.

(c) Column Y needs 140 units, of which Row B can supply 100, the rest coming from Row C. This exhausts the demands of Columns X and Y. The balance remaining in Row C of 120 units find their way through Cell CZ to Column Z.

(d) All the demand has now been satisfied and all the supply has been exhausted. The sequence of each allocation is also given.

(e) The TC of supplying the steel as per our allocation is: $(240 \times 24) + (60 \times 45) + (100 \times 30) + (40 \times 27) + (120 \times 30) = \$16,140$

Discussion

Although the allocation has been made to obtain a feasible solution, satisfying supply and demand, it is highly unlikely that this answer is an optimal solution, and that it gives us the minimum cost. It is unscientific because we have not bothered to look at the cell costs. There is of course the possibility of tinkering with the figures and reallocating the products, but for now this is all we have.

11.3.3 The Minimum-cost Cell Method

We will solve the same problem using a more rational method. As the name indicates, we should try to transport the product by the cheapest route possible, and this is done as follows:

(a) Select a cell having the lowest cost and allocate as much product as we can, using that route

(b) Continue by selecting the next cheapest route and repeat the procedure until the demand has been satisfied

Example 11.7 Solve the steel plant problem described in Example 11.6 using the minimum-cell cost method.

Solution

TABLEAU for Example 11.7

Steel plants	Warehouses			Supply
	X	Y	Z	
A	24	15 **140 (2)**	18 **100 (3)**	240, 100, 0
B	45 **140 (5)**	30	36 **20 (4)**	160, 140, 0
C	9 **160 (1)**	27	30	160, 0
Demand	300, 140, 0	140, 0	120, 20, 0	560

(a) Inspection of cell costs indicates that CX is the cheapest route. Hence, allocate the maximum quantity of 160 in Row C to Column X, leaving 140 units to be yet supplied.

(b) The cell with the next cheapest cost is AY with a unit cost of 15. If we supply 140 units from A via AY to Column Y, we will be left with 100 units in Column A.

(c) The next cell is AZ where 100 units are supplied to Column Z from Row A.

(d) The next cell to consider is BZ with a unit cost of 36 and we therefore assign 20 units for Column Z from Plant B, exhausting the demand at Z. The balance of 140 units demanded in Column X via Route BX from Row B exhausts all of the demand and supply.

The TC is $(160 \times 9) + (140 \times 15) + (100 \times 8) + (20 \times 36) + (140 \times 45) = \$12{,}360$.

Discussion

There is no question that the minimum-cost cell method is rational. Could there have been other combinations that could have lowered the cost? We cannot tell right away. Perhaps there is an optimal solution.

11.3.4 The Penalty or Vogel's Method

The penalty method makes allocations minimizing the penalty or opportunity costs. We pay a penalty because of our failure to select the best alternative. The following steps briefly describe the method.

(a) Set up the tableau as usual
(b) Calculate the penalty cost of each row and column
(c) Select the row or column with the largest penalty and allocate as much as possible to the cell having the minimum cost
(d) Eliminate rows and columns that are exhausted of their demand/supply
(e) Recalculate the penalties in rows and columns and continue with the allocation until all supply and demand have been met

We will apply the penalty method to the same problem previously solved by the NW corner method and the minimum cost cell method. Notice that the tableau has an additional row and an additional column to take care of penalties.

Example 11.8 Solve the steel plant example described in Example 11.6 using the penalty method.

Solution

TABLEAU for Example 11.8

Steel plants	Warehouses X	Warehouses Y	Warehouses Z	Supply	Penalty
A	24 **140 (2)**	15	18 **100 (3)**	240, 100,0	3
B	45	30 **140 (3)**	36 **20**	160, 20, 0	6
C	9 **160 (1)**	27	30	160, 0	18
Demand	300, 140, 0	140, 0	120, 20, 0	560	
Penalty	15, 21	12, 15	12, 18		

Calculate the penalties for rows and columns: for example, in the first row, Cell AY has the lowest cost of 15 and the next higher cost is AZ of 18, indicating a penalty of 3. This is the penalty we would pay per unit if we allocate AZ instead of AY. In a similar manner, we calculate the penalty for the other rows and columns, as shown on the matrix.

The largest penalty is in Row C, with a value of 18 and the cell with the lowest cost per unit is CX. To avoid paying this penalty, we allot the maximum we possibly can to this cell, which is

Continues

Example 11.8: *Continued*

160 units, thus exhausting the entire supply. Therefore, we can eliminate Row C from further consideration. The demand in Column X is now (300 – 160) = 140.

Recalculation of the penalties in rows and columns is necessary to determine which penalty is the highest, and match the corresponding cell with the lowest cost. These new penalties are shown in the penalty row: 21, 15, and 18. Column X has the highest penalty of 21, and the lowest cell cost is 24 for Cell AX. We can supply the maximum of 140 units from Row A to Column X, thus exhausting the demand. Eliminate Column X from further consideration. This procedure is repeated over and over again until the demand is satisfied. Let us now calculate the TC:

$$(140 \times 24) + (100 \times 18) + (140 \times 30) + (20 \times 36) + (160 \times 9) = \$11{,}520$$

Discussion

Notice that $11,520 is an improvement over the minimum cost method that gave a figure of $12,360. But how are we to know whether this answer is the optimum solution or not? This question is dealt with in the following section.

11.3.5 How Do We Determine an Optimum Solution?

From the previous examples showing the application of the NW corner, minimum cost cell, and the penalty methods, we were able to obtain feasible solutions of allocating units of products via combinations of routes, without being certain whether our solutions were optimum. Fortunately, there are some methods to evaluate our solutions, just for this objective. One of the popular methods is the modified distribution (MODI) method. An example will best demonstrate the procedure.

Example 11.9

TABLEAU 1 for Example 11.9

Warehouse / Factory	A	B	C	Supply
a	8	6 **13**	3 **2**	15
b	9	11	8 **16**	16
c	6 **4**	5 **7**	7	11
d	3 **13**	10	9	13
Demand	17	20	18	55

Continues

Example 11.9: *Continued*

Suppose we would like to transport 55 units of goods from four factories a through d to three warehouses A, B, and C, as shown in the tableau. The unit costs are given in the right-hand top corner of each cell.

Solution We have allocated the units by the low-cost method and the cost works out to be:

$$(13 \times 6) + (2 \times 3) + (16 \times 8) + (4 \times 6) + (7 \times 5) + (13 \times 3) = \$310.$$

To check if there are alternative solutions with a net lower transportation cost, we need to examine in turn each unused route. If we send one unit along Route aA, some modification has to be done to balance the supply and demand column of 17 and 15 units respectively, and one unit must be subtracted from either Route aB or aC. If we reduce one unit from Route aB, one unit must be added to Route cB to maintain the column total of 20. Also, to keep the row total of 11 units for c, one unit must be subtracted from Route cA, and this subtraction compensates for the initial modification of sending one unit by Route aA. These changes will have no effect on the feasibility of the initial allocation.

TABLEAU 2 for Example 11.9

Factory \ Warehouse	A	B	C	Supply
a	8 (+1) ⟶	6 **13 (−1)**	3 **2**	15
b	9	11	8 **16**	16
c	6 **4 (−1)** ⟵	5 **7 (+1)**	7	11
d	3 **13**	10	9	13
Demand	17	20	18	55

Revised TC = $(8 \times 1) + (12 \times 6) + (2 \times 3) + (16 \times 8) + (3 \times 6) + (8 \times 5) + (13 \times 3) = \$311$.

This is unacceptable because this means an increase of \$1.

We could have come to the same conclusion by considering the cell costs as:

$$C_{aA} - C_{cA} + C_{cB} - C_{aB} = 8 - 6 + 5 - 6 = \$1$$

where C_{aA} represents the cost of transport by that route.

Also, this inspection procedure could be repeated for each unused route in turn, to test whether we could improve the initial feasible solution.

Another way to avoid this laborious procedure is by considering shadow costs. Shadow costs are obtained by assuming that the transportation cost for all used routes is made up of two

Continues

Example 11.9: *Continued*

parts—dispatch and receiving costs. So, the dispatch costs are a, b, c, and d, while receiving costs are A, B, and C. So, the unit cost of Cell aA = C_{aA} = a + A. As per the initial feasible solution, the cost of transport of occupied cells are c + A = 6; d + A = 3; a + B = 6; c + B = 5; a + C = 3; and b + C = 8. There are six equations and seven unknowns. If A is assumed to be zero, we can find the values of the other unknowns to be a = 7, b = 12, c = 6, d = 3, and A = 0, B = −1, C = −4. These shadow prices are shown on the side and top of the matrix. The cell costs shown previously:

$$C_{aA} - C_{cA} + C_{cB} - C_{aB} \text{ can also be written in terms of shadow costs as,}$$

$$C_{aA} - (c + A) + (c + B) - (a + B) \text{ is equivalent to } C_{aA} - (a + A)$$

which is useful as a general result, because it illustrates that sending one unit along a previously unoccupied route increases the TC by the unit transport cost of the new route, minus the sum of the shadow cost for that route. Thus, if this difference is negative for any unused route, a savings in cost will be made if each unit is transferred to this route.

We can now calculate this difference for each unused route in the initial allocation as:

$C_{aA} - (a + A) = 8 - (7 + 0) = 1$
$C_{bA} - (b + A) = 9 - (12 + 0) = -3$ **we must examine this route!**
$C_{bB} - (b + B) = 11 - (12 - 1) = 0$
$C_{dB} - (d + B) = 10 - (3 - 1) = 8$
$C_{cC} - (c + C) = 7 - (6 - 4) = 5$
$C_{dC} - (d + C) = 9 - (3 - 4) = 10$

These results are entered in the top left-hand corner in the appropriate cells. Notice that a saving of $3 can be made for each unit that can be sent along Route bA, so we use this route as fully as possible to satisfy the requirements. The matrix illustrates how this is achieved. The arrows shown on the matrix indicate the modifications and illustrates that this in no way affects the feasibility of the solution.

TABLEAU 3 for Example 11.9

Shadow → Costs ↓	Warehouse / Factory	0 A	−1 B	−4 C	Supply
7	a	1 8	6 13 − x ←	3 2 + x	15
12	b	−3 9 x →	0 11	8 16 − x	16
6	c	6 4 − x ↑	5 ← 7 + x	5 7	11
3	d	3 13	8 10	10 9	13
	Demand	17	20	18	55

Continues

Example 11.9: *Continued*

Note also that the value of x should be so chosen that no allocation becomes negative and this is so when $x = 4$. The new matrix is:

TABLEAU 4 for Example 11.9

Warehouse / Factory	A	B	C	Supply
a	8	6 **9**	3 **6**	15
b	9 **4**	11	8 **12**	16
c	6 **0**	5 **11**	7	11
d	3 **13**	10	9	13
Demand	17	20	18	55

The new cost = (9 × 6) + (6 × 3) + (4 × 9) + (12 × 8) + (11 × 5) + (13 × 3) = $298

which shows that sending four units along a route whose potential saving is $3/unit reduces the TC by (3 × 4) = $12.

If we wanted to determine whether further improvement to this solution is possible, new shadow costs can be calculated for a second feasible solution, and the test procedure can be repeated.

TABLEAU 5 for Example 11.9

Shadow →	Warehouse	0	2	−1	Supply
Costs ↓	Factory	A	B	C	
4	a	4 8	6	3 **6**	15
9	b	0 9 **4**	11	8 **12**	16
3	c	3 6	5 **11**	5 7	11
3	d	3 **13**	5 10	7 9	13
	Demand	17	20	18	55

By arbitrarily assigning a value of zero to shadow cost A, the new shadow costs a, b, c, d, B, and C can be obtained by solving the following:

$$b + A = 9,\ d + A = 3,\ a + B = 6,\ c + B = 5,\ a + C = 3,\ b + C = 8$$

Continues

Example 11.9: *Continued*

This works out to a = 4, b = 9, c = 3, d = 3, B = 2, and C = -1. Also, the differences between the unit cost and the sum of the shadow costs for the unused routes are:

$$C_{aA} - (a + A) = 8 - (4 + 0) = 4$$

$$C_{cA} - (c + A) = 6 - (3 + 0) = 3$$

$$C_{bB} - (b + B) = 11 - (9 + 2) = 0$$

$$C_{dB} - (d + B) = 10 - (3 + 2) = 5$$

$$C_{cC} - (c + C) = 7 - (3 - 1) = 5$$

$$C_{dC} - (d + C) = 9 - (3 - 1) = 7$$

Since none of these values are negative, no further improvement is possible. The minimum cost allocation is $298.

The MODI method for optimizing, just described, has been applied to the NW corner results given in Example 11.6, and the optimal solution obtained, which works out to be $11,520.

Example 11.10 Find the minimum cost for the network given, starting with the NW corner solution using the MODI method.

Table E11.10a Data for Example 11.10

	D1	D2	D3	Supply
S1	24.0	15.0	18.0	240.0
S2	45.0	30.0	36.0	160.0
S3	9.0	27.0	30.0	160.0
Demand	300.0	140.0	120.0	560.0

Solution Initial solution by NW method:

Table E11.10b NW method for Example 11.10

	D1	D2	D3	Supply
S1	240.0	0.0	0.0	240.0
S2	60.0	100.0	0.0	160.0
S3	0.0	40.0	120.0	160.0
Demand	300.0	140.0	120.0	560.0

Therefore, the initial solution is $16,140.00.

Continues

Example 11.10: *Continued*

A net cost change table (or cost-improvement) can be determined for the empty cells:

Table E11.10c Net cost change for Example 11.10

	1	2	3
1	0.0	6.0	6.0
2	0.0	0.0	3.0
3	–33.0	0.0	0.0

Solution after Iteration 1:

Table E11.10d Iteration 1 for Example 11.10

	D1	D2	D3	Supply
S1	240.0	0.0	0.0	240.0
S2	20.0	140.0	0.0	160.0
S3	40.0	0.0	120.0	160.0
Demand	300.0	140.0	120.0	560.0

Solution = $14,820.00

Net cost change table:

Table E11.10e Net cost change for Iteration 1

	1	2	3
1	0.0	6.0	-27.0
2	0.0	0.0	-30.0
3	0.0	–33.0	0.0

Solution after Iteration 2:

Table E11.10f Iteration 2 for Example 11.10

	D1	D2	D3	Supply
S1	240.0	0.0	0.0	240.0
S2	20.0	140.0	0.0	160.0
S3	60.0	0.0	100.0	160.0
Demand	300.0	140.0	120.0	560.0

Solution = $14,220.00

Continues

Example 11.10: *Continued*

Net cost change table:

Table E11.10g Net cost change for Iteration 2

	1	2	3
1	0.0	–24.0	–27.0
2	30.0	0.0	0.0
3	0.0	3.0	0.0

Solution after Iteration 3:

Table E11.10h Iteration 3 for Example 11.10

	D1	D2	D3	Supply
S1	140.0	0.0	100.0	240.0
S2	0.0	140.0	20.0	160.0
S3	160.0	0.0	0.0	160.0
Demand	300.0	140.0	120.0	560.0

Solution = \$11,520.00

Net cost change table:

Table E11.10i Net cost change for Iteration 3

	1	2	3
1	0.0	3.0	0.0
2	3.0	0.0	0.0
3	0.0	30.0	27.0

Note, there are no negative numbers.

Optimal solution by MODI:

Table E11.10j MODI solution for Example 11.10

	D1	D2	D3	Supply
S1	140.0	0.0	100.0	240.0
S2	0.0	140.0	20.0	160.0
S3	160.0	0.0	0.0	160.0
Demand	300.0	140.0	120.0	560.0

Optimal solution = \$11,520.00

11.3.6 The Unbalanced and Other Transportation Problems

In most problems connected with transhipment, it would be rather rare that the supply and demand are exactly balanced. Two examples of imbalanced supply and demand are shown. In the first example, the demand is 200 units while the supply is 185, in which case a dummy supply of 15 units is introduced at zero cost (Tables 11.1 and 11.2).

Table 11.1 An example of imbalanced supply and demand

From \ To	A	B	C	Supply
A	5	6	8	52
B	3	9	7	38
C	2	6	9	95
D (dummy)	0	0	0	15
Demand	120	35	45	200

Similarly, there can be supply that exceeds demand, as shown in the matrix:

Table 11.2 Another example of imbalanced supply and demand

From \ To	A	B	C	D (dummy)	Supply
A	4	8	4	0	200
B	3	2	9	0	220
C	7	5	6	0	255
Demand	175	90	310	100	675

All three allocation methods can be used to find an initial problem solution. In the case of the NW corner or penalty method, the dummy row is treated as if it were one of the regular rows, but with the low-cost cell method it is better to leave out the dummy row until the very end.

Prohibited routes: The best way to handle the allocation of goods where prohibited routes are involved, due to government rules or road construction, is to assign an arbitrarily large unit cost of transportation to the route.

Maximization problem: In transportation problems it is possible to be confronted with tableaus giving profits instead of costs, in which case the distribution of supplies to meet demands is one to optimize (maximize) profits. An initial solution can be found by first converting the problem to a minimization case by subtracting each profit figure in the tableau from the largest. The solution is produced as before and the initial figures are used to calculate the final profit.

Degeneracy: If m represents the number of rows and n the number of columns, there must be $m + n - 1$ number of occupied cells for us to evaluate the empty cells for solution improvement. If not, we have a case of degeneracy. The way to obviate this situation is to designate one of the empty cells as an occupied cell by assigning zero quantity, taking care that this cell connects all the missing links in evaluating the empty cells.

11.3.7 The Assignment Problem

The assignment problem is a special case of a transportation problem where n resources (e.g., workers) need to be assigned to n jobs, knowing the cost of each choice. The task is to minimize the TC of the assignment. Say you wanted to assign five machinists to five machines, where the capabilities of each worker on a particular machine varied. There would be $5 \times 4 \times 3 \times 2 \times 1 = 120$ ways of assigning the workers. The Hungarian method is used for solving such problems, and the best way is to demonstrate it with the help of an example.

Example 11.11 Four workers a, b, c, and d are to be assigned four jobs A, B, C, and D and their capability to perform these jobs is understood as a measure of time needed to complete the jobs. The problem is to assign the workers the jobs so that the total time needed to finish the jobs is a minimum.

Solution

TABLEAU 1 for Example 11.11

		Jobs			
		A	B	C	D
Workers	a	30	36	42	48
	b	38	46	44	36
	c	52	34	32	38
	d	38	42	46	34

TABLEAU 2 for Example 11.11

	A	B	C	D
a	0	2	10	14
b	8	12	12	2
c	22	0	0	4
d	8	8	14	0

TABLEAU 3 for Example 11.11

	A	B	C	D
a	0	2	10	14
b	6	10	10	0
c	22	0	0	4
d	8	8	14	0

TABLEAU 4 for Example 11.11

	A	B	C	D
a	0	2	10	20
b	0	4	4	0
c	22	0	0	10
d	2	2	8	0

TABLEAU 5 for Example 11.11

	A	B	C	D
a	0	0	8	20
b	0	2	2	0
c	24	0	0	12
d	2	0	6	0

Continues

Example 11.11: *Continued*

Step 1: Deduct the smallest nonzero time in each column from all entries in that column, thus obtaining at least one zero in each column. This is shown in Tableau 2.

Step 2: Deduct the smallest nonzero time in each row from all entries in that row, only if there is no zero in that row. This is shown in Tableau 3.

Step 3: Draw the minimum number of lines that will cover all the zeros in the matrix. In Tableau 3, we notice that three lines are enough. If there are as many lines as there are rows or columns, an optimum assignment is possible. If not, further adjustments are necessary. Because there are only three lines instead of four, we need to proceed further.

Step 4: Subtract the smallest uncovered entry in Tableau 3 that is 6 from all uncovered entries. Enter the results in Tableau 4. Then add the same number, that is 6, to the elements at which the lines of Matrix 3 cross (6 in Line a and 6 in Line c), obtaining 20 and 10, respectively. Enter these results in Tableau 4 and complete the tableau.

Step 5: In Tableau 4, it is again possible to cover all zeros with only three lines, and we need to repeat Step 4, getting Tableau 5 in which four lines cover all the zeros. Hence, this is the optimal solution.

Step 6: Make the assignments, starting with the zeros, that are unique in each row or column. Notice that, in this example, two sets of assignments can be made.

Assignment I: cC, aA, bD, dB = 32 + 30 + 36 + 42 = 140 time units

Assignment II: cC, aB, bA, dD = 32 + 36 + 38 + 34 = 140 time units

Special Cases of Assignment

1. In cases of assignment, we may come across a situation where a worker cannot be assigned to a certain machine or to a certain job. In such cases these machines or jobs are marked with an M which implies that such an assignment is not permissible because the cost is prohibitive.
2. There are situations where five jobs need to be assigned to six machines. For making a minimum cost assignment, we need a square matrix for which a sixth row, known as a dummy row, is added with zeros in all the row cells. In a similar way, if there were six jobs and five machines, we could add a dummy machine column to make up a square matrix.
3. Cases can also arise in assignment problems where an assignment needs to be made, which maximizes profit (instead of minimizing costs). In such cases, we transform the profits into opportunity losses. An example of such a case is shown.

Example 11.12 A TV station has three programs that bring in high profits (in millions) as shown in the matrix. Which program should be aired at what time to reap the largest profit?

Table E11.12 Data for Example 11.12

	Program		
Hours	**A**	**B**	**C**
7 to 8 p.m.	56	54	61
8 to 9 p.m.	53	57	60
9 to 10 p.m.	58	56	57

Solution

TABLEAU 1 for Example 11.12

	Program		
Hours	**A**	**B**	**C**
7 to 8 p.m.	56	54	61
8 to 9 p.m.	53	57	60
9 to 10 p.m.	58	56	57

TABLEAU 2 for Example 11.12

	Program		
Hours	**A**	**B**	**C**
7 to 8 p.m.	5	7	0
8 to 9 p.m.	7	3	0
9 to 10 p.m.	0	2	1

TABLEAU 3 for Example 11.12

	Program		
Hours	**A**	**B**	**C**
7 to 8 p.m.	5	5	0
8 to 9 p.m.	7	1	0
9 to 10 p.m.	0	0	1

TABLEAU 4 for Example 11.12

	Program		
Hours	**A**	**B**	**C**
7 to 8 p.m.	4	4	0
8 to 9 p.m.	6	0	0
9 to 10 p.m.	0	0	2

Show the third program at 7 p.m., the second program at 8 p.m., and the first program at 9 p.m. Profit = 61 + 57 + 58 = $176 million.

11.4 DECISION ANALYSIS

There is hardly a day that passes when we are not asked to make a decision. Of course, some of the results of these decisions are inconsequential, such as wearing the wrong color of socks with a pair of shoes. Other decisions can be more painful, such as buying a used car that turned out to be a bad deal.

11.4.1 Overview

Decision analysts distinguish among three different kinds of decisions:

(a) *Decision making under certainty:* These are decisions where you think you have all the necessary information to make an intelligent choice. However, this doesn't necessarily mean that you are likely to make the right decision. In simple cases it may depend on what your personal preferences are at the time of decision making (e.g., buying a 3-piece business suit for $300). In more complicated cases, you may need additional information (e.g., buying a single-family house).
(b) *Decision making under uncertainty with constraints:* These cases are quite common where resources, budgets, time, and legal constraints pose a problem. Various algorithms and techniques are available, such as linear programming to evaluate the available data and select the best outcome.
(c) *Decision making under conflict:* In such cases, apart from information regarding the problem-situation you may have to take account of what your competitor may do. For example, if you are responsible for fixing the price of a certain product that your company is producing, you would have to consider what your competitors are likely to charge for a similar product. These decisions are dealt with by game theory.
(d) *Decision making under uncertainty with known probabilities (or under risk):* In these cases you are taking into consideration the question of chance that is likely to play a part in your decision. We are not considering our competitors. In fact, pure chance is connected with nature, like a freak thunderstorm. Decision making under conditions of uncertainty naturally takes into consideration probabilities assigned to future outcomes. This category of decision making is also known as *decision under risk*. When a manager doesn't know the probability exactly, he can use subjective estimates.
(e) *Decision making under uncertainty without known probabilities:* These cases are similar to what is described in (d), where there is more than one possible state of nature, but the probability of a particular state is not known.

11.4.2 Decision Making Under Conditions of Uncertainty

When probabilities of outcomes are unknown, the payoff for every decision and state of nature pair must be stated in the form of a matrix. The state of nature will determine the outcome of the decision. The decision maker has no control over the state of nature. Payoffs are expressed in terms of profits or costs.

Decision makers operate based on their personality and style of their operation. For example, decision makers can be optimistic, even-handed, risk-taking, or even pessimistic individuals. The following criteria are often used to describe decision types.

(a) The decision maker adopting the *maximax* criterion selects the maximum of the maximum payoffs from a matrix.
(b) If the maximum of the minimum payoffs is selected, we refer to this as the *maximin* criterion.

(c) In the minimax regret criterion (also known as the *Savage* criterion) the decision maker first creates a regret table and then chooses the payoff that contains the minimum of the maximum possible regrets.
(d) The *Hurwicz* criterion requires the decision maker to decide on a coefficient of optimism (α). Then, for each decision, we multiply the maximum payoff by α and the minimum payoff by (1 – α), and sum these to choose the decision with the largest sum.
(e) The *equal likelihood* criterion weights each state of nature equally. Here we multiply payoffs by weights for each state and choose the decision with the largest sum.

The example shown will illustrate the application of all these criteria in a common setting.

Example 11.13 A developer has 10 million dollars to invest in one of four projects: school buildings (A), water and sewage projects (B), housing (C), or transport projects (D). The amount of profit depends on the market economy and employment situation for the near future. The possible states of nature are poor economy (P), normal growth (N), good economic conditions (G). The table shows the payoff matrix. Determine the optimum action using the five decision criteria. The cell values indicate the profits.

Table E11.13a Data for Example 11.13

	State of nature (economy)		
Course of action	P	N	G
A	–6	10	31
B	–29	21	42
C	0	8	12
D	–7	6	18

Solution An examination of the payoff matrix reveals that Option D is in every way inferior to Option A (cell by cell). Therefore, we will eliminate it from further consideration.
The revised matrix is:

Table E11.13b Revised data for Example 11.13

	State of nature (economy)		
Course of action	P	N	G
A	–6	10	31
B	–29	21	42
C	0	8	12

(a) As per the maximin criterion, a decision maker who is pessimistic (or cautious) about the outlook of a decision, considers the worst case scenario that might occur and is prepared for it. Such a person chooses C.

Continues

Example 11.13: *Continued*

Table E11.13c Pessimistic decision for Example 11.13

Course of action	State of nature (economy)			Minimum of each row	
	P	N	G		
A	–6	10	31	–6	
B	–29	21	42	–29	
C	0	8	12	0	← Choose C

(b) With the maximax criterion, on the other hand, a decision maker who is optimistic about the outcome considers the best of all possible gains and expects the best from nature, chooses B.

Table E11.13d Optimistic decision for Example 11.13

Course of action	State of nature (economy)			Minimum of each row	
	P	N	G		
A	–6	10	31	31	
B	–29	21	42	42	← Choose B
C	0	8	12	12	

(c) Hurwicz Criterion: While the maximin is ultraconservative and the maximax is ultraoptimistic, the Hurwicz criterion is between the two extremes, using a weighted combination of the best and the worst consequences for each alternative. In calculating the weighted combination an optimism index (α) is used, depending on the decision makers' attitude toward the state of nature. Alpha (α) ranges between 0 and 1. The weighted combination for Alternative A_i is called the Hurwicz payoff H_i

$$H_i = \alpha M_i + (1 - \alpha) m_i$$

where M_i and m_i are the maximum and minimum gains for Alternative A_i. In this example, if we adopt α = 0.6 we get the following:

Table E11.13e Hurwicz criterion for Example 11.13

Alternative	Max gain	Min gain	Hurwicz payoff
A	31	–6	0.6(31) + 0.4(–6) = 16.2←max
B	42	–29	0.6(42) + 0.4(–29) = 13.6
C	12	0	0.6(12) + 0.4(0) = 7.2

The value of α needs to be determined by the decision maker. If α was chosen to be 0.90 the payoffs would be:

A	(0.9)(31) + (0.1)(–6)	= 27.3
B	(0.9)(42) + (0.1)(–29)	= 34.9 (max)
C	(0.9)(12) + (0.1)(0)	= 10.8

Continues

Example 11.13: *Continued*

and the choice would be Alternative B instead of A, when α was chosen to be 0.6.

(d) Minimax regret criterion: This criterion proposed by Savage illustrates the use of the concept of opportunity loss resulting from an incorrect decision. The best way to deal with this criterion is to set up a *regret* matrix:

Table E11.13f Regret matrix for Example 11.13

	State of nature (economy)			
Alternative	**P**	**N**	**G**	**Max regret**
A	6	11	11	11 ← Minimax regret
B	29	0	0	29
C	0	13	30	30
Max return	0	21	42	

Notice that if the decision maker had selected B and nature turns out to be G, he receives the maximum payoff of 42. However, if he had selected C and nature is G the cost of the mistake is (42 – 12) = 30. The best way to set up the regret table is to have a row at the bottom of the matrix with the maximum return and then subtract cell by cell, the individual cell values from the column maximum return. Then, select the maximum regret in each row and select the minimum regret from this column, as indicated. Since A is the row with the least regret, we choose Alternative A.

(e) Laplace criterion: This criterion is simple to apply. If the three states of nature are equally likely, we assign a probability of 1/3 to each cell value and find the expected value (EV).

A $(-6 + 10 + 31)(1/3) = 11.55 \leftarrow \text{max}$
B $(-29 + 21 + 42)(1/3) = 11.22$
C $(0 + 8 + 12)(1/3) = 6.66$

Choose Alternative A, because it has the highest EV.

Discussion

Notice how a decision is guided by the criterion used as demonstrated in this example. How would you decide what to do under similar circumstances? The summary of outcomes:

Maximin	Alternative C
Maximax	Alternative B
Hurwicz	Alternative A if α = 0.6; alternative B if α = 0.9
Minimax regret	Alternative A
Laplace	Alternative A

11.4.3 Decision Making Under Uncertainty with Probabilities

It is possible for a decision maker to assign preliminary values of probabilities to the state of nature, based on past records and current trends. Use is also made of two criteria, EV and expected opportunity loss (EOL). To illustrate these concepts let us take up a simple case.

Example 11.14 A developer wants to diversify his investment through construction of either apartments (A), bridges (B), or condominiums (C). He consults his economic adviser and comes up with probabilities of the state of nature for normal (N) conditions of 0.7 and depressed (D) economic conditions of 0.3 as shown in the following payoff matrix (in millions of dollars).

Table E11.14a Data for Example 11.14

	State of nature (economy)	
Alternative	**Normal (N) 0.7**	**Depressed (D) 0.3**
A	60	40
B	100	-50
C	20	10

Solution The EV of each alternative is:

$$EV(A) = (60 \times 0.7) + (40 \times 0.3) = 54$$
$$EV(B) = (100 \times 0.7) + (-50 \times 0.3) = 55 \leftarrow \text{max}$$
$$EV(C) = (20 \times 0.7) + (10 \times 0.3) = 17$$

Alternative B is the obvious winner, followed closely by A. All it means is that an average payoff of $55 million would result if this decision situation were repeated a large number of times. In a similar way, a decision maker can calculate the EOL for each decision outcome. Recall that we used the concept of regret in the minimum regret criterion (after Savage). Here, we set up a table that indicates the opportunity loss (or regret) in much the same way.

Table E11.14b Regret for Example 11.14

	State of nature (economy)	
Alternative	**Normal (N) 0.7**	**Depressed (D) 0.3**
A	40	0
B	0	90
C	80	30
Max payoff	100	40

A: EOL $(40 \times 0.7) + (0 \times 0.03) = 28$

B: EOL $(0 \times 0.7) + (90 \times 0.3) = 27$ ←minimum regret or opportunity loss

C: EOL $(80 \times 0.7) + (30 \times 0.3) = 65$

The best decision is minimizing the regret, so we choose B and this result corresponds to what we obtained from EVs.

Decision trees: It is useful to describe decision situations with the help of a decision tree, particularly in complicated cases. For example, the problem in the previous section can be converted to a decision tree. The circles and squares are the nodes, while the branches are the alternatives possible under normal and depressed conditions. Thus, the decision tree merely depicts the sequence of decision making. If we want to find the best decision, all we do is work backward from the ends of the branches toward Node 1, by calculating the EVs of the payoffs:

$$\text{EV (node 2)} = (60 \times 0.7) + (40 \times 0.3) = 54$$

$$\text{EV (node 3)} = (100 \times 0.7) + (-50 \times 0.3) = 55$$

$$\text{EV (node 4)} = (20 \times 0.7) + (10 \times 0.3) = 17$$

These expected payoffs are shown in Figure 11.1.

EV of perfect information: Referring to the example in the previous section, the question that one might consider is what additional information could the decision maker obtain regarding the future that would help in making a good decision, within certain limits? This information can be computed in terms of an EV and is called the expected value of perfect information (EVPI).

If we look at the payoff table, the most significant information that is driving our decision making is the state of nature. If we were sure of these states, we would be in good shape. One way of working at this dilemma is to argue that normal conditions prevail for 70 percent of time, while depressed conditions are likely to occur 30 percent of the time. However, nothing can ever be predicted. Based on these conditions, it is reasonable to say that each of the decision outcomes obtained using perfect information must be weighted by its respective probability:

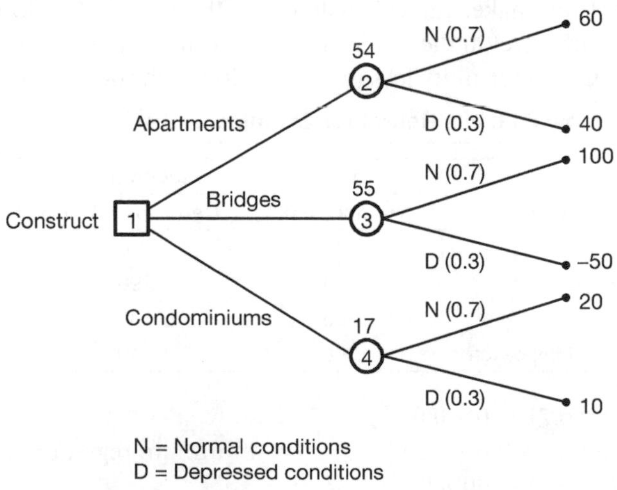

Figure 11.1 Tree diagrams for decision making

Table 11.3 Weights for data of Example 11.14

Alternative	State of nature (economy)	
	Normal (N) 0.7	Depressed (D) 0.3
A	60	40
B	100	–50
C	20	10

(100 × 0.7) + (40 × 0.3) = \$82 million (EV given perfect information). This indicates that if we had perfect information, \$82 million would be the value of the decision, by taking the maximum values from each of the columns. Now, we have already computed the value of the best decision (as calculated previously) that the EV of Alternative B is \$55 million:

$$\therefore \text{EVPI} = 82 - 55 = \$27 \text{ million}$$

This is the maximum the decision maker would pay to obtain perfect information (or near perfect information) from outside sources, such as from an economic expert, although it is doubtful if such a large amount would ever actually be paid. Incidentally, it is interesting to note that this value of EVPI of \$27 million corresponds to the EOL of \$27 million obtained earlier.

Decision making with additional information: Although there are few instances of decision makers paying huge sums of money to experts for perfect information, there are plenty of cases where decision makers have paid large sums of money to obtain additional information. For instance, geotechnical engineers have paid soil-boring experts to take additional samples of soil in addition to the ones they already have taken to make better choices on the site selection of facilities, such as electrical power stations, costing billions of dollars.

In the case of our original problem, suppose the decision maker hires an economic expert to supply him with a report with conditional probabilities for each state of nature in the future:

N = Normal economic conditions
D = Depressed economic conditions
O = Optimistic report
P = Pessimistic report

If the conditional probabilities for each report outcome, given that each state of nature is:

Probability (O/N) = 0.9; Probability (P/N) = 0.1
Probability (O/D) = 0.2; Probability (P/D) = 0.8

which means that, if the future economic conditions are given as normal, the probability of an optimistic report is p(O/N) is 0.9, and so on. The decision maker now has quite a good bit of information, although it is by no means perfect. Prior probabilities are also available:

p(N) = 0.7 and p(D) = 0.3 as noted before

With this information, the decision maker can calculate the posterior probabilities, using Bayes' rule (see Chapter 5). Looking at Figure 11.2, it is easy to compute these values:

$$p(\mathrm{N/O}) = \frac{p(\mathrm{O/N})p(\mathrm{N})}{p(\mathrm{O/N})p(\mathrm{N}) + p(\mathrm{O/D})p(\mathrm{D})} = \frac{(0.9)(0.7)}{(0.9)(0.7) + (0.2)(0.3)} = 0.913$$

In a similar way:

$$p(\mathrm{N/P}) = \frac{(0.1)(0.7)}{(0.1)(0.7) + (0.8)(0.3)} = 0.226$$

$$p(\mathrm{D/O}) = \frac{(0.2)(0.3)}{(0.9)(0.7) + (0.2)(0.3)} = 0.087$$

$$p(\mathrm{D/P}) = \frac{(0.8)(0.3)}{(0.1)(0.7) + (0.8)(0.3)} = 0.774$$

In addition, we can compute the probability of an optimistic (O) and pessimistic (P) report (see Figure 11.2).

$$p(\mathrm{O}) = (0.9)(0.7) + (0.2)(0.3) = 0.69$$

$$p(\mathrm{P}) = (0.1)(0.7) + (0.8)(0.3) = 0.31$$

As explained before, if we work backward from the nodes on the right hand side toward the left, we can compute the EVs, and these are shown over Nodes 4 through 9. We finally arrive at the EV at Node 1 of $74.0442 million, given the results of the report (see Figure 11.3).

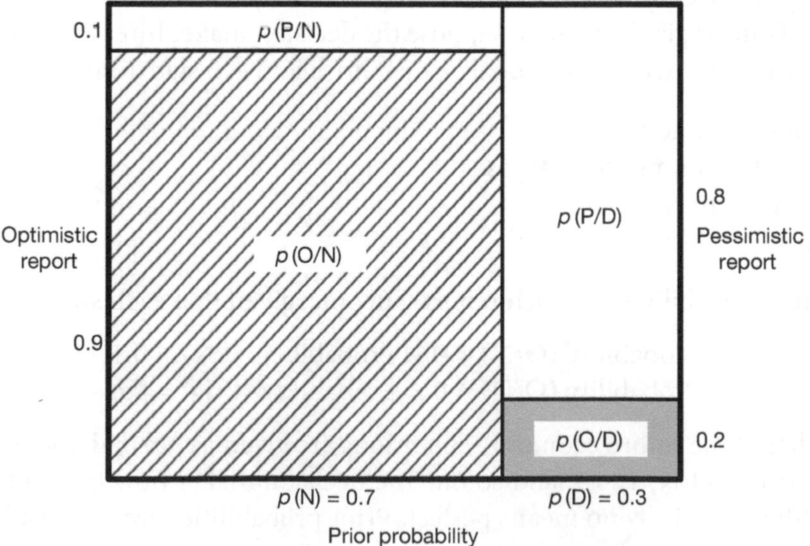

Figure 11.2 Block diagram

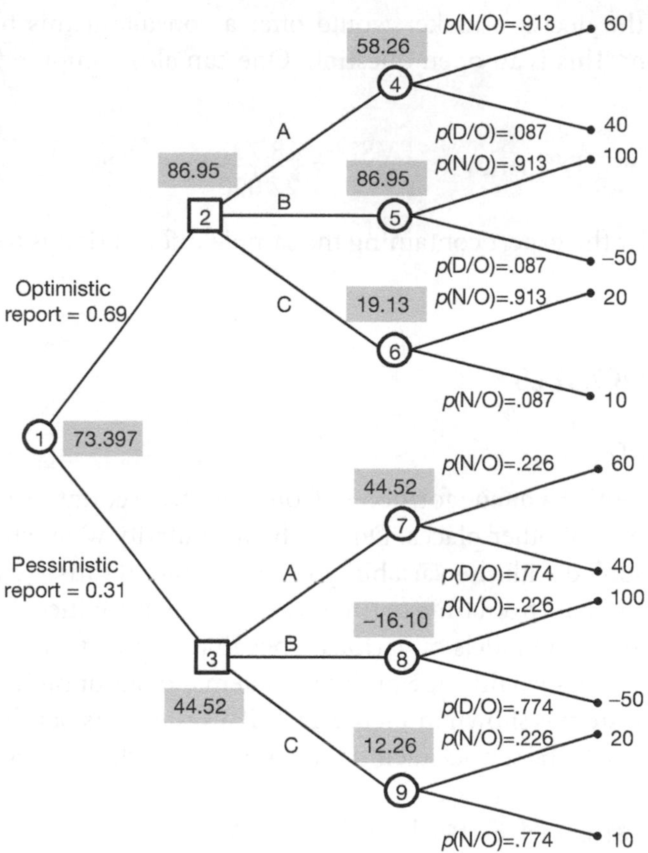

Figure 11.3 Tree diagram

A summary of our computations is:

Table 11.4 Summary of probability values

State of nature (1)	Prior probability (2)	Conditional probability (3)	Prior × conditional probabilities (4) = (2) × (3)	Posterior probabilities (5) = (4) ÷ (Σ4)
Normal (N)	0.7	p(O/N) = 0.9	p(ON) = 0.63	p(N/O) = 0.913
Depressed (D)	0.3	p(O/D) = 0.20	p(OD)=0.06	p(D/O) = 0.087
			Σ = 0.69	Σ = 1.000

Going back to our original computation, we found that the EV of our decision without any additional information was \$55 million. Now that we have obtained additional information, the EV has risen to \$73.797 million. The difference between these two figures is obviously the expected value of the sample information (EVSI).

EVSI = EV(with additional information) – EV(without additional information)
= \$73.797 – \$55.000 = \$18.797 million

Does this mean that the decision maker would offer a consultant this huge amount for the additional information? This is an open question. One can also compute the efficiency of this sample information:

$$\text{Efficiency} = \frac{\text{EVSI}}{\text{EVPI}} = \frac{18.797}{27.000} = 69.62\%$$

The report submitted by the expert containing the sample information is 69.62 percent efficient as perfect information.

11.5 QUEUING MODELS

11.5.1 Introduction

It is quite common to notice a queue formed in front of service counters in airports, hospitals, grocery stores, and scores of other places. Due to the irregularity with which customers arrive demanding service, coupled with the variability in time taken to satisfy customers, queues are likely to build up and dissipate from time to time. The mathematical theories of queues are complex but they provide us with models of various types that help us to analyze and predict how a system dealing with queues would cope or fail with demand put upon it. A general schematic diagram of a queuing system is shown in Figure 11.4. If a system has parallel service facilities, it is referred to as channels. If the service facilities are sequential, the steps are known as phases.

11.5.2 Characteristics of Queuing Systems

There are a number of operating characteristics of a queuing system. These are: (a) the probability of a specific number of customers in the system; (b) the mean waiting time for each customer; (c) the expected (mean) length of the queue; (d) the expected (mean) time in the system for each customer; (e) the mean number of customers in the system; and (f) the probability of the service facility being idle.

One of the features of a queuing system is queue discipline, that is, what happens between the moment of arrival of a customer wanting service till the time he/she leaves the system. Two of the several options are most popular: first-in-first-out (FIFO), indicating that the first customer to arrive is the first to depart, and last-in-first-out, indicating that the last customer to arrive is the first to depart. Sometimes, a customer may decide to leave the queue (balking) or he may join another queue, if he thinks he can better his chances of being served (jockeying). Queuing models are generally identified by three alphanumeric values. The first value indicates the arrival rate assumption, while the second value gives the departure rate assumption. The third value indicates the number of departure channels. We will consider four models:

1. D/D/1 queuing model that assumes deterministic arrivals as well as deterministic departures with one departure channel

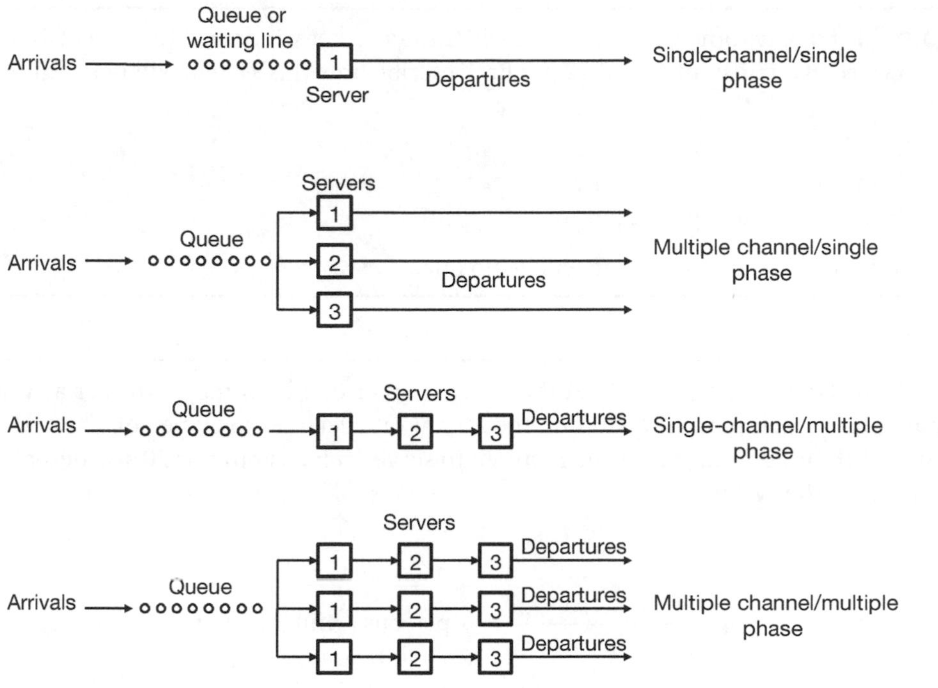

Figure 11.4 Schematic diagram of queuing systems

2. M/D/1 queuing model that assumes exponentially distributed arrival times, deterministic departures, and one departure channel
3. M/M/1 queuing model that assumes both exponentially distributed arrivals and departure times with one departure channel
4. M/M/N queuing model that is similar to M/M/1 except that it has multiple departure channels

11.5.3 Model 1 (D/D/1) Deterministic Queuing Model

An excellent starting point is the D/D/1 queuing model. This model is the simplest waiting-line model and assumes the following: (a) deterministic arrival, (b) deterministic service time, (c) single-channel server, (d) FIFO, and (e) infinite queue length. A simple queuing model's traffic intensity is important, where:

$$\text{Traffic intensity } (\rho) = \frac{\text{Mean rate of arrival } (\lambda)}{\text{Mean rate of service } (\mu)}$$

also, mean interarrival time = $1/\lambda$ and mean service time = $1/\mu$.

Thus, $\rho = \dfrac{(1/\mu)}{(1/\lambda)}$

We will examine this model working through a couple of examples.

Example 11.15 Customers arrive at a movie theater ticket window at the rate of 10 per hr and are serviced at a constant rate of 12 per hr. Describe how this system will perform.

Solution

Mean arrival rate, $\lambda = 10$ per hr
Mean service rate, $\mu = 12$ per hr

In this situation since $\lambda < \mu$ there will be no queue.

Example 11.16 Customers arrive at the ticket counter of a local movie theater at a rate of 240 persons/hr, at 5:30 p.m. After 10 min the arrival rate declines to 60 persons/hr and continues at that level for 20 min. If the time required to serve each customer is 20 sec, describe the performance of the system.

Solution

$$\lambda_1 = \frac{240 \text{ person/hr}}{60 \text{ min/hr}} = 4 \text{ persons / min}, \ t \leq 10 \text{ min}$$

$$\lambda_2 = \frac{60 \text{ person/hr}}{60 \text{ min/hr}} = 1 \text{ person / min}, \ t > 10 \text{ min}$$

$$\mu = \frac{60 \text{ person/hr}}{20 \text{ min/hr}} = 3 \text{ persons / min, for all } t$$

$\therefore$ Number of person arrivals at time $t = 4t$ for $t \leq 10$ min

And, number of person arrivals $[40 + 1(t - 10)]$ for $t > 10$ min

Also, the number of persons with tickets (departures) is $3t$ for all t.

These equations are depicted in Figure E11.16. Notice that when the arrival curve shown in the figure is above the departure curve, a queue will exist. The queue will dissipate at the time when arrival and departure curves intersect:

$$\therefore 40 + 1(t - 10) = 3t$$

$$\text{and } t = 15 \text{ min.}$$

Thus, the queue that began to form at 5:30 p.m., will disappear at 5:45 p.m. The longest queue will occur at $t = 10$ of 10 persons; the longest delay will be $13.33 - 10 = 3.33$ min:

$$\text{Total delay} = \frac{1}{2}(10 \times 10) + \frac{1}{2}(10 \times 5) = 75 \text{ persons/min}$$

Continues

Example 11.16: *Continued*

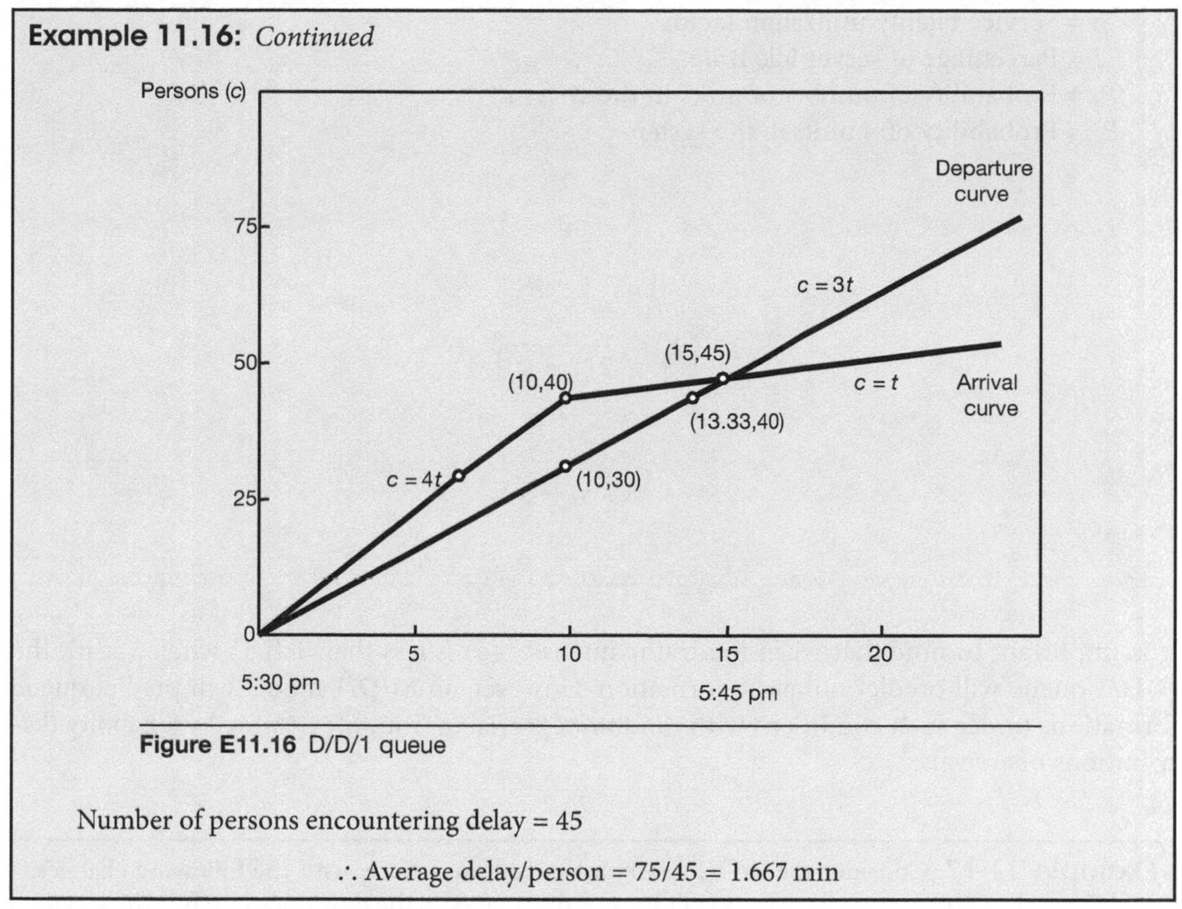

Figure E11.16 D/D/1 queue

Number of persons encountering delay = 45

$\therefore$ Average delay/person = 75/45 = 1.667 min

11.5.4 Model 2 (M/D/1)

This model assumes that the arrival times of successive units in a queue are exponentially distributed (i.e., Poisson arrivals). However, departures are deterministically distributed. Such problems are best solved mathematically. Defining traffic intensity (ρ) as the ratio of average arrivals to departures (λ/μ) and assuming ρ is less than 1, it can be shown that for an M/D/1 queue with the following notation:

λ = Mean arrival rate ($1/\lambda$ = mean time between arrivals)
μ = Mean service rate ($1/\mu$ = mean service time)
n = Number of customers (units) in the system (including those waiting and in service)
L = Mean number in the system
L_q = Mean number in the waiting line (queue length)
W = Mean time in the system
W_q = Mean waiting time (in the queue)

ρ = Service facility utilization factor
I = Percentage of server idle time
P_o = Probability of number of units in the system
P_n = Probability of n units in the system

$$\rho = \frac{\lambda}{\mu}$$

$$P_0 = 1 - \rho$$

$$L_q = \frac{\rho^2}{2(1-\rho)}$$

$$L = L_q + \rho$$

$$W_q = \frac{L_q}{\lambda}$$

$$W = W_q + \frac{1}{\mu}$$

It is important to note that when the traffic intensity (ρ) is less than 1 (i.e., when $\lambda < \mu$), the D/D/1 queue will predict no queue formation. However, an M/D/1 model will predict queue formations under such conditions with randomness arising from the assumed probability distributions of arrivals.

Example 11.17 Consider the D/D/1 example worked out before, with the following changes. The arrival rate of customers is 165 per hr over the period till the movie starts, while the service rate is 3.25 persons/min. Compute the characteristics of this system.

Solution Mean arrival rate λ = 165/6 = 2.75 persons/min (exponential)

Constant service rate μ = 3.25 persons/min; ρ = 0.846

L_q = mean number of units in the waiting line = $\frac{\rho^2}{2(1-\rho)}$ persons, or

$= \frac{(0.846)^2}{2(1-0.846)} = 2.324$ persons

L = mean number of units on the system = $L_q + \rho = 3.173$ persons

W_q = Mean time in queue = $\frac{L_q}{\lambda} = \frac{2.324}{2.75} = 0.845$ min

W = Mean time in the system = $W_q + \frac{1}{\mu} = 0.845 + 0.308 = 1.153$ min

ρ = Service facility utilization factor = 0.846 min
P_0 = Probability of no units in the system = $1 - \rho = 0.154$

11.5.5 Model 3 (M/M/1)

If we assume exponentially distributed arrival times as well as exponentially distributed departure times with one channel, we end up with a useful model with several applications. For example, if we look at the departure patterns of a movie theater booth, we find that some customers do not have the correct change or have not decided which movie to see. Under standard M/M/1 conditions, ρ is less than one and the equations for solving the characteristics of the queue are:

Mean number of units in the system: $L = \frac{\lambda}{\mu - \lambda}$

Mean number of units in the queue: $L_q = \frac{\lambda^2}{\mu - \lambda} = \frac{\rho^2}{1 - \rho}$

Mean time in the system (for each unit): $W = \frac{1}{\mu - \lambda}$

Mean waiting time: $W_q = \frac{\lambda}{\mu(\mu - \lambda)}$

Percentage of server idle time: $I = P_0 = 1 - \frac{\lambda}{\mu} = 1 - \rho$

Where P_0 = probability of no customers in the system

Example 11.18 A movie theater ticket booth has a mean arrival rate of 3 persons/min and the service rate is 4 persons/min. Calculate the characteristics of this queuing system applying the M/M/1 model.

Solution

$$\lambda = 3; \mu = 4; \rho = \frac{\lambda}{\mu} = 0.75$$

$$L = \frac{\lambda}{\mu - \lambda} = \frac{3}{4 - 3} = 3 \text{ persons}$$

$$L_q = \frac{\rho^2}{1 - \rho} = \frac{(0.75)^2}{1 - 0.75} = 2.25 \text{ persons}$$

$$W = \frac{1}{\mu - \lambda} = \frac{1}{1} = 1.00 \text{ min}$$

$$W_q = \frac{\lambda}{\mu(\mu - \lambda)} = \frac{3}{4(4 - 3)} = 0.75 \text{ min}$$

$$I = 1 - \frac{\lambda}{\mu} = 1 - \rho = 1 - 0.75 = 25\%$$

11.5.6 The Economics and Operating Characteristics of Queuing Discipline

Engineers and managers making decisions regarding the level-of-service that needs to be maintained in handling goods, vehicles, and people, often have to consider the overall savings that would accrue if an additional worker were to be used. After all, saving customers' time is highly important for a business that intends to increase its clientele. The cost trade-off relationship between the TC of the service facility and the level-of-service is shown in Figure 11.5. The objective is to minimize the TC of service and waiting time to achieve a respectable service level.

11.5.7 Model 4 (M/M/N)

If we extend Model 3 (M/M/1) by considering multiple channels, we get a more general formulation that can be applied in many cases of managerial decision making. We are still assuming Poisson arrivals and exponential service times. The mean service rate is determined by $N(\mu)$ where N is the number of services.

The following equations describe the operational characteristics of M/M/N queuing.

1. The probability of having no units in the system is:

$$P_0 = \frac{1}{\displaystyle\sum_{n_c=0}^{N-1} \frac{\rho^{n_c}}{n!} + \frac{\rho^N}{N!\left(1 - \frac{\rho}{N}\right)}}$$

   where n_c is the departure channel number

2. The probability of having n vehicles in the system is:

$$P_n = \frac{\rho^n P_0}{n!} \quad \text{for } n \le N$$

$$P_n = \frac{\rho^n P_0}{N^{n-N} N!} \quad \text{for } n \ge N$$

3. The average length of queue (in units) is:

$$L = \frac{P_0 \rho^{N+1}}{N!\, N}\left[\frac{1}{(1 - \rho/N)^2}\right]$$

4. The average time spent in the system is:

$$W = \frac{\rho + L}{\lambda}$$

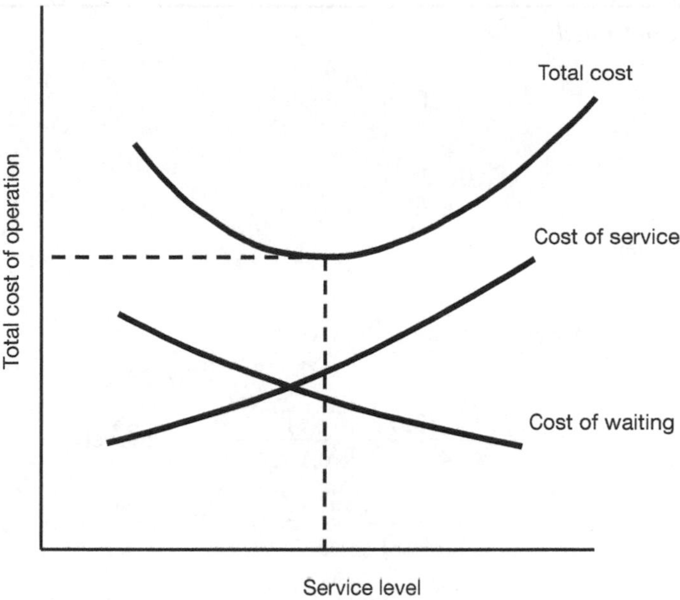

Figure 11.5 Service level vs. total cost

5. The average waiting time in the queue is:

$$W_q = \frac{\rho + L}{\lambda} - \frac{1}{\mu}$$

The probability of waiting in a queue (which is the probability that the number of units in the system, n, is greater than the number of departure channels, N) is:

$$P_{n>N} = \frac{P_0 \rho^{N+1}}{N!\, N(1 - \rho/N)}$$

Example 11.19 An entrance to a tollway has four toll booths. Vehicles arrive at an average of 1800 veh/hr and take 6.67 sec to pay their tolls. Both arrivals and departures can be assumed to be exponentially distributed. Compute the average queue length, the time in the system, and the probability of waiting in a queue.

Solution

$$\lambda = \frac{1800}{60} = 30 \text{ veh/min}$$

$$\mu = \frac{60}{6.67} = 9 \text{ veh/min}$$

Continues

Example 11.19: *Continued*

$$\rho = \frac{30}{9} = 3.33$$

$$\rho/N = \frac{3.33}{4} = 0.8333 < 1$$

For four booths open:

$$P_0 = \frac{1}{1 + \frac{3.333}{1!} + \frac{(3.333)^2}{2!} + \frac{(3.333)^3}{3!} + \frac{(3.333)^4}{4!(0.1667)}} = 0.0213$$

$$L = \frac{0.0213(3.333)^{4+1}}{4! \times 4}\left[\frac{1}{(0.1667)^2}\right] = 3.289 \text{ veh}$$

$$W = \frac{3.333 + 3.289}{30} = 0.2207 \text{ min}$$

$$P_{n>N} = \frac{0.0213(3.333)^{4+1}}{4! \times 4(0.1667)} = 0.548 \text{ min}$$

Example 11.20 Refer to Example 11.18. The administration now feels that delays to vehicles are excessive and that a fifth booth should be opened. Compute the values of P_0, L, W, and $P_{n>N}$.

Solution

$$P_0 = \frac{1}{1 + \frac{3.333}{1!} + \frac{(3.333)^2}{2!} + \frac{(3.333)^3}{3!} + \frac{(3.333)^4}{4!(0.1667)} + \frac{(3.333)^5}{5!(1 - 3.333/5)}} = 0.032$$

$$L = \frac{0.0213(3.333)^{5+1}}{5! \times 5}\left[\frac{1}{(0.333)^2}\right] = 0.659 \text{ veh}$$

$$W = \frac{3.333 + 0.659}{30} = 0.133 \text{ min}$$

$$P_{n>N} = \frac{0.032(3.333)^{5+1}}{5! \times 5(0.3334)} = 0.219 \text{ min}$$

Example 11.21 Compare the two systems with 4 and 5 booths as described in Examples 11.18 and 11.19. If the value of time is $10 per hr and the wages and cost of operating a booth is $100/hr, what would be your advice to the administration?

Solution

Table E11.21 Results for Example 11.20

		4 booths	5 booths
P_0	Probability of having no vehicles in the system	0.0213	0.032
L	Average queue length	3.287 veh	0.659 veh
W	Average time spent in the system	0.331 min	0.133 min
$P_{n>N}$	Probability of having to wait in queue	0.548	0.219

Opening a fifth booth reduces the queue length by 3.287 – 0.659 = 2.628 veh

Average time saved in the system is 0.331 - 0.133 = 0.198 min

∴ Since $\lambda = 20$ veh/min, $20 \times 0.198 = 3.96$ min

Total saving $3.96 \times 60 \times 10 =$ $2,376/hr

Net saving = $2,376 – 100 = $2,276/hr

11.6 SIMULATION

11.6.1 Introduction

Many real-world decision problems are so complex that they are difficult, if not impossible, to solve by simply applying the usual optimization models described in previous sections. In such situations, engineers and managers conduct experiments. Simulation is one experimental technique commonly used. Simulation is a modeling technique consisting of experiments to evaluate a system's behavior or response over time. It is by no means an optimizing technique, but it does allow the experimenter to examine the problem for a possible solution. In short, to simulate is to imitate a real-world situation. For example, if a store manager using three cashier servers wanted to increase the efficiency of his outfit by reducing the length of the queues, he could simulate this situation and come up with a solution.

11.6.2 Random Numbers

Most simulations make use of random numbers. They can be generated by using a calculator, but statisticians have generated scores of pages of these numbers that have no discernable pattern, thus making them random. You can generate random numbers by using a spinner with a pointer that is marked from 0 to 99 and, thus, capable of producing a random number from 0 to 99. Flipping coins and rolling dice can also be used if necessary.

11.6.3 Simulations Using Known Probabilities

We will now demonstrate the use of simulation through some simple examples. First, empirical (actual) data is collected from the site of operation. Second, a probability distribution is developed and an interval of random numbers is assigned to each class of the distribution. Third, random numbers derived from tables or otherwise, are applied to derive the results and lastly these are interpreted. A number of simple examples follow.

Example 11.22 Trucks arriving for unloading earth vary in length from 20 to 30 ft. The time needed to unload a 20 ft truck is 12 min, while a 30 ft takes 22 min. Using a uniform distribution, simulate the time needed to unload 6 trucks, on a random basis adopting the following random numbers 20, 31, 98, 24, 01 and 56. Assume that we don't know the distribution of the length of trucks.

Solution

Let a = time needed for a 20 ft truck to unload = 12 min

and b = time needed for a 30 ft truck to unload = 22 min

RN = random number as a percentage

Simulated value = $a + (b - a)(\text{RN}) = 12 + 10(\text{RN})$

1. $12 + 10(0.20) = 14.0$ min
2. $12 + 10(0.31) = 15.1$ min
3. $12 + 10(0.98) = 21.8$ min
4. $12 + 10(0.24) = 14.4$ min
5. $12 + 10(0.01) = 12.1$ min
6. $12 + 10(0.56) = \underline{17.6 \text{ min}}$

95.0 min

Average time per truck = 95/6 = 15.83 min

Certainly, 15.83 min is not the average time needed for unloading 6 trucks, but if we simulated 100 trucks we would possibly get a more realistic answer.

Example 11.23 A professor spends time with students based on normally distributed times with a mean of 20 min and a standard deviation of 5 min. Using random numbers find the average time he takes for 5 students, adopting the following normally distributed random numbers −0.25, 1.13, 0.35, 0.75, 2.09.

Continues

Example 11.23: *Continued*

Solution Let μ = 20 min; σ = 5 min; and RN = random number

Simulated value = μ + σ(RN) = 20 + 5(RN)

20 + 5(−0.25) = 18.75
20 + 5(1.13) = 25.65
20 + 5(0.35) = 21.75
20 + 5(0.75) = 23.75
20 + 5(2.09) = 30.45
120.35 min

Average time spent with each student = 24.07min

This answer is based on a sample of only 5 students. For realistic results, hundreds of students would need to be taken into account.

Example 11.24 Airplanes arrive at a small airport, described by a Poisson distribution (see Chapter 6) with a mean of three planes per hour. Simulate the number of plane arrivals for an 8-hr shift.

Solution First, set up a table connecting plane arrivals using the cumulative Poisson distribution.

$$P(x) = \sum \frac{\lambda^x e^{-\lambda}}{x!} \text{ where } \lambda = 3$$

Table E11.24 Cumulative probability for plane arrivals

Plane arrival	0	1	2	3	4	5	6	7	8	9	10
Cum. *P*(*x*) Prob.	0.050	0.199	0.423	0.647	0.815	0.916	0.966	0.988	0.996	0.999	1.0

Next, select eight random numbers in sequence and match the plane arrivals to each of them:

853	540	985	903	266	373	920	164	
5	3	7	5	2	2	6	1	= 31 planes

Example 11.25 A life insurance salesman has kept good records for the past year regarding his performance. There is a 50 percent chance that when a client walks into his office there is a genuine interest in buying a policy. However, this by itself does not always end up in a sale. Fifty percent of the time there will be no sale, 1/3 of the time it will result in a sale of $100,000 and 1/6 of the time in a sale of $200,000. Using a simulation of 20 cases, determine the probability of a sale and the

Continues

Example 11.25: *Continued*

expected policy value. After you have completed the simulation, show your results in the form of a decision tree and compare your results with the theoretical tree. What is the EV of your sale?

Solution There are two stages to this problem. In the first stage, there is a 50 percent chance that a prospective client will be interested in talking about buying a policy. In the second stage, the person who shows an interest will decide what he will do. Assign random numbers accordingly.

Table E11.25a Two stages in Example 11.25

	Probability	Cum. probability	RN
Stage 1			
Interest	0.5	0.5	00 – 49
No interest	0.5	1.00	50 – 99
Stage 2			
No interest	0.50	0.50	00 – 49
Sale of $100,000	0.33	0.83	50 – 82
Sale of $200,000	0.17	1.00	83 – 99

Simulation of 20 insurance calls:

Table E11.25b Simulation results for Example 11.25

		Interest			Interest			
Trial	**Stage 1 R. No.**	**Yes**	**No**	**Stage 2 R. No.**	**No**	**10^5**	**2×10^5**	**Value**
1	46	X		59		X		10^5
2	16	X		72		X		10^5
3	86		–	–				–
4	25	X		45	X			0
5	03	X		05	X			0
6	62		–	–				–
7	23	X		32	X			0
8	36	X		91			X	2×10^5
9	94		–	–				–
10	70		–	–				–
11	12	X		15	X			0
12	75		–	–				–
13	41	X		38	X			0
14	69		–	–				–
15	34	X		84			X	2×10^5
16	48	X		71		X		10^5
17	02	X		22	X			0
18	24	X		64		X		10^5
19	95		–	–				–
20	12	X		90			X	2×10^5
Result		13	7		6	4	3	10×10^5

Continues

Example 11.25: *Continued*

For the theoretical tree:

$$\left(\frac{1}{4}+\frac{1}{6}+\frac{1}{12}\right)+\left(\frac{1}{2}\right)=1.00$$

$$(25\% + 16.7\% + 8.3\%) + 50\% = 100\%$$

$$\text{No sale} = 0.25 + 0.50 = 75\%$$

$$\text{Sale} = 0.167 + 0.083 = 25\%$$

$$\therefore \text{Probability of selling } \$10^5/\text{given a sale} = \frac{0.167}{0.167+0.083} = 0.668$$

$$\text{Probability of selling } \$2 \times 10^5/\text{given a sale} = \frac{0.083}{0.167+0.083} = 0.332$$

$$\text{Total} = 1.00$$

EV of policy = $(0.668 \times 10^5) + (0.332 \times 2 \times 10^5) = \$133{,}200$

For the simulated tree (see Figure E11.25):

$$\left(\frac{6}{20}+\frac{4}{20}+\frac{3}{20}\right)+\left(\frac{7}{20}\right)=1.00$$

$$(30\% + 20\% + 15\%) + 35\% = 100\%$$

$$\text{No sale} = 0.30 + 0.35 = 65\%$$

$$\text{Sale} = 0.20 + 0.15 = 35\%$$

$$\therefore \text{Probability of selling } \$10^5/\text{given a sale} = \frac{0.20}{0.20+0.15} = 0.5714$$

$$\text{Probability of selling } \$2 \times 10^5/\text{given a sale} = \frac{0.15}{0.20+0.15} = 0.4286$$

$$\text{Total} = 1.00$$

EV of policy = $(0.5714 \times 10^5) + (0.4286 \times 2 \times 10^5) = \$142{,}860$

Continues

Example 11.25: *Continued*

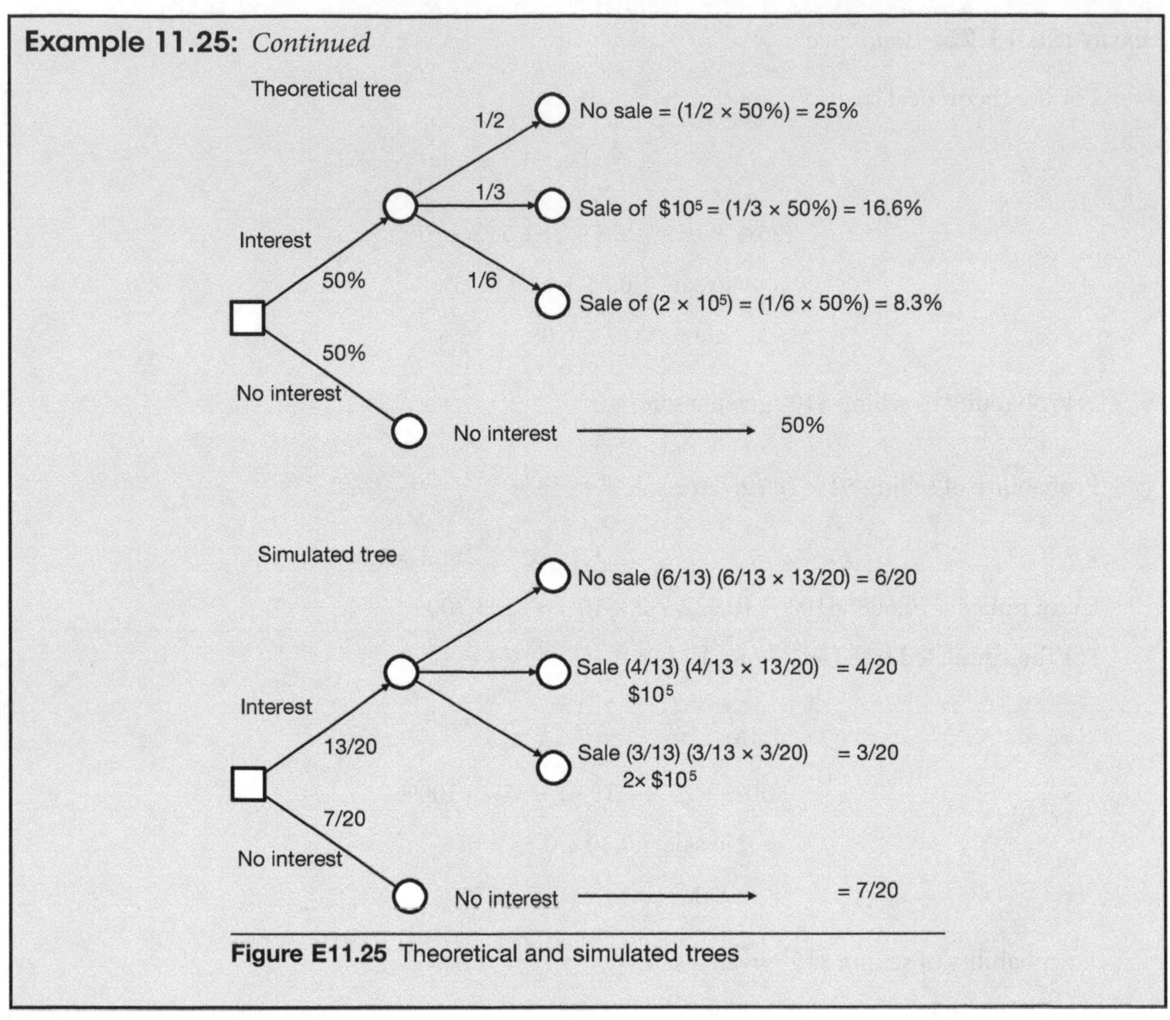

Figure E11.25 Theoretical and simulated trees

Example 11.26 A one-man repair shop handles jobs as shown in the table:

Table E11.26a Data for Example 11.26

Time between jobs	Probability	Job time	Probability
10	0.10	8	0.15
20	0.15	16	0.30
30	0.35	24	0.25
40	0.20	32	0.20
50	0.15	40	0.10
60	0.05		
	1.00		1.00

Continues

Example 11.26: *Continued*

Simulate the repair shop performance for 10 jobs and compute the following:

(a) The average turnaround time per job

(b) The average number of jobs waiting to be done

(c) The idle time of the repair man

Solution Assign random numbers to time between jobs:

Table E11.26b Random numbers for time between jobs

Arr. interval prob.	Probability	Cum. prob. P_x	RN
10	0.10	0.10	00–09
20	0.15	0.25	10–24
30	0.35	0.60	25–59
40	0.20	0.80	60–79
50	0.15	0.95	80–94
60	0.05	1.00	95–99

Assign random numbers to repair time:

Table E11.26c Random numbers for repair time

Repair time (Y)	Probability (P_y)	Cum. prob.	RN
8	0.15	0.15	00–14
16	0.30	0.45	15–44
24	0.25	0.70	45–69
32	0.20	0.90	70–89
40	0.10	1.00	90–99

Table E11.26d Summary of results for Example 11.26

Job	r_1	Arr. int.	Arr. clock	Enter service clock	Waiting time	Length of queue	r_2	Service time	Dep. clock	Time in service
1	19	20	20	20	0	0	65	24	44	24
2	51	30	50	50	0	0	17	16	66	16
3	63	40	90	90	0	0	85	32	122	32
4	37	30	120	122	2	1	89	32	154	34
5	76	40	160	160	0	0	71	32	192	32
6	34	30	190	192	2	1	11	8	200	10
7	27	30	220	220	0	0	10	8	228	8
8	59	30	250	250	0	0	87	32	282	32
9	08	10	260	282	22	1	08	8	290	30
10	89	50	310	310	0	0	42	16	326	16
					26	6		72		219

Continues

Example 11.26: *Continued*

$$\text{Average turnaround time} = \frac{172}{10} = 17.2 \text{ min}$$

$$\text{Average waiting time} = \frac{26}{10} = 2.6 \text{ min}$$

$$\text{Average number of jobs waiting} = \frac{6}{10} = 0.6 \text{ jobs}$$

11.7 MARKOV ANALYSIS

Markov analysis is a technique used by engineers and managers to forecast future trends. The objectives of Markov analysis are to provide probabilistic information about a situation using the results of a just-previous experiment. For example, the probability of your car needing a major repair next year might depend on how many major repairs it had last year. Let us demonstrate Markov analysis by working out an example.

Example 11.27 A gas station manager surveyed 900 customers who frequently fill gas at his pumps. Of those, 300 bought gas in the first week of March, while 600 did not. It is expected that in the second week that 75 percent of those who bought gas will return, in addition to 20 percent of those who did not buy gas. What is the expected number of purchasers and nonpurchasers in the second week? What is likely to happen in subsequent weeks?

Solution
We can express this question in matrix form:

$$(300, 600)\begin{bmatrix} .75 & .25 \\ .20 & .80 \end{bmatrix}$$

Customers = (300 × 0.75) + (600 × 0.20) = 225 + 120 = 345

Noncustomers = (300 × 0.25) + (600 × 0.80) = 75 + 480 = 555

Total customers = 900

This result can be expressed:

$$(300, 600)\begin{bmatrix} .75 & .25 \\ .20 & .80 \end{bmatrix} = (345, 555) \text{ for the second week}$$

For the third week the expected results are:

$$(345, 555)\begin{bmatrix} .75 & .25 \\ .20 & .80 \end{bmatrix} = (370, 530)$$

Continues

Example 11.27: *Continued*

And, for the fourth week the expected results are:

$$(370, 530)\begin{bmatrix} .75 & .25 \\ .20 & .80 \end{bmatrix} = (384, 516)$$

If one works at the figures of purchasers and nonpurchasers, it is evident that the number of purchasers is increasing, but at a slower rate as the weeks pass. Evidently the sales will reach a steady state in due time.

Example 11.28 Reexamine the problem in Example 11.27. What would the sales be in a steady state?

Solution

$$\text{Let } X = (x, y)$$

Where:

$$x = \text{purchasers}$$

$$y = \text{nonpurchasers}$$

$$\text{and } X = (x + y)$$

$$(x, y)\begin{bmatrix} .75 & .25 \\ .20 & .80 \end{bmatrix} = (x, y)$$

and $0.75x + 0.20y = x$ and $0.25x + 0.80y = y$ from which we find $x = 0.80y$; and since $x + y = 900$

$$\therefore (400, 500)\begin{bmatrix} .75 & .25 \\ .20 & .80 \end{bmatrix} = (400, 500)$$

In the steady state, there will be 400 buyers per wk and 500 nonbuyers. Let us interpret this transition matrix.

Table E11.28 Transition matrix

		Next period	
		P	*N*
This period	*P*	0.75	0.25
	N	0.20	0.80

The first row of numbers tells us what proportion of customers who buy gas in the first period will still buy gas in the next period. Thus, the 0.75 in Row P, Column P, means that of the

Continues

Example 11.28: *Continued*

customers who buy gas in the first period, 75 percent will still be buying gas in the next period. The second row gives us the same information about those who did not buy gas in the first period. This information could be put in the form of difference equations (with this period subscripted 1 and the next Period 2).

$$P_2 = 0.75P_1 + 0.20N_1$$

$$N_2 = 0.25P_1 + 0.80N_1$$

and this is exactly what we did in the examples.

11.7.1 Characteristics of Markov Analysis

Now that we have solved some problems, we notice that Markov analysis is applicable to systems that exhibit probabilistic movement from one state to another, over time and this probability is known as a *transition probability*. The set of all transition probabilities when expressed in matrix form is called a *transition matrix*. For instance, (in Example 11.27):

$$\begin{bmatrix} .75 & .25 \\ .20 & .80 \end{bmatrix}$$

is a transition matrix. These matrices have the following characteristics:

(a) The probabilities for a given beginning state sum to 1.0
(b) The probabilities apply to all parties in the system
(c) The probabilities are constant over time
(d) The states are independent over time

If **M** is any square matrix and **X** is a vector such that **XM** = **X**, then **X** is called a fixed vector for **M**. The term fixed vector is also called *stationary vector* or *Eigen vector* by some authors. In our example (400, 500) is a fixed vector for **M**, where

$$\mathbf{M} = \begin{bmatrix} 0.75 & 0.25 \\ 0.20 & 0.80 \end{bmatrix}$$

Also, if **M** is a Markov matrix, and **X** is a *probability vector*, such that **XM** = **X**, then **X** is called a *fixed probability vector* for **M**. If a Matrix **M** has a fixed Vector **X**, it is possible to find a fixed probability vector for **M** by dividing **X** by the sum of the elements in **X**. Thus:

$$\begin{bmatrix} .75 & .25 \\ .20 & .80 \end{bmatrix} = \left(\frac{400}{900}, \frac{500}{900}\right) = \left(\frac{4}{9}, \frac{5}{9}\right) = (0.444, 0.556)$$

Example 11.29

Show that the Matrix $\mathbf{M} = \begin{bmatrix} 0.75 & 0.25 \\ 0.20 & 0.80 \end{bmatrix}^n$ is equal to $\begin{bmatrix} 0.444 & 0.556 \\ 0.444 & 0.556 \end{bmatrix}$.

Solution

$$\mathbf{M}^2 = \begin{bmatrix} 0.61 & 0.39 \\ 0.31 & 0.69 \end{bmatrix} \quad \mathbf{M}^3 = \begin{bmatrix} 0.54 & 0.45 \\ 0.37 & 0.63 \end{bmatrix} \quad \mathbf{M}^4 = \begin{bmatrix} 0.50 & 0.50 \\ 0.40 & 0.60 \end{bmatrix}$$

$$\mathbf{M}^5 = \begin{bmatrix} 0.45 & 0.55 \\ 0.42 & 0.58 \end{bmatrix} \quad \mathbf{M}^6 = \begin{bmatrix} 0.444 & 0.556 \\ 0.444 & 0.556 \end{bmatrix}$$

These results tally with:

$(300, 600)(\mathbf{M}) = (345, 555)$ Second week

$(300, 600)\mathbf{M}^2 = (370, 530)$ Third week

$(300, 600)\mathbf{M}^3 = (384, 516)$ Fourth week

and eventually with:

$(300, 600)\mathbf{M}^6 = (400, 500)$ when the system reaches a steady state.

Example 11.30 Refer to Example 11.27. Using a decision tree, determine the probabilities of a customer purchasing or not purchasing gas in Week 3, given that the customer buys in the present month. Summarize the resulting probabilities in a table.

Solution The situation is:

Table E11.30a data in Example 11.30

	Next week	
This week	**Buy**	**Not buy**
Buy	0.75	0.25
Not buy	0.20	0.80

Figure E11.30 shows the two decision trees from which the following conclusions can be drawn

Table E11.30b Probability values for Example 11.30

Starting state	Prob. of buy	Prob. of not buy	Sum
Buy	0.625	0.3875	1.00
Not buy	0.310	0.6900	1.00

Continues

Example 11.30: *Continued*

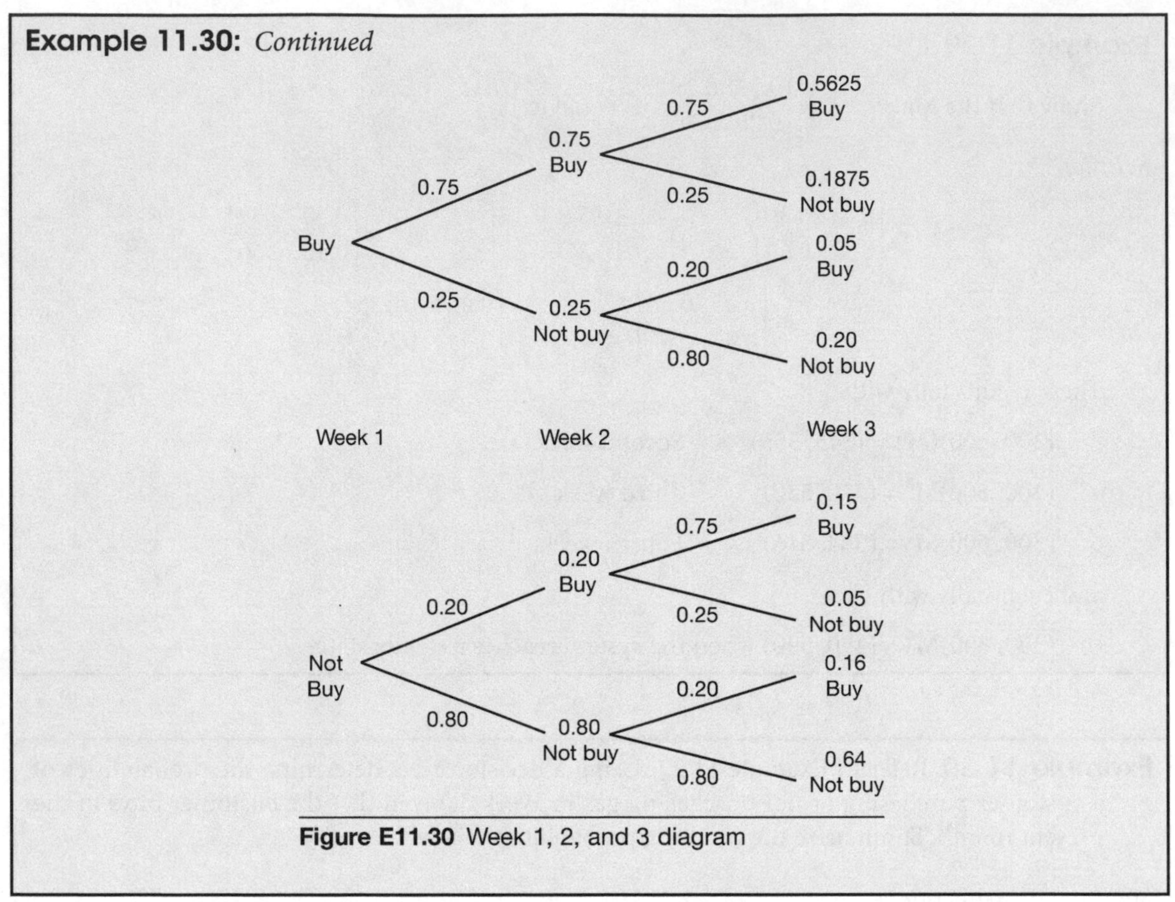

Figure E11.30 Week 1, 2, and 3 diagram

11.7.2 Special Transition Matrices

We have seen that the use of Markov analysis can be used to determine the probability of being in a given state at some future time period. To find this probability we successively multiply the transition matrix by itself *n* times. When such a condition exists, the constant state probabilities are called steady-state probabilities.

If we want to generalize long-run probabilities, we must consider three kinds of Markov chains: regular chains; absorbing chains; and cyclical chains.

Regular chains: We have already looked at these chains before. In a regular chain, all states communicate with one another. For example, Matrix P, describes three states, W = well, S = sick for 1 wk, and V = sick for 2 wk.

Table 11.5a An example of regular chains

		Period 2		
		W	**S**	**V**
Period 1	W	0.75	0.25	0
	S	0.50	0	0.50
	V	0.30	0	0.70

In this example, for instance, there are three zero entries. But the two-period matrix has no zeros.

Table 11.5b An example of regular chains

		Period 3		
		W	**S**	**V**
Period 1	W	0.6875	0.1875	0.1250
	S	0.5250	0.1250	0.3500
	V	0.4350	0.0750	0.4900

As noted, regular chains will lead eventually to an equilibrium distribution. This property may be useful when we are predicting the behavior of entire populations rather than the probabilistic movements of individuals.

Absorbing chains: In these cases, there are one or more absorbing states that the individual cannot leave once it is entered. To visualize what is meant by an absorbing state, examine the well/sick/very sick matrix given before and modify the very sick row to dead (*D*). If this matrix is modified to the following:

Table 11.6 An example of absorbing chains

		Period 2		
		W	**S**	**D**
Period 1	W	0.80	0.19	0.01
	S	0.50	0.47	0. 03 = Q
	D	0	0	1

Notice that this is not a regular matrix because a person cannot go from being dead to well or sick no matter how much time passes. Here, death is an absorbing state. Another example is a pollutant that may be transformed into a totally harmless substance. The opposite of an absorbing state is a transient state, which is the regular chain.

Cyclical chains: Sometimes an individual is trapped in a cyclical pattern that he cannot escape. For instance a judge may rotate his visitation to his four offices located in different parts of the state.

Table 11.7 An example of Markov cyclical chains

		N	S	E	W
	N	0	1	0	0
Period 1	S	0	0	1	0
	E	0	0	0	1
	W	1	0	0	0

In summary, with regular Markov chains one could draw two conclusions: (1) in the long-run, the probability of being in a particular state approaches equilibrium and is independent of the state that the individual is initially and (2) equilibrium probabilities may be interpreted as the percent of time spent in each state over the long-run. With absorbing chains, we are not bothered by equilibrium as much as we are interested in knowing how quickly it will be trapped. With cyclical chains, the interest lies in finding how many periods are needed before completing a rotation.

EXERCISES

1. A car salesman would like to forecast the demand for Toyota trucks for next year based on sales for this year. His records indicate the following (Table P11.1):

Table P11.1

Month	Sales
January	10
February	12
March	15
April	16
May	20
June	22
July	20
August	15
September	12
October	12
November	10
December	9

(a) Compute a 3-mo MA forecast for April of the current year through January of the next year

(b) Assigning weights of 2.0, 1.0, and 0.5 to the months in sequence, starting with the most recent month and forecast his sales for the month of January next year

(c) Compare the two forecasts using MAD and write your comments
(d) Compute the exponentially smoothed forecast ($\alpha = 0.3$) for January next year
(e) Conduct a least-square trend forecast for January and February of next year

2. For the last 8 yr, there have been a high number of accidents on a 20-mi stretch of freeway as shown in Table P11.2:

Table P11.2

Year	Accidents
1	40
2	45
3	35
4	42
5	50
6	58
7	49
8	65

(a) Compute a 3-yr MA forecast for Years 4 to 9
(b) Compute a 3-yr weighted average forecast with weights of 3, 2, and 1
(c) Compute the exponentially smoothed forecast with $\alpha = 0.3$ for the accident data
(d) Using least-squares trend analysis forecast the likely accidents in the next 2 yr

3. A movie theater chain would like to know the patronage for their shows depending on the ticket price. In the last year, a record was kept for seven ticket rates and the corresponding patronage (Table P.11.3).

(a) Set up a simple regression analysis equation for this data connecting ticket price with patronage (Y).
(b) What patronage would be expected for tickets at \$4.50 and \$7.00?

Table P11.3

Ticket price	Patronage (in 100s)
\$3.00	52
\$3.50	48
\$4.00	47
\$5.00	40
\$5.50	24
\$6.75	18

4. The manager of the movie chain is curious to know whether predictions of patronage can be made based on rating scores of movies. She has six cases, shown in Table P11.4.

Table P11.4

Rating	Patronage (100s)
5	52
6	48
8	47
6	40
5	24
8	18

Set up a regression equation and predict what patronage may be anticipated with a score of 7.

5. If in the previous problem, you consider the excellence ratings score given (see Table P11.5) to films shown, as well as the price, how would your answer change?

Table P11.5

Price ($)	Patronage (100s)	Rating
3.00	52	5
3.50	48	6
4.00	47	8
5.00	40	6
5.50	24	5
6.75	18	8

6. A large shipping company needs to supply material from three sources (a, b, and c) to four sources (A, B, C, and D) for unit costs shown in the matrix in Table P11.6.

Table P11.6

To	A	B	C	D	
From					Supply
a	50	75	30	45	120
b	65	80	40	60	170
c	40	70	50	55	110
Demand	100	100	100	100	400

(a) Find the initial feasible allocation using the NW corner, least-cost cell, and the penalty methods and compute the TC

(b) Using the solutions found in (a), find the optimal solution

7. Three production plants at Locations I, II, and III supply steel goods to warehouses at A, B, and C as shown (Table P11.7) and have unit costs in dollars:

Table P11.7

	A	B	C	Supply
I	20	28	16	40
II	24	20	24	60
III	16	24	20	80
Demand	80	60	40	180

(a) Determine initial allocation using the NW corner, low-cost cell, and penalty methods, and find the TC of shipment
(b) Use the initial allocation and find the optimal solution

8. A manager has worked out a transhipment problem as shown in the tableau (Table P11.8).

Table P11.8

Factory	A	B	C	Supply
a	24	20	12	12
b	8	30	6	8
c	18	14	M	6
d	22	16	12	16
e	0	0	0	4
Demand	18	10	18	46

(a) Does this tableau represent an unbalanced shipment problem that has been rectified?
(b) Is this a degenerate solution? How would you correct this problem?
(c) Do you notice a prohibited route?
(d) Work through this problem by applying the NW corner, low-cost cell, and the penalty methods.
(e) Apply any method you know to find the optimum solution.

9. Three companies X, Y, and Z are bidding on three jobs A, B, and C. X can do 100 jobs/wk and charges $6 for A, $5 for B, and $7 for C. Y can do 120 jobs and charges $9 for A, $6 for B and $10 for C. Z can do 100 jobs and charges $8 for A, $10 for B, and $9 for C. If 100 jobs of Type A, 80 jobs of Type B, and 90 jobs of Type C are to be done, how would you assign jobs to each company using the NW corner, low-cost cell, and the penalty methods? Refine your answers by applying one of the optimization methods.

10. An electrical appliance can be produced at the rate of 100 per da throughout the year, but because of seasonal demand, only 40 per da are sold during the first three quarters of the year, but 160 per da are sold during the last quarter of the year. Storage costs amount to \$200 per quarter and the expenses of production can rise at a rate of \$100 per quarter. If the production costs are \$4000 during the first quarter:
    (a) Set up these data as a transportation problem
    (b) Find the production schedule that gives the minimum cost
11. Find the minimum cost for supplying material as given in Table P11.11 with unit costs in each cell. Start with the minimum cell cost solution and then use the MODI method to refine your answer.

Table P11.11

	D1	D2	D3	Supply
S1	24	15	18	240
S2	45	30	36	160
S3	9	27	30	160
Demand	300	140	120	560

12. Find the minimum cost starting with Vogel's solution for Exercise 11 and then use the MODI method to refine your answer.
13. Four machinists are to be assigned to four jobs. The time taken by each job is shown in Table P11.13. Assign the machinists to the job in such a way that the total sum of the time needed is a minimum.

Table P11.13

Machine operator	1	2	3	4
A	20	23	10	12
B	7	20	9	12
C	13	15	13	20
D	16	16	12	10

Determine the assignment of machines to each operator to minimize the total number of hours and the optimal solution.

14. The manager of a shoe store has five salesmen who have varied capabilities to work in four stores. Their past records indicate that their sales profits in (1000s) are as shown in the matrix in Table P11.14. Assign them to the four stores to bring in the highest overall profit.

Table P11.14

Store salesman	1	2	3	4
A	88	93	92	91
B	84	80	91	93
C	91	88	95	85
D	97	90	90	95
E	85	90	75	80

15. Refer to Exercise 14. After the manager makes the assignment as per the matrix, Salesmen A, B, C, and D feel that they do not want to be assigned to Stores 2, 4, 3, and 1 respectively for personal reasons. Reassign the salesmen and indicate the difference in total profits because of the new assignment.

16. If a large building, currently provided with the minimum fire protection, is not protected against fire, the total damage could be as much as $3 million in case of a fire, whose probability of occurrence is 60 percent. However, a fire protection engineer has estimated that if sophisticated equipment worth $0.75 million is provided, there is a probability that there would be no damage in case of a fire. (1) What are the states of nature, the alternatives, and their probability? (2) Construct a matrix and decision tree using this information and indicate if the sophisticated equipment should be installed.

17. A person with expensive equipment needs to replace its motor. He finds a used motor at a second hand outfit for $100, but finds that the testing is not reliable. Seventy percent of the time it tests good motors as good and 40 percent of the time, it tests bad motors as good. From past experience, a used motor is good 20 percent of the time. The only other alternatives are to either order a brand new motor for $300, which is available only after 6 mo at a loss of $100, or buy one from ready stock for $450. (a) Sketch a decision tree showing all options. (b) What is the best strategy?

18. A manufacturing company can produce three Machines A, B, and C for third world countries depending on good, stable, and unstable conditions. The payoff table (Table P11.18) indicates profits/losses in millions per year.

Table P11.18

	Market conditions		
Machine	Good (0.2)	Stable (0.7)	Unstable (0.1)
A	240	140	-60
B	120	80	40
C	70	60	60

(a) What would be the best machine to manufacture?
(b) Compute the opportunity loss table.
(c) Draw a decision tree diagram.
(d) What consulting fee would the company be willing to pay to obtain perfect information regarding market conditions to gain the best profit?

19. The manufacturing company described in Exercise 18 is not satisfied with the results of its analysis and would like to hire a marketing company that knows about third world countries. Their reports indicate that there is a 0.6 probability of a positive report given good conditions, a 0.3 probability of a positive report given stable conditions and a 0.1 probability of a positive report given unstable conditions. There is a 0.9 probability of a negative report given unstable conditions, a 0.7 probability given stable conditions, and a 0.4 probability given good conditions. Using Bayes' theorem and a decision tree, determine the best strategy the company can follow and the maximum sum of consulting fees that could be paid to the consulting company.

20. A company manufacturing bicycle parts has just acquired a new machine to produce axles. It was found that when the machine was set up correctly, there was a 10 percent chance of the item being defective; but if the machine was set up incorrectly, the chances of the item being defective rose to 40 percent. The information currently available is that the machine is set up incorrectly 50 percent of the time. What is the probability that the machine is set up incorrectly, if the sample is found to be defective?

21. A professor spends about 10 min with each student, although 12 min is scheduled for each student to be just on time. What is the average number of students waiting to be served, the average waiting time for a student and the mean time a student spends in the system (i.e., waiting and being served)? (Use D/D/1.)

22. In Example 11.16 worked out in the text, the service rate $\mu = 2.5$ persons/min.

(a) Describe the performance of the system, all the factors remaining the same.
(b) Draw a neat sketch of the system. (Use D/D/1.)

23. A local carwash station has an automated system requiring 5 min to wash a car plus 1 min to wipe it dry. Customers arrive at the rate of 8 veh/hr. If the system operates as per M/D/1, compute the characteristics of this operation.

24. The local airport in Walla Walla has a single runway with one traffic controller. It takes a plane 10 min to land and clear the runway with landing following an M/M/1 regime. According to Federal Aviation Administration rules, the traffic controller must have at least 15 min rest time every hour. If two planes arrive per hour, then

(a) What is the average number of planes that have to wait to land?
(b) What is the average time a plane must be in the air before it can land?
(c) Will the airport have to hire a second traffic controller to follow FAA rules?

25. (a) During the first few days of every semester, a student advisor takes 3 min to sign off on each student's schedule. Unfortunately he is the only authority to approve such schedules and when the arrival rate of students is 18 per hr, his work becomes pretty hectic. Compute the characteristics of this system.

    (b) The advisor feels that if some graduate student help is available his advising would go faster. If 5 min is a reasonable time for a student to wait, the help of one graduate student will reduce the waiting time by 1/2 min, how many graduate students should be hired? (Use M/M/1.)

26. A hardware store operates a single-server queuing system (with one cashier and one checkout counter) 12 hr per da. This system has the following details (1) infinite calling population (2) first come/first serve, (3) M/M/1 system, (4) 20 customers/hr arrive at the checking counter queue, and (5) 25 can be checked out. Compute P_0, L, L_q, W, W_q and V.

27. Refer to problem 26. The manager of this store feels that if the queue is divided into two equal parts and a second checkout counter is operated, there would be greater efficiency, despite the cost of operating a second counter costing \$10/hr. Compute the characteristics of this proposal and estimate the probable savings per week, assuming the value of time of customers equal to \$10 per hr (because time spent in the queue represents a loss to the store).

28. Refer to problem 26. The manager of the store feels that it may be a good idea to hire a helper at \$5/hr. He tries out this arrangement and finds that because of a reduction of delay the arrival rate has correspondingly climbed to 38 customers/hr. Compute the characteristics of this arrangement, assuming customer time is worth \$10/hr and estimate the savings per week to the store.

29. Refer to problem 26. The manager of the store would like to try having two separate servers (counters) to speed up the queue. If the mean arrival and service rate remain as before and the cost of operating an additional counter is \$10/hr, compute the characteristics of this system and estimate the savings per week to the store.

30. A printing shop has two machines or equal speed serving customers. The arrival interval of customers is distributed as in Table P11.30.

Table P11.30

Arrival interval (*A*)	Probability (*P*)
2 min	0.2
4 min	0.5
6 min	0.3
Service time (*S*)	**Probability (*P*)**
5 min	0.3
6 min	0.5
7 min	0.4

Simulate these operations for 5 customers.

31. A CPM/PERT project shown consists of the following activities and probabilities attached. For example, Task (1 – 2) can be completed in 5 wk with a probability of 0.2 or in 6 wk with a probability of 0.8. Simulate this network three times, determine the critical path, and compute the expected time of completion (Figure P11.31 and Table P11.31).

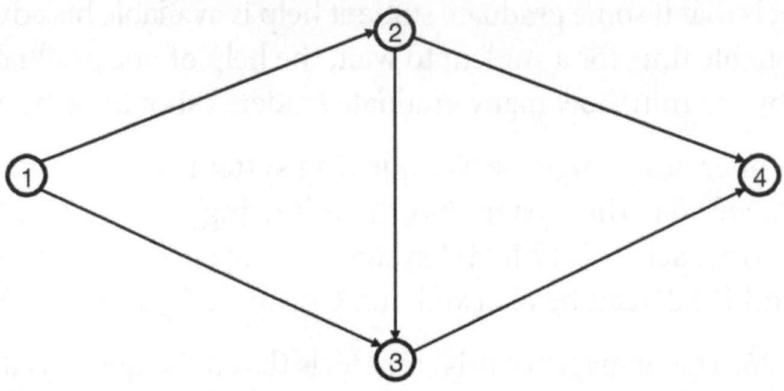

Figure P11.31 A CPM/PERT project

Table P11.31

Activity	Duration (*x*), weeks	Probability, *p* (*x*)
1 – 2	5	0.2
	6	0.8
1 – 3	9	0.4
	10	0.6
2 – 3	8	0.6
	9	0.4
2 – 4	3	0.8
	4	0.2
3 – 4	7	0.5
	9	0.5

32. If the trucks in the Example 11.22 worked out in the text vary between 20 and 40 ft, with a 40-ft truck taking 35 min to unload, redo the simulation for 10 trucks.

33. If the professor in the Example 11.23 spends 15 min each on an average with a standard deviation of 4 min, rework the problem and simulate for 10 students, assuming normal distribution.

34. The airplanes in Example 11.24 are now arriving in the airport at 4 per hr. Simulate for a 10-hr shift.

35. A computer repair shop consists of an engineer and one machine. Clients come in to the shop at 9 a.m. at the following time interval (Table P11.35a).

Table P11.35a

Time	Interval
6	0.05
12	0.10
18	0.20
24	0.35
30	0.20
36	0.10

However, the problems are such that repairs can only be checked and done either by the engineer, or the mechanic, or sometimes by both. The distribution is as follows:

Engineer only 0.30
Mechanic only 0.60
Both engineer and mechanic 0.10

Simulate this operation for 10 customers assuming the following distribution (Table P11.35b).

Table P11.35b

Engineer	*p*(*x*)	Mechanic	*p*(*y*)	Both	*p*
10 min	0.2	20 min	0.3	10	0.1
20 min	0.4	30 min	0.5	30	0.7
30 min	0.4	40 min	0.2	50	0.2

36. A customer generally buys two brands of shoes P and Q changing brands as shown in Table P11.36:

Table P11.36

		Next period	
		P	Q
This period	P	0.55	0.45
	Q	0.60	0.40

(a) Determine the probabilities that a customer will buy shoe P or shoe Q in Period 3 in the future, given that the customer bought P or Q in this period.
(b) Draw decision trees to illustrate your answer.
(c) What would be the situation in the steady state?

37. An engineer has observed a particular machine's record regarding breakdown, as shown in the matrix (Table P11.37), where O indicates that the machine operates and B indicates that there is a breakdown.

Table P11.37

		Next day	
		O	**B**
This day	O	0.4	0.6
	B	0.8	0.20

(a) Assuming that the machine operates on Day 1, determine the probabilities that the machines will be operating or down on Day 2 through Day 6
(b) Determine the steady state probability for this transition matrix

38. A village of 20,000 people has a labor force of 10,000. In any particular month, there is a large number (E) who are gainfully employed and a small proportion who are unemployed (U). It has been observed that 95 percent of those employed in one year are still employed the following year. While 45 percent of the currently unemployed find jobs and are employed in the next year, and these percentages hold true year-after-year, put down this information in the form of a transition matrix and interpret.

39. A housing complex manager has observed that tenants pay their rents as per the transition matrix given in Table P11.39:

Table P11.39

		Next month	
		Pay	**Not pay**
This month	Pay	0.9	0.1
	Not pay	0.8	0.20

(a) If a renter did not pay his rent this month, what is the probability that she will not pay in any of the next 3 mo?
(b) What is the steady state of this matrix?

40. It is said by sociologists that in general people tell the truth with a probability p such that $0 < p < 1$. Then for a rumor (that could be either true or false) the probability that a person is likely to tell the truth can be represented by (Table P11.40):

Table P11.40

		What everyone told his friend	
		True	**False**
What everyone was told	True	p	$1-p$
	False	$1-p$	p

Using this as a base and assuming p = 0.9, what is the probability that the fourth person who hears the rumor is hearing the truth?

41. In one of the Washington State counties it was found that in 1998, 0.07 of the rural population moved to the city and 0.04 of the city population moved to the rural area. (a) Describe this in the form of a Markov matrix; (b) What is the fixed probability vector of this matrix? (c) If this trend continued in future years what proportion of the population will end up in the city?

42. The manager of a college fund finds that of those members who did not donate last year, there is a strong probability that 10 percent of them will contribute a small gift this year. Of those that gave small gifts last year 20 percent will not give at all but 10 percent will give large gifts this year. Also, of those who gave large gifts last year, 10 percent will skip a contribution this year. Twenty percent will give small gifts and the rest will continue to give large gifts.

    (a) Describe the donation pattern by a Markov matrix
    (b) If this pattern continues, what fraction of the alumni can eventually be expected each year to contribute large, small, and zero gifts

Systems Thinking 12

12.1 INTRODUCTION

Systems thinking is the conscious use of concepts captured in the word *system* to order our thoughts. It is a way of thinking about a problematic situation using ideas to understand its complexity and also to make use of the particular concept of wholeness to order our thoughts (Checkland, 1999). Systems thinking emerged in the early twentieth century and continued to transform itself in a series of steps, stages, or waves for dealing with problems encountered in engineering, management, city planning, and social systems. Toward the end of World War II, a sophisticated set of mathematical tools and techniques began to be formulated and applied for solving engineering and management problems based on the scientific method. Over the fifty years or more since they were developed, these tools have become known by various labels, including operations research (OR), management science, systems analysis, and systems engineering, depending on who was using them and in what context. Many of these tools and techniques are described in Chapters 10 and 11, under the title of Hard Systems Methodology (HSM).

Although HSM is still popular and extensively used in problem solving, particularly in engineering, management, and planning, over the past 30 years it has been widely recognized that this methodology is inadequate for dealing with ill-structured problems of the real-world, where the means (or techniques) and the ends (or goals) are not clearly defined. Another criticism leveled against HSM is that it does not address the human component present in almost all managerial and engineering problems. The recognition of complexity present in all but the simplest systems is another factor that HSM is not able to cope with. These and other weaknesses became progressively obvious to practitioners (especially planners and managers dealing with societal problems) who felt that changes to HSM were necessary. Considering that HSM is deterministic and/or stochastic in nature, with the principal goal of reaching an optimal solution, these criticisms did not come as a big surprise. Naturally, over the years, systems thinking has undergone a major transformation, resulting in the development of a range of methodologies for use in engineering, management, and planning. The decision-making process has been greatly enriched by this transformation (Jackson, 2000; Midgley, 2000; Rosenhead and Mingers, 2001).

The purpose of this chapter is to broaden the view of those readers interested in learning about other paradigms besides the hard systems methodologies described in Chapters 10 and 11. Considering HSM as the first wave of systems thinking, this chapter describes three other paradigms of inquiry that have evolved: (1) soft systems thinking; (2) critical systems thinking; and (3) multimodal systems thinking. In Chapter 13 some selected case studies based on the paradigms spelled out in this chapter are described.

Before proceeding, we would like to draw the attention of the reader to the introductory remarks we made in Chapter 10 regarding methodology, methods, and techniques. To put these terms in perspective, methodology concerns itself with the study of the principles of the use of methods, in the sense that it sets out to describe and question the methods that may be employed in a particular activity under investigation. Methodology is, therefore, a higher-order term than methods, procedures, models, tools, and techniques (Jackson, 2000).

12.2 SYSTEMS THINKING

12.2.1 The Nature of Systems

The term *system* has been in use for such a long time that most people have lost the significance of the word. Systems have come to be associated with anything and everything, from anthropology to zoology, and from communication to computation. With such a wide range of associations, it is best to start by defining the word. In an intuitive sense a system is an entity that maintains its existence and functions as a whole through the interactions of its parts. While there are differing definitions of a system, the one common strand through all these definitions is that it is a set of elements, relationships and procedures to achieve a specific purpose. A system may be a part of another system and at the same time contain several smaller systems. In Chapter 1 we described the hierarchy of systems to be found in the world, ranging from simple static structures to living sociocultural systems. We also explained the connections between systems thinking and how it could lead to the development of systems theory. Systems thinking is a *framework of thought* that helps us to deal with complex things in a holistic way. Systems thinking looks at the *whole* and the *parts*, as well as the *connections* between the parts, examining the whole in order to understand the parts is the essence of systems thinking. It is thus the opposite of reductionism (the idea that the whole is simply the sum of its parts). A collection of parts that do not connect is not a system; it is simply a *heap* or a pile of things.

It will be evident to the reader that there are a number of ways of classifying systems. We can have living and nonliving systems, hard and soft systems, simple and complex systems and the list can go on and on. Systems thinking is also associated with two ways of looking at systems—*systematic* and *systemic*. Here, systematic refers to the systematic (or sequential) procedures adopted to achieve a predetermined objective, while systemic refers to holistic thinking about the system or the problem situation embedded in the system.

The difference between simple and complex systems is interesting. It is the *relationships* and the *interactions* among parts that makes it a simple or a complex system and not just the number of parts in a system. For example, an aircraft engine may look complex to a layman, but an aeronautical engineer would classify it as a simple system, although it looks complicated. Similarly, some *Sudoku* puzzles may look complex to beginners, but this complexity is a matter of detail; through trial and error, a unique solution can be found. In the case of complex systems, however, the parts of the system can relate and interact with each other in a myriad of ways, many of which may be unknown. Complex systems, therefore, cannot be predicted like simple systems because we cannot tell how the parts will connect and how these connections will affect the whole. Two players playing a game of chess is a good example of a complex system. How the move of one player will affect her opponent is not known. Further details regarding the nature of systems are given throughout this chapter.

12.2.2 System of Systems Thinking

When scientific knowledge advances, to the extent that there is a discrepancy between theory and practice, there is a paradigm shift, according to the eminent scientific historian Thomas Kuhn (1962). Such paradigm shifts have also occurred with systems thinking. The four paradigms of systems thinking described in this chapter are:

1. Hard systems thinking (HST)
2. Soft systems thinking (SST)
3. Critical systems thinking (CST)
4. Multimodal systems thinking (MST)

Chapters 10 and 11 have already dealt with HST, but for the sake of a more complete perspective, we will take another look at this paradigm.

HST or functionalist approaches: Though there is a wide diversity in the techniques embraced by HST, they all have certain common characteristics. First, they are essentially goal-seeking strategies using quantitative tools for achieving an optimal or near-optimal solution. Second, they need a clear definition of ends and the optimal means for achieving those ends. This characteristic is a handicap when a messy and complex situation has to be dealt with, which is inevitable in nearly all engineering and planning projects. And third, they are best suited for tackling problems that don't involve human activity systems.

SST or interpretive approaches: This is a form of systemic thinking that understands reality as the creative thinking of human beings. It takes into consideration social reality as the construction of people's interpretation of their experiences and works with the aspirations of people's views, interpretations, and intentions. Although there are quite a number of soft systems methodologies that have been employed since the 1970s, we describe four that have been extensively used: (1) Ackoff's Interpretive Planning; (2) Checkland's Soft Systems Methodology

(SSM); (3) Senge's Fifth Discipline; and (4) the Strategic Options Development and Analysis (SODA).

CST or emancipatory approaches: While many practitioners have hung on to and made good use of both HST and SST, it became obvious to practitioners that emancipatory interests for dealing with inequalities, such as power and economic differences in society, were not being adequately considered by SST. As a result, CST emerged in the 1990s to address these inequalities. Werner Ulrich, a Swiss planner inspired by Churchman, made a breakthrough by operationally addressing this problem (Jackson, 2000).

MST: The most recent addition to the family of systems thinking is MST, and it has recently been adopted in Europe. Developed by de Raadt (1995) and his colleagues in Sweden, MST uses as many as 15 performance indicators for questioning the validity of decisions made by planners and policy-makers. Many of these performance indicators cover issues of sustainability, environmental, and ethical issues (Eriksson, 2003).

12.3 HARD SYSTEMS THINKING

12.3.1 Preamble

HST concentrates on objectivity. From an engineering, management and planning perspective, these approaches aim to gain control over a situation, using professional knowledge to achieve that goal. It is assumed that the relationship between cause and effect can be determined, since constraints are firm and goals are unambiguous (Rosenhead and Mingers, 2001). Hard systems methodologies can be conveniently classified into three distinct classes: (1) systems analysis; (2) systems engineering; and (3) OR. They all have certain commonalities that are not difficult to detect.

12.3.2 Systems Analysis

Systems analysis is defined as an approach for systematically examining the costs, effectiveness and risks of alternative policies or strategies—and designing additional ones if those examined are found wanting (Jackson, 2000). The development of systems analysis is associated with the RAND Corporation, a nonprofit organization in the consulting business. Developed in response to wartime needs, this methodology gained prominence in the mid-1960s when it was adopted for cost-benefit analysis. Later, in the 1970s, the International Institute for Applied Systems Analysis was established to research possible areas of expanding the application of systems analysis. It came up with a seven-step procedure for effective systems analysis:

1. Formulating the problem
2. Identifying, designing, and screening alternatives
3. Building and using models for predicting consequences for selecting an alternative

4. Comparing and ranking alternatives
5. Evaluating the analysis
6. Decision and implementation
7. Evaluating the outcome

The successful application of systems analysis is contingent on a well-defined problem situation along with the availability of feasible alternatives from which to choose.

12.3.3 Systems Engineering

This branch of HST is defined as: *the set of activities that together lead to the creation of a complex, man-made entity and/or procedure, and the information flows associated with its operations* (Checkland, 1999). The following sequence is commonly used:

1. Problem definition
2. Choice of objectives
3. Systems synthesis
4. Systems analysis
5. Systems selection

12.3.4 Operations Research

OR was originally developed to tackle logistical problems encountered during World War II. A large variety of quantitative techniques have since been developed for use in the civilian sector of manufacturing, production, planning, and services. The step-by step procedure is:

1. Formulate the problem
2. Construct a mathematical model to represent the problem under study
3. Derive a solution from the model
4. Test the model and the corresponding solution derived
5. Establish controls over the solution
6. Implement the solution, provided a budget can be worked out

The three strands of the hard systems paradigm have greatly aided engineers, planners, and managers in problem solving. All three strands emphasize formulating an objective, to handle the problem situation successfully. Once the objective is identified, the appropriate solution can then be picked from a range of alternatives using a systematic approach. However, this means-ends approach runs into trouble when there is an element of complexity attached to the objectives and to the means themselves. The complexity in these cases can give rise to new issues and further complexity that needs to be resolved. The conventional techniques of problem solving traditionally have been unable to cope with such issues. These problems gave an impetus for the development of a new paradigm that would be able to deal with the complexity and plurality embedded in problems with multiple and unclear means and ends. This new focus resulted in the *soft systems paradigm*, along with a wide range of methodologies developed to tackle complex problem situations encountered in dealing with living systems.

12.4 SOFT SYSTEMS THINKING

12.4.1 Preamble

SST places its emphasis based on the views and inputs of people who are themselves involved in the process. In contrast to HST, its primary area of concern is perception, values, beliefs, and interests of stakeholders (attributes that are not taken into account by HSM unfortunately). It also recognizes that if managers are to work successfully, they have to be cognizant that multiple perceptions of reality need to be considered in dealing with problems situated in pluralistic environments. Although SST comes in a wide variety of forms, we will describe four of the more common ones used in practice. To set the stage for soft systems thinking, a few ideas on the path from optimization to learning are given next.

12.4.2 The Path from Optimization to Learning

Paradigms of inquiry: It is a well-established fact that while physicists and chemists perform laboratory experiments for investigating a phenomenon, the results of their experiments are repeatable and can be publically tested which is a characteristic of the scientific method. Indeed, this scientific method can be described as being based on at least three fundamental principles that characterize it and lend power and authority: (1) reductionism; (2) repeatability; and (3) refutation (Checkland, 1999). Scientists, in general, select a small portion of the world for investigation and carry out experiments over and over again until they are satisfied with their results. These results are then added to the body of existing knowledge. Thus, scientific knowledge is accumulated, unless a particular body of knowledge is refuted by another scientific experiment. This quality of replicability of experimental results stems from the fact that the phenomenon under investigation must be homogenous through time.

For a number of compelling reasons, the social sciences from the outset sought to copy the methods and practices established by scientists in the natural sciences, but unfortunately, this practice led to various problems. As is well-known, there is an ontological unity assumed in the scientific method in the sense that all objects in the universe—whether they be inert, living, conscious, or rational beings—are taken to be fundamentally and qualitatively the same. However, fortunately there has been a growing awareness that the social world is qualitatively different from the natural world, and, as a consequence, it was soon realized that the methodological assumptions of the scientific method were untenable when they were applied to social problems. Thus, it is not surprising that new systems methodologies, predominantly in real-world problem solving, were developed (Oliga, 1988).

System typologies: Checkland (1999) describes three system typologies: (1) natural, (2) physical, and (3) human activity. The first two fall in the category of hard systems, where the well-known methodologies of the scientific method and systems engineering have been and continue to be successfully applied. Essentially, the hard systems approach defines the objectives to be achieved, and then engineers the system to achieve them. The third system type, the human activity system, is usually messy, ill-defined, and cannot be described by its state, in which case the analyst must concede to its purposeful activity, human values, and nonphysical relationships. This is so because human activity systems can be expressed only as perceptions of people who attribute meaning to what they perceive. There is, therefore, no testable account of a human activity system, only possible accounts, all of which are valid according to a particular world-view (or *Weltanschauung*). In contrast to hard systems engineering, SSM does not seek to mechanically design a solution as much as it orchestrates a process of learning. We will elaborate on these differences in hard and soft systems later in this chapter.

Action research (AR): We have already seen that the scientific method that has been practiced by the natural scientists, based on testing hypothesis is extremely powerful. However, when scientists, engineers, planners, and managers began to apply the scientific method (and the hard systems methodologies that followed) to social and human systems, the results were generally unsatisfactory and, at times, a total disaster. As a consequence, the question that constantly cropped up was: If the scientific method and hard systems methodologies are not truly applicable to social and human activity systems because they are not homogenous through time, thus making replicability impossible, what else can be done? This was the context in which Action Research (AR) emerged. The roots of AR date back at least to the time of Kurt Levin (1890-1947), a psychologist who, with his band of researchers, became interested in human group dynamics, particularly from the point of view of bringing about change in society. They immersed themselves in human activity problems and pursued them till the problem unfolded through time. In essence, the principal objective of AR becomes the change process itself.

In an introductory chapter written in the *Handbook of Action Research* (2001), Peter Reason and Hilary Bradbury describe AR as a participatory, democratic process concerned with developing practical knowledge, grounded in a participatory worldview. It seeks to bring together action and reflection, theory and practice, in participation with others, and in finding practical solutions. A wider purpose is to increase the wellbeing of people and communities through economic, political, psychological, equitable, and environmental action within the ecology of the planet.

The crucial elements of the AR process are: (1) it is a collaborative process between researchers and people involved in the problem situation; (2) it is a process of critical inquiry; (3) it is a process that focuses on social practice; and (4) it is a deliberate process of reflective learning.

Checkland and Holwell (1998) have provided a set of rich pictures to draw the differences in the processes of the scientific method (implying hypothesis testing) and that of AR. Figure 12.1 shows the hypothesis-testing process (as part of the scientific method) applicable in

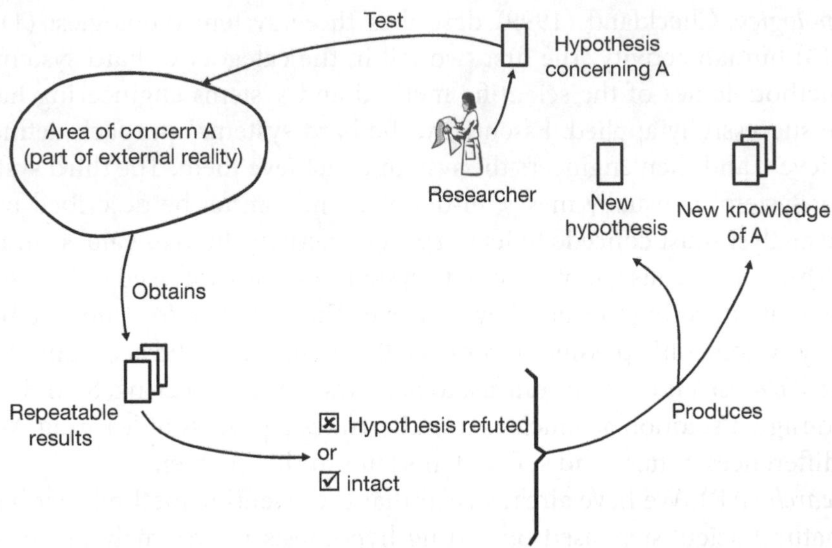

Figure 12.1 The hypothesis-testing research process of natural science (scientific method)

investigations in the natural sciences and HSM. This shows how a researcher wants to test a hypothesis about a perceived reality and produces a new hypothesis or new knowledge. In contrast, Figure 12.2, shows a generalized research or investigation process that can be adopted, irrespective of whether it's applied to natural, physical, or social systems. Here, F is a framework of ideas used in a methodology M to investigate an area of interest A. Using the methodology may teach us not only about A but also about the adequacy of F and M. However, we have to modify F, M, and even A when we are considering AR, because unlike systems engineering (that is *systematic*) we are now dealing with *systemic* thinking, and the *learning* process of a new M (or soft systems thinking), based on a systemic F that is then used for a new attack on the original A. This emphasis on the change in F, M, and A in which the investigator or researcher becomes immersed and involved in the flux of the real-world problem situation is probably the hallmark of AR. Figure 12.3 shows how the cycle of AR in human activity systems works. Many of the concepts will become clearer after reading Checkland's soft systems methodology (SSM) given in the next section.

12.4.3 Checkland's Soft Systems Methodology (SSM)

SSM emerged as a result of dissatisfaction with the limitation of traditional HSM, in that the original intention of a holistic, interdisciplinary, experiential discipline of addressing problems occurring in human activity systems was being betrayed. Partially as a result of this feeling, Peter Checkland and his colleagues at Lancaster University in the United Kingdom began an AR program designed to extend systems ideas to ill-structured management problems. Their aim was to produce a systems methodology capable of dealing with problems that had a human content (Jackson, 2000).

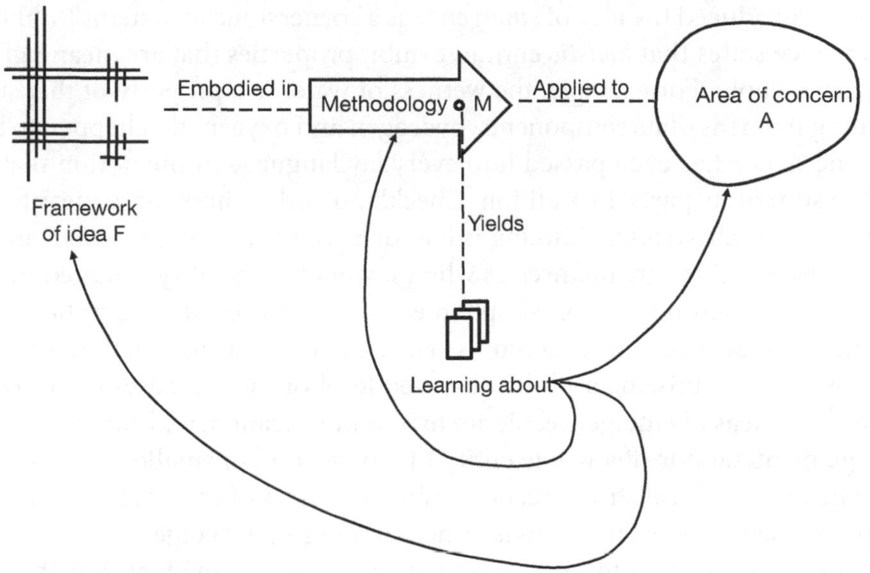

Figure 12.2 Elements relevant to any piece of research

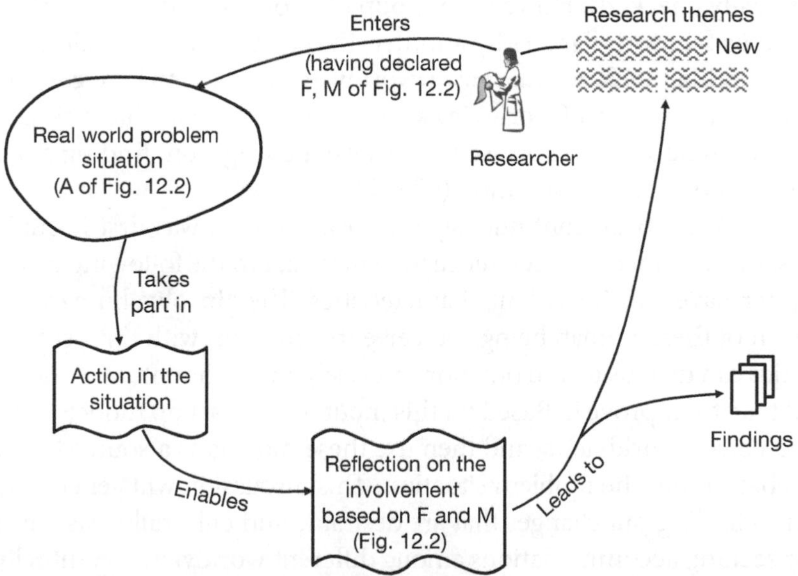

Figure 12.3 The cycle of action research in human situations

Checkland introduced the idea of *emergence* as a cornerstone in systems thinking. The principle of emergence states that *holistic* entities exhibit properties that are meaningful only when attributed to the whole. For example, the wetness of water is a property of that substance that has no meaning in terms of its components, hydrogen and oxygen, that happen to be gases. This concept of emergence has even passed into everyday language in the notion that the whole is more than the sum of its parts. In addition, Checkland added three more ideas to assemble the core concepts on which systems thinking is based: *hierarchy*, *communication*, and *control*. Hierarchy means that entities are themselves wholes, where each entity is nested inside the other (like Russian dolls); communication simply means the transfer of information. In the formal systems models, the decision-making process ensures that controlled action is taken in light of the system's purpose or mission and the observed level of the measures of performance. Collectively, then, the ideas of emergence, hierarchy, communication, and control provide the basic systems image or metaphor of a whole entity that may contain smaller wholes or be part of a larger whole in a hierarchical structure, possessing processes of communication and control, as well as adapting itself to strive in an environment undergoing change.

In applying systems ideas to human activity systems, it is evident that the most difficult part is the learning component. For example, in the case of natural or physical systems, accounts of real-world manifestations are publicly testable (if it is said that a car has four wheels, this fact can be easily checked). However, a group of people could be described as terrorists by one observer and as freedom fighters by another. Every observation could be considered valid according to a particular *Weltanschauung* or worldview. Indeed, the concept of human activity systems is to consider a variety of worldviews while looking at a problem situation. The essence of SSM, therefore, can be simply expressed as a way of getting from finding out about a problem situation to taking action in that situation (Checkland, 1999).

Checkland's SSM has been continuously improved since it was first formulated in the early 1970s. It can best be described with reference to Figure 12.4 in the following way. Almost all problems we encounter have two important characteristics. They involve human beings who have a purpose, and each of these human beings perceive the problem with different worldviews. These two characteristics can help us to find out more precisely about the nature of the problem and what steps we should take to improve it. Based on this input, we can set up models of purposeful action reflecting the perceived worldviews, and then use these models as a source to help us structure a discussion or debate about the problem situation. This discussion will generate further questions that would lead to finding out changes that are desirable and culturally feasible. The idea is to set up a process for seeking accommodations among different worldviews, eventually leading to possibly a worldview for improving the situation that different people would be willing to accept. This process that has just been described constitutes a *learning cycle*. It could also be looked on as a process of *group learning*. The best results can be expected when this learning cycle is carried out by people in the group and not left to outside experts, although such experts or professionals (engineers, planners, etc.) can help in the process, without interfering or coercing. The changes and/

or improvements recommended by the group can now be implemented, unless further discussion and debate should be necessary, in which case the group can go back and revisit the steps again. In this sense SSM's learning cycle can be seen as never-ending. It offers a way of continuously managing any complex human problem situation by encouraging the understanding and cognizance of multiple perspectives, thus bringing rigor to the processes of analysis, debate, and *taking action to improve* (Checkland and Poulter, 2008).

It is obvious that SSM developed because the traditional methods of systems engineering, based on defining goals and objectives, simply did not work when applied to messy, ill-structured, real-world problems, particularly related to human activity systems. Over the years Checkland has crystallized SSM, and his general framework is illustrated in Figure 12.5. Regarded as a whole, Checkland's SSM is a learning tool that uses the system's ideas to organize four basic mental processes: Perceiving (Stages 1 and 2), predicting (Stages 3 and 4), comparing (Stage 5), and determining needed changes and actions (Stages 6 and 7). The output and

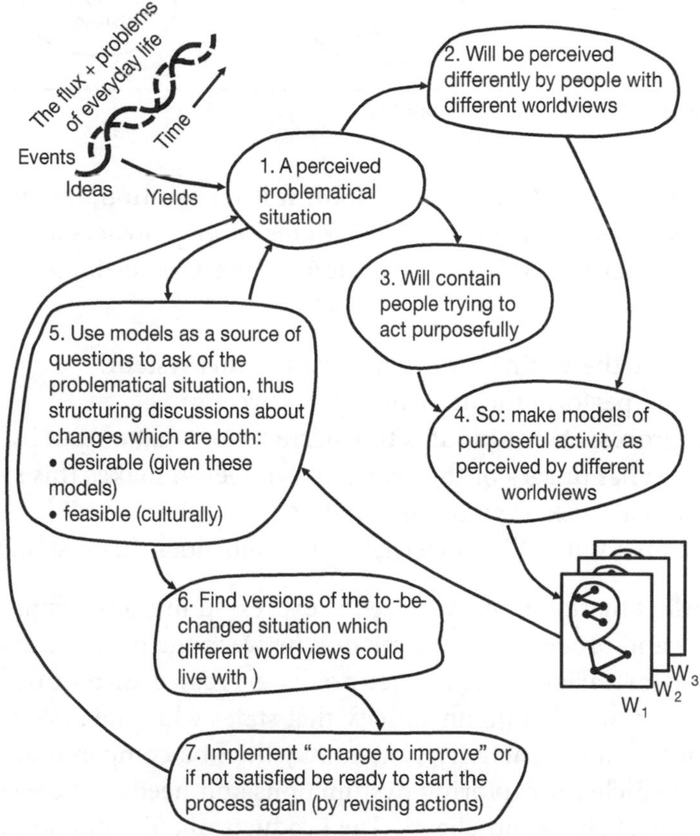

Figure 12.4 Cycle of learning for action

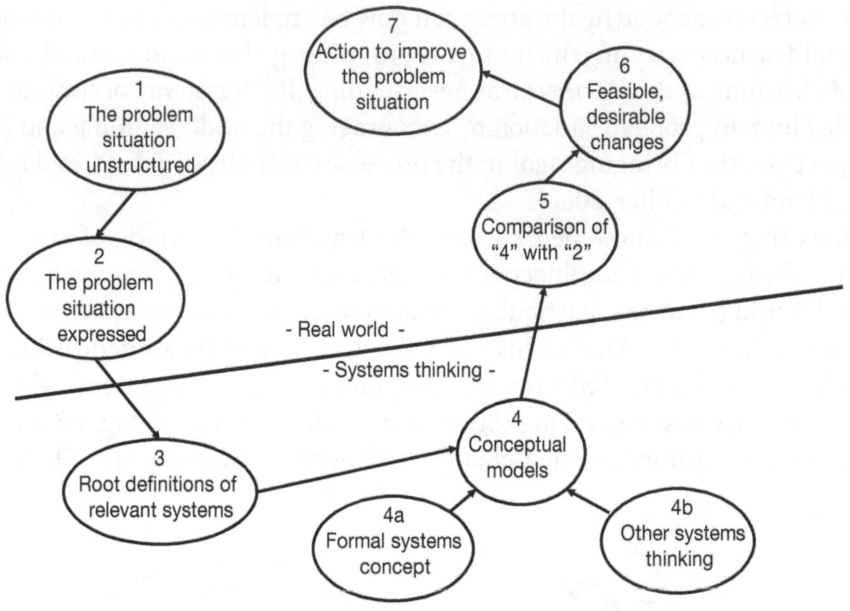

Figure 12.5 Checkland's soft systems methodology

utility of SSM is quite different from that of HSM. It is recognition, learning, and insight. Six elements—customers, actors, transformation, *Weltanshauung*, owners, and the environmental constraints—help to structure SSM. The definitions of these six elements, that form the mnemonic CATWOE are:

> **C**ustomers: Who are the victims or beneficiaries of this system?
> **A**ctors: Who would perform the activities of this system?
> **T**ransformation process: What input is transferred into what output by this system?
> ***W****eltanschauung*: What images of this world (worldviews) makes this system meaningful?
> **O**wner: Who could abolish or stop this activity?
> **E**nvironmental constraints: What external constraints does this system take as given?

In SSM the real-world situation to be analyzed is expressed in nonsystems language using the concepts of structure and process, plus the relation between the two. This constitutes a relevant system and encapsulates various specific viewpoints expressed in root definitions (RDs). An RD is a concise description of a human activity that states what the system is. From the RD a conceptual model of the human activity system may then be compared to the real world. The model is the formal vehicle for exploring dysfunctions and needed changes in the real world, involving both systems analysts and clients. The products of SSM should provide the basis for needed changes and such changes can fall into three categories: structural changes, procedural changes, and attitudinal changes. The entire process is done interactively between clients and key informants.

A comparison of HSM with Checkland's SSM is appropriate at this stage and this is shown in Table 12.1. The main difference is as follows: HSM considers goal-seeking to be an adequate model of human behavior and relies heavily on the language of problems and solutions to deal with problems. SSM does not consider goal-seeking to be an adequate model for representing and solving human activity problems. SSM is therefore relevant to arguing and debating about real-world problems, and not merely models of the world. And this posture leads to learning and not just to optimizing. This results in the language of issues and accommodations rather than mere solutions. SSM is also a process of managing, where managing is interpreted very broadly as a process of achieving organized action. It is evident that in most real-world problematic situations the applications of a combination of both HSM and SSM would be appropriate to structure a debate in which a spectrum of needs, interests and values could be teased out and discussed (Khisty, 1995). Such a combination embraces the best attributes of two methodologies. Similar combinations can be made with other methodologies given in this chapter.

12.4.4 Ackoff's Interactive Planning

Spanning over a period of more than 40 years, Russell Ackoff's work has made a significant impact on many fields in management and organization. Only the very basic outline of his work is provided here and a good reference of his classic writings on management is *Ackoff's Best*

Table 12.1 Hard systems versus soft systems methodologies

Attribute	HSM	SSM
Orientation	Systematic goal-seeking	Systemic learning
Roots	Simplicity paradigm	Complexity paradigm
Beliefs	Systems can be engineered Models are of the world Closure is needed Finding solutions to problems	System can be explored Models are intellectual constructs Inquiry is neverending Finding accommodation to issues
Human content	Nonexistent	High
Questions	How?	What and how?
Suitability	Well-structured problems	Ill-structured problems
Advantages	Uses powerful methods but needs professionals to run the programs	Available to owners and practitioners
Disadvantages	Not transparent to the public	Transparent and understood by public
Principles	Reductionism	Participants part of the research inquiry
	Replicability	Allows reflective learning
	Refutation possible	Process is recoverable
	Results homogenous through time	Results may not be homogenous through time

(1999). Readers working their way through Chapter 2 will have come across Ackoff's strategies for tackling problems encountered in their professional life: absolving, resolving, optimizing, or dissolving.

Ackoff offers three operating principles for interactive planning:

Participative implies that the process of planning is more important than the plan itself and that the process needs to be enriched by the participation of those who will be affected by the plan. Effective planning cannot be done to or for an organization; it must be done by it. Therefore, the role of the planner is not to plan for others, but to encourage participants and stakeholders to plan for themselves.

Continuous implies that the planning must account for any unexpected changes. For example, we carry a spare tire in our cars because we assume we could have a flat tire, although we can't predict when we may have one. Organizations and their environments change continually over time and such change needs to be monitored continuously.

Holistic has two parts—coordination and integration. The principle of coordination asserts that all parts of an organization at the same level should be planned for simultaneously and interdependently, while the principle of integration asserts that planning done independently at any level of an organization cannot be effective, unless all levels are planned for simultaneously and interdependently. When coordination and integration are combined the holistic principle is achieved.

Ackoff also spells out five phases of Interactive Planning as:

- *Formulation* of a messy situation determines what problems and opportunities face the organization planned for, how they react, and what obstructs or constrains the organization doing something to correct them. The output of this phase takes the form of a *reference scenario*.
- *Ends planning* determines what is wanted by means of an idealized redesign of the system planned for. Goals, objectives, and ideals are extracted from this design. Comparison of the reference scenario and the idealized redesign identifies the gaps to be closed or narrowed by the planning process.
- *Means planning* entails closing or narrowing the gaps by selecting or inventing appropriate courses of action, practices, projects, programs, and policies.
- *Resource planning* assesses what types of resources will be required and when, and how they will be made available.
- *Implementation and control* determines who is to do what, when it is to be done, and how the assignments and schedules will be executed to produce effective performance. These five phases of Interactive Planning can usually be taken up simultaneously according to the convenience of the project (Ackoff, 1999).

12.4.5 Senge's Fifth Discipline

Peter Senge studied system dynamics at MIT, developing his own style of systems thinking. He brought systems thinking to the forefront of management and leadership studies through his influential book *The Fifth Discipline* (Doubleday, 1990). Senge applies his ideas in the context of organizational learning, where systems thinking is the fifth of five learning disciplines: personal mastery, mental models, shared vision, team learning, and systems thinking. A brief account of each of his five disciplines are:

1. *Personal mastery* means learning to develop one's own proficiency as a life-long journey, without dominating others. It creates an organizational environment that encourages all its members to develop goals and purposes they wish to choose and adopt.
2. *Mental models* are conceptual structures in the mind that drive cognitive processes of understanding. They influence people's actions because they mold people's appreciation of what they see. They also help us to reflect on, clarify, and improve our internal pictures of the world.
3. *Shared vision* builds a sense of commitment in a group by developing shared images of the future we seek to create, and the practices by which we hope to get there. While personal vision is one's individual ideas, shared vision is a collective commitment.
4. *Team learning* transforms conversational and collective thinking skills, so that groups can reliably develop the intelligence and the ability that is greater than the sum of individual talents. Team learning creates a resonance that helps the team to succeed in achieving the goals that it sets.
5. *Systems thinking* is a way of thinking about a problem, and a language for describing the understanding, the forces, and the interrelationships that shape the behavior of systems. It helps to change systems more effectively, and to act more in tune with larger processes of the natural and economic world.

Senge has developed a pragmatic style of systems thinking that has become popular over the last fifty years. He has identified eleven *laws of the fifth discipline* that are integral to developing learning organizations. Most of them point to the counterintuitive behavior of complex systems. His eleven laws are:

1. Today's problems come from yesterday's solution. Most solutions merely shift the problem from one part of a system to another.
2. The harder you push, the harder the system pushes back, because in living systems we generally don't understand the feedback mechanism.
3. Behavior grows better before it grows worse. One must therefore look at both the costs and benefits in the short as well as in the long term.
4. The easy way out leads back in. Using systemic thinking is essential for getting the total picture.

5. The cure may be worse than the disease.
6. Faster is slower, because systems tend to have optimal rates of growth and, when growth becomes excessive, problems arise.
7. Cause and effect are not closely related in time and space, particularly in complex systems.
8. Small changes can produce big results, but the areas of highest leverage are often the least obvious, because small changes can sometimes bring large long-term benefits. They can also create long-term disbenefits.
9. You can have your cake and eat it too, but not at once.
10. Dividing an elephant in half does not produce two small elephants, because living systems have integrity. Besides, their character depends on the whole. This applies to organizations too.
11. There is no blame. Systems thinking shows us that there is no outside and no inside of a problem, because you and your problem are part of a single solution. The cure lies in your relationship with your enemies.

According to Senge, complexity can be of two types: (1) detail complexity consisting of many variables and (2) dynamic complexity where cause and effect are not close in time and space, and therefore obvious interventions do not produce expected outcomes. As a result, Senge developed a language of systems thinking that describes complicated interrelationships and patterns of change in a simple way, relying on three concepts that can be seen as the building blocks of systems thinking. These concepts are: *reinforcing* or *amplifying feedback* (e.g., snowballing effect as it rolls downhill); *balancing* or *stabilizing feedback* (e.g., where a system is brought back to its original goal through a self-correcting mechanism); and *delays* that usually go undetected. Reinforcing and balancing feedback are to be found in nearly every organization, but usually go undetected. Delays occur when the impact of feedback processes take longer to come through than expected.

Four examples, describing Senge's method, are:

Balancing process: Tom gets into the shower at his hotel, which has a sluggish faucet (a fact he doesn't know about). He starts to adjust the water temperature to his liking, but because of the delayed feedback, he finds that he is adjusting and readjusting the faucet to correct his settings. This balancing process takes time and creates some frustration, all because of the delay in the feedback process. Examples of this kind occur all the time, such as the time-lag between ordering materials for construction and receiving them. Figure 12. 6a shows the balancing process.

Limits to growth: Mike, a highly motivated long-distance runner, makes rapid progress toward increased speed and endurance in the first few months of his effort. However, this initial spiral of success creates unforeseen adverse effects, manifested in his physical health. He is forced to slow down his effort to maintain whatever he has achieved thus far and achieve some kind of balance, as shown in Figure 12. 6b. There are many such examples in the real world where there is a limit to the growth process.

Shifting the burden: It has been found that in many cases short term solutions not only fail to solve the problem but make it worse and also make it difficult to eliminate the fundamental problem. In our daily lives there are cases when we impatiently resort to a cheap temporary fix in home improvement projects, instead of doing a thorough investigation and then spending a little more money to rely on a permanent repair. Alcohol, drugs, and tranquilizers are well-known fixes of drowning our problems, but the problem only becomes worse, resulting in serious health issues. For example, Bob, a night-shift factory worker, suffers from stress at work. He has several options to improve his situation, by reducing his work-load or changing to a day-shift. However, these changes are not easy for him to make because it may take some time to do so, and there can be delays that may occur that are difficult to predict. He is not willing to wait and turns to drugs and alcohol to forget his problems. Bob's condition becomes progressively worse because shifting the burden to drugs results in his becoming an addict, a condition that is worse than his original problem. Figure 12.6c illustrates this case (Jackson, 2000).

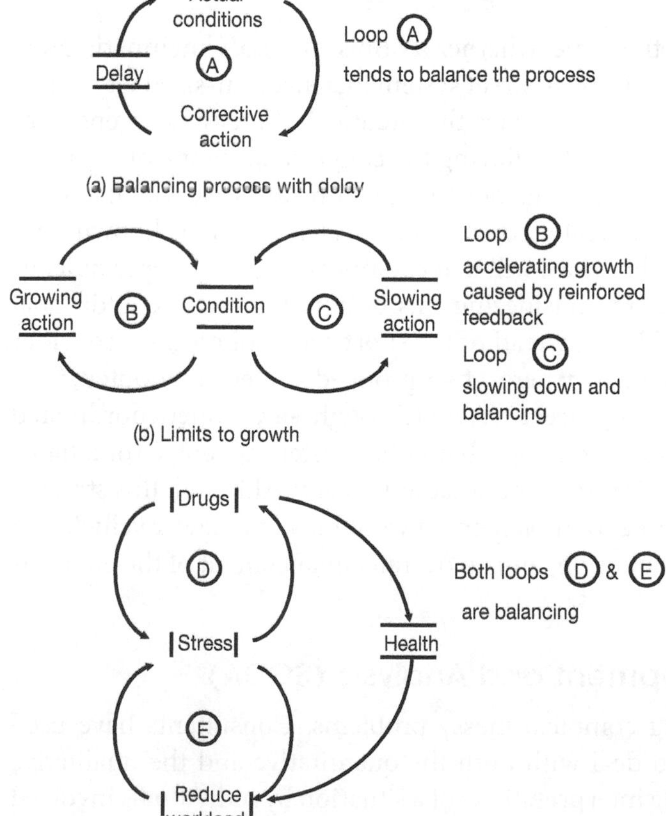

Figure 12.6 Three of the common stepping stone examples suggested by Senge: (a) balancing process with delay, (b) limits to growth, and (c) shifting the burden

Tragedy of the commons: Cases of the tragedy of the commons occur all the time. It begins in most instances with people benefitting individually by sharing a common resource, for example, a park or a freeway or a shopping mall. But sooner or later the amount of activity grows too large for the commons to support, and the facility deteriorates or breaks down. In essence, the value of a public facility decreases for everybody when it is overused.

This tragedy has been studied carefully by planners for the last two or three decades in the case of transportation systems. Our huge highway system, for instance, was put together with the objective of providing higher mobility coupled with increased accessibility through a well designed system of freeways, arterials, and local streets. But whenever there is congestion on this system we tend to add lanes and expand this extensive highway system still further, and in doing so we endanger the quality of life and the ecological sustainability of modern society. Furthermore (and ironically), this very expansion results in more traffic congestion, drastically reducing mobility and accessibility, thereby lowering business productivity, increasing fuel consumption, increasing pollution, and adversely affecting safety, and the list goes on and on. It is estimated that all these adverse effects cost a staggering amount of about $200 billion per year (Khisty and Zeitler, 2001).

All across North America the story is the same, whether it's Boise, Idaho, Cincinnati, Ohio, or Los Angeles, California. Let's take a case of a busy street system in a medium-sized city where there is severe traffic congestion. Figure 12.6d illustrates the situation where the city engineer considers the short-term strategy, for dealing with relieving the congestion on one of its major streets, by adding lanes. If we look at Loop 1, and consider adding more lanes, we will, in course of time, end up with perhaps the same if not greater congestion, because it's a well-known phenomenon that adding lanes to a congested highway, will make it more congested pretty quickly. This phenomenon is indicated in Loop 2. In considering this option, there is the additional environmental degradation that is inevitable. If instead of the short-term solution we consider the long-term solution as shown in Figure 12.6e, where a two-pronged strategy is adopted, first, by improving the traffic capacity of the existing street system (through, say, proper coordinated signals, etc.), (see Loop 1), and second, by introducing a better bus system coverage (or a light-rail system) (see Loop 3), we will be able to improve the situation greatly. Although this strategy would undoubtedly take a little bit more time to implement, it would result in fewer vehicles on the street system, less congestion, less air pollution, and an overall enhancement of the environment (see Loop 4).

12.4.6 Strategic Options Development and Analysis (SODA)

SODA is a method designed for tackling complex, messy problems. Consultants have used it with much success because it is able to deal with both the quantitative and the qualitative aspects of a problem. SODA focuses on the interpretations of a situation by individuals involved with the problem. Personal constructs are elicited and drawn by means of cognitive maps, in a

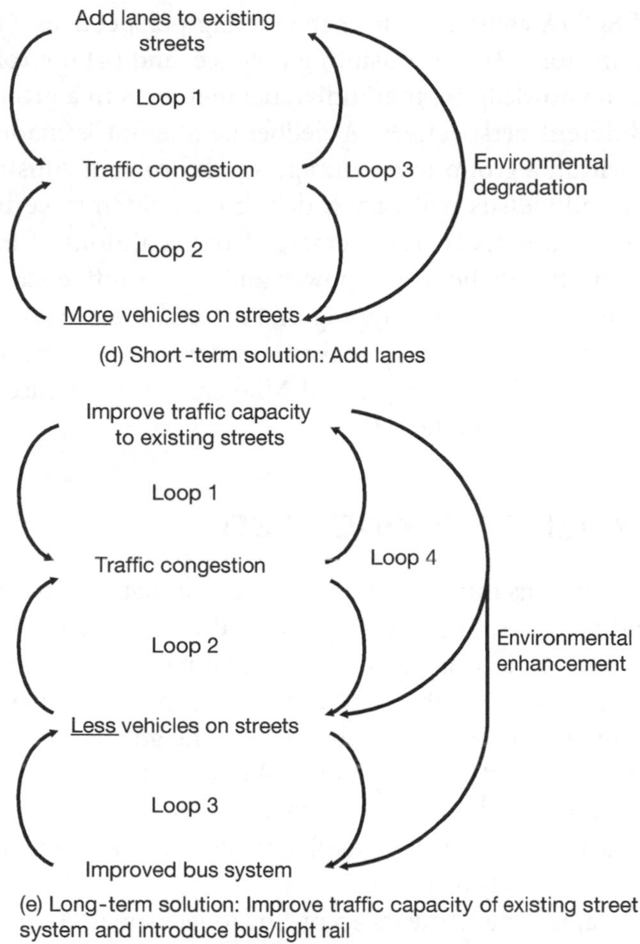

Figure 12.6 (d) Short-term and (e) Long-term solutions

participatory fashion during interviews. The objective of this model is to facilitate a consensus and commitment among members toward common goals and issues.

A consultant is likely to resort to a SODA approach *only* if some or all of the following conditions prevail: First, the consultant is personally interested in the practical aspects of social psychology and cognitive psychology, that is, in being explicit and reflective about managing a social process and in analyzing the potential tensions among members of the problem solving team. Second, the consultant is able to relate personally to a small number of significant persons in the group or organization. Third, the consultant is able to work on problems, proceeding from broad concepts to specific commitments. And finally, a consultant is more interested in designing and managing problem solving workshops rather than in researching the characteristics of the problem.

The framework of SODA consists of four interacting perspectives: (1) the individual, (2) the nature of the organization, (3) the consulting practice, and (4) the role of technology. The approach is subjective, acknowledging that individual members in a group will likely view the same problem from different perspectives. A deliberate attempt is made to include different roles and experiences within a group to encourage a rich problem construct. The methods of communication among individuals will vary widely from written to verbal to body language. Organizational direction and goals typically emerge from coalitions of individuals within the organization, particularly those who wield power and try to influence the decision-making process. SODA makes use of cognitive mapping that records the process of exploring options. Visualizing and drawing the issues simply makes it easier to see the whole picture for group decision-making (Jackson, 2000; Rosenhead and Mingers, 2001). A case study describing the application of SODA is given in Chapter 13.

12.5 CRITICAL SYSTEMS THINKING (CST)

The idea behind critical systems thinking (CST) is to encourage professionals, such as administrators, managers, and planners, to get closely involved with citizens while making decisions. Citizen involvement is therefore considered an essential part of the decision process, so that citizens get first-hand knowledge of the implications of a proposed plan of action for, say, a new bridge or improvements to a water supply project. The goal is to present the plan in such a way that citizens can gain reflective competence about what the professionals are proposing and to participate in the planning process as it proceeds toward planning and implementation. Although this kind of participation is also possible in SSM, citizen participation is not emphasized, nor is it done as exhaustively as in CST.

CST embraces five major commitments according to Jackson (2000): The first commitment is critical awareness that comes from closely examining the assumptions and values of the design proposed. The second is social awareness that involves the recognition of social pressures and consequences of actions proposed. Third, and most importantly, is a dedication to human emancipation, seeking to achieve for all individuals the maximum development of their potential. The fourth commitment is the informed use of systems methodologies in practice. And the final commitment is the complementary and informed use of systems methodologies of all varieties of the systems approach.

Of the many systems thinkers who have contributed to the development of CST, the one who has done most to popularize it is Werner Ulrich, a student of C. West Churchman. His style of CST is termed *Critical Heuristics of Social Systems Design*. Ulrich (1998), in a working paper presented at the University of Lincolnshire in the United Kingdom, proposed that everybody should be encouraged to effectively participate collectively in decision-making on public projects, irrespective of status or rank. In addition, citizens and management need to understand each other's roles, responsibilities, and rights. CST aims to overcome the disadvantage citizens have of not having the knowledge-skills that experts have. CST particularly helps those citizens from poorer sectors of society who may not have the influence that the wealthier and more powerful sectors wield. The critical core of the systems thinking should remind us that of all our claims to knowledge, understanding, and rationality, the systemic nature of the system is all important.

In formulating CST, Ulrich draws on the philosophy of a well-known German philosopher, Jürgen Habermas (1984) concerning the nature of rationality. Habermas believes rationality is achieved through dialogue and free questioning and debate. For example, when decision makers communicate with the public they should be intelligible, they should make sure that what they are communicating is true, and they should be sincere without having any intention of deceiving anybody.

Ulrich has proposed a checklist of twelve questions framed for users of CST, that is, by those involved in a project, including administrators, managers, and citizens. They question the justification of a project's normative contents or dispute the underlying judgment that supports the project. Table 12.2 spells out the twelve questions. Notice that each one of them has to be answered both in the *is* and *ought-to-be* mode.

Table 12.2 Checklist of critically heuristic boundary questions

Sources of motivation

(1) Who is (ought to be) the *client*? That is, whose interests are (ought to be) observed?

(2) What is (ought to be) the *purpose*? That is, what are (ought to be) the consequences of the inquiry of design?

(3) What is (ought to be) the *measure of improvement*? That is, how can (should) we determine whether and in what way the consequences, taken together, constitute an improvement?

Sources of power

(4) Who is (ought to be) the *decision maker*? That is, who is (ought to be) in a position to change the measure of improvement?

(5) What *resources* and other conditions of success are (ought to be) controlled by the decision maker? That is, what conditions of success are (should be) controlled by the decision-making body?

(6) What conditions are (ought to be) part of the *decision environment*? That is, what conditions does (should) the decision maker not control (e.g., from the viewpoint of those not involved)?

Sources of knowledge

(7) Who is (ought to be) considered an *expert*? That is, who should be involved as researcher, planner, or consultant?

(8) What *expertise* is (ought to be) brought in? That is, what is (should) count as relevant knowledge or know-how, and what is (should be) its role?

(9) Who or what is (ought to be) assumed to be the *guarantor*? That is, where do (should) those involved seek some guarantee that their findings or proposals will be implemented and will secure improvement?

Sources of legitimation

(10) Who is (ought to be) *witness* to the interest of those affected but not involved in the inquiry or design process? That is, who argues (should argue) the case of those who cannot speak for themselves but may be concerned, including the handicapped, the unborn, and nonhuman nature?

(11) To what extent and in what way are those affected given (ought they be given) the chance of *emancipation* from the premises and promises of those involved? That is, how do we treat those who may be affected or concerned but cannot argue their interests?

(12) What *worldview* is (ought to be) determining? That is, what are (should be) the visions of improvement of both those involved and those affected, and how do (should) we deal with differing visions?

12.6 MULTIMODAL SYSTEMS THINKING

MST is the most recent contribution to the family of systems thinking. It was developed by J. D. R. de Raadt in the 1990s based on the philosophy of Herman Dooyeweerd, As many as 15 performance measures or modalities are utilized in this analysis: numerical, spatial, kinematic, physical, biotic, sensitive, logical, historical, informatory, social, economic, aesthetic, juridical, ethical, and creedal. These indicators along with corresponding sample questions are given in Table 12.3. Figure 12.7 shows these performance indicators. Notice that as each performance indicator forms the foundation for the one immediately above it, all of them are bonded together as a whole in the human experience and can be both qualitatively and quantitatively measured. An important point to note is that those indicators, such as spatial and numeric, (toward the bottom of Figure 12.7) are less normative than those at the top, such as ethical and juridical (Khisty, 2006).

The sample questions suggested in Table 12.3 illustrate how MST can be used as a framework for critically examining performance indicators with input from citizens involved with a project. Naturally, such involvement with citizens will evoke further questions from citizens affected critically by the project under scrutiny.

Table 12.3 Multimodal systems thinking

Performance indicator	Clarifying sample questions
Creedal	Are the right things being done in the short and long term?
Ethical	Are the planning and implementation morally correct and ethical?
Juridical	Are the planning and implementation just and fair?
Aesthetic	Is the plan aesthetically satisfying?
Economic	Are the resources being used optimally?
Social	Are the social needs of the people respected and accounted for?
Informatory	Have all jurisdictions been dully represented and considered?
Historical	Have lessons from the past (good and bad) been fully considered?
Logical	Are all the results of the models used reliable, logical, and realistic?
Sensitive	Is the system (and subsystems) sufficiently robust to handle changes?
Biotic	Have the concepts of sustainability (for air, water, soil, etc.) been taken care of?
Physical	Has the system taken the best advantage of the topography and soil?
Kinematic	Has the movement of people, vehicles, and goods been designed for safety, comfort, convenience, economy, and sustainability?
Spatial	Have the land-use pattern and distribution of activities been designed for the health, safety, and convenience of the people?
Numerical	Has the quantitative analysis been done using the best methods available and based on the most reliable data collected?

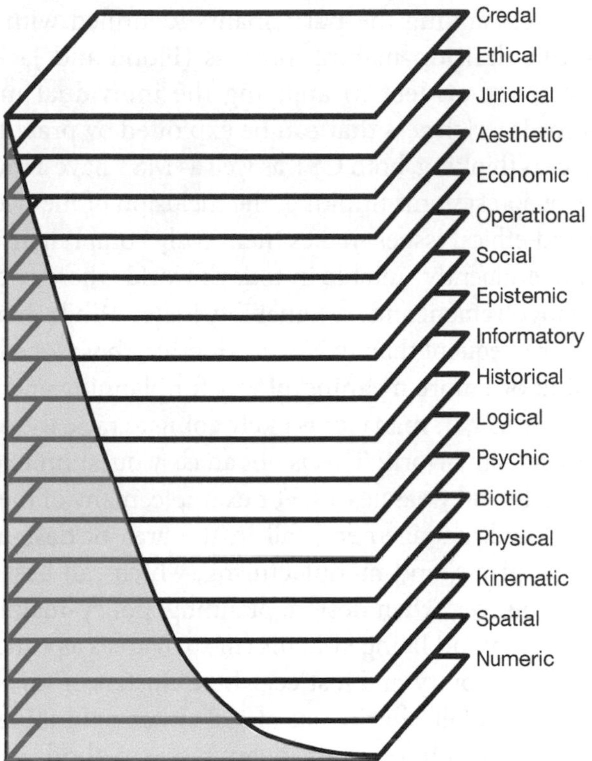

Figure 12.7 Performance indicators used in multimodal methodology

12.7 SUMMARY

Readers who have worked through this chapter will have discovered that it covers a vast terrain. It describes the latest developments in systems thinking over the last 30 years for dealing with complex planning, management, and organizational problems of today. These developments have resulted in forms of inquiry and analysis that are different from the HST that engineers and managers adopted after World War II and that are still being used to this day.

In essence, HST is based on the assumption that an efficient means of achieving a known and defined end is evident, and that once this end is specified the problem can be formulated, modeled, and worked out quantitatively. HST assumes that an orderly world exists where its methods and techniques can be employed, through rational procedures, to optimize an objective function (Flood and Jackson, 1991).

In contrast, SST seeks to explore working with the diverse perceptions of observers, having multiple views of a problem situation that can be examined and debated. Where necessary, it is assumed that the help of experts can be sought. The aim is to include the views of problem-owners and other stakeholders for finding possible ways to move forward. This involvement encourages a learning process among the participants identified with the problem situation, and leads to a collaborative decision-making process (Flood and Jackson, 1991). Although, contrasts and congruencies are evident in applying the individual methods in soft systems thinking, the diversity affords a richness that can be exploited by practitioners.

Compared to soft systems thinking, both CST as well as MST have a much wider agenda. In addition to all the attributes of soft systems thinking, the inclusion of the two important dimensions of human emancipation and ethical issues makes them truly comprehensive and gives them their greatest strengths. They are eminently suitable in today's world where engineers and planners are heavily involved in citizen involvement and sustainability issues. While their applications in Europe have been far wider and more frequent than in North America, those who have used CST and MST claim that they cannot think of a more meaningful way for planning and management.

The systems engineer, planner, or manager is likely going to raise the obvious question: Which style of thinking should I use in my work? This is not an easy question to answer in the sense that each style has its own strengths and weaknesses. For example, many of the problems encountered, particularly in engineering and management, fall in the area of designing physical systems—bridges, highways, chemical plants, and manufacturing, which can best be done, at least in the initial stages, using HSM. However, when design, planning, policy-making, and implementation have to deal with human activity and living systems (in such areas as education, safety, rehabilitation, public involvement, sustainability, and justice, where clusters of world views and perceptions are needed), the choice obviously falls for the use of soft systems thinking, critical systems thinking, and multimodal thinking, in addition to the quantitative methods of hard systems thinking.

There appears no alternative but to resort to as wide a range of systems thinking styles as is possible for coping with problems arising in practice. The spirit of this dilemma is captured most appropriately by Donald Schön (1990). He reminds us that most problems, once they are identified, must be properly formulated to make sense of their complexity and messy situation before thinking of how to tackle them. He illustrates this dilemma by comparing the problem solving process to represent a practitioner sitting on high ground overlooking a swamp. The high ground represents simple, technical problems where technical rationality and rigorous quantitative methods can be applied, very much in line with what HST offers. On the other hand, the swamp below is where the messy, complex, and difficult problems lie that are highly important, relevant, and that involve living systems. The practitioner has two options: (1) remaining on the high ground and solving the problem technically, quantitatively, and rigorously, using HSM or (2) descending to the swamp where the real problems lie, but where HSM does not apply. The swamp is where the wicked problems reside.

REFERENCES

Ackoff, R. L. (1999). *Ackoff's Best: His Classic Writings on Management*, New York, NY: John Wiley and Sons.

Checkland, P. B. (1999). *Systems Thinking Systems Practice*, Chichester, UK: John Wiley and Sons.

Checkland, P. B. and S. Holwel (1998). "Action Research: Its Nature and Validity," *Systemic Practice and Action Research*, 11:1, 9-21.

Checkland, P. B., and J. Poulter (2008). *Learning for Action*, Chichester, UK: John Wiley and Sons.

de Raadt, J. D. R. (1995). "Expanding the Horizon of Information Systems Design," *Systems Research* 12, 185-199.

Eriksson, D. M. (2003). "Identification of Normative Sources for Systems Thinking," *Systems Research and Behavioral Science*, 20: 475-488.

Flood, R. L., and M. C. Jackson, eds. (1991). *Critical Systems Thinking: Directed Readings*, Chichester, UK: John Wiley and Sons.

Habermas, J. (1984). *The Theory of Communicative Action*, Boston, MA: Beacon Press.

Jackson, M. C. (2000). *Systems Approach to Management*, New York, NY: Kluwer Academic/Plenum Publishers.

Khisty, C. J. (1995). "Soft Systems Methodology as Learning and Management Tool," *Journal of Urban Planning and Development*, American Society of Civil Engineers, 121:3, 91-107.

——— (2006). "A Fresh Look at the Systems Approach and an Agenda for Action," *Systemic Practice and Action Research*, 19:1, 3-25.

Khisty, C. J. and U. Zeitler (2001). " Is Hypermobility a Challenge for Transport Ethics and Systemicity?" *Systemic Practice and Action Research*, 14:5, 597-613.

Lewin, K. (1947). "Frontiers in Group Dynamics11: Channels of Group Life - Social Planning and Action Research, *Human Relations* 1, 143-153.

Midgley, G. (2000). *Systemic Intervention*, New York, NY: Kluwer Academic/Plenum Publishers

Oliga, J. C. (1988). "Methodological Foundations of Systems Methodologies," *Systems Practice*, 1:1, 87-112.

Reason, P. and H. Bradbury, eds. (2001). "Introduction." P. Reason and H. Bradbury, *Handbook of Action Research*, London: Sage Publication.

Rosenhead, J. and J. Mingers, eds. (2001). *Rational Analysis for a Problematic World Revisited*, 2nd ed., Chichester, UK: John Wiley and Sons.

Senge, P. M. (1990). *The Fifth Discipline: The Art and Practice of the Learning Organization*, New York, NY: Doubleday.

Schön, D. (1987). *Educating the Reflective Practitioner: Toward a New Design for Teaching and Learning in the Professions*, San Francisco, CA: Jossey-Bass.

Ulrich, W. (1998). "Systems Thinking as if People Mattered: Critical Systems Thinking for Citizens and Managers," Working Paper #23, Lincoln School of Management, Lincoln, UK.

EXERCISES

1. How would you compare the three strands of systems analysis, systems engineering, and OR? If you were proposing to build a small house for a client, how would you make use of these three strands when building the house? Would you think of using any other techniques to improve the situation?

2. How and why did the path from the use of optimization, so widespread in engineering and management, slowly transform to one of learning?

3. Mrs. Smith, the owner of a newly built house, wants to put up a fence around her house, which is located in a beautiful part of the city. Although she is paying for the fence herself, it appears that the neighborhood housing association has an influence on her decision to select the type and color of the fence. She resents the interference but is willing to attend the association's meeting to argue her case. Formulate the RD and draw appropriate rich pictures to illustrate this problematic situation using Checkland's SSM.
4. What role does action research play in the planning and management contexts?
5. What are the basic differences between CST and multimodal thinking? Draw up a table showing these differences.
6. Based on your own reading and the information given here, write a short history of the evolution of systems thinking.
7. Peter Senge's Fifth Discipline has been widely used in thousands of cases in practically every discipline. What are the striking features of this style of SST in comparison to Checkland's SSM?

Systems Thinking: Case Studies 13

(With contributions from Dr. P. S. Sriraj, University of Illinois at Chicago.)

13.1 INTRODUCING THE CASE STUDIES

This chapter describes a number of case studies that can provide readers with examples of how real-world problems, encountered today and tomorrow, can be tackled using a blend of appropriate forms of analyses. Making decisions in today's world is not easy for the simple reason that the dynamics of this world are not as clear as many imagine and because complexity and uncertainty are inevitable in nearly every situation.

The various methodologies and approaches regarding systems thinking have been illustrated and discussed in previous chapters. However, the hard systems thinking described in Chapters 10 and 11, and systems thinking covered in Chapter 12, are particularly pertinent for appreciating what is contained in this chapter. Four case studies are presented. They are important in the sense that they illustrate how an experienced professional chooses to tackle problems that occur in the course of the day.

The first case study deals with the application of robustness analysis to a transportation project that involves high costs coupled with high uncertainties. It is an alternative approach to the commonly used hard systems methodology (HSM) of applying the *Maximax*, *Maxmin*, or *Hurwicz* criteria. The second case study describes a real-life situation where Checkland's soft systems methodology (SSM) and multimethodology are applied to the Chicago region. While the main thrust of this study uses SSM, the performance indicators of multimethodology are also applied. The third case study describes a real-life situation drawn from the Chicago Transit Authority (CTA), using strategic options development and analysis (SODA). The fourth and last case study deals with an unusual situation connected with the Bhopal gas tragedy that occurred in December 1984. Although this is an ex post facto study, and therefore prescriptive, it demonstrates how Checkland's SSM would have helped to understand this crisis situation more effectively and more efficiently had it been applied at the time of the incident.

This is an appropriate place to examine crisis situations vis-à-vis project management. Crisis situations can differ in magnitude and severity, in terms of the damage they can cause to human life and the environment. The severity of the situation usually dictates what action needs to be taken depending on the resources and time available. Accordingly, the four case studies

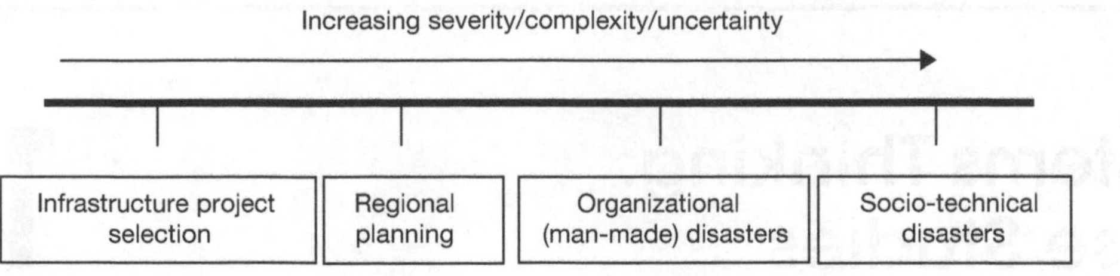

Figure 13.1 Scale of crisis and complexity indicator

described in this chapter are lined up in Figure 13.1 based on the increasing degrees of complexity, severity, and uncertainty involved.

In reality a case study illustrates how a professional, such as an engineer, planner, or manager, handled a particular project or problem under circumstances that were then prevailing. However, there is no guarantee that the procedure, method, or methodology used by a professional is necessarily the best, because the timing, expertise, technology, and know-how are constantly changing. Neither can it be assumed that a similar outcome would result if some equally knowledgeable professional followed exactly the same methodology in another case resembling the one under examination. These case studies should, therefore, be considered examples of good practice rather than templates for replication.

13.2 CASE STUDY 1: TRANSPORTATION PROJECT SELECTION USING ROBUSTNESS ANALYSIS

13.2.1 Introduction

This case study is associated with decision making for a highway project. In the case of projects in the private sector, a planner takes risks into account by selecting an alternative, because there is no guarantee that the alternative will lead to the desired outcome. However, decision makers in the public sector have to justify using taxpayers' dollars and, hence, do not have the option of taking undue risks. In such cases, uncertainty in public project selection is dealt with in several other ways. One such method is robustness analysis. Robustness analysis provides an approach to the structuring of transportation problems where uncertainty is high and where sequential time-phased decision making is necessary. The technique emphasizes the need to make early decisions in such a way as to preserve many future options that currently seem attractive. In other words, the strength of robustness analysis is on exploiting flexibility in phased transportation project planning.

This case study explains the application of robustness analysis with the help of an example. It illustrates how a decision maker can keep a variety of options open, that may become viable in the future, by selecting a particular alternative at the present time.

13.2.2 Background: Robustness and Debility

Decision makers generally use well-known operations research procedures under conditions of uncertainty when the knowledge of probabilities associated with outcomes is not known. Chapter 11 gives details of these procedures, such as the Hurwicz's criterion. The significance of uncertainty associated with a decision depends on the cost of reversing a commitment once made. In transportation projects, the cost is in the millions of dollars and, hence, when a decision is made it is hard to reverse it because of the horrendous costs involved. This implies that there should be some kind of flexibility built into any decision-making process, enabling decision makers to alter their definition of what is *best* at the present. This is especially true with transportation investment projects that depend directly on the geography and the development of the area. In such cases, the uncertainty is with respect to the population growth and the allocation sequence of projects for the entire region. These problems can be tackled by adopting robustness analysis that enables a planner to help preserve a wide range of options in the future that seem attractive now.

Rosenhead (1980) says that *robustness analysis is a way of working which focuses attention on the possibilities (not probabilities) inherent in a situation*. The concept of debility is defined as the number of unsatisfactory end-states still attainable after an initial decision, expressed as a ratio of all such end-states. A low debility score as opposed to a high robustness score is preferred with respect to any alternative. A low debility score indicates there is lower risk for undesired outcomes. In some cases, the debility matrix may be quite useful. A simple example is presented to illustrate robustness analysis.

13.2.3 Road Construction Using Robustness

A medium-sized city that is rapidly growing is planning to construct a new airport and expanding the existing road network to connect the city to the proposed airport. The city has developed a preliminary alignment of the proposed network and plans to get this project completed in a sequential manner.

The city is looking at a planning horizon of 20 years for the investment projects for the nine network links illustrated in Figure 13.2. The estimates of travel demand are uncertain. Three scenarios are generated, one for each of the three possible states in the future. The first scenario represents the original projection of operating costs when the development starts in and around Node 1. The second corresponds to a state of economy that leads to the operating costs being 20 percent more than the original projection with the development pattern centered on Node 6. The third corresponds to a state of economy that results in operating costs 20 percent less than the original projection with the development pattern centered on Node 2.

The objective is to select links that need to be constructed early on in the sequence. The results of this exercise are analyzed to bring out the advantages of using robustness analysis. The costs associated with each link for each of the scenarios are shown in Table 13.1.

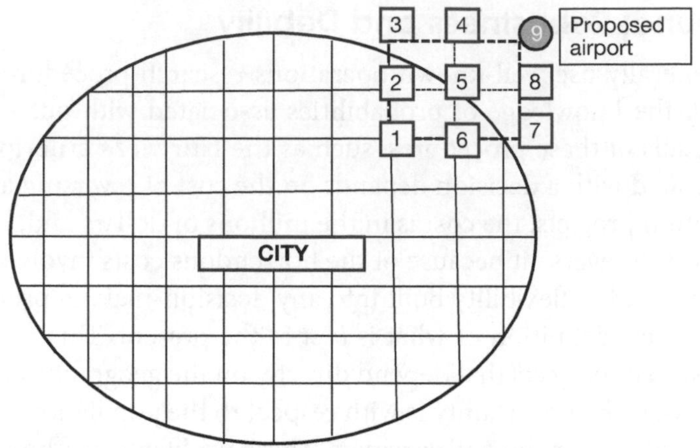

Figure 13.2 Proposed road expansion

Table 13.1 Link costs for each future (millions of dollars)

Link	Scenario 1	Scenario 2	Scenario 3	No. of times link is part of a plan	Robustness*
1-2	250	300	200	14	14/27
1-6	400	480	320	11	11/27
2-3	200	240	160	8	8/27
2-5	500	600	400	9	9/27
3-4	400	480	320	8	8/27
4-5	250	300	200	12	12/27
4-9	400	480	320	11	11/27
5-6	250	300	200	10	10/27
5-8	500	600	400	12	12/27
6-7	400	480	320	10	10/27
7-8	200	240	160	10	10/27
8-9	200	240	160	16	16/27

*Figures obtained based on calculations from Table 13.2

Robustness analysis: Using the network shown in Figure 13.2 a set of nine alternative plans are prepared for each scenario. It should be noted that the plans need to be exhaustive in nature and unique to the project at hand. The robustness of a particular link is the frequency with which that link appears as a component of these plans. In other words, the robustness of a link is the proportion of the 27 plans that include that particular link. The details of each plan are listed in Table 13.2. Note that each set of plans for Scenarios 1, 2, and 3 naturally begin with Nodes 1, 6, and 2, respectively. The costs for each plan are merely the sum of the costs of the individual links that are a part of each plan.

Table 13.2 Details of plans for each scenario

Scenario 1			Scenario 2			Scenario 3		
Plan	**Nodes**	**Cost**	**Plan**	**Nodes**	**Cost**	**Plan**	**Nodes**	**Cost**
A	1, 2, 5, 8, 9	1450	A	6, 1, 2, 3, 4, 9	1980	A	2, 3, 4, 5, 8, 9	1240
B	1, 2, 3,4, 9	1250	B	6, 1, 2, 5, 4, 9	2160	B	2, 3, 4, 5, 6, 7, 8, 9	1520
C	1, 6, 7, 8, 9	1200	C	6, 1, 2, 5, 8, 9	2220	C	2, 1, 6, 7, 8, 9	1160
D	1, 6, 5, 4, 9	1300	D	6, 1, 2, 3, 4, 5, 8, 9	2580	D	2, 1, 6, 5, 8, 9	1280
E	1, 6, 5, 8, 9	1350	E	6, 5, 4, 9	1080	E	2, 1, 6, 5, 4, 9	1240
F	1, 2, 5, 4, 9	1400	F	6, 5, 8, 9	1140	F	2, 5, 4, 9	920
G	1, 2, 5, 6, 7, 8, 9	1800	G	6, 7, 8, 9	960	G	2, 5, 8, 9	960
H	1, 2, 3, 4, 5, 6, 7, 8, 9	2150	H	6, 7, 8, 5, 4, 9	2100	H	2, 1, 6, 7, 8, 5, 4, 9	1920
I	1, 2, 3, 4, 5, 8, 9	1800	I	6, 7, 8, 5, 2, 3, 4, 9	3120	I	2, 5, 6, 7, 8, 9	1240

The robustness score for each of the links is calculated as outlined earlier and the results are tabulated in Table 13.1. A look at the robustness scores shows that the most robust link is 8-9, with the highest robustness score of 16 out of a possible 27. This is to be expected because the final destination is the airport. Since there are only two links that lead to the airport (4-9 and 8-9), these two appear as part of most plans. Apart from Link 8-9, the next most robust link is Link 1-2 with a robustness of 14 out of a possible 27. This means that Link 1-2 is included in 14 of the 27 plans for the three scenarios. Therefore, Link 1-2 is the most robust and the candidate for construction in the first phase of the project (Sriraj, 1999).

A look at the robustness score for Link 2-3 or 7-8 indicates that these links have a lower number of options for the scenario when compared to the robust Link 1-2. The robustness scores calculated are good for the first stage of construction. A similar procedure can be followed for the other stages of construction. This method will give flexibility to a planner who can pick the project that needs to be constructed, depending on the situation, without any extra cost. This is the versatility of robustness in the sense that it allows for decisions to be changed as, and when, the situations arise.

13.2.4 Algorithm for Robustness

In summary, an algorithm for conducting a robustness analysis for any long-range planning process involving sequential decision making is:

1. Set a budget
2. List all proposed projects and their capital and operating costs
3. Generate scenarios of demand
4. Draw up a set of plans for each scenario
5. Find the number of times a particular project appears as part of the plan

6. Determine the robustness for the project (as demonstrated in the example)
7. Select the project that is the most robust subject to budget constraints being met

13.2.5 Discussion and Summary

Uncertainty is embedded in most long-range planning processes. Planners have attacked such problems by using conventional techniques that do not take into account the uncertainty associated with the problem. By doing so, public policy makers sometimes may commit the funding for projects that are not high priority. Robustness analysis tries to eliminate this uncertainty by picking that project that keeps the most number of options open for the future. This by itself does not guarantee the elimination of uncertainty involved in such long-range planning, but it does help in limiting the uncertainty from affecting a majority of the decisions adversely. Robustness analysis will be useful to decision makers especially in developing countries that do not have the luxury of a healthy economy to invest in projects with limited future options.

The algorithm for robustness analysis talks about the need to draw up a set of plans for each future scenario that is taken into consideration. It is evident then that the robustness scores are only as good as the plans from which they are derived. Thus, if the plans for each scenario do not encompass all the relevant information, it is likely that the process will not be successful. This problem can be overcome by doing an in-depth analysis for each scenario. The time and money spent for such an analysis will be infinitesimally small compared to the huge costs of the projects.

In summary, the advantages of robustness analysis are:

- It is technically very simple
- It is easy to adopt because of its simplicity
- It provides insight for tackling daunting problems without any complex procedural analysis

At the same time, some of the concerns about robustness analysis are:

- The process is simple enough to be taken for granted and not taken seriously
- The robustness score is an indicator of flexibility and one needs to look at other factors involved with the plans before deciding on the one to implement (e.g., the debility score)
- It entails making a number of plans that may take time
- The outcome is sensitive to the number and content of the plans

13.3 CASE STUDY 2: CHECKLAND'S SOFT SYSTEMS METHODOLOGY AND MULTIMETHODOLOGY APPLICATION TO THE CHICAGO REGION

We examine the applicability of Checkland's SSM in the context of a regional planning problem in a real-world case study involving regional/infrastructure planning for low-income sectors of society. The problem of the accessibility of low-income populations to jobs has, for the last few decades, caused concern for planners and decision makers, especially in urban areas. Studies have shown that there is no single solution that can solve the problem situation and improve the accessibility of low-income individuals. Experts have advocated various solutions to this spatial mismatch phenomenon. While the onus for this lies within the transportation sector, the need, however, is for a more integrated, cross-sectoral approach that would involve other sectors such as housing and economic development as well. This research was the first step toward identifying the data needs of this complex, cross-sectoral exercise designed to identify feasible solutions to the spatial mismatch in Chicago.

13.3.1 Background: Mixing and Matching

There is a significant gap in major metropolitan areas between the residences of low-income population and the location of suitable jobs for this population (Arnott, 1997). At the same time, research has also shown that the solution does not lie within the transportation sector alone. This is especially true in Chicago where a majority of the low-income population lives in the near-south and near-west sides of the city, with the entry-level jobs suitable for this population spread throughout the region. While the transportation network is adequate, there are other potential solutions that need to be explored, especially from the housing and economic development perspectives. It is in this context that the SSM approach is applied to bring about a synergistic solution to a complex problem. By involving the stakeholders in the process of developing the system, it not only makes the system receptive to the needs of the stakeholders, it also makes the developer understand the intricacies and the underlying layers of issues and complexities with the help of the stakeholders.

13.3.2 The Chicago Region Characteristics and Root Definitions

The participants in the interview were members of the Community Mobility Task Force (CMTF) of the Chicago Area Transportation Study, the metropolitan planning organization for the region. The interviews were conducted over a period of two months with each participant interviewed separately. The interviews were semistructured and were guided with the help of a set of 25 questions about the role of information, and the use of information systems in the stakeholders' organization.

The organizations interviewed for this exercise were all public agencies involved in some form with one of the three sectors—transportation, housing, and economic development. The problem situation being addressed is the plight of the low-income population and the lack of their accessibility to a suitable job. The difficulty in getting to various activities by residents and the random pattern observed in the locations of places of residence create a mismatch in a large urban area. Researchers have attempted to bridge this gap between jobs and home locations to make the low-income population's transition to the labor force easier. There is a need for planners and decision makers to break out of the silo approach and look at more holistic solutions.

It is against this backdrop that the various agencies representing the three pertinent sectors of transportation, housing, and economic development in the Chicago region were approached.

The interviews and discussions are couched in SSM to facilitate synergies between the various agencies and to foster better interaction between the various sectors. The agencies interviewed and included in this process were:

- Center for Neighborhood Technology (CNT)
- Pace Suburban Bus Service
- Metra (commuter rail)
- DuPage County Human Services
- Work, Welfare, and Families (WWF)
- The Chicago Department of Transportation

The rich pictures, root definitions (RDs), and the conceptual models are all developed as part of the SSM process and the results highlighting the synergy are then tabulated. The views of the stakeholders were captured based on their reflections about the transportation, housing, and economic development systems.

The rich picture (Figure 13.3) for this problem situation shows that the low-income population does not have the accessibility to get to entry-level jobs, given that they are living in one

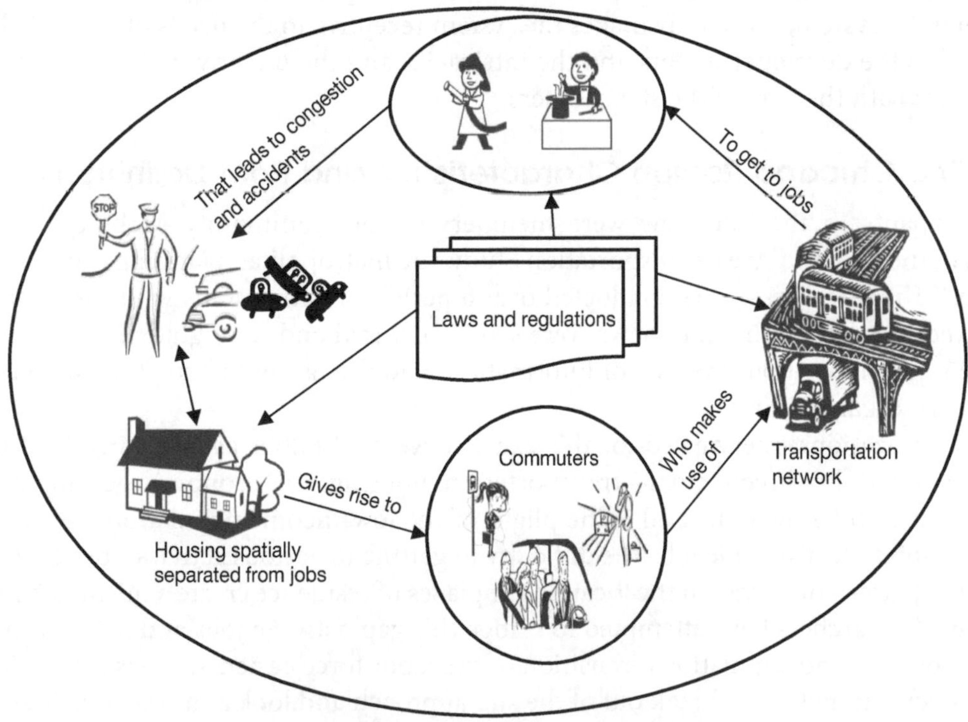

Figure 13.3 Rich picture depicting the problem situation

part of the region (six-county northeastern Illinois region) and the entry-level jobs (employers who are part of the economic development) in the region are located spatially elsewhere. Also, the transportation network serving to bridge the gap between the origins and destinations in the form of highways and transit does not help. The rich picture also highlights the lack of affordable housing in proximity to job locations or vice versa. The same argument holds true for the transportation network as well.

The stakeholders and their responses are categorized according to the sector they represent resulting in conceptual models for the transportation, housing, and economic development/ community organization systems.

The Transportation System

This system represents transit planners, transportation officials, and others involved with the transportation network in the region. The RD for this system is outlined:

Root definition: A comprehensive network of roads and transit services that caters equitably and fulfills the need of the population in the region in a cost-effective, environmentally friendly, and sustainable manner.

- Clients—The traveling public
- Actors—Transportation planners, engineers, and decision makers
- Transformation—To transform the transportation network/system from being an efficient system to one that is equitable, sustainable, and efficient that caters to the various strata of society, including the low-income population group that typically seek entry-level jobs with start times outside of the peak-hours
- *Weltanschauung*—A transportation system that caters to all sections of society
- Owners—The state Department of Transportation, the transit providers in the region, and the local and regional government
- Environment—The population of the region, the bureaucracy surrounding the transportation environment, and the physical layout of the region including the land use and the housing system

The Housing System

The housing system consists of representatives from the state/county agencies dealing with housing and related issues as well as the community organizations that have an affiliation toward promoting the cause of the low-income population.

Root Definition: A fair and affordable housing system that is located strategically closer to the transportation service as well as within reach of economic development and daycare services in the region.

- Clients—Low-income population
- Actors—Housing and Urban Development administrators for the region, urban planners of the local government, the regional government, builders, developers, and employers

- Transformation—To enhance the region's housing availability to cater better to the needs of the target population by ensuring better coordination between the transportation and economic development sectors
- *Weltanschauung*—A housing system that is receptive to the needs of the low- and middle-income population in the region
- Owners—The regional housing authority, the departments of human services, community organizations, and the local government
- Environment—The transportation network along with the economic development in the region represent the environment in which the housing industry works

The Economic Development System

The economic development system is comprised of representatives from agencies espousing the cause of welfare families, as well as the agencies that deal with employers and developers in the region.

Root definition: A flourishing economy that is reflected by a happy and contented community that earns a living and has shelter and access to jobs.

- Clients—The population in the region as well as other parts of the world that depend on the region's products and services.
- Actors—The employers in the region, the workforce boards, and the human services agency
- Transformation—To transform an industry that is vibrant and healthy and provides adequate job opportunities for every individual in the region
- *Weltanschauung*—A profitable industry that works in tandem with the transportation and housing sectors to address the needs of the low-income population
- Environment—The jobs in the region along with the businesses that have to co-exist in a regional environment, including the housing and transportation systems.

13.3.3 Multimodal Performance Indicators

The conceptual models (Figures 13.4, 13.5, and 13.6) outline the necessary tasks needed to bring about the transformation desired in the RD and the CATWOE mnemonic (introduced in Chapter 12). As can be seen from the three composite conceptual models, all the sectors are in agreement that there is a need to move out of the silo approach and on to a more holistic and inclusive approach toward addressing the needs of the low-income population in a region.

The salient features of the three conceptual models are discussed next, followed by an evaluation of the tasks outlined with the help of a set of performance indicators. The conceptual model of the transportation system brings to the fore the various tasks necessary to enable the transformation outlined in the RD and with the help of the CATWOE mnemonic. The stakeholders from the three sectors focused on how they would address the situation to make

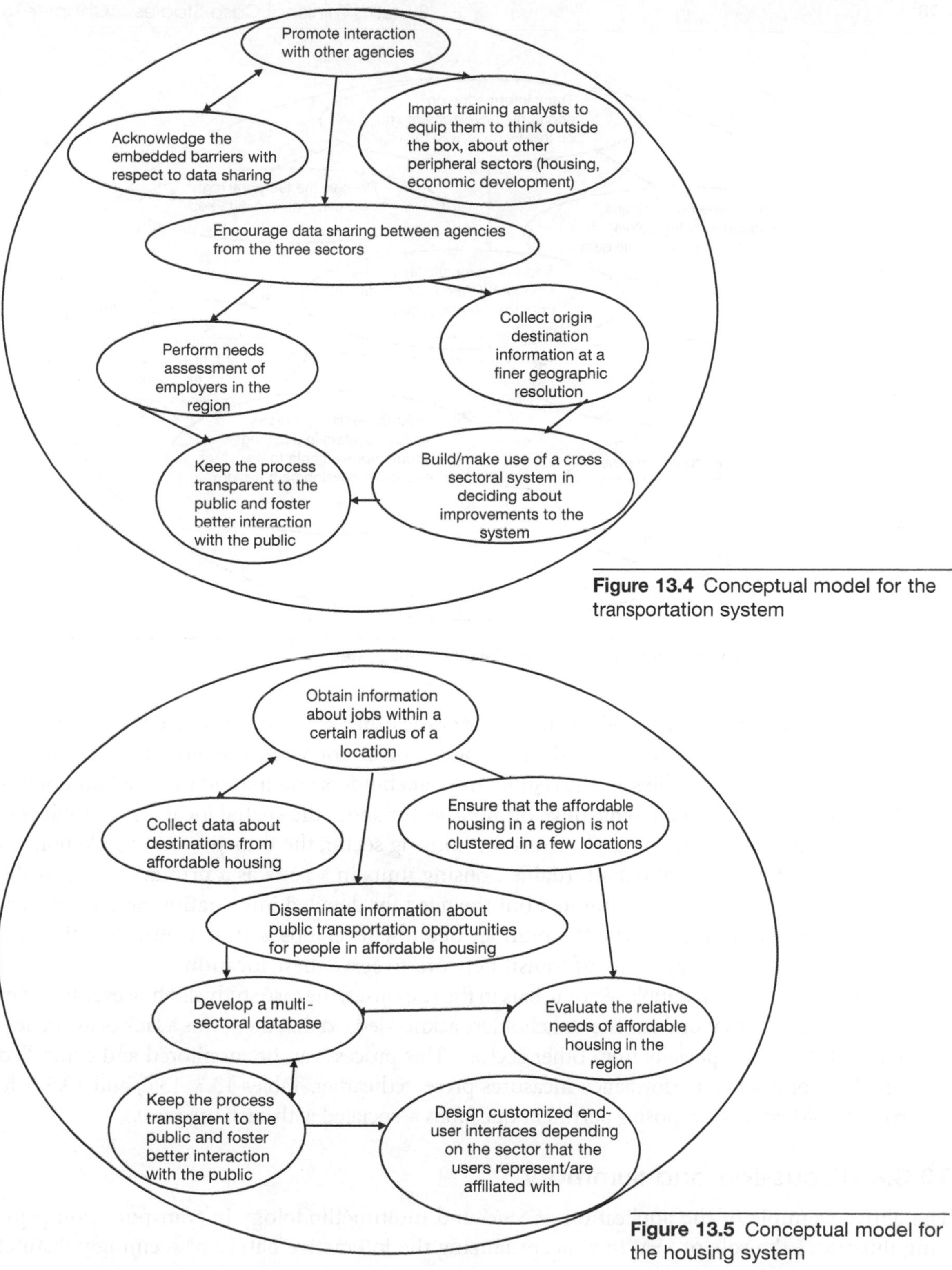

Figure 13.4 Conceptual model for the transportation system

Figure 13.5 Conceptual model for the housing system

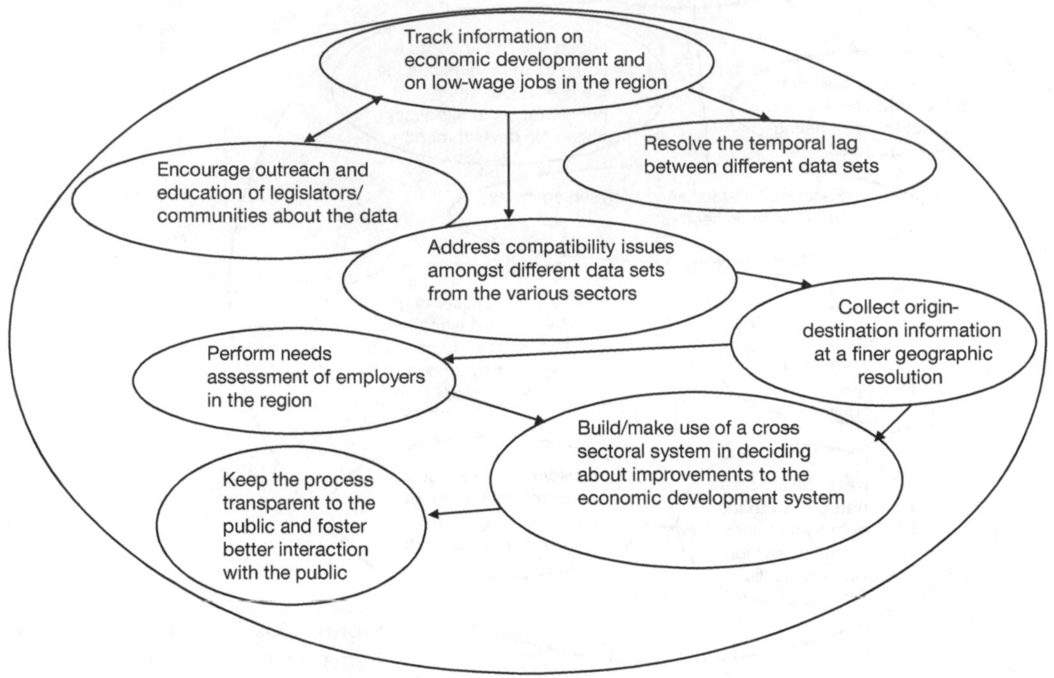

Figure 13.6 Conceptual model for the economic development system

it better and in doing so, identified the needs from the other sectors that would enable them to achieve their goal. Thus, in the case of the transportation sector, while the typical input was that of origin-destination trip tables for the region, the stakeholders understood the need for obtaining information from the economic development sector about the spatial location of employers and openings in the region. Similarly, from the housing sector, the transportation stakeholders identified the information about affordable housing units in a zone as a primary need. At the same time, the stakeholders also pointed out the need for detailed information on factors such as the location of day care centers, the number of entry-level jobs within a certain radius of a neighborhood, and the availability of transit network to serve these locations.

While there was a perceptible sectoral-bias in the responses, the case study did however flesh out a core area of need—that of data. The stakeholders acknowledged that there was a lack of awareness of the available data, especially from other sectors. This process can be monitored and controlled with the help of a set of performance measures presented earlier. Tables 13.3, 13.4, and 13.5 help control the SSM process by posing pertinent questions associated with each measure.

13.3.4 Discussion and Summary

The above example of the application of SSM and multimethodology in transportation planning illustrates the role of SSM in conceptualizing the interactive nature of seemingly distinct

Table 13.3 Performance indicators for the transportation system

Performance indicator	Question	Verdict on the current situation
Creedal	Are the right things being done in the short and long term?	Yes, the formation of the CMTF is a step in the right direction
Ethical	Are the planning and implementation morally correct and ethical?	Yes
Juridical	Are the planning and implementation just and fair?	Yes, as far as the transportation system is concerned, there is no evidence to prove otherwise
Aesthetic	Is the plan aesthetically satisfying?	No, not at present
Economic	Are the resources being used optimally?	Yes
Social	Are the social needs of the people respected and accounted for?	Yes
Informatory	Have all jurisdictions been duly represented and consulted?	Yes
Historical	Have lessons been learned from the past?	The transportation system authority has been aware of past history and has acted accordingly
Logical	Are all the results of the models used reliable and logical?	Yes, the integration of housing and economic development into the transportation sector is acknowledged by stakeholders as logical
Sensitive	Is the system sufficiently robust to handle changes?	The transportation system is a mature system in Chicago and can make the transition in a smooth manner
Biotic	Have the concepts of environmental sustainability been taken care of?	Yes
Physical	Has the system taken advantage of the conditions?	No
Kinematic	Has the movement of people, goods, and vehicles been designed for accessibility and environmental sustainability?	Yes
Spatial	Have the spatial attributes been designed for the health and safety of the people?	No
Numerical	Have quantitative methods been developed based on the data?	Yes

problems that can be presented holistically. In this particular case, the stakeholders were presented with the problem of accessibility for low-income individuals in the six-county Chicago region. The stakeholders were also encouraged to think outside the box in terms of data needs and barriers. The rich picture depicts the holistic nature of the problem situation from a bird's-eye view and brings out the interconnected nature of seemingly disconnected activities. The conceptual models help to crystallize the thoughts of the stakeholders and highlight the necessary steps to ameliorate the situation. The set of performance measures serve to keep a check

Table 13.4 Performance indicators for the housing system

Performance indicator	Question	Verdict on the current situation
Creedal	Are the right things being done in the short and long term?	No, there is a need for more affordable housing
Ethical	Are the planning and implementation morally correct and ethical?	Yes
Juridical	Are the planning and implementation just and fair?	Yes, the housing authority and the Department of Human Services ensure that a fair and equitable housing solution is presented
Aesthetic	Is the plan aesthetically satisfying?	No, not at present
Economic	Are the resources being used optimally?	Yes
Social	Are the social needs of the people respected and accounted for?	Yes, but it is difficult to ensure social needs are accounted for
Informatory	Have all jurisdictions been duly represented and consulted?	Yes
Historical	Have lessons been learned from the past?	There is a debate about affordable housing among planners and decision makers, albeit without any consensus
Logical	Are all the results of the models used reliable and logical?	Yes, the integration of housing and economic development into the transportation sector is acknowledged by stakeholders as logical
Sensitive	Is the system sufficiently robust to handle changes?	The affordable housing segment can survive only with changes that incorporate other sectors into the mix
Biotic	Have the concepts of sustainability been taken care of?	No
Physical	Has the system taken advantage of the conditions?	No
Kinematic	Has the movement of people, goods, vehicles been designed for accessibility and sustainability?	No, it is more often than not dictated by availability and zoning than accessibility
Spatial	Have the spatial attributes been designed for the health and safety of the people?	No
Numerical	Have quantitative methods been developed based on the data?	Yes

on the whole process and provide adequate feedback with respect to achieving the various goals and objectives. Users must exercise caution when making use of these performance measures in evaluating problem situations. Depending on the problem situation, one might need to address the questions differently to ensure that the essence of the question is retained without losing focus.

Table 13.5 Performance indicators for the economic development system

Performance indicator	Question	Verdict on the current situation
Creedal	Are the right things being done in the short and long term?	Yes, the formation of the CMTF is a step in the right direction with representation from WWF and CNT
Ethical	Are the planning and implementation morally correct and ethical?	Yes
Juridical	Are the planning and implementation just and fair?	No, there is no evidence to prove that employers either look to settle in an impoverished neighborhood or provide access for low-income population
Aesthetic	Is the plan aesthetically satisfying?	Yes
Economic	Are the resources being used optimally?	Yes
Social	Are the social needs of the people respected and accounted for?	No, not necessarily from the economic development perspective
Informatory	Have all jurisdictions been duly represented and consulted?	No, typically the public are not taken into confidence
Historical	Have lessons been learned from the past?	No
Logical	Are all the results of the models used reliable and logical?	Yes
Sensitive	Is the system sufficiently robust to handle changes?	The economic development in a region as large as the Chicagoland area can handle changes without problems
Biotic	Have the concepts of sustainability been taken care of?	Yes, but not from the perspective of the low-income population
Physical	Has the system taken advantage of the conditions?	No
Kinematic	Has the movement of people, goods, and vehicles been designed for accessibility and sustainability?	Yes, however, the economic development sector has not done all it could to improve the condition of the low-income population
Spatial	Have the spatial attributes been designed for the health and safety of the people?	Yes
Numerical	Have quantitative methods been developed based on the data?	Yes

This case study has introduced the concept of SSM and has, with the help of an example, illustrated the strength and use of this methodology in a wide setting that has been traditionally viewed as being very quantitative. The use of SSM helps provide structure to seemingly disjoint and complex problem situations, such as the case study discussed. The use of the control mechanism in the form of performance indicators in conjunction with SSM is new to the SSM arena and serves to enrich the handling of the problem situation, as illustrated in this study.

13.4 CASE STUDY 3: USING SODA FOR THE CHICAGO TRANSIT AUTHORITY STUDY

13.4.1 Introduction

Public agencies are increasingly required to involve stakeholders in their planning and decision-making processes. Many transit agencies invest significant resources in terms of staff, time, and money to realize their stakeholder involvement efforts. This causes them to hold practical reservations about stakeholder involvement programs, believing they increase costs, create delay, open the door to emotional considerations and self-interest, and ignite controversy rather than consensus (Burby, 2003). Despite these reservations, stakeholder involvement is often a legally required activity that affects multiple departments within a transit organization. This is because agencies are agents of stakeholders and must understand and incorporate the desires of stakeholders into their planning and engineering activities for effective outcomes.

Overall, a substantial gap exists between public agencies and external groups that their perspectives have not been understood and considered. This case study is an attempt to better understand this conceptual gap by systematically assessing internal and external stakeholder perceptions of the success of a recent stakeholder involvement program for a rail line rehabilitation project of the CTA.

The outcome of stakeholder participation processes should have a genuine impact on policy and be seen to do so. One of the main complaints about public participation methods is that they are often used simply as a means to legitimize predetermined decisions or give the appearance of consultation without any genuine intent of acting on their recommendations (Rowe and Frewer, 2000). The end result is a distrust of public agencies by the public. To combat this, the public should be able to see what is going on and how decisions are being made. Such transparency is likely to reduce public suspicions about the agency. If any information needs to be withheld from the public it would seem important to admit what is being withheld and why.

If disconnections exist in either the process or in acceptance, it tends to result in a gap between the expectation of the stakeholders and in the strength of the message from the agency. It is our hypothesis that most public involvement exercises are left with a gap between the agency and stakeholder expectations on one or both counts. To test this hypothesis we used the CTA's Brown Line Rehabilitation Project as a case study.

13.4.2 Background: Brown Line Rehabilitation Project

CTA is the transit provider in the city of Chicago and parts of suburban Cook County in Northeastern Illinois. The rail network of CTA consists of eight lines and this study analyzes the Brown Line. Since 1998, ridership has increased on the Brown Line by 21 percent, making it the highest rate of growth in CTA's rail system. The CTA began introducing operational changes in the mid-1990s to accommodate increasing demand on the Brown Line. Despite service changes, persistent crowding on Brown Line platforms and in rail cars continued to negatively affect the rail transit experience for passengers.

In 1997, the CTA began to plan a significant rehabilitation project for this line. This was a large-scale construction project with substantial impacts throughout some of the most densely built neighborhoods of Chicago's north side. The major objectives of this rehabilitation project were to lengthen all station platforms between Chicago and Kimball stations to accommodate eight-car trains and rehabilitate rail infrastructure and stations to meet American Disabilities Act accessibility requirements in all stations. Prior to the rehabilitation project, all Brown Line stations outside of the loop could only accommodate six-car trains, with the exception of Merchandise Mart, Fullerton, and Belmont stations, making it one of only two CTA rail lines that could not accommodate eight-car trains (Mattingly et al., 2010).

During this process, a Brown Line Task Force was set up to serve as a major conduit for communication among key internal and external stakeholders throughout this project. The task force met on a monthly basis to supplement rather than supplant the CTA communications platform to the community. According to the literature, public involvement should not be a one-size-fits-all approach, as no one approach constitutes good participation, rather this depends on the project-specific circumstances (Bickerstaff et al., 2002; Grossardt et al., 2003). On the Brown Line Rehabilitation Project, the CTA's outreach and involvement approach catered to the needs and characteristics of the demographic profile of the project impact area and those of the elected officials and communities. This area tends to be very highly educated and organized and is savvy with computers and technology. The CTA was asked to respond to a higher level of engagement and to provide different types of information throughout the process, which they attributed to higher levels of education and activism within this community. The CTA was aggressive in sending the message out to the community about the anticipated delays and the provision of alternative bus service, and the scheduled construction activities. In their opinion, they responded with what they deemed very intense outreach involving a wide range of outreach activities ranging from aldermanic briefings to distributing flyers to awareness raising media to community meetings. They understood their process to be extremely effective. However, not all external groups agreed with this sentiment.

13.4.3 SODA Application

The purpose of this study was to measure and assess internal and external stakeholder perceptions and expectations of a meaningful and effective involvement effort in public transit planning. This study methodology was centered upon three sets of structured interviews with distinct internal and external stakeholder groupings for CTA outreach and involvement. The interviews served to identify the gaps between the process and the acceptance, if any, of stakeholder involvement. This process, with theoretical underpinnings in cognitive mapping, made use of SODA, a qualitative approach steeped in public participation theory through the software, Decision Explorer. The method was systemic in nature and looks at the system as a whole rather than taking the reductionist approach of breaking it into parts. The interviews with each stakeholder grouping carried a parallel set of questions based upon the key findings of the literature review including assessing the extent to which the process enhanced the effectiveness and quality of the plan, increased public acceptance of the plan through use of local knowledge, built public trust in the CTA, and assured cost-effective decision making. The hypothesis enveloping each of these concepts was that there exists a gap between the process and the acceptance.

Two sets of structured interviews were conducted with internal and external CTA stakeholders on the Brown Line Rehabilitation Project. Internally within the CTA, the outreach and involvement process for this project has been heralded as a success. The major objective of this case study was to assess the extent to which there exists a gap between the CTA's conception of success for public outreach and involvement in this case and the public's conception of success for the CTA's outreach and involvement efforts in the Brown Line Rehabilitation Project.

Ten internal CTA employees were interviewed who were involved in the Brown Line Rehabilitation Project from a range of departments, including Planning, the President's Office, Rail and Bus Operations, Construction, and Customer Communications. Nine external stakeholders were interviewed including elected officials, University representatives, community and neighborhood group representatives, Chambers of Commerce, and business owners.

Upon the completion of the interviews, each interview was digitally transcribed. Based on direct quotes pulled from each interview transcription in response to the same question sets, a gap analysis of internal and external understandings of the outcome of the stakeholder involvement process was performed. *Decision Explorer* cognitive mapping software was used to complete this gap analysis. Decision Explorer allows for structured analysis of qualitative data by creating a model of interlinked ideas using a cognitive map. The map consists of short phrases, or concepts, gathered directly from the interviews, whose relationships are indicated by the causal links drawn between them. Together, concepts and links form a model. The term *cognitive map* refers to the data set as a whole. The benefit of this cognitive mapping software is that it maintains the richness of the data by managing its complexity, rather than having to rely on weaker, generalized statements to summarize the interview data.

Comparative models were created to assess the internal versus external perspectives on the success of the Brown Line Rehabilitation outreach and involvement process with respect to the internal and external goals. This was an iterative process to verify the inter-coder reliability of the links between concepts. Beyond mapping, Decision Explorer contains advanced techniques used to analyze relationships in the interview findings. The software produces two types of statistics: central scores and potency scores. The central score represents the level of influence that a concept has in relation to other concepts in the model, while the potency score reflects the extent to which a peripheral concept contributes to the higher order central concepts. Central and potency scores were calculated from cognitive maps developed for external and internal stakeholders. In this way we are able to assess the conceptual gaps that exist between internal and external stakeholders about the success of the Brown Line Rehabilitation stakeholder involvement process. For analysis purposes, the top five central and potency scores were selected and analyzed to assess the gaps between internal and external perceptions of the success of this rail rehabilitation project. This method provided a robust, systematic technique to measure and analyze stakeholder perceptions using qualitative data. The software was used to construct the cognitive map for the external stakeholders to the agency (See Figure 13.7) as well as the internal stakeholders (See Figure 13.8). These cognitive maps reflected the opinions of the two groups pertaining to the agency's goals for public involvement.

13.4.4 Findings

The findings demonstrate systematic gaps between the internal and external CTA stakeholders' perceptions of the overall goals of the stakeholder involvement process. These gaps occur when

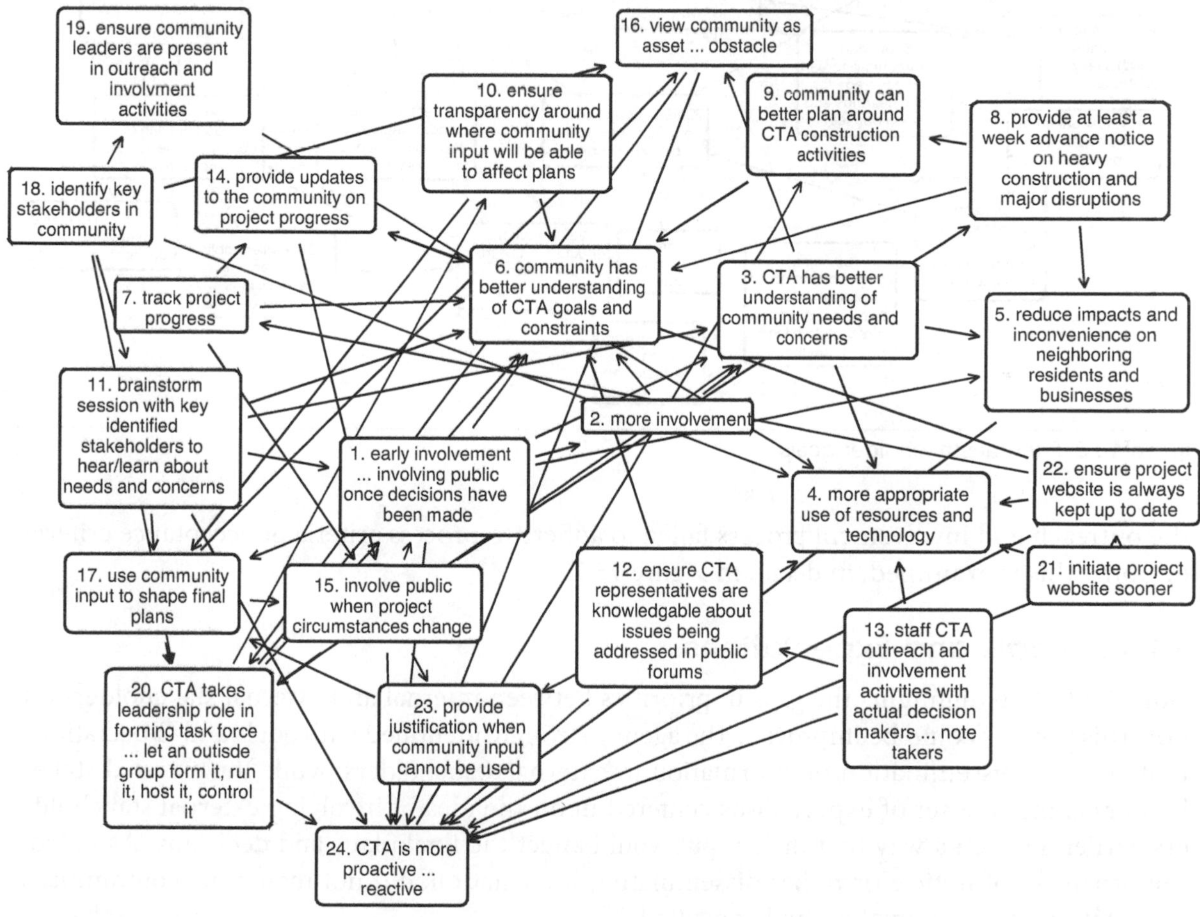

Figure 13.7 External stakeholder goals

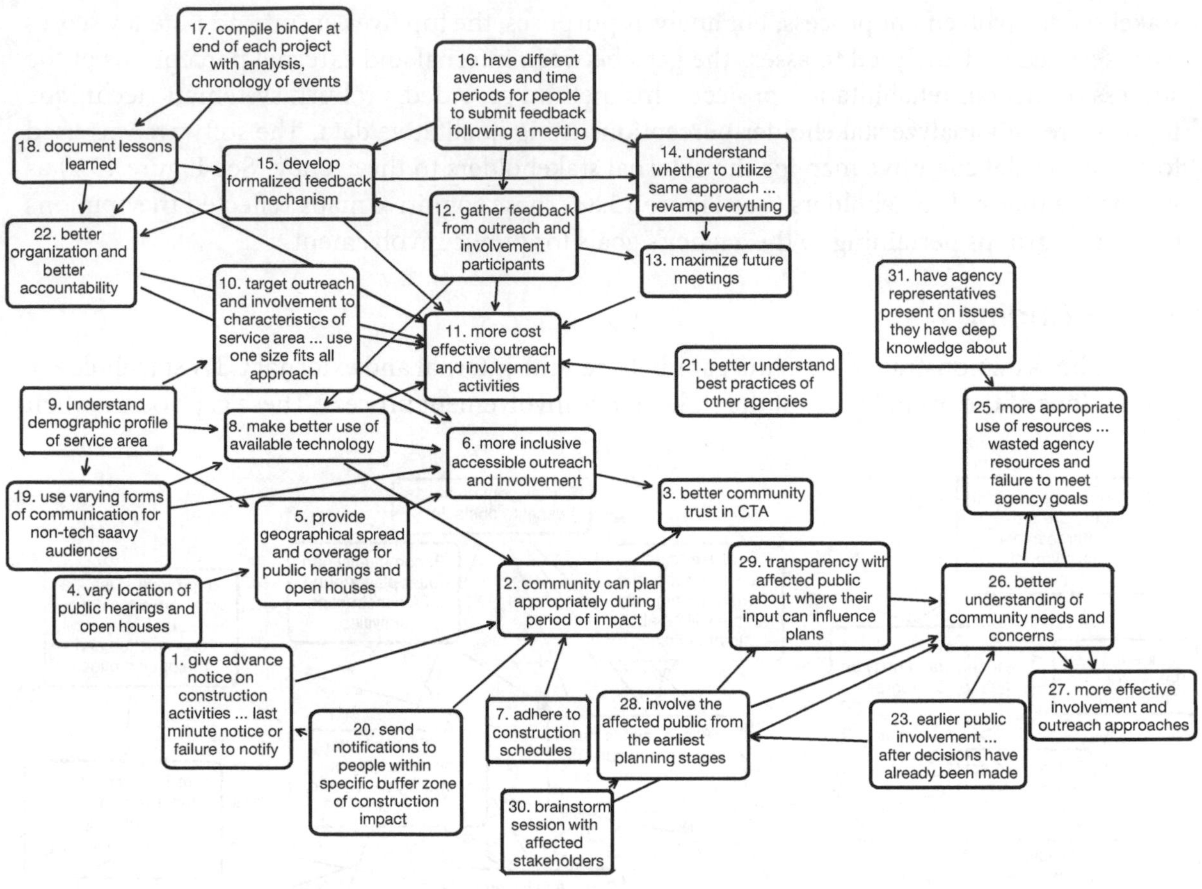

Figure 13.8 Internal stakeholder goals

the outreach and involvement process failed to adhere to process criteria or acceptance criteria or both. This is examined, in detail, in Figure 13.7.

13.4.5 Goals and Expectations

Gap: Table 13.6 highlights the gaps in priorities between internal and external stakeholders on how this process could be improved. The agency strongly identified with goals and expectations that advance dissemination of information to external stakeholders, while the external stakeholders clung to a set of expectations centered upon the idea of involving external stakeholders earlier, in such a way that their input would affect the final plans and decisions. As noted, the provision of notic,e or rather disseminating information, did not match the community's expectation of being involved and consulted.

Analysis: The significant gaps in these analysis scores were very revealing about the distinct set of expectations internal and external stakeholders carry into such a process. While

Table 13.6 Comparison of central and potency scores: internal vs external

Central scores: Goals and expectations	
Internal CTA perception	Gather feedback from outreach and involvement participants
	More cost-effective outreach and involvement activities
	Make better use of available technology
	More inclusive, accessible outreach and involvement
	Better organization and more accountability
External stakeholder perception	CTA is more proactive rather than reactive
	CTA takes leadership role in forming task force rather than let outside group form it, run it, host it, control it
	Community has better understanding of CTA goals and constraints
	Early involvement rather than involving public once decisions have been made
	Use community input to shape final plans
Potency scores: Goals and expectations	
Internal CTA perception	Give advance notice on construction activities rather than last minute notice or failure to notify
	Community can plan appropriately during period of impact
	Better community trust in CTA
	Vary location of public hearings and open houses
	Provide geographical spread and coverage for public hearings and open houses
External stakeholder perception	Early involvement rather than involving the public once decisions have been made
	More involvement
	CTA has better understanding of community needs and concerns
	More appropriate use of resources and technology
	Reduce impacts and inconveniences on neighboring residents and businesses

the external stakeholders were clamoring for more involvement, earlier involvement, proactive agency leadership, ultimately building toward the expectation that their input was used to shape the final plan, the agency asserted their concern for more cost-effective outreach and involvement activities.

13.4.6 Discussion and Summary

The purpose of this case study was to qualitatively measure and assess gaps in internal and external perceptions of success of an outreach and involvement program. We found gaps between stakeholder and agency perceptions of the success in this process. Conducting stakeholder involvement is very costly and time consuming, and, as these study findings indicate, it is fraught with the difficulty of balancing external stakeholder expectations of the process with meeting agency goals and objectives. These findings have several clear implications for how transit agencies can improve the effectiveness of their efforts and close this gap between stakeholder expectations and agency perceptions in future projects by appropriate use of resources.

13.5 CASE STUDY 4: CRISIS MANAGEMENT USING SOFT SYSTEMS METHODOLOGY FOR BHOPAL GAS TRAGEDY

13.5.1 Introduction

Crisis management and planning has developed into an important, fundamental and invaluable tool for the very existence and survival of an organization in view of the disasters that have occurred by nature or by corporate mismanagement. Disasters in this century have been caused either by natural forces, terrorist acts, or by corporate mismanagement. Well-known examples include the Bhopal chemical spill, Exxon-Valdez oil spill, crashes of Value-Jet and TWA flights, and the Oklahoma City bombing. These examples, except for the Oklahoma City bombing, fall under the category of corporate disasters, resulting in severe damage not only to the organization but also to the community and the environment. These and other cases of a similar nature have now come to be termed as sociotechnical disasters (Richardson, 1994).

One of the earliest definitions of crisis management came from the public relations industry, and was defined as the successful management of public and stockholder opinion in the midst of a corporate disaster (Braverman, 1997). The word crisis is defined as an unstable situation of extreme danger or difficulty (Webster's, 1997). In the context of organizations, a crisis is defined as an event that can destroy or affect an entire organization (Mitroff et al., 1996). Such events are usually characterized by tremendous loss of resources—both human and material. When such an event occurs, usually without any warning, most organizations find themselves in a situation where they are typically not ready to comprehend and cope with the magnitude of the event. Typically this is the main reason for the large-scale destruction of life and property that comes to be associated with many disasters and crisis situations. Researchers have been able to instill in managers and organizations that crises—major or minor—are no longer to be treated as isolated, episodic events. Instead, the need is for crisis management to be included in the daily routine of managers and the top brass of organizations.

In the past, crises were handled in a haphazard and random manner by trying to isolate the crisis from the rest of the organization. But when the systemic nature of crises was established, organizations soon found themselves facing a problem of much bigger magnitude. This called for changes in the practice of crisis management by harnessing the effectiveness of systems methodologies in such situations. Systems methodologies, emanating from systems thinking have been developed by various researchers to tackle problems of varied magnitude and complexity (Flood and Jackson, 1991). It is the objective of this case study to make use of some of these methodologies to conduct a post-crisis audit and to develop a crisis management program for the city of Bhopal in the wake of the chemical spill at Union Carbide India Limited (UCIL).

13.5.2 Crisis Management

The term crisis management, in the corporate sector, refers to the successful management of public and stockholder opinion in the midst of a disaster. This definition can easily be extended to cover not just socio-technical disasters, but also natural disasters, with the emphasis being on the successful management and coordination of various entities affecting an event. Incidents that have occurred over the past decade or so, have only served to remind corporations in particular, and society at large, of the need to be prepared. The rate and frequency at which these incidents have occurred since the early 1970s also bolster the need for a sound crisis management team (Richardson, 1994). Many of these types of incidents have been recognized as being organizationally induced sociotechnical disasters. They occur as a result of a combination of human, technological and organizational systems failure. Such disasters usually result in massive economic and social costs to say nothing of the large-scale damage to human life and potential disrepute for the organization concerned.

There are two prevailing views and opinions about these crisis situations; the first is a simple systems viewpoint, where individuals are held responsible for the disasters. The second considers disasters as stemming from the complex interaction between the various parts of an organization and the environment surrounding it in terms of people, infrastructure, competitors, and so on. It is systemic in nature and takes into account the interaction of the different parts of the system. This approach needs to be encouraged and adopted by organizations across the world, to better equip them to face any untoward emergencies and/or disasters. The current practice of crisis management, up until a few years back, has been to adopt the simple systems approach, find a few culprits and provide piecemeal relief to those affected. Neither is this a permanent fix nor do organizations learn from their mistakes, with the result that they are condemned to possibly repeat them again.

Multinationals like Exxon and Union Carbide, which took to this approach, have learned their lesson the hard way—after thousands of lives were lost and severe damage done to the environment. Organizations have also learned that it would behoove them to be prepared for crisis rather than merely reacting to it. The complexity of these crisis situations escalates when considering the problems embedded in the indigenous cultural settings of a developing country. This case study shows how an organization and its stakeholders could adopt crisis-preparedness through systems methodologies.

In a real sense, the complexity of the problem determines which methodology to apply. The UCIL situation in Bhopal was one such case where there were different stakeholders and many different complex issues tied to a central theme—that of the gas leak at the premises of the factory.

13.5.3 The Bhopal Gas Tragedy

On December 3, 1984, a poisonous gas leaked from the UCIL plant in Bhopal, India due to gross negligence and faulty design. This leak of Methyl Iso-Cyanate (MIC) left about 10,000 people dead and more than 300,000 people disabled, blinded and injured. At the time of the accident, Union Carbide was the 37th largest company in the United States with more than 100,000 employees in over 40 countries. Its branch in India (UCIL) was the 21st largest company in that country with more than 10,000 employees.

The facility started manufacturing MIC from 1979. MIC is one of the intermediate chemicals used in the manufacture of pesticides. It is a dangerous chemical, lighter than water but twice as heavy as air. It has the ability to react with most substances. This was one of the reasons for the disaster on the night of December 3, 1984. A large amount of water got into the MIC storage tank and was not detected until the MIC started reacting with the water. Before any preventive action could be taken, about 40 tons of MIC had escaped into the air in two hours, spreading eight kilometers away from the factory.

Aftermath: Thousands of people were killed in their sleep or as they fled in terror, and hundreds of thousands were injured or affected. People who survived death were left with lingering disabilities and diseases. Exposure to MIC resulted in damage to the eyes and lungs and also resulted in respiratory ailments such as chronic bronchitis and emphysema, and a number of other diseases. Despite all these effects, Union Carbide claimed that MIC was a *mild throat and ear irritant*.

Shortcomings: Listed here are a few of the pertinent reasons for the disaster: (1) Unmotivated top management of UCIL coupled with the hiring of personnel who lacked expertise in the chemical industry, (2) reduction in personnel, leading to a compromise in security and safety, (3) boom in the city population resulting in inadequate infrastructure facilities, leading to residential neighborhoods located near the facility. These factors were instrumental in the scale of the tragedy growing to huge proportions. The lack of any trained personnel in the top and middle management led to the inadequate storage facilities for the MIC. Also, this resulted in lack of proper training for employees to handle a crisis. There were not enough people to monitor the plant, which subsequently led to the leak of MIC for two hours before detection. Lack of adequate personnel resulted in technical malfunctions and human errors. Several of the safety valves were either removed or did not work. The emergency refrigeration unit was under repair. The audio emergency signal to warn people in the vicinity of the plant was turned off. Cast-iron pipes were used instead of special stainless steel pipes. Protective slip blinds were not used before washing pipelines (Bowonder and Miyake, 1988). These were some of the failures stemming from management.

The general population of Bhopal in the near vicinity of the facility was not aware of the potential dangers that could result from such a disaster. The public's impression about the factory was that it produced harmless pesticides for agriculture and that was about all the common man knew about Union Carbide and its production. The local authorities were equally ignorant about the effects the chemicals could cause. Naturally, this led to widespread panic once the incident surfaced. The authorities asked the people to flee the city, while the most prudent thing to do would have been to lie low on the ground. The health community treated only the obvious symptoms, such as itching or watery eyes, and did not try to get to the root of the problem. The poor infrastructure facilities did nothing to help either. This resulted in a breakdown of communications in the fog of the battle. The local authorities did not have a contingency plan either to tackle such an emergency. Also, the cultural differences between the developed world and a developing country like India were a major factor overlooked by the parent company in West Virginia when the plant was set up in India. All these reasons added up to the tragedy growing to such huge and unmanageable proportions.

13.5.4 Application of SSM

To analyze and explore this problem situation from a post-audit standpoint, Checkland's SSM is applied. The process of SSM includes a series of steps that helps to weave through the maze of complexity encountered in this situation. The problem situation is analyzed and a rich picture depicting the various stakeholders and the interconnectedness between them is shown in Figure 13.9. The rich picture identifies the presence of three systems that are considered for further debate. They are (1) the management system of UCIL, (2) the political decision-making system of Bhopal, and (3) the community satisfaction system. These three systems are further explored. The use of SSM helps in critically analyzing these systems and identifying the most relevant activities that need to be attended to in the near future. The exploration of each of these systems is centered on six elements (CATWOE) that ensure that the nature of the process is systemic.

13.5.4.1 Formulation of Root Definitions

The RD of the system under observation helps in modeling the system, with the assistance of the CATWOE mnemonic (Checkland and Tsouvallis, 1996). An RD defines a set of activities,

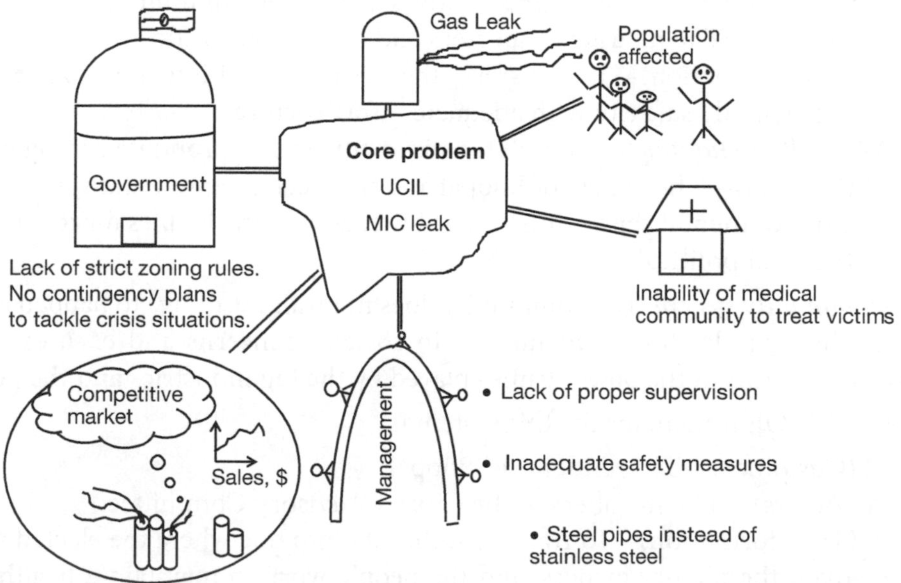

Figure 13.9 Rich picture for exploring and making decisions

including a transformation process, and always represents a particular worldview. It is an idealized representation of what the system should be. The three systems of the Bhopal gas tragedy are explored below, beginning with the formulation of RDs. A rich picture about the existing situation is used to understand the inter-relationships and the underlying complexities (See Figure 13.9)

1. *The management system of UCIL:* The policy-making body of the organization ensures smooth day-to-day operation of the factory in order to keep the organization ahead of competition and at the same time cares for the safety of the workers and the community.

   The CATWOE mnemonic used for developing the conceptual model of this system is:

   C (Customer)—Employees, customers, and community
   A (Actors)—Management
   T (Transformation)—To transform the management into a crisis-prepared system
   W (*Weltanschauung*)—An organization that cares for its employees and the community
   O (Owners)—Union Carbide, United States
   E (Environmental constraints)—Population of Bhopal, other competitors in the pesticide industry, and the bureaucratic machinery of the government

2. *The political decision-making system:* The government maintains a good coordination with the industries within its region and enforces strict zoning ordinances and ensures good infrastructure for the overall safety of its citizens.

   The CATWOE mnemonic in this situation is:

   C (Customers)—Elected representatives and council of ministers
   A (Actors)—The industry inspectors and other public officials
   T (Transformation)—To transform the city of Bhopal into a well zoned and environmentally safe city with adequate infrastructure
   W (*Weltanschauung*)—An industrialized city with appropriate zoning ordinances
   O (Owners)—The people of Bhopal and the elected representatives
   E (Environmental constraints)—The bureaucracy, the lobbies for the industry, and the local political parties

3. *The community system:* The community does not wait for things to happen and instead forces the issue by taking an interest in societal concerns and ensures it does not become a victim in the power games played by the big industries and the politicians.

   The CATWOE mnemonic for this system is:

   C (Customer)—The residents of Bhopal
   A (Actors)—The members of the Citizen Advisory Committee
   T (Transformation)—To transform the city into one where the elected representatives, the factory owners, and the people work to maintain a healthy environment

W (*Weltanschauung*)—A safe and pollution free city
O (Owners)—The people of Bhopal
E (Environmental constraints)—Population, lack of proper infrastructure, and poor health facilities

13.5.4.2 Conceptual Models

The conceptual models for each of these systems were developed based on these RDs. The purpose of a conceptual model is to accomplish what has been defined in the RD. While the RD is a representation of the system as is, the conceptual model is an account of *what the system must do in order to be the system named in the definition* (Checkland and Tsouvallis, 1996). The conceptual model is constructed by using the minimum number of verbs necessary to carry out the tasks needed to transform the existing situation to match the idealistic RD (Flood and Jackson, 1991).

The conceptual models constructed for systems 1, 2, and 3 are as shown in Figures 13.10, 13.11, and 13.12, respectively.

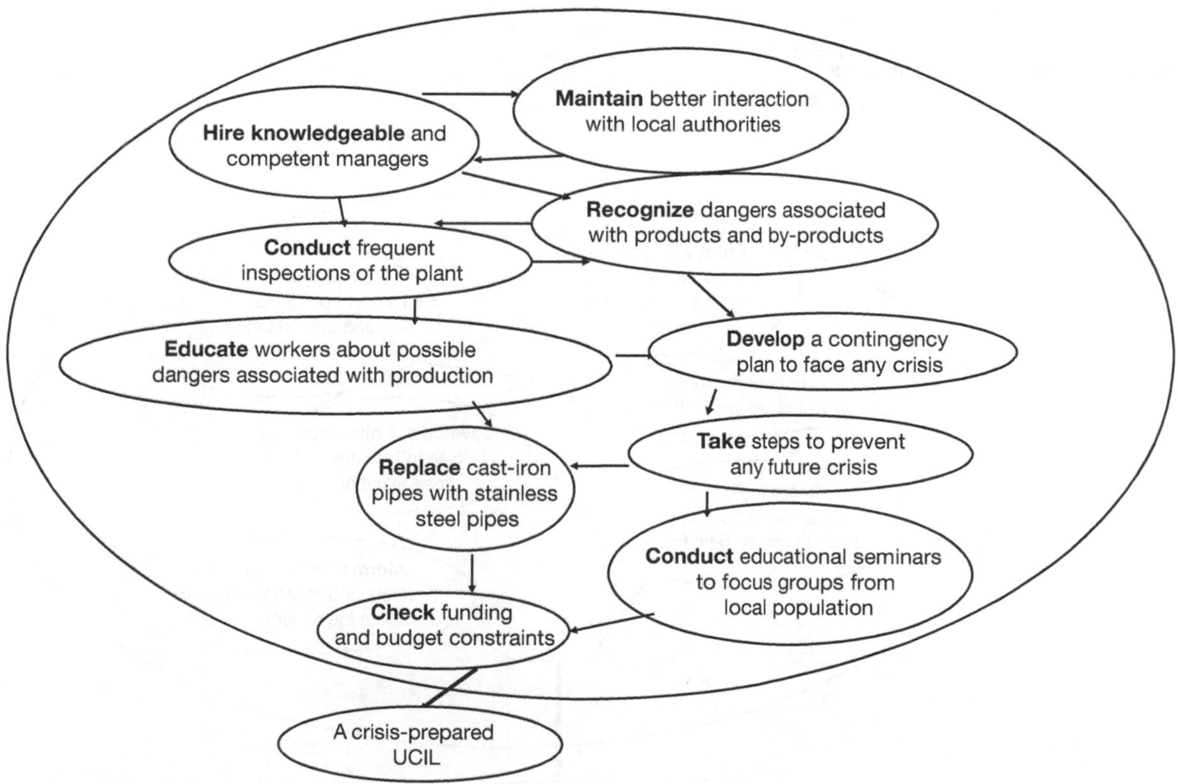

Figure 13.10 Conceptual model of the management system of UCIL

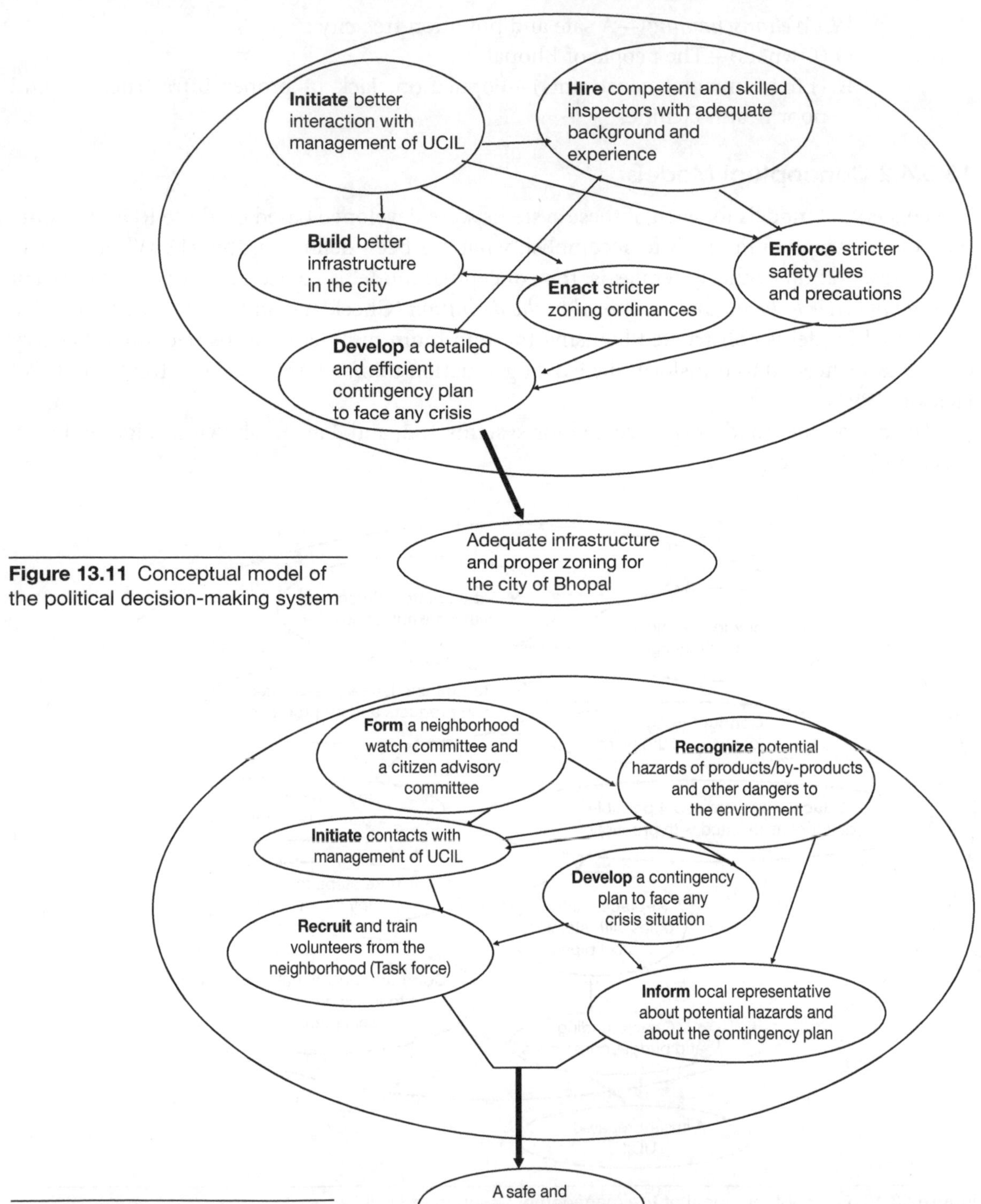

Figure 13.11 Conceptual model of the political decision-making system

Figure 13.12 Conceptual model of the community system

13.5.4.3 Comparison of Conceptual Models and Root Definitions

The conceptual models were then compared with the RDs to identify the changes that needed to be made to the real situations to bring about improvements to the problems. This process of comparison is tabulated in Tables 13.7 through 13.9, for the three systems. The activities suggested in the conceptual models are critically analyzed to check for their existence in real life situations. Then, the means to achieve these suggested ends were explored along with their potential effectiveness (Sriraj and Khisty, 1999).

13.5.5 Lessons Learned and Actions to Be Taken

The comparison of the conceptual models and the RDs of Systems 1, 2, and 3, as tabulated in Tables 13.7, 13.8, and 13.9 respectively, give a clear picture of the necessary steps that need to be taken to effect the transformations identified in the CATWOE mnemonics. It is abundantly evident that a significant percent of the blame lies with the management of the UCIL. Thus, it becomes the responsibility of the management of UCIL to take the first step toward correcting the situation. This, they could have achieved by hiring competent, well-qualified managers. As outlined earlier, the managers at UCIL did not have either the foresight or the technical excellence to prevent or handle a situation of such a magnitude. Even the handful of technically qualified people did not stay for long with the organization, partly because they were being paid less than what the market was offering. Thus, it becomes the responsibility of the upper management to attract and retain qualified and talented people within its organizational structure. Well-qualified managers in turn will strive to make sure that the company is prepared to face any crisis situation. They will do so by paying attention to detail in the day-to-day management of the factory and its workers and by helping to form policies with a clear vision for the future. They will also help in drawing up an effective contingency plan.

The other deficient area was in the skills of plant operators, many of whom did not have the proper training and skills required for handling sophisticated equipment. Management should

Table 13.7 The management system of UCIL

Activity (1)	Existence (2)	How is it done? (3)	How? Judgment? (4)
1. Competent managers	No	Appoint a committee and draw guidelines for selection	Competent managers will keep the work force prepared and well informed, which will minimize any potential danger.
2. Contingency plan	No	Form a brain trust with representatives from the various sectors that are likely to be affected in a crisis	The efficient handling of a crisis with minimal loss to life and property
3. Overcome technological deficiency	Yes	Replace cast-iron pipes with steel pipes, check in water supply	Successful treatment of MIC without any major incidents
4. Ignorance of work force	Yes	By conducting training sessions with experts	By the awareness of the workforce to potential dangers and their preparedness

Table 13.8 The political decision-making system

Activity (1)	Existence (2)	How is it done? (3)	How is it judged? (4)
1. Competent inspectors	No	Stop arbitrary transfers of inspectors and introduce a better screening process for selection.	By keeping tabs on the length of stay at a particular location for a given inspector
2. Better infrastructure	No	Build better roads, improve communication links, and improve medical facilities	A well prepared medical community and a good public communication system
3. Stricter zoning	No	Remove residential neighborhoods located near factories and vice versa	By ensuring a safe distance between residential neighborhoods and industrial areas
4. Contingency Plan	No	Form a think-tank of experts from the field of planning, sociology, and other high-tech areas to come out with a well-thought-out plan	By the ability of the government and its infrastructure to react to adverse situations

Table 13.9 The community system

Activity (1)	Existence (2)	How is it done? (3)	How is it judged?(4)
1. Neighborhood watch committee	No	By selecting people from the neighborhood with civic sense and social commitment	By the active involvement of the committee with respect to issues that might affect the community, and by the existence of neighborhood watch patrols
2. Initiate and maintain contacts with local representatives	Partial	By holding monthly citizen participation meetings in the presence of the people's representative	Same as above

provide training courses in safety and equipment handling on a periodical basis. The operators, maintenance foremen and other workers were not trained for handling emergencies and this in turn caused a delay in making decisions once the events started unfolding. This was partly because of the absence of a formal contingency plan and because the organization was not crisis-prepared. A contingency plan should be formulated with proper input from the management and feedback from the workforce.

Any contingency plan will only be as successful as the people handling it and the resources at their disposal. It is the responsibility of the government to provide the basic amenities needed for the community and to make sure that the necessary infrastructure is available to help the community. The gas leak seriously exposed the inadequacies in terms of the infrastructure in the city of Bhopal. Also, the medical community was not given all the details about the nature of the leak and the possible remedies. In the absence of information, the doctors could do nothing but treat the symptoms instead of attending to the patients in a systemic manner.

The deficiencies in the organization and the government were just a part of the larger problem—that of illiteracy, poverty and underdevelopment. Bhopal is a medium-sized city and soon after independence the policies of the government set the city on the path toward rapid industrialization. This attracted more people toward the city, thus making the existing infrastructure facilities inadequate. The government did not do an adequate job of promoting the virtues of education and providing facilities for people to learn.

A crisis management program for the city of Bhopal is outlined in the following section and includes not only the organizations at risk, but also the different stakeholders who stand to benefit from such a program. This program is especially applicable in developing countries like India where the population density is high and the resources are scarce.

13.5.6 The Crisis Management Program

An attempt has been made at developing a crisis management program for the city of Bhopal to enable it to be prepared in case of disasters of a similar nature. The program includes both the government and the potential sources of such disasters. The organization should be required to have complete knowledge of all its products and their effects, if they are exposed to the environment. This detail would help in deciding the course of action to be taken to prevent any repetition of such an incident. It also becomes the responsibility of the employees who operate the plant and handle these chemicals to ensure their proper maintenance. The organizations can ensure this by providing incentives for the maintenance supervisors and their teams. Crisis-prepared organizations learn from each crisis, and emerge stronger and healthier than before.

The government, for its part, should improve the infrastructure facilities with respect to the medical community and also pass stricter zoning ordinances. This is easier said than done, especially in a country like India where the population density is very high. This problem of zoning can be tackled by laying the onus on the organization by making transport facilities available to the employees, to and from work, a practice not uncommon in India. This will ensure that people do not live in the immediate vicinity of potentially dangerous zones. It should be made mandatory for any organization dealing with chemicals to have a crisis management plan. The government should also make sure that proper medical facilities are provided for and necessary communication links available to everyone. This again is part of a bigger problem—that of population, which is the root-cause for most problems in such developing countries. The communication aspect has been improving slowly over the past decade or so and will soon become a nonfactor.

The people should make sure that they are well informed about their surroundings and the factories they are working for. This is a two-way street where there should be proper coordination between the organizations and the citizens. The people should form citizens groups that can take up these societal issues either with the organizations or with the local elected representatives. Thus, it is evident from this analysis that it is not just the responsibility of a single organization in terms of dealing with sociotechnical disasters. Instead, it is the responsibility of everyone associated with the source to make sure that the community as a whole is prepared.

13.5.7 Discussion and Summary

This case study has critically analyzed the issues surrounding the chemical spill and the subsequent gas leak from the premises of the UCIL, Bhopal. Conventional methods of dealing with crises have not been successful in preparing the organizations for future crisis events and the learning process associated with such events has been grossly underestimated in the past. With the help of systems methodologies like SSM, this case study has highlighted the systemic nature of such events and has also shown its usefulness in preparing a crisis management plan for a city or an organization or a community. The process should be further explored using a variety of scenarios and making use of a menu of methodologies to arrive at a template of tasks that need to be performed by the organizations and its stakeholders in order to be crisis-prepared. Current literature on this topic deals only with the crisis-preparedness of the organization and does not focus on the different stakeholders who are privy to the functionality of the organization. These stakeholders are most likely to be the affected parties in the event of a crisis. Further research is needed to explore the use of systems methodologies to help communities deal with such disasters in a better manner, with minimal damage to living beings and their environment.

It should be emphasized, at this point, that the synthesis and use of these soft methodologies in the planning and day-to-day management of organizations is not meant as a slight to the conventional hard methodologies that have been in practice for so long. Instead, the two strands should be used in a judicious manner to integrate the best of both these views. It is the responsibility of the management to incorporate the use of these new and emerging paradigms into its daily functioning. By doing so, the management can ensure that a systemic outlook toward problem situations is imbibed into its components and workforce. This will make it easier for the organization to be systemic in its thinking when a crisis situation does occur. It would behoove an organization to expose its workforce to these tools by holding invited seminars from experts in the field of systems thinking and crisis management, and providing short courses with training exercises for its employees. City managers, engineers, and planners also need to be exposed to these exercises in order for the community to benefit from systemic thinking (Khisty, 1996).

On the other side of the coin, the government in such developing countries should do everything under its power to educate its citizens about the need to preserve and sustain the environment. This can be done only if the people are educated enough to understand such issues. With the media becoming a potent force in disbursing information, more and more critical issues affecting the lives of the citizenry should be brought to the forefront. This will spur the citizens to seize the proverbial bull by the horn, instead of waiting for the government to assist them in each and every endeavor.

13.6 CONCLUDING REMARKS

A wide variety of case studies have been described in this chapter that indicate the diversity of applications possible in dealing with infrastructure improvement by utilizing the tools and procedures detailed in previous chapters. Readers who have gone through these case studies may have noticed some commonalities in these applications. A few pertinent ones are:

Whereas HSM aims at achieving objectives that are spelled out before the analysis has begun, the SSM and the other soft systems methods are concerned with learning and inquiring about a complex and uncertain situation. Managing is interpreted as a process leading to organized action. SSM and the other soft systems methods assume that individuals and groups will perceive a problem in different ways and expect that decision makers take correspondingly different actions into consideration. That is why participative democracy is necessary.

The role of public input in decision making is therefore vital in the sense that planning, designing, evaluating, and policy making cannot be adopted and implemented without popular support.

Without rigorous evaluation, current practices may continue to waste resources and may fail to promote larger societal goals. A systemic and systematic evaluation not only assesses the effectiveness and efficiency of any planning effort, but also provides feedback to help in improving programs.

In this day and age, issues connected with sustainability and ethics are vital, and these issues have been addressed in some of the case studies included in this chapter. Readers may think hard about how to tackle these types of problem situations and others that they may encounter with the help of the methods illustrated in this chapter. However, attention should be paid to the fact that the case studies presented in this chapter do not include all of the details pertaining to the thinking of the stakeholders or all of the realities in which they are situated. Thus, readers are advised to exercise a judicious approach toward applying these methods to tackle complexity.

REFERENCES

Ackoff, R. L. (1974). *Redesigning the Future*, New York, NY: John Wiley and Sons.

Albrechts, L. (2002). "The Planning Community Reflects on Enhancing Public Involvement. Views from Academics and Reflective Practitioners," *Planning Theory & Practice*, 3:3, 331-347.

Arnott, R. (1997), "Economic Theory and the Spatial Mismatch Hypothesis," *Working Papers in Economics*, Boston, MA: Boston College.

Arnstein, S. (1969). "A Ladder of Citizen Participation," *Journal of the American Planning Association*, 35:4, 216-224.

Beer, S. (1985). *Diagnosing the System for Organisations*, Chichester, UK: John Wiley.

Bickerstaff, K. and G. Walker (2001). "Participatory Local Governance and Transport Planning," *Environment and Planning* A 33:3, 431-451.

Bickerstaff, K., R. Tolley, and G. Walker (2002). "Transport Planning and Participation: the Rhetoric and Realities of Public Involvement," *Journal of Transport Geography*, 10:1, 61-73.

Bowonder, B. and T. Miyake (1988). "Managing Hazardous Facilities: Lessons from the Bhopal Accident," *Journal of Hazardous Materials*, vol. 19, 237-269.

Braverman, M. (1997). Are You Crisis-prepared? *Crisis Management Group, Inc.*, Maryland.

Brown Line Customer Survey (2005). *Market Research Department*, Chicago Transit Authority, Chicago.

Burby, R. (2003). "Making Plans That Matter: Citizen Involvement and Government Action." *Journal of the American Planning Association*, 69:1, 33-49.

Checkland, P. (1981). *Systems Thinking, Systems Practice*, Chichester, UK: John Wiley.

Checkland, P. and C. Tsouvalis (1996). "Reflecting on SSM: the Link Between Root Definitions and Conceptual Models," *Working Paper No.5, The Center for Systems & Information Sciences*, University of Humberside. Hull, United Kingdom.

Churchman, C. W. (1968). *The Systems Approach*, New York, NY: Delacorte Press.

Falcocchio, J. (2004). "Performance Measures for Evaluating Transportation Systems: Stakeholder Perspective," *Transportation Research Record: Journal of the Transportation Research Board*, 1895, 220-227.

Flood, R. L. and M. C. Jackson (1991). *Creative Problem Solving: Total Systems Intervention*, John Wiley & Sons. Chichester, United Kingdom.

Glass, J. J. (1979). "Citizen Participation in Planning: the Relationship between Objectives and Techniques," *Journal of the American Planning Association*, 45:2, 180-189.

Grossardt, T., K. Bailey, and J. Brumm (2003). "Structured Public Involvement: Problems and Prospects for Improvement." *Transportation Research Record: Journal of the Transportation Research Board*, 1858, 95-102.

Innes, J. and D. E. Booher (2004), "Reframing Public Participation: Strategies for the 21st Century," *Planning Theory & Practice*, 5:4, 419-436.

Irvin, R. and J. Stansbury (2004). "Citizen Participation in Decision-making: Is it worth the effort?" *Public Administration Review*, 64:1, 55-65.

Jackson, M. C. (1988). "Systems Methodologies as Complementary Tools for Managing Situational Complexity," *Transactions of the Institute of Measurement and Control*, 10:3, 155-160.

Khisty, C. J. (1996). "Education and Training of Transportation Engineers and Planners Vis-à-vis Public Involvement," *Transportation Research Record 1552*, Washington, D.C., National Academy Press, 171-176.

Mattingly, M., P. S. Sriraj, E. Welch, and B. Bhojraj (2010). "Measuring and Assessing the Perceptions of Success in a Transit Agency Stakeholder Involvement Program," *Transportation Research Record 2174*, Washington, D.C., 89-98.

Mitroff, I. I., Pearson, C. M., and K. L. Harrington (1996). *The Essential Guide to Managing Corporate Crises: A Step-by-Step Handbook for Surviving Major Catastrophes*, Oxford University Press, Oxford, United Kingdom.

O'Connor, R., M. Schwartz, J. Schaad, and D. Boyd (2000). "State of the Practice: White Paper on Public Involvement. Transportation in the New Millenium," *Committee on Public Involvement in Transportation A1D04*. Washington, D.C.

Prevost, D. (2006). "Geography of Public Participation: Using Geographic Information Systems to Evaluate Public Outreach Program of Transportation Planning Studies," *Transportation Research Record: Journal of the Transportation Research Board*, 1981, 84-91.

Richardson, B. (1994). "Socio-technical Disasters: Profile and Prevalence," *Disaster Prevention and Management*, 3:4, 41-69.

Rowe, G. and L. J. Frewer (2000). "Public Participation Methods: A Framework for Evaluation," *Science, Technology and Human Values*, 25:1 3-29.

Sriraj, P. S. (1999). "Dealing with Complexity and Uncertainty in Urban Planning using Systems Intervention," *Doctoral Dissertation*, Illinois Institute of Technology, Chicago, IL.

Sriraj, P. S. and C. J. Khisty (1999). "Crisis Management and Planning Using Systems Methodologies," *ASCE Journal of Urban Planning and Development*, 125:3, 121-133.

Szyliowicz, J. (2002). "Measuring the Effectiveness of Public Involvement Approaches, *TR News*, May-June, 220, 35-38. Retrieved from http://onlinepubs.trb.org.

Ulrich, W. (1983). *Critical Heuristics of Social Planning: A New Approach to Practical Philosophy*, Berne, CH: Paul Haupt.

Webster's Revised Unabridged Dictionary (1997). Plainfield, NJ: G&C Merriam.

Weeks, J. (2002). "Public Involvement by Minorities and Low-income Populations: Removing the Mysteries," *TR News*, 220, 25-31. Retrieved from http://onlinepubs.trb.org.

Sustainable Development, Sustainability, Engineering and Planning

14

14.1 INTRODUCTION

The American Society of Civil Engineers (ASCE) defines sustainability as *a set of environmental, economic and social conditions in which all of society has the capacity and opportunity to maintain and improve its quality of life indefinitely without degrading the quantity, quality or the availability of natural, economic and social resources* (ASCE, 2009). The concept of sustainable development is not new. The history of life on earth has also been a history of interaction between living things and their natural environment (Carson, 1962) that has, at times, been in better balance than at others. Several ancient civilizations grappled with sustainable development and sustainability issues. Where some were successful at identifying and addressing what was critical to sustain their development, others failed and entire civilizations were lost (Chambers et al., 2000). In recent history, however, sustainable development has become a global priority and, subsequently, a policy objective for several nations, beginning in the 1980s with the release of a global blueprint for sustainable development—*Our Common Future*. The blueprint, developed by the United Nations' World Commission on Environment and Development (WCED), offered what has become the most widely cited definition of sustainable development: *development that meets the needs of present generations without compromising the ability of future generations to meet their own needs* (WCED, 1987). Several operational definitions of sustainable development have been developed based on this conceptual definition, often as the starting point for communities (i.e., municipalities, nations, etc.), agencies, and other entities interested in addressing sustainability both systemically and systematically. Table 14.1 gives examples of operational definitions of sustainable development and sustainability relative to transportation systems.

14.1.1 Important Sustainability Issues for the Engineering Community

The world's population passed the seven billion mark in 2011. While the world's population took over two centuries to double between the 18th and 20th centuries, it tripled between 1950 and 2000 with unprecedented global population growth. Other demographic changes are unprecedented. The year 2005 is the midpoint of a decade that spans three unique and important transitions in the history of humankind. Before 2000, young people (i.e., under 60 years) outnumbered old people; in 2000, old people outnumbered young people for the first time

Table 14.1 Operational definitions of transportation for sustainable development

Source	Definition
New Zealand Ministry of Transport	Sustainable transport is about finding ways to move people, goods, and information in ways that reduce its impact on the environment, economy, and society. Some options include: (1) using transport modes that use energy more efficiently, such as walking or cycling and public transport; (2) improving transport choice by increasing the quality of public transport, cycling and walking facilities, services, and environments; (3) improving the efficiency of our car use, such as using more fuel-efficient vehicles, driving more efficiently, avoiding cold starts, and car pooling; (4) using cleaner fuels and technologies; (5) using telecommunications to reduce or replace physical travel, such as teleworking or teleshopping; (6) planning the layout of cities to bring people and their needs closer together and to make cities more vibrant and walkable; and (7) developing policies that allow and promote these options, such as the New Zealand Transport Strategy (2008).
Centre for Sustainable Transportation (Project funding: Centre for Sustainable Transportation and the Government of Canada: Environment Canada and Transport Canada)	A sustainable transportation system is one that: (1) allows the basic access needs of individuals and societies to be met safely and in a manner consistent with human and ecosystem health and with equity within and between generations; (2) is affordable, operates efficiently, offers choice of transport mode, and supports a vibrant economy; (3) limits emissions and waste within the planet's ability to absorb them, minimizes consumption of nonrenewable resources, reuses and recycles its components, and minimizes the use of land and the production of noise (2003).
Organization of Economic Cooperation and Development (Environment Directorate)	Environmentally sustainable transportation is transportation that does not endanger public health or ecosystems and that meets needs for access consistent with the (1) use of renewable resources at below their rates of regeneration and (2) use of nonrenewable resources below their rates of regeneration (1999).
PROSPECTS: Developing Sustainable Urban Land Use and Transport Strategies: Methodological Guidebook: Procedures for Recommending Optimal Sustainable Planning of European City Transport Systems (Project Funding: European Commission's Energy, Environment and Sustainable Development Programme)	A sustainable urban transport and land use system: (1) provides access to goods and services in an efficient way for all inhabitants in the urban area; (2) protects the environment, cultural heritage, and ecosystems for the present generation, and (3) does not endanger opportunities of future generations to reach at least the same welfare level as those living now, including the welfare they derive from their natural environment and cultural heritage (2003).
The Sustainable Transportation Action Network, The Urban Environmental Management Research Initiative, Global Development Research Center	Sustainable transportation concerns systems, policies, and technologies. It aims for the efficient transit of goods and services and sustainable freight and delivery systems. The design of vehicle-free city planning, along with pedestrian and bicycle-friendly design of neighborhoods is a critical aspect for grassroots activities, as are telework and teleconferencing. It is more about accessibility and mobility than about transportation (Accessed Nov 2010).

in the history of the world. Until approximately 2007, rural dwellers always outnumbered urban dwellers; from 2007 onward, urban dwellers have outnumbered rural dwellers (Cohen 2005). Demands for natural resources have followed suit with a growing number of major cities (i.e., cities with over one million people), megacities (i.e., cities with over 10 million) and megaregions (i.e., contiguous regions of major cities and megacities). Global sustainability and sustainable development priorities have been articulated, including climate change mitigation and adaptation, water scarcity, and the end of carbon-based energy. In developing regions, lack of access to basic sanitation, clean water, basic health care, food security and other basic human development issues are important priorities for sustainable development. The eight United Nations' Millennium Development Goals (see Table 14.2) were adopted by 170 heads of states and governments in 2005 to reverse poverty, hunger, disease, and other deplorable conditions affecting billions of people, by the year 2015. As urbanization continues with growing demands for improved quality of life, engineers have the opportunity and challenge of reengineering cities to be more sustainable.

The Vision of Civil Engineering in 2025, a roadmap for the Civil Engineering profession, in the US recognizes these pressures and expands the role of civil engineers in the future to include (ASCE 2007):

- Planners, designers, constructors, and operators of society's economic and social engine—the built environment
- Stewards of the natural environment and its resources
- Innovators and integrators of ideas and technology across the public, private, and academic sectors
- Managers of risk and uncertainty caused by natural events, accidents, and other threats
- Leaders in discussions and decisions shaping public environmental and infrastructure policy

Table 14.2 The United Nations' millennium development goals

Goal 1: Eradicate extreme poverty
Goal 2: Achieve universal primary education
Goal 3: Promote gender equality and empower women
Goal 4: Reduce child mortality
Goal 5: Improve maternal health
Goal 6: Combat HIV/AIDS, malaria, and other diseases
Goal 7: Ensure environmental sustainability
Goal 8: Develop a global partnership for development

14.1.2 ASCE Code of Ethics and Sustainable Development

The first of the seven fundamental canons of the ASCE Code of Ethics requires engineers to hold paramount the safety, health, and welfare of the public and to strive to comply with the principles of sustainable development in the performance of their professional duties. Engineering for sustainable development is, therefore, an ethical responsibility for civil engineers. Engineering for sustainable development can sometimes present ethical dilemmas requiring engineers to balance their responsibility for improving human welfare with their fidelity to clients who are their main sources of remuneration. Increasingly, engineers will be required to assume leadership roles to educate their clients and contractors about the benefits of sustainable engineering that may call for trading off short-term costs against long-term benefits. More and more engineering companies are demonstrating that they can take a leadership role in adapting to be more sustainable while simultaneously developing a competitive edge in their business, attracting clients who are willing to pay for higher quality products and services with fewer risks in the long run (Anderson, 1999).

14.1.3 Sustainability and Systems Thinking

The principles of sustainability and systems thinking are closely linked. Taken together they contribute core values that form the basis for reshaping our planet. In contrast to the mechanistic Cartesian view of the world, the systems view emerging from general systems theory can be characterized as systemic, organic, holistic, complex, and ecological. In this sense the world is not seen as a machine made up of a multitude of parts, but is pictured as one indivisible, dynamic whole whose parts are essentially interrelated, interdependent, and can be understood only as patterns of a complex process. The principles of systems thinking are covered in Chapter 12, and readers are urged to go through it to more fully appreciate the topics dealt with in this chapter.

Planners, engineers, and other professionals engage in sustainable development by using their skills and creativity for improving man-made and human-activity systems. Collectively, these systems, while functioning within the limits of the natural environment, aim to enhance the social quality of life and the economy of communities. In this context, systems thinking plays a vital role in sustainable development, focusing on the relationship among parts forming a connected whole. For the most part professionals have to deal with complex systems consisting of a large number of different parts having an enormous number of possible connections between parts, and with multiple feedback loops, where small changes in the parts may make substantial differences in the outcome.

Civil engineering infrastructures (i.e., man-made support systems) involve planning, designing, and implementation of physical systems as well as human-activity systems. This involvement demands both systemic (i.e., holistic) and systematic (i.e., procedural or step-by-step) methodologies. Typically, hard systems methodology are most appropriate for the design of physical systems, while for human-activity systems there are a wide variety of methodologies available, ranging from soft systems to critical systems to multimodal systems. Sustainable development requires very careful consideration of the methodological choices available, recognizing that not only does such development impact humans beings but also nonhuman beings. These issues may require ethical and emancipatory considerations to be addressed, in addition to what soft systems thinking can handle by itself. Details of hard, soft, critical, and multimodal methodologies are given in Chapter 12.

14.1.4 System Interconnections and Interdependencies

The triple bottom line (TBL) model for sustainable development, widely embraced as a useful construct to guide sustainable development, communicates some of the systems' interconnections and interdependencies that are at play as we pursue engineering for sustainable development. Figures 14.1a and 14.1b show two different frameworks for the TBL concept: the Venn diagram and Russian dolls models.

TBL thinking relates to preserving natural environment systems while advancing the development of the social and economic systems of communities. While the Venn diagram model tends to place all three systems at the same priority level, the Russian dolls model distinguishes among the three systems, recognizing the natural environment system primarily in which all other social and economic activities are conducted. The social system is recognized next, out of which an economic system is developed and maintained. These systems all interact with one another and are interdependent. A community with a failing natural environment system can expect to see associated impacts that may limit their ability to improve the community quality of life or further develop the community's economic competitiveness. Improper thinking about economic development may result in negative impacts on the natural environment, which can ultimately affect the economic competitiveness of societies. For example, the indiscriminate logging of rainforests for timber products has sometimes resulted in the destruction of natural

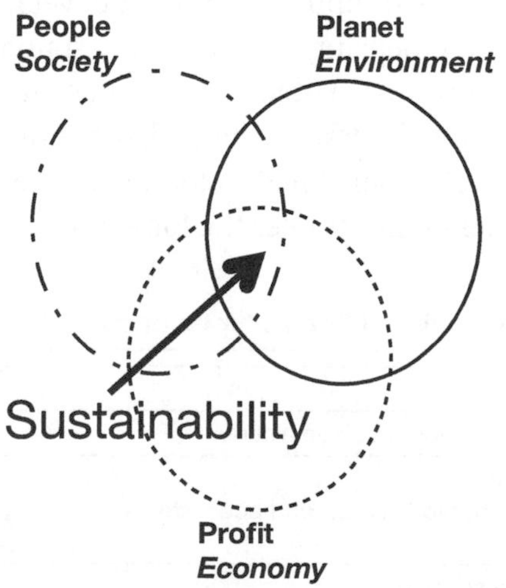

Figure 14.1a Triple bottom line—Venn diagram model

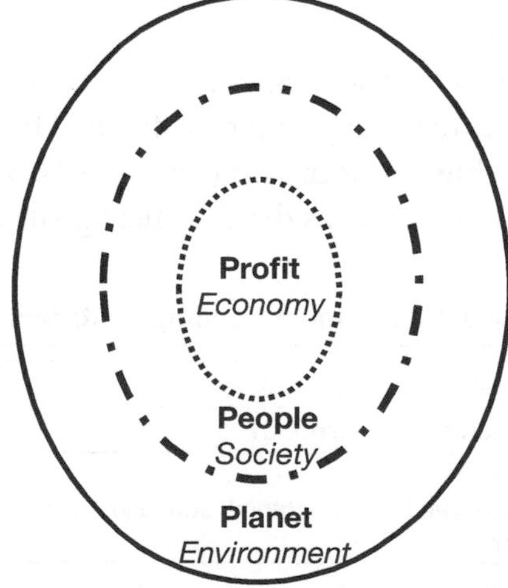

Figure 14.1b Triple bottom line—Russian dolls model

habitats of various species, increased soil erosion, and ultimately the loss of the rainforest that supports the livelihood of communities in various ways.

Modern societies sometimes strive for better balance, understanding how their civil infrastructure systems can be interdependent, and recognizing the need to incorporate these interdependencies in designing for more resilient infrastructure. The Northeast blackout of 2003 for example, the largest blackout in North American history, demonstrates the risks and costs of cascading failures in interdependent systems. The event was caused by a combination of human error and equipment failures. It contributed to at least 11 deaths, left over five million people without power for up to two days and cost an estimated $6 billion (Minkel, 2008). The blackout affected parts of the Northeast and Midwestern United States and Ontario, Canada. Lack of power meant the lack of air conditioning, access to banking systems, inability to use computer systems, and a loss of water pressure in some areas where pumps lost power. It also meant the shutdown of railroad services in Amtrak's northeast corridor, shut down of passenger screenings at airports, shut down of some regional airports, many gas stations unable to pump fuel due to lack of electricity, shut down of many oil refineries on the east coast, and the disruption of cellular communication devices, radio stations and cable-TV stations, and large numbers of factories. Viewed and managed as complex systems, the sustainability of these interdependent systems includes their resilience, reliability and risks—all of which are important performance parameters that must be monitored and managed effectively.

14.1.5 Strong Versus Weak Sustainability

Several types of capital are essential for human development: human, natural, technological, manufactured, and financial capital are all examples of much needed inputs for human development and progress (Hawken et al., 1999). *Our Common Future*, a blueprint for attaining sustainability, developed by the U.S. National Research Council Board on Sustainable Development in 1999, identifies the different types of capital that ought to be developed, versus sustained, in the quest for sustainable development (see Table 14.3). The report frames sustainable development in terms of reconciling society's development goals with its environmental limits over the long term.

Table 14.3 Sustainable development: Sustaining versus developing different types of capital

Sustainable Development	
What should be developed?	What should be sustained?
People Child survival, life expectancy, education, equity, and equal opportunity	Nature Earth, biodiversity, and ecosystems
Economy	Life support
Wealth, productive sectors, and consumption	Ecosystem services, resources, and environment
Society	Community
Institutions, social capital, states, and regions	Cultures, groups, and places

(Source: *Our Common Journey*, U.S. National Research Council, 1999)

The concepts of strong and weak sustainability arise out of transformations between natural and man-made forms of capital. Strong sustainability requires that natural capital stocks be held constant. Weak sustainability, on the other hand, allows for the substitution of equivalent human-made capital for depleted natural capital (Wackernagel and Rees, 1996). Many economists believe that weak sustainability is good enough. According to this view, society is sustainable provided that the aggregate stock of manufactured and natural assets is not decreasing. In contrast, strong sustainability recognizes ecological services that are unaccounted for and life-support functions performed by many forms of natural capital, and the considerable risk associated with their irreversible loss. Strong sustainability therefore requires that natural capital stocks be held constant independently of human-made capital. As population and material expectations (or demand) are rising, capital stocks should actually be enhanced to maintain community quality of life, all else being equal. In other words, per capita stocks must be increased to maintain present levels of community quality of life unless there are unforeseen changes in population growth, lifestyles or technology to mitigate for the increasing numbers of people.

14.1.6 Shallow and Deep Ecology

The distinction between shallow and deep ecologies was originally made in the early 1970s by the Norwegian philosopher and mountaineer Arne Naess (Sessions, 1995). It has now been accepted as a useful terminology to refer to the major division within contemporary environmental thought. Shallow ecology is anthropocentric. It views humans as above or outside of nature, as the source of all value, and ascribes only instrumental or use value to nature. Of course, this perspective goes with the domination of nature. Naturally, the ethical framework associated with shallow ecology is no longer adequate to deal with some of the major problems of today, most of which involve threats to nonhuman forms of life.

On the other hand, deep ecology does not separate humans from the natural environment, nor does it separate anything else from it. It does not see the world as a collection of isolated objects but rather as a network of phenomena that are fundamentally interconnected and interdependent. Deep ecology recognizes the intrinsic values of all living beings and views humans as just one particular strand in the web of life. This implies a corresponding ecologically oriented ethics that confers inherent value on all forms of life. A deep ecological system, in essence, is holistic and seeks to view all processes in that light (Sessions, 1995).

The short-term shallow approach stops short before the ultimate level of fundamental change, often promoting technological fixes, (e.g., recycling, increased automotive efficiency, and export-driven mono-cultural organic agriculture) based on the same consumption-oriented values and methods of the military-industrial economy. The long-range deep approach involves redesigning our whole system, based on values and methods that truly preserve the ecological and cultural diversity of natural systems.

14.2 MODELS OF SUSTAINABLE DEVELOPMENT AND SUSTAINABILITY

Theoretical concepts and models of sustainable development and sustainability provide operational frameworks that can be used in engineering planning, design and evaluation. Seven useful models of sustainable development and sustainability are discussed in the sections that follow:

1. The TBL Framework
2. The IPAT Model
3. The Ecological Footprint (EF) Model
4. The Triaxial Representation of Technological Sustainability (TRTS)
5. The Quality of Life/Natural Capital Model (QOL/NC)
6. The Sustainability Footprint (SF) Model
7. The True Sustainability Index (TSI)

14.2.1 The Triple Bottom Line Framework

The Triple Bottom Line (TBL) framework identifies environmental, social and economic capital as the three pillars of sustainability, also known as the three-legged stool of sustainability. There are various frameworks (see Figures 14.1a and 14.1b, for example) being used today; however, the essential premise of the TBL is that humans have access to different types of capital—environmental, social and economic—and that they must manage these various forms of capital in a way that ensures that communities can continue to improve the quality of their lives indefinitely. The basic TBL framework (see Figure 14.1a) is concerned with keeping capital transformations in a state where there is a balance between the quality of the environment, and available economic resources and social well-being. The Russian dolls TBL framework (see Figure 14.1b) has the same concerns but prioritizes the importance of the environment higher than social well-being or economic activity, recognizing that the environment is the most critical constraint to social and economic development. Similarly, the society is prioritized higher than the economy indicating that the economic system is developed within the social system.

When addressing planning, design, and performance monitoring of civil infrastructure systems over their life cycles, a fourth fundamental dimension is considered for system sustainability: the facility's functional performance (see, for example, Deakin, 2001; Zietsman and Rilett, 2002; Jeon and Amekudzi, 2005). There is emerging consensus that for sustainability, a civil infrastructure system such as a transportation system must achieve the minimum levels of functional performance (e.g., providing certain minimum levels of mobility/accessibility, travel time reliability and safety) for its users while interacting favorably with the natural environment, contributing to improved quality of life (including social equity) within the community, and enhancing the community's economic competiveness.

14.2.2 The IPAT Model

Developed in the 1970s, the IPAT model, focuses on ecological sustainability, emphasizing the relationship between the environmental impact, number of consumers, their affluence (or level of consumption) and the technological efficiency of delivering a particular product or service (Erlich and Holdren, 1971, 1972). The environmental impact of a service or product is characterized as a product of the population, its affluence and level of technological development as shown:

$$I = P \times A \times T$$

where:

I is the Impact
P is the Population
A is Affluence
T is Technology

As an example, the amount of fuel used to travel a certain distance depends on the overall demand for travel, the mode of transportation and the efficiency of that form of travel. To achieve ecological sustainability, the impact needs to be within the natural limits imposed by the Earth's carrying capacity. Table 14.4 shows an application of the IPAT model to evaluating the sustainability impacts of travel in two communities, reinforcing the fact that technological and behavioral improvements are both important to achieve progress toward sustainability. Policies and regulations, together with enforcement, can be used to promote such changes in behavior.

14.2.3 The Ecological Footprint Model

The Ecological Footprint (EF) model was developed by Wackernagel and Rees in the 1990s as an ecological accounting tool to enable estimation of the resource consumption and waste assimilation requirements of a defined human population or economy in terms of a corresponding productive land area. The EF is the amount of land and water required to support the living standards of a given population. The EF of a person is calculated by considering all of the biological materials consumed and all of the biological wastes generated by that person in a given year. All these materials and wastes are then individually translated into an equivalent number of global hectares. Measured EFs are compared to the biocapacity that is the total amount of area available to generate resources and absorb wastes. When human demand exceeds available biocapacity, this is referred to as overshoot. For ecological sustainability, the EFs of communities and individuals must not exceed available biocapacity. Table 14.5 shows the Ecological Footprints and productivity of various energy sources (Wackernnagel and Rees, 1996). The footprint varies inversely as the productivity of an energy source.

Table 14.4 The IPAT model: Comparison of two cities

IPAT Example—A Comparison of Two Cities

Blacktop and Parktown are two small towns of 1000 people each. Blacktop, suffering from urban sprawl, is a place where all residents own fuel-efficient cars (15 km per liter on the average) and travel an average of 20,000 km per year. In Parktown, good public transport and well-planned amenities mean that only half of the inhabitants need a car. Those that do own a car do not need it so often and therefore replace it less frequently. Each car owner drives 10,000 km per year on the average. However, their older model vehicles are less fuel efficient (10 km per liter) than those in Blacktop.

We measure the environmental impact of both small cities in terms of fuel consumption, using the IPAT formula.

Blacktop

P = 1,000 (Number of cars owned)

A = 20,000 (Number of km traveled per yr)

T = 1/15 (liter of fuel required per km)

I = 1,000 * 20,000 * (1/15) = 1,334,000 liter per yr (that corresponds to 3,150 tn of CO_2 per yr)

Parktown

P = 500 cars

A = 10,000 km per yr

T = 1/10 liters per km

I = 500 * 10,000 * (1/10) = 500,000 liter per yr (1,181 tn of CO_2 per yr)

Thus, despite the fact that Blacktop has more fuel-efficient vehicles (technology), the residents have a greater CO_2 emissions impact than those of Parktown due to both the number owning cars (population) and their lifestyles that involve significant travel (affluence or lifestyle)

(Adapted from Chambers et al., 2000)

Table 14.5 Ecological footprints and productivity of various energy sources

Energy source	Footprint for 100 Gigajoules per yr (in hectares)	Productivity (in Gigajoules per hectare per yr)
Fossil fuel		
Ethanol approach	1.25	80
CO_2 absorption approach	1.0	100
Biomass replacement approach	1.25	80
Hydro-electricity (average)	0.1	1,000
Lower course	0.2-0.67	150-500
High altitude	0.0067	15,000
Solar hot water	0.0025	Up to 40,000
Photovoltaics	0.1	1,000
Wind energy	0.008	12,500

(Adapted from Wackernagel and Rees, 1996)

14.2.4 Triaxial Representation of Technological Sustainability

The Triaxial Representation of Technological Sustainability (TRTS) framework was developed by Pearce and Vanegas in the 1990s, using Thermodynamic Laws as the foundation. Humans, by their activities that expend energy, tend to increase the earth system's entropy (or degree of disorder). In all systems, entropy increases with the expenditure of energy. In the earth system, the potential exists for energy received from the sun to exceed the amount of energy lost as thermal radiation: the difference is called the *solar energy budget*. The solar energy budget can be used to offset entropy increases, resulting from transformations of matter and energy within the earth system, by fixing more carbon in the earth system in the gradual and long-term development process of carbon-based fuels. The requirements for sustainability are the conservation of matter and energy, and stable entropy. As long as we consume less energy than that supplied by the solar energy budget, sustainability is theoretically possible. For any system to be sustainable there must be no net loss of the sum total of matter and energy circulating within the system. In addition, the state of entropy within the system must be stable for the system to exist indefinitely (Pearce, 1999).

Based on these foundations, the TRTS model is a three-dimensional framework that may be used for planning, project design, and evaluation. The framework measures the sustainability impact of a system, service or product by evaluating the stakeholder satisfaction (a measure of social quality of life) it produces in comparison with its impact on the resource base (i.e., demand for natural resources) and ecosystem (i.e., waste assimilation requirements). Figure 14.2 depicts that plans, projects, or services that are considered more sustainable will be found

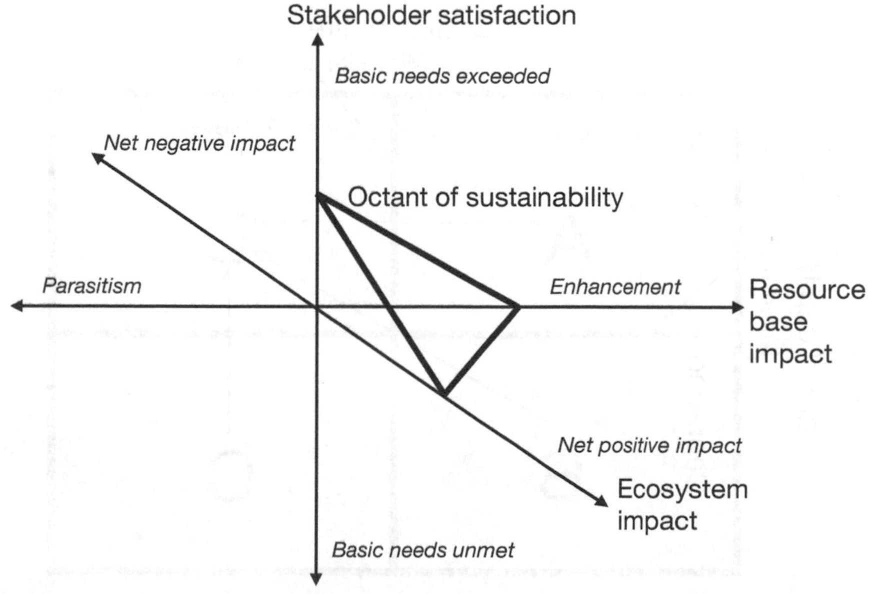

Figure 14.2 Triaxial representation of technological sustainability

in the octant of sustainability where stakeholder satisfaction is met or exceeded, with a neutral or positive impact on the resource base and ecosystem, inferring that technological developments are occurring within the carrying capacity of the environment. This framework explicitly includes human and environmental capital measures.

14.2.5 The Quality of Life/Natural Capital Model

The QOL/NC model was presented by Chambers, Simmons, and Wackernagel in 2000. This model assumes that the primary objective of sustainability is to achieve satisfying lives for all while staying within the bounds of nature. It is one of the few models that distinguish between sustainable development and sustainability. Shown in Figure 14.3, the QOL/NC model describes sustainable development as the process or pathway to the state of sustainability, and differentiates between higher and lower levels of sustainability, once the state of sustainability has been reached.

The QOL/NC model describes four zones that can be used to classify communities or other entities according to their positions with respect to sustainability (see Figure 14.3). Zone A represents a situation where nature is protected but even the most basic quality of life objectives are not met. An example of this is a community that has been denied access to its natural resources and is therefore unable to feed, shelter, or clothe itself, that is, wildlife conservation projects where the development needs of the locals are ignored. Zone B represents a situation where the environment is being degraded and yet the QOL is low (e.g., poor public health, lack of access to sanitation and clean drinking water, etc.). Societies where poverty is endemic and results in

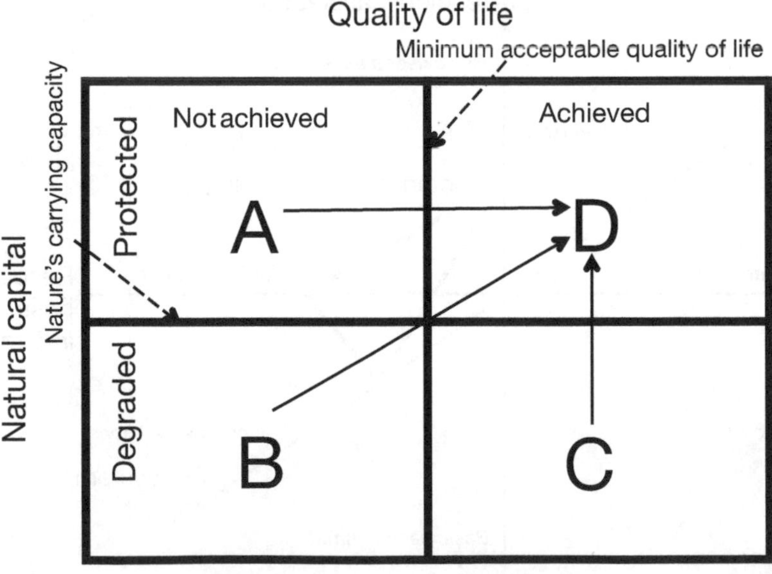

Figure 14.3 Quality of life/natural capital model

further degradation of the quality of life and the natural environment are Zone B communities. Several economically developing communities fall into this zone. Zone C represents a situation where people enjoy a satisfying QOL but natural capital is not being adequately protected. Many economically developed societies fall into this category. In Zone D, both key goals of sustainability are satisfied: a high QOL is achieved without degrading the environment either at the local or global level. Some ecovillages fall into this category. Identified by the United Nations as a model village development and winner of the 1997 world prize in zero emissions from the UN Zero Emissions Research Initiative, Gaviotas, located in Colombia's hostile eastern savannahs (the llanos) has demonstrated that social and ecological well-being are compatible partners. Founded in 1971 by Paolo Lugari, Gaviotas is a small village of a few hundred that has used wind generators to solve the problem of energy and hydrophonics to address the problem of rain-leached soils. In addition, the village is embarking successfully on restoring the indigenous rainforest. From the onset, Gaviotas was conceived as a place that would offer a high quality of life, where schooling, housing, healthcare and food are free, and there is no poverty and no crime.

Once the state of sustainability is reached, a community may continue to develop to higher levels of sustainability where higher QOL levels are achieved at less and less impact to the natural environment (Chambers et al., 2000). This model indicates that there are different ideal sustainable development pathways as shown in Figure 14.3, and that Zone B countries and other communities can leapfrog the historic environmental burdens of development through development that is more sustainable.

14.2.6 The Sustainability Footprint

The sustainability footprint (SF) model was developed by Amekudzi, Khayesi and Khisty in the late 2000s as an evaluation framework based on the three key types of capital fundamental to sustainability and development (2009) and expanded in 2011. Developed primarily for evaluating the impact of civil infrastructure on developing sustainably, the SF model builds on the TFST, the EF, and the QOL/NC to create a four-dimensional evaluation framework (see Figure 14.4) that evaluates sustainable development progress as the impact of infrastructure system performance on community quality of life, ecological capital and economic capital, over time. The SF is defined as the rate of change of some measure of civil infrastructure system performance related to the quality of life of system users as a function of the life cycle environmental and economic costs and benefits associated with attaining that system performance (Amekudzi et al., 2011).

The model can be formulated as shown. The SF of some entity C (e.g., a municipality, metropolitan area, nation, etc.) between two finite points in time, t = i and t = i + n can be formulated as a vector of quantities as shown:

$$SF_{C_{t=i}}^{t=i+n} = \left(\mathrm{d/dt}\left(\frac{z_c}{x_c y_c}\right), \mathrm{dZ_c/dt}, \mathrm{dY_c/dt}, \mathrm{dX_c/dt}, \mathrm{Z_c}, \mathrm{Y_c}, \mathrm{X_c} \right)$$

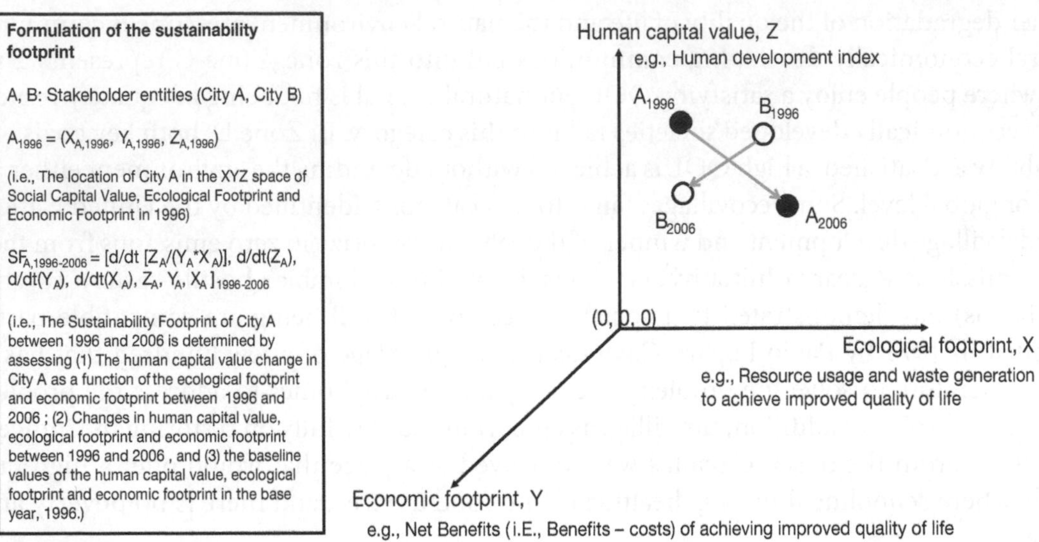

Figure 14.4 Sustainability footprint model

where:

$SF^{t=i+n}_{C_{t=i}}$ is the SF of City C between time t = i and t = i + n (where n is some finite amount of time);

Z_c is the system performance of the infrastructure related to the quality of life or some other measure of community quality of life of the city's inhabitants (a subset of which is the system's users);

X_c is the wastes generated and output to the environment, and resources used or inputs taken from the environment by the city, associated with the infrastructure system, and

Y_c is the economic net benefits, (i.e., economic benefits less costs), of the city associated with the infrastructure system.

The model indicates that infrastructure systems, services, and products that have the greatest positive impact on the quality of life of their users while minimizing life cycle costs to the environment, and maximizing associated life cycle net benefits, are the most superior from the viewpoints of sustainable development and sustainability.

14.2.7 The True Sustainability Index

The TSI evaluates how an organization contributes to the improvement or deterioration of economic, environmental and social conditions at the local, regional or global level. Simply reporting on trends in individual performance will fail to respond to this underlying question that is at the heart of sustainable development and sustainability (Center for Sustainable Organizations, 2009). The TSI frames the sustainability performance of an organization as a function of

Table 14.6 True sustainability index taxonomy

TBL* **AOIs****		**Environmental**	**Social**	**Economic**
Natural capital		X***		
Anthro capital	Human capital		X	X
	Social capital		X	X
	Constructed capital		X	X

(Adapted from Center of Sustainable Organizations, 2009)
*TBL: Triple bottom line
**AOIs: Areas of impact
***Hatch marks indicate areas where metrics are needed to fully reflect organizational impacts on vital capitals for each of the three bottom lines.

impacts on vital capitals in the world that people rely on for well-being: natural capital, human capital and constructed capitals (called anthro capitals). A quotient construct, the denominators represent normative impacts by organizations on the carrying capacity of vital capitals and the numerators represent the actual impacts by organizations on the same things. These vital capitals are evaluated in the context of the TBL capitals. The index consists of 15 sustainability metrics that link the vital capitals to the TBL capitals. Metrics include climate, air, water, solid waste assimilation (environmental); human health, social institutions, constructed capital, climate change mitigation (social); and livable wages, business ethics, economic institutions and economic infrastructure/material goods (economic). The environmental bottom line metrics are intended to cover all major areas of environmental impact and assume that human well-being is tied to ecological well-being. Social bottom line metrics address organizational impacts on people and society and measure impacts on all three types of anthro capital. Economic bottom line metrics focus on impacts on the broader economy and are tied to impacts on human economic well-being (McElroy, 2009). Table 14.6 shows the TSI taxonomy.

14.3 PLANNING AND DESIGNING FOR SUSTAINABLE DEVELOPMENT AND SUSTAINABILITY

The planning-design process begins with visioning a future we are trying to achieve, distilling goals and objectives that can guide decision making, and determining performance measures and targets that will lead to the achievement of said vision. The next phase involves scoping out alternative plans to achieve articulated visions and evaluating them based on the preselected performance measures. Such planning processes involve understanding the tradeoffs associated with the alternative scenarios and identifying the most superior plans in the context of the multiple criteria being assessed (e.g., functional performance, environmental impact, social impact and economic impact). Figure 14.5 illustrates the transportation planning and project development process (adapted from Meyer and Miller, 2000).

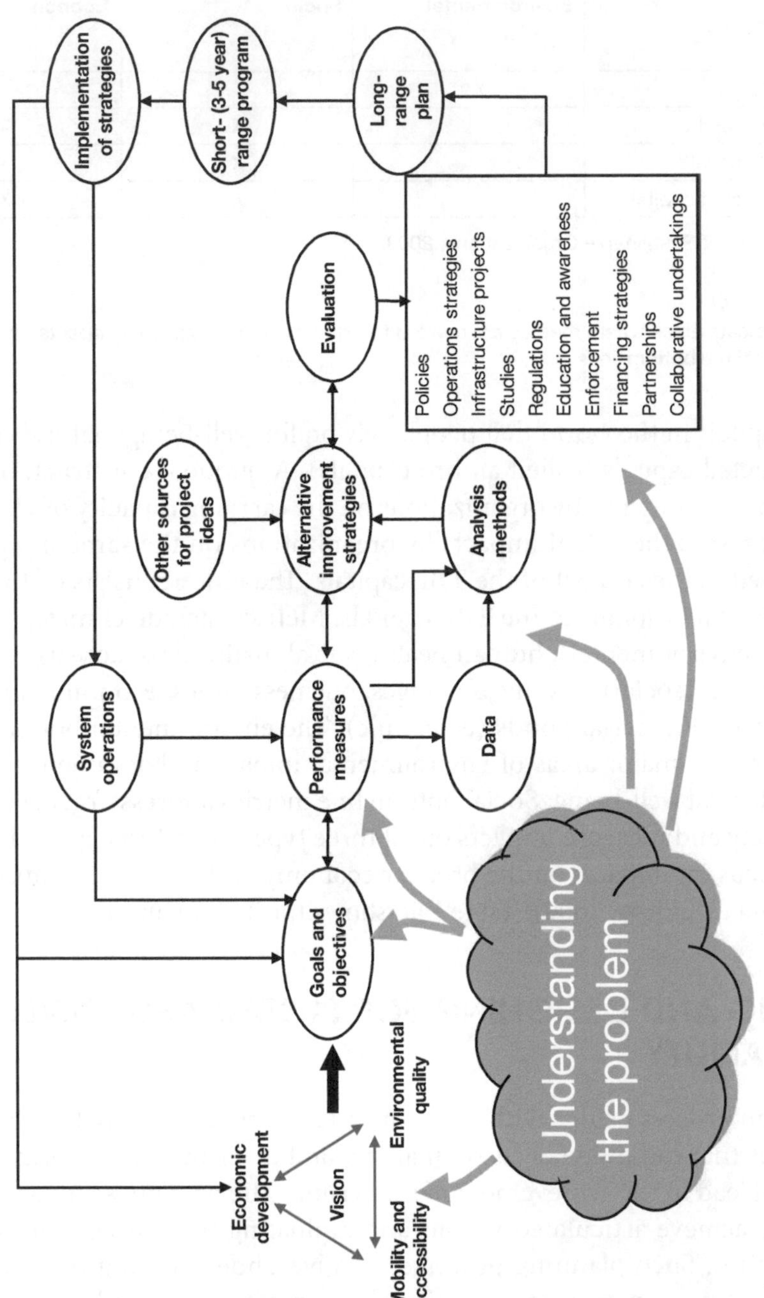

Figure 14.5 Transportation planning framework (Adapted from Meyer and Miller, 2000)

14.3.1 Planning and Project Development Methodologies for Sustainable Development and Sustainability

Planning and design for sustainable development and sustainability is a multiple criteria decision making (MCDM) process. MCDM refers to methods and procedures by which concerns about multiple conflicting criteria can be formally incorporated into the decision-making process. Several good applications of MCDM in infrastructure planning and project development point to the fact that there are seldom optimal alternatives but rather dominant or superior alternatives where the alternatives dominate on several of the decision criteria being used, and oftentimes decision makers must consider trade-offs where some alternatives are stronger on various criteria but weaker on others (Jeon et al., 2010; Zietsman et al., 2006).

It is also the case that sometimes sustainability planning and project development decisions are framed with one of the TBL capital as the dominant criterion or objective, and the other criteria as secondary objectives or constraints in the decision-making problem. For example, a Zone 2 community is likely to frame a sustainable development plan with the main objective of improving quality of life or economic capital and with the natural environment as a constraint, given that the key issues in these communities relate to the failure to achieve basic quality of life standards (e.g., lack of access to sanitation, clean water, good transportation, etc.). On the other hand, a community in Zone 3 would likely frame a sustainability plan in terms of achieving environmental objectives, such as greenhouse gas reduction targets, given that the critical issues in these communities tend to be more environment related. Independent of which criterion or criteria are selected as predominant, sustainability plans, policies, programs, and projects must always be considered in the context of the other key sustainability parameters. This indicates that, if they are framed as single-objective problems, then the other nonpredominant criteria must be considered as constraints. If they are framed as multiple criteria problems, then the predominant issue or issues will be given the highest weight while the other criteria will receive lower weights reflecting that they are secondary criteria in the decision-making process.

In addition to the models and methodologies previously discussed, there are several other approaches and methodologies being applied in infrastructure decision-making contexts that can be used in planning and project development for sustainable development and sustainability. Examples include scenario planning, backcasting, performance-based planning, context-sensitive solutions (CSS) asset management, strategic environmental assessment (SEA), environmental impact assessment (EIA), equity analysis, and health impact assessment (HIA). Table 14.7 gives a summary of descriptions and applications of such methodologies.

Table 14.7 Examples of methodologies applicable to sustainability decision making

Tool	Description	Example applications
MCDM methods	Multiple criteria decision-making methodologies are used to incorporate multiple criteria, sometimes conflicting, in decision-making processes. They can help identify dominant alternatives, where there are alternatives that are best with respect to all the decision criteria being used (e.g., functional performance, environmental quality, equity, and economic development measures). They can also help with evaluating tradeoffs among various alternatives where there are no dominant alternatives but alternatives that are better on certain criteria and worse off with respect to the other criteria under consideration (Jeon et al., 2010; Zietsman et al, 2006).	Evaluation of plan or project alternatives for sustainability or sustainable development
Scenario planning	Scenario planning develops alternative plausible scenarios for the future, scopes out alternative plans, evaluates them against these alternative futures and selects the plan that will achieve predetermined goals most significantly independent of the scenario that actually transpires. This tool is helpful for addressing uncertainties that are prevalent when we attempt to forecast the future in planning. Scenarios first emerged following World War II as a method of military planning. One of the original applications was by the Royal/Dutch Shell Company in the 1970s to evaluate ways to remain competitive in various oil supply scenarios (Schwartz, 1996; Federal Highway Administration, 2008).	Development of infrastructure and sustainability plans, programs, and policies
Backcasting	Backcasting is an analytical tool that recasts the decision-making environment to have more control over the future by deciding on the desired status of selected critical factors (e.g., related to livability, the environment, and economy, for example). Policies are then developed and implemented to promote technological innovation as well as behaviors to achieve the desired future state (Federal Highway Administration, 2011; Barrella and Amekudzi, 2011).	Development of sustainability plans, programs, and policies
Performance-based planning	Performance-based planning develops clear performance goals, measures and time-sensitive targets that guide the planning, project development and evaluation processes. Planning goals, objectives and performance targets are made transparent and the planning agency reports to the general public and political decision makers on progress toward these targets (National Cooperative Highway Research Program, 2000).	Development of sustainability plans, programs, and policies
Context sensitive solutions (CSS)	CSS (Context Sensitive Solutions) is a collaborative, interdisciplinary approach that involves all stakeholders to develop a transportation facility that fits its physical setting and preserves scenic, aesthetic, historic and environmental resources, while maintaining safety and mobility. CSS is an approach that considers the total context within which a transportation improvement project will exist (Federal Highway Administration, 2010).	Project development

Tool	Description	Example applications
Asset management	Asset management is a systematic process of maintaining, upgrading and operating physical assets cost effectively. It combines engineering principles with sound business practices and economic theory, and it provides tools to facilitate a more logical, organized approach to decision making (Federal Highway Administration, 1998). Asset management relates to making decisions to optimize the performance of facilities and systems over their life cycles. It involves collecting condition data on facilities, developing models to estimate performance and allocating resources to optimize system performance.	Budget setting, project prioritization for program development
Environtmental impact assessment (EIA)	EIA is a tool used to evaluate the impacts of projects on the environment (natural and human). It involves evaluating how projects impact the natural environment, public health, and other sociocultural resources.	Project development
Strategic environmental assessment (SEA)	SEA is a methodology for incorporating environmental considerations into policies, plans, and programs. It is a systems-level EIA.	Development of policies, plans, and projects
Equity analysis	Equity analysis is a tool used to evaluate the distribution of benefits and burdens of projects, plans and policies to determine if they are equitably distributed over the various populations within a community. Equity analyses are sometimes conducted as part of EIAs.	Development of plans and projects; input into EIAs
Health impact assessment (HIA)	HIA is commonly defined as a combination of procedures, methods, and tools by which a policy, program, or project may be judged as to its potential effects on the health of a population, and the distribution of those effects within the population (World Health Organization, 1999).	Inputs for sustainability plans; evaluation of health impacts of large-scale projects; may be used to address health issues in EIAs
Life cycle assessment (LCA)	LCA is a technique used to assess the environmental aspects and potential impacts associated with a product, process, or service by (a) compiling an inventory of relevant energy and material inputs and environmental releases; (b) evaluating the potential environmental impacts associated with identified inputs and releases; and (c) interpreting the results to help make a more informed decision (United States Environmental Protection Agency, 2010).	Inputs for sustainability plans and EIA; evaluation of environmental impacts of large-scale projects

REFERENCES

Amekudzi, A., C. J. Khisty, and M. Khayesi (2009). "Using the Sustainability Footprint Model to Assess Development Impacts of Transportation Systems," *Transportation Research Part A: Policy and Practice*, 43:3, 339-348.

Amekudzi, A. A., J. M. Fischer, M. Khayesi, C. J. Khisty, and S. O. Asiama (2011). "Risk-Theoretical Foundations for Setting Sustainable Development Priorities: A Global Perspective," *Proceedings of the 2011 Annual Conference of the Transportation Research Board*, Washington, D.C.

ASCE (2009), Board Approved Definition of Sustainability, Cited from *ASCE Committee on Sustainability Meeting Agenda*, November 15, 2010, 1st Green Highways and Streets Conference, Denver, CO, United States.

Anderson, R. (1999). *Mid-Course Correction: Toward a Sustainable Enterprise: The Interface Model.* White River Junction, VT: Chelsea Green Publishing Company.

Barrella, E. and A. Amekudzi (2011). "Using Backcasting for Sustainable Transportation Planning," *Proceedings of the 2011 Annual Conference of the Transportation Research Board*, Washington, D.C.

Carson, R. (1962). *Silent Spring*, New York, NY: Houghton Mifflin.

Center for Sustainable Transportation, Definition of Sustainable Transportation, Accessed at http://www.centreforsustainabletransportation.org/ in December 2010.

Center for Sustainable Organizations (2009), *A Strawman Specification for a Comprehensive Triple Bottom Line True Sustainability Index (for Organizations) consisting of full-quotient, Context-Based Metrics*, Version, 2.4.

——— (2009). *The True Sustainability Index.* Accessed at http://www.sustainableorganizations.org/true-sustainability-index.html in December 2010.

Checkland, P. B. (1984). "Systems Thinking in Management: The Development of Soft Systems Methodology and Its Implications for Social Sciences," *In* H. Ulrich and G. J. B. Probst, eds., *Self-Organization and Management of Social Systems: Insights, Promised, Doubts and Questions*, Berlin, Germany: Springer-Verlag.

Chambers, N., C. Simmons, and M. Wackernagel (2000). *Sharing Nature's Interest. Ecological Footprints as an Indicator of Sustainability*, London/Sterling, VA: Earthscan.

Deakin, E. (2001). "Sustainable Development and Sustainable Transportation: Strategies for Economic Prosperity, Environmental Quality, and Equity," *IURD Working Paper Series*, Working Paper 2001-03, University of California at Berkeley, Institute of Urban and Regional Development.

Erlich, P. and J. Holdren (1971). "Impacts of Population Growth," *Science*, vol. 171, 1212-1217.

Erlich, P. and J. Holdren (1972). One-Dimensional Ecology, *Bulletin of Atomic Scientists*, 28:5, 16, 18-21.

FHWA (1998). *Asset Management Primer*, United States Department of Transportation, Washington, D.C.

——— (2008). *About Scenario Planning, Planning—Planning, Environment and Realty.* Accessed at www.fhwa.dot.gov/planning/scenplan/about.htm in January 2011.

——— (2010). *FHWA and Context Sensitive Solutions (CSS).* Accessed at http://www.fhwa.dot.gov/context/ in December 2010.

——— (2011). *Transportation Planning for Sustainability Guidebook.* Developed by Georgia Institute of Technology (Adjo A. Amekudzi, Michael D. Meyer, Catherine L. Ross, and Elise Barrella) for the Federal Highway Administration.

Hawken, P., A. Lovins, and H. Lovins (1999). *Natural Capitalism: Creating the Next Industrial Revolution*, New York, NY: Little, Brown.

Jeon, C. M. and A. A. Amekudzi (2005) "Addressing Sustainability in Transportation Systems: Definitions, Indicators and Metrics," *ASCE Journal of Infrastructure Systems*, 11:1 31-50.

Jeon, C. M., A. A. Amekudzi, and R. Guensler (2010). "Evaluating Transportation System Sustainability: Atlanta Metropolitan Region," *International Journal of sustainable Transportation*, 4:4 227-247.

Meyer, M. D. and E. J. Miller (2000). *Urban Transportation Planning: A Decision-Making Approach*, New York, NY: McGraw-Hill.

Minkel, J. R. (2008). "The 2003 Northeast Blackout—Five Years Later," *Scientific American*, August 13. Accessed at www.scientificamerican.com in January 2011.

Ministry of Transport (New Zealand) (2008).The *New Zealand Transport Strategy 2008*. Accessed at http://www.transport.govt.nz/ourwork/keystrategies/new-zealand-transport-strategy/ in December 2010.

National Cooperative Highway Research Program, Transportation Research Board (2000). *Guidebook on Performance-Based Planning*. Report prepared by Cambridge Systematics.

National Research Council (1999). Board on Sustainable Development. *Our Common Journey*.

Pearce, A. (1999). *Sustainability and the Built Environment: A Metric and Process for Prioritizing Improvement Opportunities*, Ann Arbor, MI: UMI Dissertation Services.

Pearce, A. and J. A. Vanegas (2002). "Defining Sustainability for Built Environment Systems," *International Journal of Environmental Technology and Management*, 2:1/2, 94-113.

Sessions, G. (1995). *Deep Ecology for the 21st Century*, Boston, MA: Shambala Publications.

Schwartz, P. (1996). *The Art of the Long View: Planning for the Future in an Uncertain World*. Doubleday, New York/London/Toronto/Sydney/Auckland.

Sustainable Transportation Action Network, Global Development Research Center. Urban Environmental Management. Sustainable Transportation, Accessed at www.gdrc.org/uem/sustran/sustran.html in November 2010.

United States Environmental Protection Agency (2010). *Life Cycle Assessment (LCA)*. Accessed at http://www.epa.gov/nrmrl/lcaccess/ in January 2011.

Wackernagel, M. and W. Rees (1996). *The Ecological Footprint, Reducing Human Impact on the Earth*, Gabriola Island, BC: New Society Publishers.

WCED, United Nations (1987). *Our Common Future*, London, UK: Oxford University Press.

World Health Organization (1999). *Gothenburg Consensus Statement*. Accessed at the United States Centers for Disease Control and Prevention Health Impact Assessment page at http://www.cdc.gov/healthyplaces/hia.htm in January 2011.

Zietsman, J. and R. Rilett (2002) *Sustainable Transportation: Conceptualization and Performance Measures*, Southwest Region University Transportation Center.

Zietsman, J., L. R. Rilett, and S. Kim (2006). Transportation Corridor Decision Making with Multiattribute Utility Theory, *International Journal of Management and Decision Making*, 7:2/3, 254-266.

EXERCISES

1. There are several definitions of sustainable development and sustainability, some conceptual and others operational. Explain the differences between conceptual and operational definitions of sustainable development and sustainability. Explain why someone is likely to find several operational definitions of sustainable development and sustainability in different communities or in the same community at different times.

2. The city council of a large metropolitan area will be meeting to consider alternative sustainability plans for the city. Assume you have recently been employed as an intern with the city's Public Works Department. The director of the Public Works Department has assigned you to research existing models for evaluating sustainability plans to ensure that the

city can track progress toward their sustainability goals. Which of the seven sustainability models described in this chapter would you recommend and why? Assess the advantages and disadvantages of using each model to determine your answer.

3. Outline and discuss the sustainable development and sustainability priorities in developed and developing regions of the world.
4. Three water supply systems are being evaluated for a megacity. As a systems engineer, what data will you collect to evaluate and compare the sustainability of the three systems? Develop a table showing the different types of data you would collect and explain how you would use the data to determine the most sustainable of the three alternatives.
5. The Ecological Footprint (EF) database contains data on the natural environment accounts of various nations around the world (www.footprintnetwork.org). The United Nations Human Development Index (HDI) database contains data on the quality of life of communities in different nations around the world. Using the EF and HDI data, evaluate and compare sustainable development and sustainability in each of the following countries: United States; Norway; China; South Korea; Ghana; and Haiti. Explain the limitations of your analysis and outline the data you would collect to improve your analysis.

Case Studies in Engineering and Planning for Sustainability

15

15.1 INTRODUCTION

This chapter presents examples of applications of sustainability principles to the engineering and planning of real-life facilities, systems, products, or services. The first case study presents Georgia-based Interface, Inc. and its rise to leadership in the application of sustainability principles to manufacturing. The world's largest manufacturer of modular carpet, Interface has embarked on a journey to environmental sustainability where the company will have zero negative impact on the natural environment, while maintaining and growing in economic competitiveness and their ability to satisfy customers. The second case reviews New Zealand's Transport Strategy, developed to implement sustainability principles in the nation's transportation system and designed to progressively engineer a sustainable transportation system. The third case demonstrates the application of multiple criteria decision making to identify superior plan alternatives and analyze trade-offs by considering the functional system performance, environmental, economic, and social impacts of infrastructure investments. Together, these three cases illustrate how the principles of sustainability can be applied to guide the planning, design, and development of facilities, systems, and services.

15.1.1 Interface's Approach to Sustainability

> "For the first 21 years of Interface's existence, the company gave no serious thought to what we were taking from or doing to the Earth. We did what was necessary—we abided by all the necessary laws and regulations—and focused, like most companies on running a growing business."
>
> "In 1994, while preparing remarks on Interface's environmental plans for a company meeting, Interface founder and Chairman, Ray Anderson, read Paul Hawken's *The Ecology of Commerce*—an experience Ray has described as an epiphany, a *spear to the chest* awakening to the urgent need to set a new course toward sustainability for Interface."
>
> —www.interfaceglobal.com/Sustainability/Our-Journey.aspx. Accessed February 11, 2011

15.1.2 Interface's Seven-front Approach to Sustainability

Interface is the worldwide leader in design, production, and sales of environmentally responsible modular carpet for the commercial, institutional, and residential markets, and it is a leading designer and manufacturer of commercial broadloom. The company is headquartered in Atlanta with annual sales worldwide on the order of several hundred millions of dollars. In 1994, Interface founder Ray Anderson realized there was a critical need to set a new direction for his company to be more environmentally sustainable. Having operated in the petroleum-intensive industry of carpet manufacturing since 1973, he had grown his company to be highly successful by all traditional measures. However, in the midst of preparing for a speech to an Interface task force on the company's environmental vision, he read Paul Hawken's *The Ecology of Commerce* and had an epiphany about the real impact of his company on the natural environment. This epiphany marked a turning point in his company. He challenged his company to rethink its focus and pursue a bold vision for the future: "to be the first company that, by its deeds, shows the entire industrial world what sustainability is in all its dimensions: people, process, product, place and profits—and in doing so, becomes restorative through the power of influence." Table 15.1 shows the vision and mission statement of Interface.

Over 15 years into their journey and a decade away from the set year for achieving Mission Zero (i.e., zero negative impact on the environment), Interface is simultaneously pursuing three paths to sustainability: (1) innovative solutions for reducing their environmental

Table 15.1 Vision and mission statement of Interface

Vision
To be the first company that, by its deeds, shows the entire industrial world what sustainability is in all its dimensions: people, process, product, place, and profits—by 2020—and in doing so we will become restorative through the power of influence.
Mission
Interface will become the first name in commercial and institutional interiors worldwide through its commitment to people, process, product, place, and profits. We will strive to create an organization wherein all people are accorded unconditional respect and dignity; one that allows each person to continuously learn and develop. We will focus on product (that includes service) through constant emphasis on process quality and engineering that we will combine with careful attention to our customers' needs so as always to deliver superior value to our customers, thereby maximizing all stakeholders' satisfaction. We will honor the places where we do business by endeavoring to become the first name in industrial ecology, a corporation that cherishes nature and restores the environment. Interface will lead by example and validate by results, including profits, leaving the world a better place than when we began, and we will be restorative through the power of our influence in the world.

(Source: www.interfaceglobal.com/Sustainability/Our-Journey/Vision.aspx. Accessed January 2011)

footprint; (2) new ways to design and make products; and (3) an inspired and engaged culture—all addressing the three basic elements of sustainability: the environment, the economy, and society. Ray Anderson and his company understood from the beginning that sustainability had to be approached from a systems perspective or a whole company approach. The systems-based approach (see Figure 15.1) touches operations and manufacturing, guides senior management and associates' decision making, and influences the company's relationships with customers, suppliers, and the entire web of commerce in which the company conducts business (www.interface.com).

Over the years, this commitment to sustainability has reached all parts of Interface's business and evolved into a shared mission—Mission Zero. Mission Zero is the company's mission to attain zero negative impact on the environment by the year 2020. Recognizing that attaining sustainability is a journey, and not an easy one, Interface has dedicated a team to focus solely on driving, sustaining, and measuring their progress toward sustainability. The journey that began in educating themselves and teaching other organizations the lessons they have learned has propelled Interface to become a recognized leader in sustainable business worldwide.

Ray Anderson and his company's resolve to change the way they did business (see Figure 15.2) evolved into a plan over time that would guide this journey to sustainability. This plan, referred to as *Climbing Mount Sustainability*, involves seven fronts or faces of *Mount Sustainability*. Ray Anderson describes how he and his company have begun to climb each of the seven fronts of sustainability, aiming for the symbolic point at the top representing zero environmental footprint or zero negative impact on the environment.

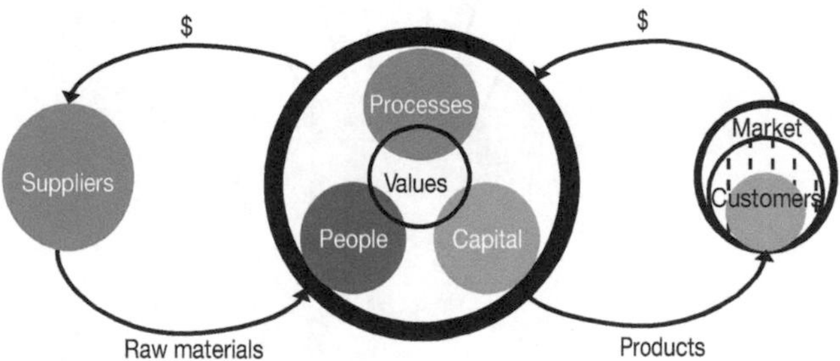

Figure 15.1 The interface model: a systems approach to achieving sustainability (adapted from Anderson, 1999)

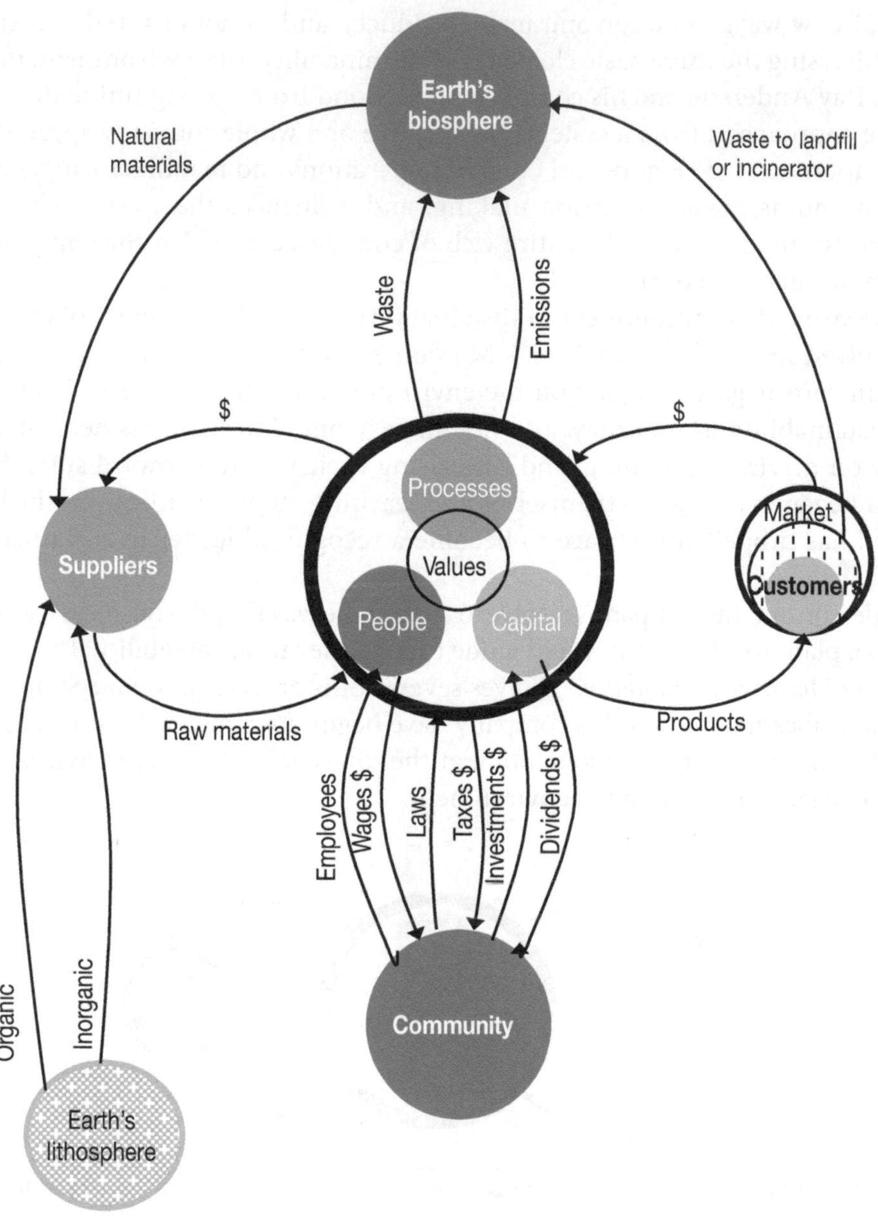

Figure 15.2 Typical company of the 20th century (adapted from Anderson, 1999)

Interface's seven fronts of sustainability are:

1. *Eliminate waste:* Eliminate the concept of waste (see Figure 15.3). This front involves eliminating all forms of waste in every area of business. For Interface this involves redesigning products and processes to reduce and simplify the amount of resources used

so that material waste will no longer be waste but, instead, will be remanufactured into new resources, providing technical nutrients for the next cycle of production.

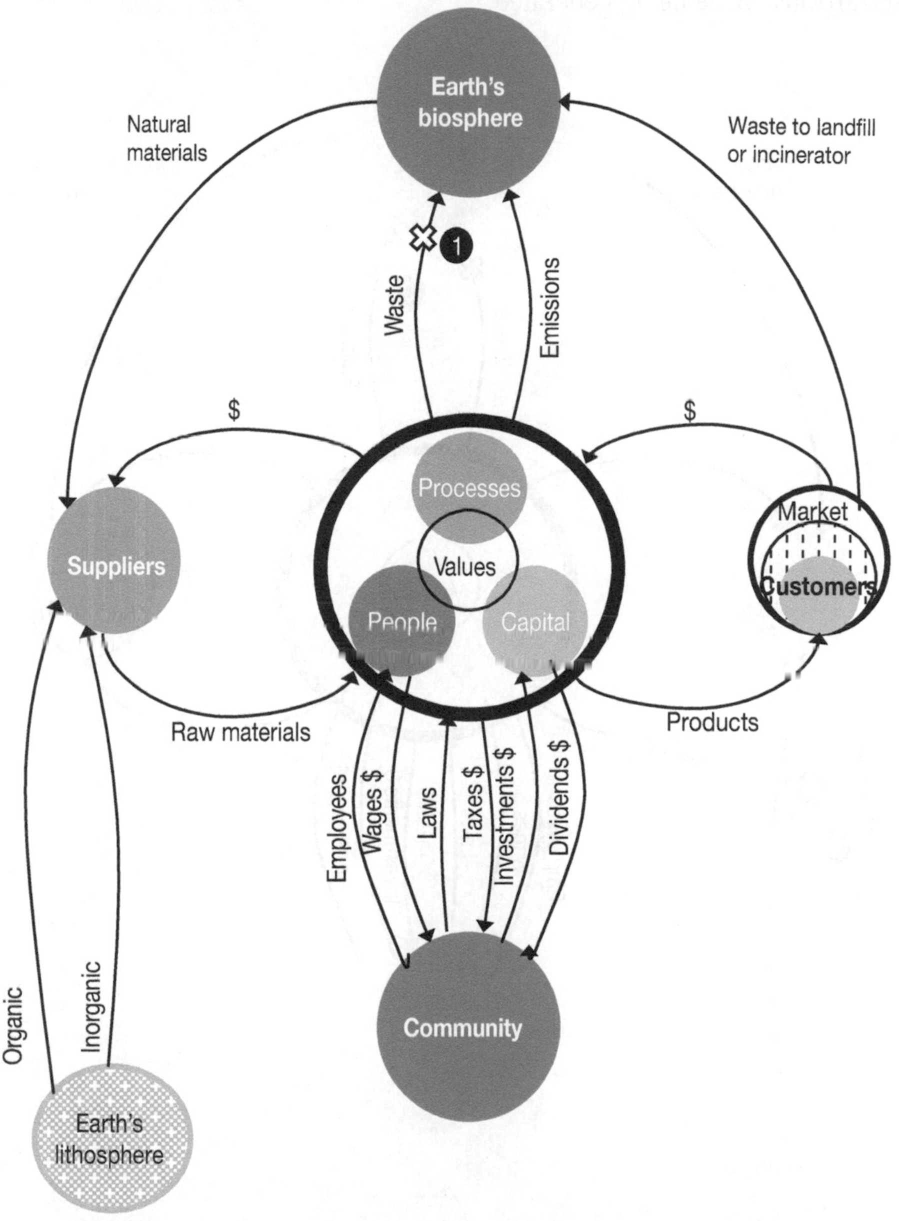

Figure 15.3 Front 1: Achieving zero waste (adapted from Anderson, 1999)

2. *Benign emissions:* Analyze emissions with the goal of emitting only harmless waste (see Figure 15.4). This front involves eliminating toxic substances from products, vehicles,

and facilities. Interface is moving determinedly toward eliminating all its emissions into the ecosphere and striving to create factories with no smokestacks, effluent pipes, or hazardous waste being generated.

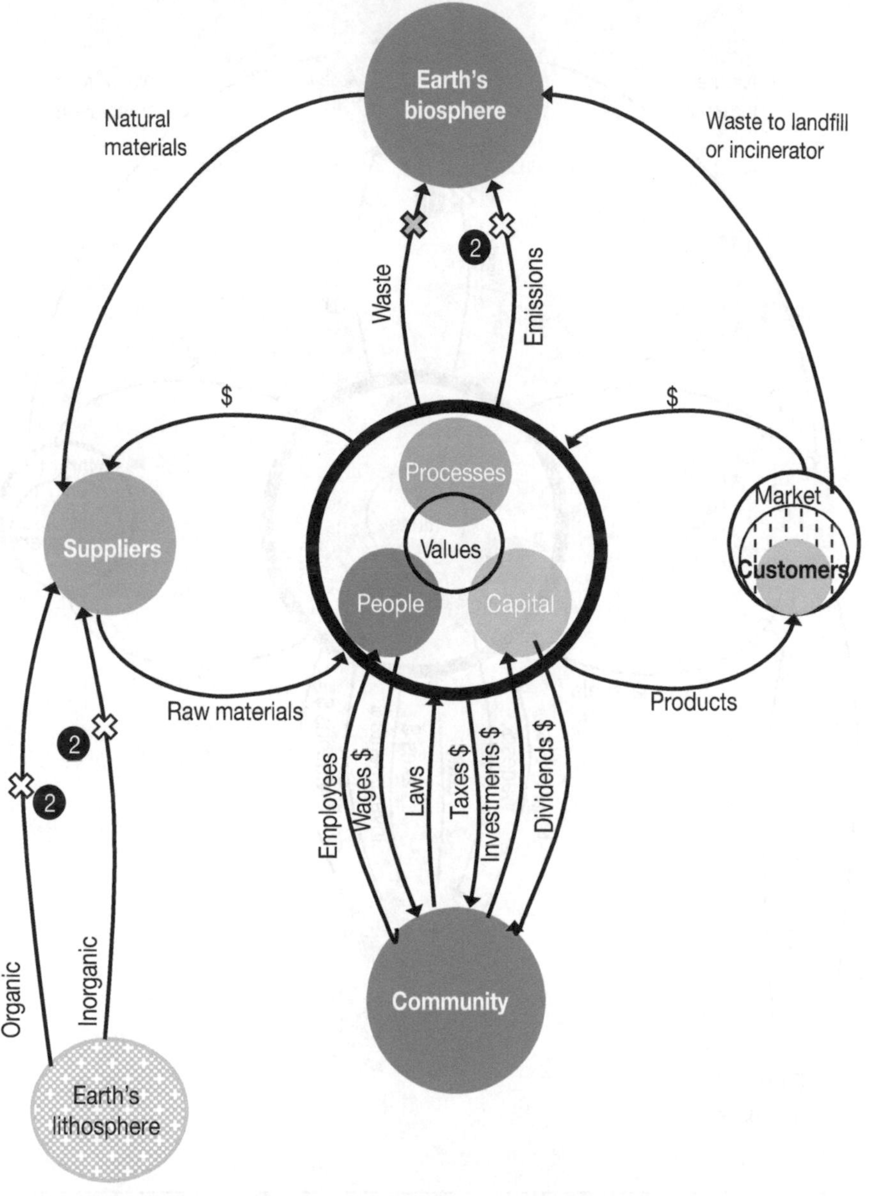

Figure 15.4 Front 2: Eliminating harmful emissions (adapted from Anderson, 1999)

3. *Use renewable energy:* Reduce energy use and use renewable sources (see Figure 15.5). On this front the goal is to operate facilities with renewable energy resources such as

solar, wind, landfill gas, biomass, geothermal, tidal and low-impact/small-scale hydroelectric or nonpetroleum-based hydrogen. Interface seeks to ensure that, by 2020, all fuels and electricity used to operate their manufacturing, sales, and office facilities will be from renewable resources.

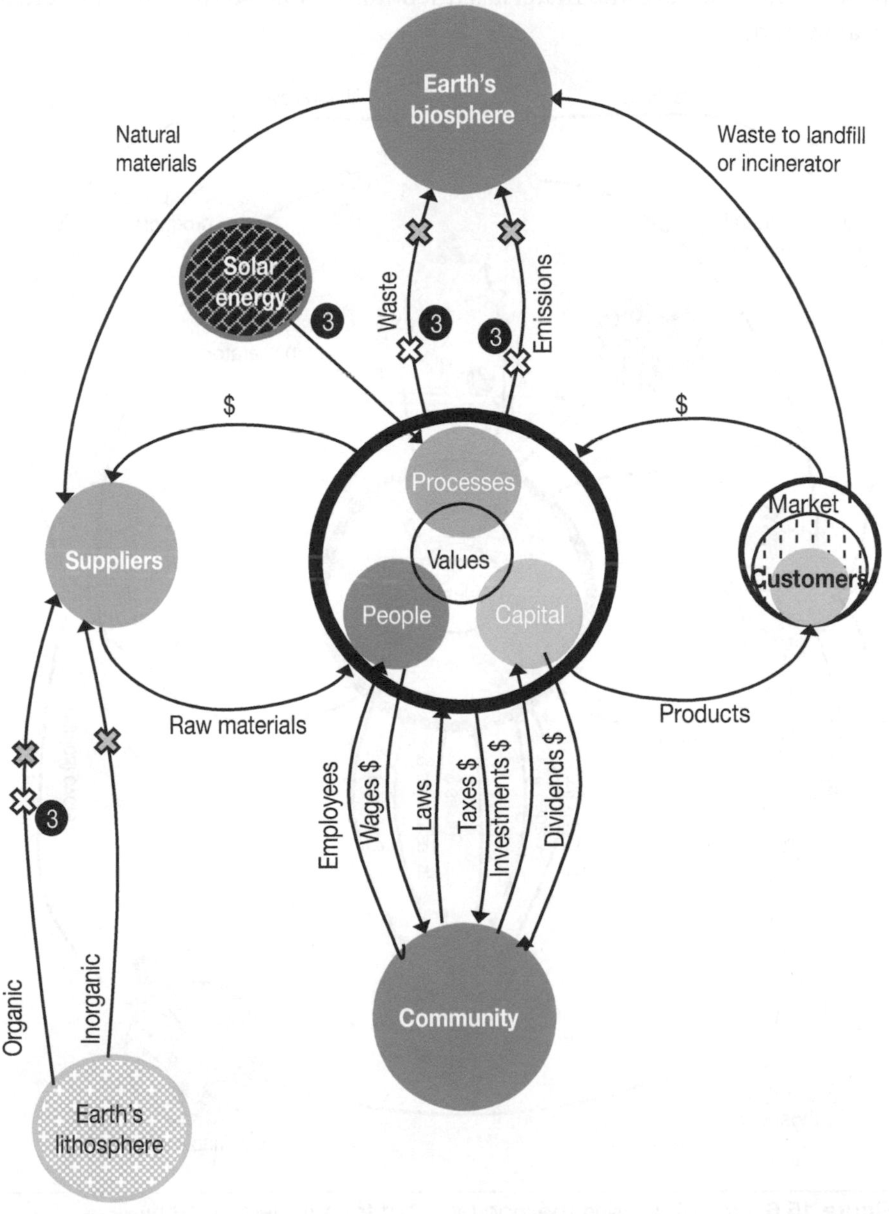

Figure 15.5 Front 3: Using renewable energy (adapted from Anderson, 1999)

4. *Close the loop:* Redesign processes and products to create cyclical material flows (see Figure 15.6). This front involves the redesign of processes and products to close the technical loop using recovered and bio-based materials. Interface is redesigning its processes and products to recycle synthetic materials, convert waste into valuable raw materials, and keep organic materials uncontaminated so they may be returned to their natural systems.

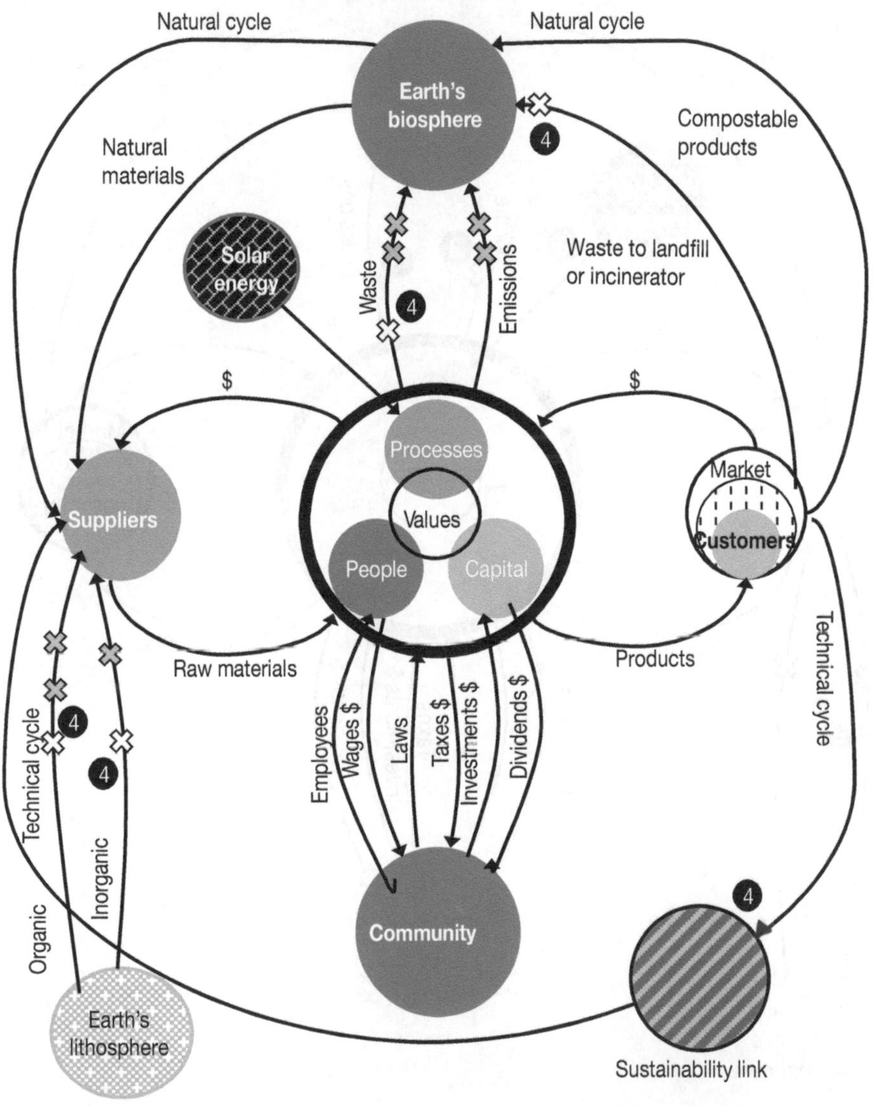

Figure 15.6 Front 4: Closing the loop (adapted from Anderson, 1999)

5. *Use resource-efficient transportation:* Reduce transportation of products and people (see Figure 15.7). This front focuses on transporting people and products efficiently to eliminate waste and emissions. Interface is working to make their transportation more ecologically efficient and participating in voluntary partnerships focused on reducing pollution and greenhouse gas (GHG) emissions and offsetting carbon dioxide emissions associated with their travel, including employee commutes.

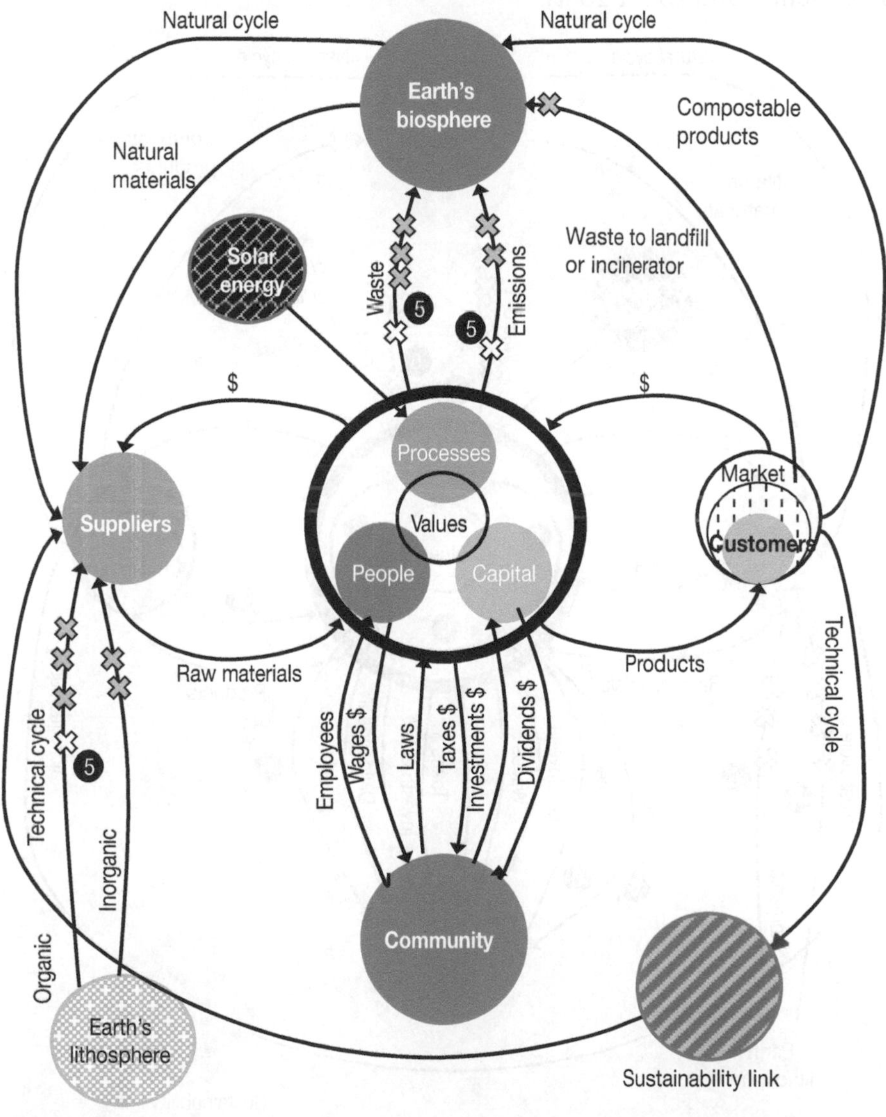

Figure 15.7 Front 5: Resource-efficient transportation (adapted from Anderson, 1999)

6. *Sensitize stakeholders:* Create a community that understands the functioning of natural systems and our impact on them (see Figure 15.8). This front involves cultivating a culture that uses sustainability principles to improve the lives and livelihoods of all of Interface's stakeholders—employees, partners, suppliers, customers, investors, and communities. Interface believes that when stakeholders fully understand sustainability and the challenges that lie ahead, they will come together into a community of shared environmental and social goals.

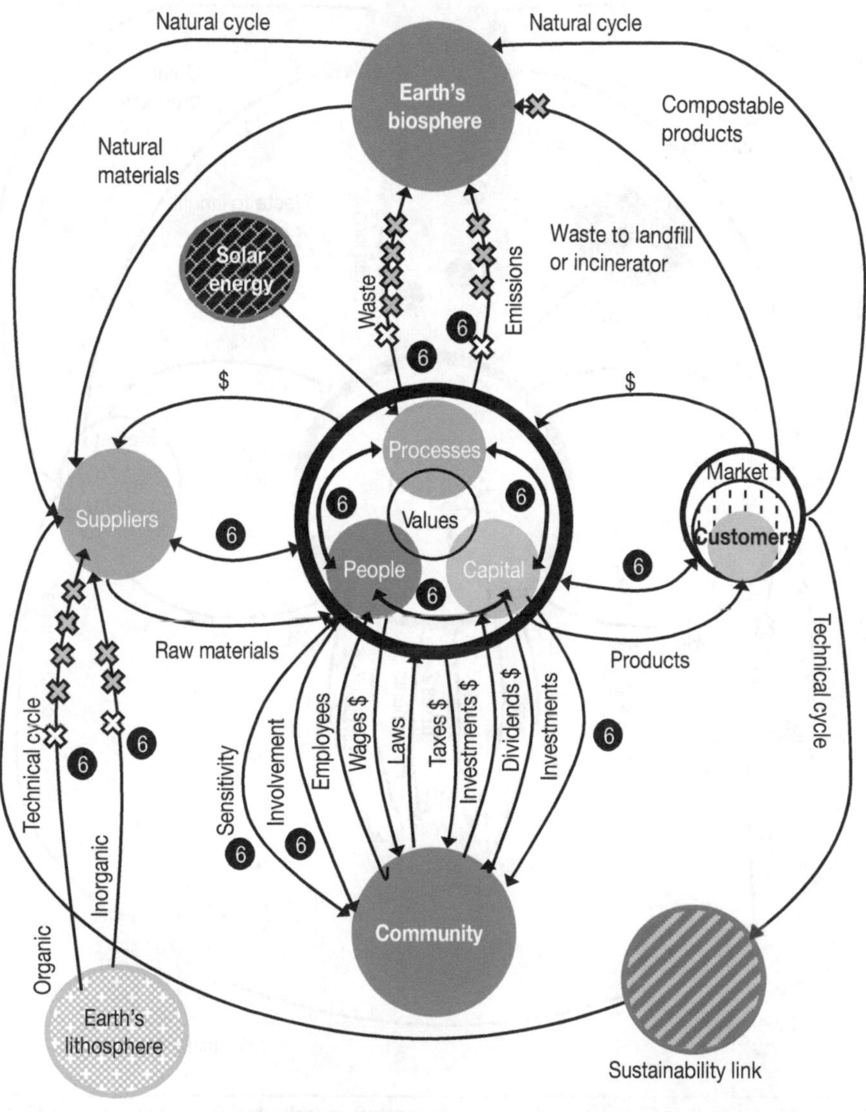

Figure 15.8 Front 6: Sensitizing stakeholders (adapted from Anderson, 1999)

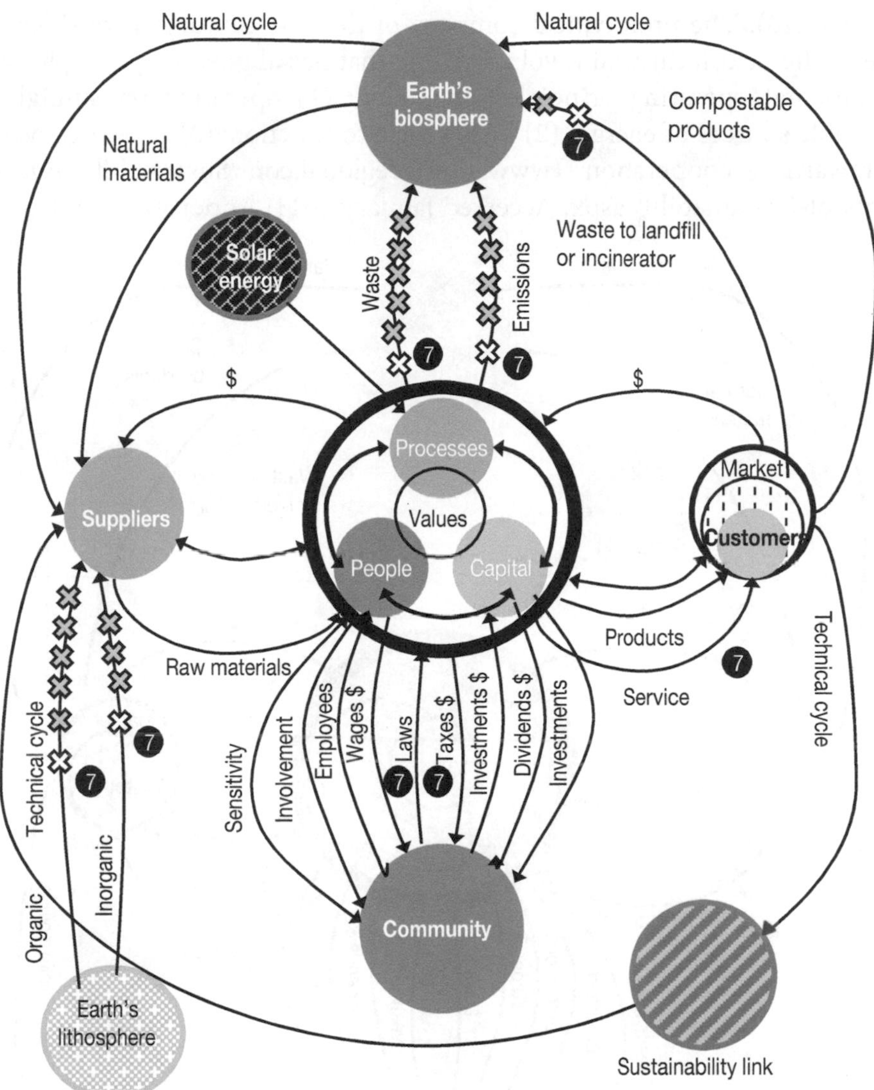

Figure 15.9 Front 7: Redesigning commerce (adapted from Anderson, 1999)

7. *Redesign commerce/create the true service economy:* Deliver service and value instead of materials (see Figure 15.9). This front involves creating a new business model that demonstrates and supports the value of sustainability-based commerce. Interface is creating new methods of delivering value to customers, changing its purchasing practices, and supporting initiatives to bring about market-based incentives for sustainable commerce. The company has taken leadership in developing a model to chart their own evolution to a prototypical model of a 21st century sustainable business (see

Figure 15.10). The prototypical company of the 21st century is considered the enterprise of the next industrial revolution, one that has aligned its principles with nature's fundamental operating principles, including: (1) operating on sunlight and other renewable sources of energy; (2) fitting form to function; (3) recycling everything; and (4) rewarding cooperation. (www.interfaceglobal.com/Sustainability/Our-Journey/7-Fronts-of-Sustainability.aspx. Accessed January 2011; Anderson, 1999)

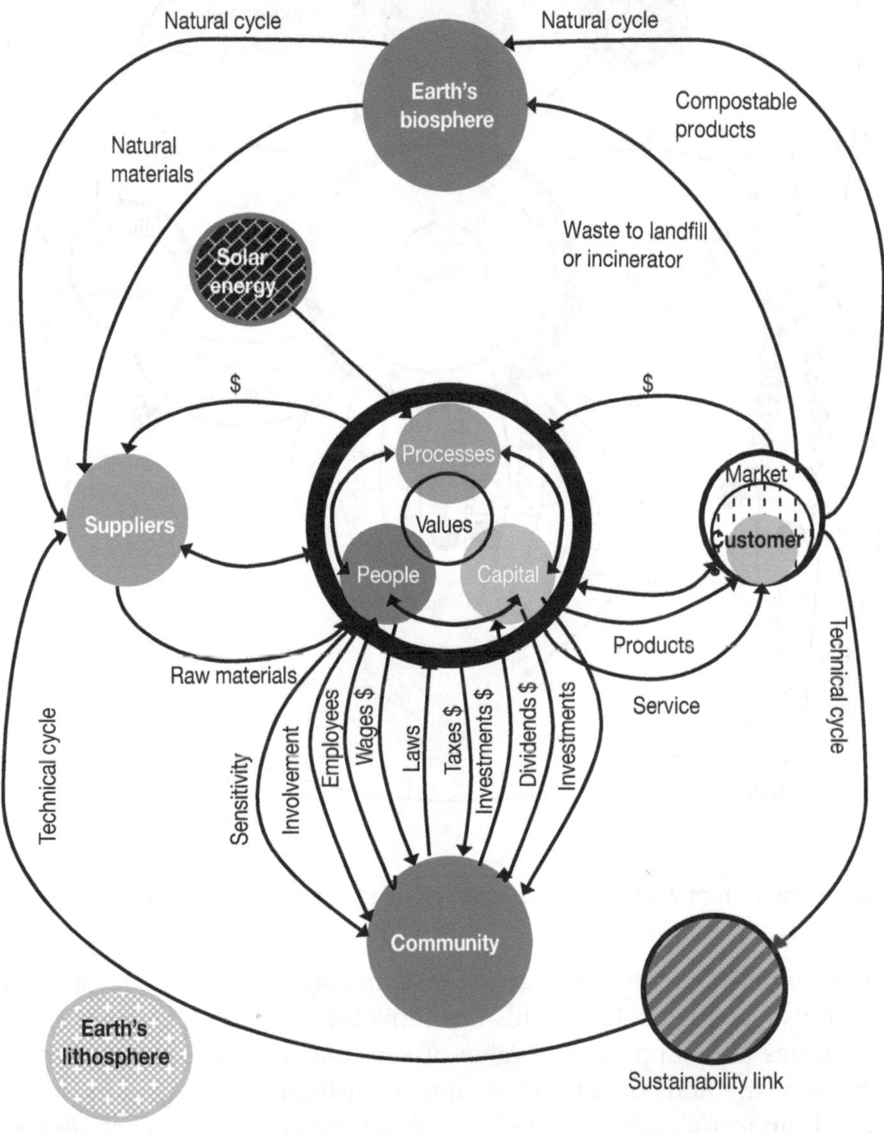

Figure 15.10 Prototypical company of the 21st century (adapted from Anderson, 1999)

15.1.3 Measuring Progress Toward Sustainability

Interface uses measurable environmental, economic, and social achievements to track their progress toward Mission Zero. The company established a metric system in 1996 to enable them to understand the impacts of their processes and products on the environment and return on investment and to drive improvement. This transparent system tracking their progress is also a means of educating others and engaging their stakeholders to join them on their mission. Because Interface values credibility, they support the development of verification and certification standards. Interface's 2009 energy ecometrics are summarized in Table 15.2.

The company has cited the growing scarcity of natural resources and the high likelihood that the government will come forward with a cap-and-trade plan in the United States as important incentives for reducing energy use and carbon intensity; such activities make good business sense. CEO Dan Hendrix has described sustainability as paving the way to innovation with process innovations such as the reduction (and in some cases the elimination of) wet painting through pattern tufting with product breakthroughs and other best selling products inspired by biomimicry and made possible through new tufting technology. "On the manufacturing floor, small changes like optimizing line speed and insulating equipment to minimize heat loss have made a big cumulative impact" (Interface, 2010). Several other changes have allowed the company to reduce its reliance on nonrenewable energy sources such as propane and natural gas, while green electricity purchases, use of landfill gas, and on-site photovoltaic arrays have boosted the use of renewable energy resources.

15.1.3.1 Verification and Certification

For credibility, Interface uses third-party verification, meaning that an independent reviewer assesses or audits their progress statements for accuracy. Interface participates in voluntary reporting programs with built-in verification. For example, to ensure the members accurately report their annual GHG emissions, the Chicago Climate Exchange hires independent auditors to review and verify data submitted by member companies. Certifications tend to be more involved than verifications. In a certification process an independent organization is hired to evaluate the product or company against a prescribed set of criteria. Certifications can be used to distinguish certain products and companies from each other, and can help advance sustainability by providing manufacturers with a detailed set of criteria toward which they can specify

Table 15.2 Interface 2009 metrics—environmental, economic, and social

Sustainability measure	Recorded change
Energy for carpet manufacture (ecometric)	Reduction by 43% since 1996
GHG emissions (ecometrics)	Reduction of 44% in absolute terms since 1996
Net sales (economic and social metrics)	Increase of 27%

(Source: www.interfaceglobal.com/Sustainability/Progress-to-Zero.aspx. Accessed January 2011)

or design. The credibility of different verifications and certifications can vary widely. Interface supports consensus-based, multi-stakeholder developed, nonproprietary standards and participates in these processes. (www.interfaceglobal.com/Sustainability/Progress-to-Zero/Circle-of-Influence.aspx. Accessed January 2011)

15.1.4 Concluding Remarks: Interface's Journey to Sustainability

The case study of Interface demonstrates the potential for manufacturing companies to create economic wealth and improve the social quality of life of their customers with less and less negative impact on the natural environment. Interface experience highlights the importance of leadership, self-education and the education of others in embarking on a successful journey to sustainability. Undertaking such a journey is not without short-term costs associated with reengineering processes and products and educating self and others, but can produce long-term benefits, however, that can ultimately outweigh the initial costs. A certain amount of risk is involved in taking leadership in moving toward sustainability because of the uncertainties involved in changing processes that have already proven to be economically beneficial in the traditional paradigm of manufacturing. However, as Interface has demonstrated, reengineering a company to have reduced negative impacts on the environment can also increase the competitive advantage of the company in the long run, especially when customers start to demand products that are not harmful to the environment, which they are doing more and more.

15.2 THE NEW ZEALAND TRANSPORTATION STRATEGY

The New Zealand Transport Strategy (NZTS) 2008 is a good example of a national transportation plan that is based on sustainability principles with a framework for tracking progress toward sustainability goals. Transportation planning is required by law in a number of countries. It is usually conducted by an agency authorized to provide for a community's (i.e., nation, city, metropolitan area) transportation needs, such as the Ministry of Transportation. A transportation plan outlines a community's long-range vision for transportation and articulates how this vision will be progressively achieved.

The New Zealand Ministry of Transport released their first plan for sustainable transportation in 2008: the NZTS. It provides direction for all parts of the transport sector: road, rail, maritime, and aviation. An update of the 2002 transport strategy, the NZTS outlines a vision for an environmentally sustainable transportation system, defines clear objectives, outlines a set of targets for the transport sector, and articulates actions to achieve these targets over the next 30 years. The plan also includes a mechanism for periodic review and the revision of the original framework as progress is made or new data becomes available for various indicators. Although the NZTS is nonstatutory, it is supported by a statutory document known as the Government Policy Statement on Land Transport Funding. The NZTS performance measurement and reporting framework includes a procedure to monitor progress toward the targets in the Transport Strategy and Government Policy Statement. In this way a tool has been provided for evaluating the effectiveness of the current policy and for guiding future decisions. The plan also provides for accountability by making the monitoring framework publicly available on the Web and through the publication of an annual progress report, including current trends. These plans are part of a series of changes to the transport sector outlined in the Land Transport Management Amendment Act of 2008 (NZTS, 2008).

15.2.1 Defining Sustainable Transportation

All serious attempts to address sustainability issues through infrastructure planning and development begin with the development of a contextually relevant definition of sustainability or sustainable development. New Zealand's initiative is no different. The New Zealand Ministry for the Environment defines sustainable transport as follows: *Sustainable transport is about finding ways to move people, goods and information in ways that reduce its impact on the environment, the economy, and society.* Some options include:

- Using transport modes that use energy more efficiently such as walking or cycling and public transport
- Improving transport choice by increasing the quality of public transport, cycling and walking facilities, services and environments
- Improving the efficiency of our car uses such as using more fuel-efficient vehicles, driving more efficiently, avoiding cold starts, and car pooling
- Using cleaner fuels and technologies
- Using telecommunications to reduce or replace physical travel such as teleworking or teleshopping
- Planning the layout of our cities to bring people and their needs closer together and to make cities more vibrant and walkable
- Developing policies that allow and promote these options (e.g., the New Zealand Transport Strategy).

15.2.2 Defining a Vision, Objectives, and Targets

The NZTS establishes a 30-year vision for transportation in 2040: *People and freight in New Zealand have access to an affordable, integrated, safe, responsive and sustainable transport system.* The long-range vision reflects the fact that many transport investments have long-term implications and take time to achieve change. The vision was developed to be consistent with national transport priorities. It is supported by five objectives:

1. Ensuring environmental sustainability
2. Assisting economic development

3. Assisting safety and personal security
4. Improving access and mobility
5. Protecting and promoting public health

A comprehensive range of performance targets has been developed to measure progress toward each objective and, hence, track the nation's progress toward their overall vision for transportation. Fifteen targets were included in the 2008 strategy with plans to add more targets as data sources are identified and relevant measures created for them. The strategy includes at least one indicator for each target. As shown in Table 15.3, the NZTS targets address environmental, economic, and social quality of life goals in addition to the functional performance of the transportation system.

15.2.3 A Multi-sector Approach

Transportation, being a means to an end, has an impact on various sectors of a nation. Transportation facilitates the movement of people and freight and as such can make a broad contribution to the government's priorities that affect the nation's sustainable development. The system effects of transportation make it possible to address sustainability in multiple sectors using

Table 15.3 NZTS targets

Ensuring environmental sustainability • Half per capital GHG emissions from domestic transport by 2040 (relative to 2007 per capita emissions) • Increase coastal shipping's share of interregional freight to 30% of tonne-kilometers by 2040 • Increase rail's share of freight to 25% of tonne-kilometers by 2040 • Become one of the first countries in the world to widely use electric vehicles • Reduce the kilometers traveled by single occupancy vehicles, in major urban areas on weekdays, by 10% per capita by 2015 compared to 2007 • Reduce the rated carbon dioxide (CO_2) emissions per kilometer of combined average new and used vehicles entering the light vehicle fleet to 170 grams CO_2 per kilometer by 2015 with corresponding reduction in average fuel used per kilometer • Increase the area of Crown transport land covered with indigenous vegetation
Assisting economic development For identified critical routes: • Improve reliability of journey times • Reduce average journey times
Assisting safety and personal security • Reduce road deaths to no more than 200 per annum by 2040 • Reduce serious injuries to no more than 1500 per annum by 2040
Improving access and mobility • Increase use of public transport to 7% of all trips by 2040 (i.e., from 111 million boardings in 2006-2007 to more than 525 boardings in 2040)
Protecting and promoting public health • Reduce the number of people exposed to health-endangering noise levels from transport • Reduce the number of people exposed to health-endangering concentrations of air pollution in locations where the impact of transport emissions is significant

transport strategies, and vice versa. In New Zealand, for example, the NZTS is coordinated with the New Zealand Energy Strategy (NZES) and the New Zealand Energy Efficiency and Conservation Strategy (NZEECS), both released in 2007. The NZES sets the strategic direction for the energy sector to contribute to New Zealand's future prosperity and sustainability. It responds to the challenges of providing enough energy to meet the needs of a growing economy, maintain the security of the supply, and reduce GHG emissions (from electricity produced by burning fossil fuels). The NZEECS is the government's action plan to maximize energy efficiency and renewable energy. It includes transport actions, some of which have been incorporated into the NZTS. The NZES incorporates the government's decision to halve per capita GHG emissions from domestic transport by 2040, relative to 2007 per capita GHG emissions. It sets out five action areas for transport, in addition to emissions trading:

1. Managing demand for travel
2. Shifting to more efficient and/or lower-impact means of transport
3. Improving fuel efficiency of the vehicle fleet
4. Developing and adopting future fuels
5. Ensuring the security of short-term oil supplies and a diverse supply of transport fuels

The NZTS builds on the work already undertaken in the transport-related components of the NZES and NZEECS. Several other national and local government strategies are impacted on or affected by transport policy including the New Zealand Housing Strategy, the New Zealand Health Strategy, the New Zealand Disability Strategy, the New Zealand Positive Aging Strategy, the New Zealand Injury Prevention Strategy, the New Zealand Urban Design Protocol, the New Zealand Waste Strategy, Regional Growth Strategies, Regional Land Transport Strategies, Long Term Council Community Plans and other strategies and programs developed by local government (NZTS 2008).

15.2.4 Implementation: Turning Strategy into Action

To be implemented, transportation plans or strategies must guide investments in the transportation system. They must guide the development, prioritization and implementation of projects to facilitate achievement of the set targets. The NZTS provides the high-level framework for transport decision making over a period of thirty years. However, the transport needs of different parts of New Zealand vary and require the development and application of context-sensitive solutions to address local conditions effectively. Implementation will also reflect that priorities change over time. The government will develop a comprehensive action plan based on the NZTS that will identify the necessary actions to be taken, assign the actions to various entities that will be responsible and accountable for executing them, and for the timing of the various actions in order to implement this strategy successfully. The government is responsible for working with various transport sector stakeholders in a collaborative, accountable and evidence-based manner with respect to achieving the NZTS objectives. The government is also responsible for monitoring and evaluating progress, and reviewing the strategy at regular intervals to ensure it responds to changing circumstances. Over time, as different actions or projects are prioritized and implemented to achieve the strategy objectives, the nation's vision of transportation will be achieved. Examples of projects include public transit developments and enhancements to offer alternative modes and reduce congestion on various portions of the highway network, operation projects to improve safety, roadway widening and traffic operation improvements to improve traffic flow, and preventative and routine maintenance to preserve the physical system.

15.2.5 Monitoring, Reporting, and Review

Any serious efforts in planning and project development include monitoring, reporting, and the management of performance. Monitoring and reporting are tools for the responsible agency to demonstrate its stewardship of the infrastructure system and accountability to the general public, and provide a formal means for tracking progress toward the predetermined objectives. The NZTS has been prepared to shape the future of transport in New Zealand to 2040. It will, therefore, guide the actions of the government and of Crown entities responsible for different aspects of regulating transport and providing infrastructure. The Ministry of Transport in particular will be accountable for the delivery of the strategy by 2040, working with many other organizations such as local authorities and private transport providers, as well as the system users—collectively known as stakeholders. Monitoring and publishing of performance for the NZTS will be accomplished by means of a Transport Monitoring Indicator Framework (TMIF). The outputs from the TMIF are accessible on the Web (via the New Zealand Ministry of Transport website) and are updated and published on an annual or more frequent basis. Table 15.4 shows an example of how objectives, targets and indicators are aligned to ensure that the overall vision for transportation gets implemented, and Table 15.5 shows the full set of indictor sets for the TMIF. Performance is monitored, reported and reviewed periodically to help the agency to progressively tailor and fine-tune decisions to ensure that performance targets are achieved.

15.2.6 Concluding Remarks: New Zealand's Transportation Strategy

New Zealand's transportation strategy is a good example of one that was developed to help a community continue to achieve economic competitiveness and a higher social quality of life while reducing its negative impacts on the natural environment and society. The strategy demonstrates good planning principles with clear linkages between the vision, goals, and objectives, and also between the objectives, performance measures, and targets. As projects are selected to achieve the predetermined targets, the country should measure progress toward sustainability. The strategy demonstrates the importance of viewing the achievement of sustainability

Table 15.4 Example of related objective, target, and indicator

Objective	Ensure environmental sustainability
Target	Reduce the kilometers traveled by single-occupancy vehicles, in major urban areas on weekdays, by 10% per capita by 2012 compared to 2007
Indicator	Distance per capita traveled in single-occupancy vehicles, in major urban areas, on weekdays

Table 15.5 Indicator sets for New Zealand transport monitoring indicator framework

Indicator set	Description
Transport volume	Tracks the use of the transport system, including motorized and nonmotorized modes. Also shows the age and composition of the fleet.
Network reliability	Describes the reliability of the transport network, congestion, and travel time.
Freight and the transport industry	Includes information on freight movements (domestic and international). Also includes share of the transport and storage industry of the Gross Domestic Product of New Zealand and transportation workforce information.
Access to the transport system	Shows how accessible the transport system is to a range of transport users. It includes indicators relating to the affordability of transport, social connectivity, and access to motor vehicles, travel perceptions, and accessibility of public transport.
Travel patterns	Shows the use of various transport modes, including active modes such as walking and cycling, for everyday journeys, for example, to work and school.
Transport safety and security	Shows how the transport system is performing with respect to safety in terms of transport-related deaths, injuries, accidents, and the social cost of accidents. It also includes personal security, resilience, and the security of the transport system.
Public health effects of transport	Shows how transport contributes to the noise levels and air quality that impact public health.
Infrastructure and investment	Shows infrastructure investment and the size and quality of transport infrastructure.
Environmental impact of transport	Includes climate change emissions and information on energy and land use, water quality, and waste management.
Transport-related price indices	Includes data on transport-related prices, including fuel and construction prices.

(Adapted from www.transport.govt.nz/ourwork/TMIF/Pages/default.aspx. Accessed February 2011)

as a systems issue in the sense that the transportation sector influences, and is influenced by, other sectors, (e.g., land use, energy, and public health). An effective strategy cannot ignore these linkages but has to be designed to be sensitive to related sectors and also needs to be fairly comprehensive. Ultimately, interlinked systems cannot be sustainable in isolation because of their linkages to, and impacts on, one another. Like Interface's approach to sustainability in the private sector, which seeks to involve all stakeholders in the supply and demand chain, public agency efforts must involve stakeholder agencies and customers to be ultimately successful in making substantive progress toward sustainability.

15.3 SUSTAINABILITY EVALUATION OF TRANSPORTATION PLAN ALTERNATIVES USING MULTIPLE ATTRIBUTE DECISION MAKING METHODOLOGY

Planning and designing for sustainability involves making decisions that involve multiple criteria: system performance, environmental, economic, and social quality of life considerations. Multiple attribute decision making (MADM) refers to making preference decisions (i.e., evaluation, prioritization, selection) over the available alternatives that are characterized by multiple, usually conflicting, attributes (Hwang and Yoon, 1981). MADM is a useful tool for evaluating planning and design alternatives with respect to achieving sustainability because it offers a framework for assessing alternatives based on multiple sustainability criteria, identifying the dominant alternative or superior alternatives, and assessing tradeoffs between or among the alternatives when there is no dominant alternative.

A visualization tool, the Sustainability Diamond, was developed by Jeon et al., (2010) based on MADM. Together with the Sustainability Diamond, MADM can be applied to evaluate infrastructure plans, policy packages, or project alternatives based on multiple sustainability criteria or parameters. The Sustainability Diamond can be used as a composite index to combine multiple performance measures into one value that reflects how well each alternative performs in contributing to municipal, regional, or national sustainability goals. These tools have been applied to compare three different transportation and land use scenarios for the Atlanta Metropolitan Region: Baseline 2005, Mobility 2030 (the adopted regional transportation plan) and Aspirations 2030 (financially unconstrained version of the regional transportation plan). This application develops an index of measures for transportation system effectiveness, environmental integrity, economic development, and social equity. The study:

- Identifies key decision criteria for the project, plan, or policy under consideration.
- Selects appropriate performance measures for each decision criterion.
- Populates the measures with data.
- Normalizes and weights the measures according to their relative importance in decision making.
- Develops weighted aggregate indices for system performance, environmental integrity, economic development, and social equity.
- Develops a comprehensive sustainability index based on the four component indexes (functional performance, economic, environmental, and social quality of life impacts). The four sustainability indexes for each alternative can be plotted for comparison and the resulting Sustainability Diamond assessed to help decision makers determine the dominant alternative or superior alternatives, or assess tradeoffs among alternatives where there is no dominant alternative for achieving sustainability.

15.3.1 MADM Approach for Sustainability Evaluation of Plan Alternatives

Five steps were used to develop the sustainability index and visualization for three metro Atlanta transportation and land use plan alternatives:

Step 1: Identify sustainability issues or goals

The long-range regional transportation plan for metropolitan Atlanta, Mobility 2030, has the following goals:

- Improving accessibility and mobility
- Maintaining and improving system performance and preservation
- Protecting and improving the environment and quality of life
- Increasing safety and security

The goals cover transportation system effectiveness (i.e., functional performance) and the three basic elements of sustainability:

- Economic: economic efficiency and development as well as financial affordability
- Environmental: environmental integrity, natural resources, and system resilience
- Sociocultural: social equity, safety and human health, and quality of life.

Step 2: Define relevant performance measures for each goal

To measure how well the three planning scenarios met the sustainability goals, 27 performance measures were identified with at least one performance measure for each goal. The list of measures was pared down to 11 because data was not readily available for all the measures. Table 15.6 shows the final list of goals and performance measures used to evaluate each component of sustainability.

Table 15.6 Sustainability goals and performance measures

Dimension of sustainability	Goals	Performance measures
Transportation system effectiveness	A1. Improve mobility	A11. Average freeway speed
	A2. Improve system performance	A21. Veh/mi traveled per capita
Environmental sustainability	B2. Minimize air pollution	B21. VOC* emissions B23. NO_x† emissions
Economic sustainability	C1. Maximize economic efficiency	C12. Total time spent in traffic
Social sustainability	D1. Maximize equity	D12-1. Equity of VOC exposure D12-2. Equity of NO_x exposure
	D2. Improve public health	D21-1 Exposure to VOC emissions D21-2 Exposure to NO_x emissions

*VOC = Volatile organic compounds
†NO_x = Nitrogen oxides

Step 3: Analyze and quantify the impact of different plans

Constructing the composite sustainability index starts with developing an index for each of the four components of sustainability. Each performance measure is assigned a raw value (see Table 15.7) that is then divided by the minimum or maximum value of said performance measure to create a normalized value. Raw values are developed from data collected for the various measures through estimates developed by models, projections from historical data, and so on. The raw values are normalized so that they are unitless and can be added together. For example, for the measure of freeway speed (A11), higher values are more desirable; therefore each raw value is divided by the highest speed to create normalized values. On the other hand, for the pollution measure (B2), lower levels of pollution are desirable. The minimum pollution is therefore divided by each raw value to create normalized values. Table 15.8 summarizes the normalized values for each component measure of sustainability.

Step 4: Construct composite sustainability index by using criteria and criteria weights

The next step is to assign weights to the measures that reflect the relative importance of each regional sustainability goal. Assigning weights is a subjective process that can be performed using various methods and requires consensus of policymakers. Table 15.8 shows the weights assigned to the four component sustainability criteria and their associated performance measures, reflecting the relative levels of importance of these criteria and measures to the decision maker. An index is calculated as the weighted average of the performance measures:

$$\text{INDEX} = \sum (\text{NORMALIZED VALUE} \times \text{WEIGHT}) \tag{15.1}$$

Table 15.7 Raw values for selected performance measures

Performance measures	Unit	Baseline 2005	Mobility 2030	Aspirations 2030
A11. Average freeway speed	miles per hour	42.17	42.21	42.21
A21. Vehicle miles traveled per capita	miles per person	35.04	31.75	31.75
B21. VOC emissions	ton per day	118.33	53.38	53.38
B23. NO_x emissions	ton per day	209.64	38.33	38.33
C12. Vehicle hours traveled per capita	minute per person	9.26	8.95	8.95
D12-1. Equity of VOC exposure (spatial)	Social equity index	19.10	23.45	23.45
D12-2. Equity of NO_x exposure (spatial)	Social equity index	20.02	23.56	23.60
D12-3. Equity of VOC exposure (income)	Income equity index	10.74	55.95	427.17
D12-4. Equity of NO_x exposure (income)	Income equity index	9.57	54.97	364.93
D21-1. Exposure to VOC emissions	Human impact index	1354.56	467.48	4134.47
D21-2. Exposure to NO_x emissions	Human impact index	2269.79	318.92	2766.65

(Adapted from Jeon et al., 2010)

The comprehensive sustainability index is calculated in the same way as the component sustainability indexes. Each sustainability component is assigned a weight based on the community's priorities. The comprehensive index is therefore the weighted average of the component sustainability indexes. Table 15.9 shows the component and comprehensive sustainability indexes for the three metro Atlanta land use and transportation plans.

Step 5: Using Sustainability Diamond to Evaluate Trade-offs

The Sustainability Diamond (see Figure 15.11) is a visualization tool to help identify dominant alternatives and evaluate trade-offs between and among alternatives, where there are no clearly dominant alternatives, based on the multiple sustainability decision criteria. The four component sustainability criteria are found along the four dimensions of the diamond and plots are made of the relative values of each alternative plan along each dimension (i.e., system effectiveness, environmental impact, economic impact and social impact).

Table 15.8 Normalized values and criteria weights for selected performance measures

			Normalized values		
Criteria weights	**Performance measures**	**Weights**	**Baseline 2005**	**Mobility 2030**	**Aspirations 2030**
A. Transport (35%)	A11. Average freeway speed	0.67	1.000	0.896	0.896
	A22. Veh/mi traveled per capita	0.33	0.906	1.000	1.000
B. Environmental (20%)	B21. VOC emissions	0.50	0.451	1.000	1.000
	B23. NO_x emissions	0.50	0.183	1.000	1.000
C. Economic (10%)	C12. Veh/hr traveled per capita	1.00	0.967	1.000	1.000
D. Social (35%)	D12-1. Equity of VOC exposure (spatial)	0.12	0.815	1.000	1.000
	D12-2. Equity of NO_x exposure (spatial)	0.12	0.848	0.998	1.000
	D12-3. Equity of VOC exposure (income)	0.12	1.000	0.192	0.025
	D12-4. Equity of NO_x exposure (income)	0.12	1.000	0.174	0.026
	D21-1. Exposure to VOC emissions	0.26	0.345	1.000	0.113
	D21-2. Exposure to NO_x emissions	0.26	0.141	1.000	0.115

(Adapted from Jeon et al., 2010)

Table 15.9 Component and comprehensive sustainability indexes for metro Atlanta plan alternatives

Sustainability indexes	Baseline 2005	Mobility 2030	Aspirations 2030
Environmental	0.317	1.000	1.000
Social	0.566	0.804	0.306
Transportation effectiveness	0.972	0.972	0.972
Economic	0.967	1.000	1.000
Comprehensive	0.698	0.906	0.731

(Adapted from Jeon et al., 2010)

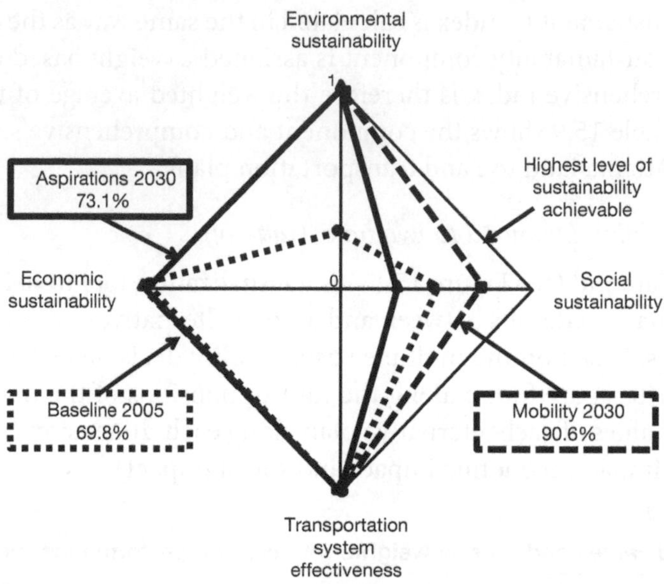

Figure 15.11 The sustainability diamond visualization tool: relative effectiveness of plan alternatives (adapted from Jeon et al., 2010)

15.3.2 Applying MADM and the Sustainability Diamond: Identifying Superior Plans

The sustainability visualization tool helps a decision maker evaluate the relative effectivencss of alternative plans (or projects) based on the index values calculated for each plan (see Table 15.9). The plan with the highest value in any of the four dimensions defines the maximum obtainable value for the set of plans being evaluated. This way plan alternatives can be evaluated based on their relative effectiveness. The comprehensive index is useful for identifying dominant alternatives and the component indexes are useful for evaluating trade-offs between and among alternatives. Figure 15.11 shows that Mobility 2030 is the dominant alternative; and while Aspirations 2030 is a better plan in terms of the natural environment, Baseline 2005 is a better plan for improving the social quality of life. A tool like the Sustainability Diamond can be used iteratively to develop desirable plans from the perspective of these multiple sustainability criteria (Jeon et al., 2010; FHWA, 2011).

15.3.3. Concluding Remarks: Applying MADM and Visualization for Selecting Plan Alternatives

The case study on applying MADM and visualization tools in selecting superior plan alternatives highlights the importance of examining the basic sustainability criteria separately as well as collectively, when evaluating plan (or project) alternatives, in order to understand which plans are weaker and stronger in various aspects of sustainability and evaluate these existing trade-offs to identify the optimal plan or superior plans (where no optimal plan exists). This methodology evaluates the relative effectiveness of alternatives with respect to functional performance and the three basic sustainability criteria (i.e., the natural environment, the economy, and social quality of life); however, it does not include any objective criteria for determining minimum thresholds for sustainability. Such data would have to be included to determine if the most attractive alternative among the various competing alternatives actually meets minimum thresholds for sustainability. Minimum thresholds of sustainability can be determined by assessing the extent to which the use of environmental capital is within the limits of a system's carrying capacity and the extent to which economic capital is being developed and used to improve the quality of life of communities.

15.4 CONCLUSION

This chapter presents three case studies that demonstrate how the concept of sustainability can be made operational in both public and private sector contexts. The first case study examines how the largest carpet manufacturing company in the world, Interface, has taken a leadership role in moving toward environmental sustainability and defining the prototypical company of the 21st century as one that creates economic wealth and improves the quality of life of its customers without negatively impacting the natural environment. The second case study presents the New Zealand Transportation Strategy, a detailed long-range plan for improving the sustainability of the transportation system while enhancing its positive impacts and reducing its negative impacts on the economy and social quality of life. The third case study presents an application of multiple attribute decision making (MADM) methods and visualization techniques to evaluate competing transportation and land use plan alternatives in a major metropolitan area. Collectively, these examples demonstrate how the principles of sustainability can be made operational in a wide variety of contexts and illustrate the benefits to be gained from such initiatives.

REFERENCES

Anderson, R. (1999). *Mid-Course Correction: Toward a Sustainable Enterprise: The Interface Model*, White River Junction, VT: Chelsea Green Publishing.

Hwang, C. L. and K. Yoon (1981). *Multiple Attribute Decision Making: Methods and Applications, Lecture Notes in Economics and Mathematical Systems*, New York: Springer-Verlag.

Jeon, C. M., A. A. Amekudzi, and R. Guensler (2010). "Evaluating Transportation System Sustainability: Atlanta Metropolitan Region," *International Journal of Sustainable Transportation*, 4:4, 227-247.

Federal Highway Administration (2011) *Transportation Planning for Sustainability Guidebook*. Prepared by A. Amekudzi, M. Meyer, C. Ross, and E. Barrella, Georgia Institute of Technology, January.

Interface (2010). Press release: *Energy Use Reduced Nearly One-Half; Sales up 27 Percent. Thirteen years in, Interface, Inc. has Reduced its Footprint While Growing the Business*, April 20.

——— *Interface's values are our guiding principles*. Accessed at www.interfaceglobal.com/Sustainability/Our-Journey/Vision.aspx in January 2011.

——— *Our Progress*. Accessed at www.interfaceglobal.com/Sustainability/Progress-to-Zero.aspx in January 2011.

——— *The Seven Fronts of Mount Sustainability*. Accessed at http://www.interfaceglobal.com/Sustainability/Our-Journey/7-Fronts-of-Sustainability.aspx in January 2011.

——— *Toward a More Sustainable Way of Business*. Accessed at www.interfaceglobal.com/Sustainability.aspx in January 2011.

——— *Verification and Certification: Third Party Involvement*. Accessed at www.interfaceglobal.com/Sustainability/Progress-to-Zero/Circle-of-Influence.aspx in January 2011.

Ministry of the Environment, New Zealand, Sustainable transport definition. Accessed at www.mfe.govt.nz/issues/transport/sustainable/ in February 2011.

Ministry of Transport, New Zealand (2008). *New Zealand Transport Strategy*. Accessed at www.transport.govt.nz/ourwork/Documents/NZTS2008.pdf in November 2010.

——— *Transport Monitoring Indicator Framework*. Accessed at www.transport.govt.nz/ourwork/TMIF/Pages/default.aspx in February 2011.

Yoon, P. K. and H. Ching-Lai, (1995). *Multiple Attribute Decision Making, An Introduction, Quantitative Applications in the Social Sciences*, Thousand Oaks/London/New Delhi: Sage Publications.

EXERCISES

1. Identify a carbon footprint calculator online and calculate your annual carbon footprint. Identify three actions that you can take to reduce your footprint significantly that would be acceptable to your lifestyle and pocketbook.

2. Obtain historic data for Interface or another manufacturing engineering or other engineering company, select an appropriate model from the models presented in Chapter 14, and evaluate the progress of the company toward sustainability. Comment on your results.

3. Classify the eight indicator sets in the New Zealand Transport Monitoring Indicator Framework into system performance, environmental, economic, and social impact indicators. Do some of the indicators fall into more than one category? Explain why.

4. Using data from the Texas Transportation Institute's Urban Mobility Report, compare and contrast the sustainability levels and trends of two major metropolitan areas in the United States, using an appropriate model. Discuss the limitations of your comparison and explain what other types of data you would collect to improve your answer.

5. Using ecological footprint data from the Global Footprint Network and the Human Development Index data from the United Nations' Human Development Index database, both available online, evaluate and compare the sustainability of a low- and high-income country. Which country is more sustainable? Why?

6. Select a large-scale infrastructure system, such as the Boston Central Artery Tunnel, the Three Gorges Dam, or the Golden Gate Bridge, and sketch a plan of how you would go about evaluating the sustainability of this system. Be sure to identify all the important data you need to collect and potential sources for the data.
7. The mayor of a city would like to give a sustainability award to the agency that has contributed most significantly to advancing the city toward sustainability in the past year. Agencies such as the Department of Public Works, the Department of Transportation, the Environmental Protection Agency, and the Department of Economic Development are being considered for the award. What criteria would you suggest to the mayor to use in determining the best agency? Explain why.

APPENDIX

Table A.1 Standard Normal Probability Values $\Phi(s)$

s	.00	.01	.02	.03	.04	.05	.06	.07	.08	.09
0.0	0.5000	0.5040	0.5080	0.5120	0.5160	0.5199	0.5239	0.5279	0.5319	0.5359
0.1	0.5398	0.5438	0.5478	0.5517	0.5557	0.5596	0.5636	0.5675	0.5714	0.5753
0.2	0.5793	0.5832	0.5871	0.5910	0.5948	0.5987	0.6026	0.6064	0.6103	0.6141
0.3	0.6179	0.6217	0.6255	0.6293	0.6331	0.6368	0.6406	0.6443	0.6480	0.6517
0.4	0.6554	0.6591	0.6628	0.6664	0.6700	0.6736	0.6772	0.6808	0.6844	0.6879
0.5	0.6915	0.6950	0.6985	0.7019	0.7054	0.7088	0.7123	0.7157	0.7190	0.7224
0.6	0.7257	0.7291	0.7324	0.7357	0.7389	0.7422	0.7454	0.7486	0.7517	0.7549
0.7	0.7580	0.7611	0.7642	0.7673	0.7704	0.7734	0.7764	0.7794	0.7823	0.7852
0.8	0.7881	0.7910	0.7939	0.7967	0.7995	0.8023	0.8051	0.8078	0.8106	0.8133
0.9	0.8159	0.8186	0.8212	0.8238	0.8264	0.8289	0.8315	0.8340	0.8365	0.8389
1.0	0.8413	0.8438	0.8461	0.8485	0.8508	0.8531	0.8554	0.8577	0.8599	0.8621
1.1	0.8643	0.8665	0.8686	0.8708	0.8729	0.8749	0.8770	0.8790	0.8810	0.8830
1.2	0.8849	0.8869	0.8888	0.8907	0.8925	0.8944	0.8962	0.8980	0.8997	0.9015
1.3	0.9032	0.9049	0.9066	0.9082	0.9099	0.9115	0.9131	0.9147	0.9162	0.9177
1.4	0.9192	0.9207	0.9222	0.9236	0.9251	0.9265	0.9279	0.9292	0.9306	0.9319
1.5	0.9332	0.9345	0.9357	0.9370	0.9382	0.9394	0.9406	0.9418	0.9429	0.9441
1.6	0.9452	0.9463	0.9474	0.9484	0.9495	0.9505	0.9515	0.9525	0.9535	0.9545
1.7	0.9554	0.9564	0.9573	0.9582	0.9591	0.9599	0.9608	0.9616	0.9625	0.9633
1.8	0.9641	0.9649	0.9656	0.9664	0.9671	0.9678	0.9686	0.9693	0.9699	0.9706
1.9	0.9713	0.9719	0.9726	0.9732	0.9738	0.9744	0.9750	0.9756	0.9761	0.9767
2.0	0.9772	0.9778	0.9783	0.9788	0.9793	0.9798	0.9803	0.9808	0.9812	0.9817
2.1	0.9821	0.9826	0.9830	0.9834	0.9838	0.9842	0.9846	0.9850	0.9854	0.9857
2.2	0.9861	0.9864	0.9868	0.9871	0.9875	0.9878	0.9881	0.9884	0.9887	0.9890
2.3	0.9893	0.9896	0.9898	0.9901	0.9904	0.9906	0.9909	0.9911	0.9913	0.9916
2.4	0.9918	0.9920	0.9922	0.9925	0.9927	0.9929	0.9931	0.9932	0.9934	0.9936
2.5	0.9938	0.9940	0.9941	0.9943	0.9945	0.9946	0.9948	0.9949	0.9951	0.9952
2.6	0.9953	0.9955	0.9956	0.9957	0.9959	0.9960	0.9961	0.9962	0.9963	0.9964
2.7	0.9965	0.9966	0.9967	0.9968	0.9969	0.9970	0.9971	0.9972	0.9973	0.9974
2.8	0.9974	0.9975	0.9976	0.9977	0.9977	0.9978	0.9979	0.9979	0.9980	0.9981
2.9	0.9981	0.9982	0.9982	0.9983	0.9984	0.9984	0.9985	0.9985	0.9986	0.9986
3.0	0.9987	0.9987	0.9987	0.9988	0.9988	0.9989	0.9989	0.9989	0.9990	0.9990
3.1	0.9990	0.9991	0.9991	0.9991	0.9992	0.9992	0.9992	0.9992	0.9993	0.9993
3.2	0.9993	0.9993	0.9994	0.9994	0.9994	0.9994	0.9994	0.9995	0.9995	0.9995
3.3	0.9995	0.9995	0.9995	0.9996	0.9996	0.9996	0.9996	0.9996	0.9996	0.9997
3.4	0.9997	0.9997	0.9997	0.9997	0.9997	0.9997	0.9997	0.9997	0.9997	0.9998

Note: $\Phi(s) = P(S \leq s) = \frac{1}{\sqrt{2\pi}} \int_{-\infty}^{s} e^{-s^2/2} ds$ and $\Phi(-s) = 1 - \Phi(s)$

Table A.2 Critical values of t-distribution

ν	Significance level, α (see example for one-sided and two-sided cases)									
ν	0.20	0.15	0.10	0.05	0.025	0.01	0.005	0.0025	0.001	0.0005
1	1.376	1.963	3.078	6.314	12.71	31.82	63.66	127.3	318.3	636.6
2	1.061	1.386	1.886	2.920	4.303	6.965	9.925	14.09	22.33	31.60
3	0.978	1.250	1.638	2.353	3.182	4.541	5.841	7.453	10.21	12.92
4	0.941	1.190	1.533	2.132	2.776	3.747	4.604	5.598	7.173	8.610
5	0.920	1.156	1.476	2.015	2.571	3.365	4.032	4.773	5.893	6.869
6	0.906	1.134	1.440	1.943	2.447	3.143	3.707	4.317	5.208	5.959
7	0.896	1.119	1.415	1.895	2.365	2.998	3.499	4.029	4.785	5.408
8	0.889	1.108	1.397	1.860	2.306	2.896	3.355	3.833	4.501	5.041
9	0.883	1.100	1.383	1.833	2.262	2.821	3.250	3.690	4.297	4.781
10	0.879	1.093	1.372	1.812	2.228	2.764	3.169	3.581	4.144	4.587
11	0.876	1.088	1.363	1.796	2.201	2.718	3.106	3.497	4.025	4.437
12	0.873	1.083	1.356	1.782	2.179	2.681	3.055	3.428	3.930	4.318
13	0.870	1.079	1.350	1.771	2.160	2.650	3.012	3.372	3.852	4.221
14	0.868	1.076	1.345	1.761	2.145	2.624	2.977	3.326	3.787	4.140
15	0.866	1.074	1.341	1.753	2.131	2.602	2.947	3.286	3.733	4.073
16	0.865	1.071	1.337	1.746	2.120	2.583	2.921	3.252	3.686	4.015
17	0.863	1.069	1.333	1.740	2.110	2.567	2.898	3.222	3.646	3.965
18	0.862	1.067	1.330	1.734	2.101	2.552	2.878	3.197	3.610	3.922
19	0.861	1.066	1.328	1.729	2.093	2.539	2.861	3.174	3.579	3.883
20	0.860	1.064	1.325	1.725	2.086	2.528	2.845	3.153	3.552	3.850
21	0.859	1.063	1.323	1.721	2.080	2.518	2.831	3.135	3.527	3.819
22	0.858	1.061	1.321	1.717	2.074	2.508	2.819	3.119	3.505	3.792
23	0.858	1.060	1.319	1.714	2.069	2.500	2.807	3.104	3.485	3.767
24	0.857	1.059	1.318	1.711	2.064	2.492	2.797	3.091	3.467	3.745
25	0.856	1.058	1.316	1.708	2.060	2.485	2.787	3.078	3.450	3.725
26	0.856	1.058	1.315	1.706	2.056	2.479	2.779	3.067	3.435	3.707
27	0.855	1.057	1.314	1.703	2.052	2.473	2.771	3.057	3.421	3.690
28	0.855	1.056	1.313	1.701	2.048	2.467	2.763	3.047	3.408	3.674
29	0.854	1.055	1.311	1.699	2.045	2.462	2.756	3.038	3.396	3.659
30	0.854	1.055	1.310	1.697	2.042	2.457	2.750	3.030	3.385	3.646
40	0.851	1.050	1.303	1.684	2.021	2.423	2.704	2.971	3.307	3.551
50	0.849	1.047	1.299	1.676	2.009	2.403	2.678	2.937	3.261	3.496
60	0.848	1.045	1.296	1.671	2.000	2.390	2.660	2.915	3.232	3.460
120	0.845	1.041	1.289	1.658	1.980	2.358	2.617	2.860	3.160	3.373
∞	0.842	1.036	1.282	1.645	1.960	2.326	2.576	2.807	3.090	3.291

Example: With $\nu = 8$, at $\alpha = 0.05$ in a one-sided case, $(1 - \alpha) = 0.95$ and $t_\alpha = 1.860$. For a two-sided case, at $\alpha = 0.05$, we use $\alpha/2 = 0.025$. Thus, for $\nu = 8$, $t_{\alpha/2} = 2.306$.

Table A.3 Critical values of the chi-square distribution

	Significance level, α (see note below)									
ν	0.99	0.975	0.95	0.90	0.10	0.05	0.025	0.01	0.005	0.001
1	—	0.001	0.004	0.016	2.706	3.841	5.024	6.635	7.879	10.827
2	0.020	0.051	0.103	0.211	4.605	5.991	7.378	9.210	10.597	13.815
3	0.115	0.216	0.352	0.584	6.251	7.815	9.348	11.345	12.838	16.268
4	0.297	0.484	0.711	1.064	7.779	9.488	11.143	13.277	14.860	18.465
5	0.554	0.831	1.145	1.610	9.236	11.070	12.833	15.086	16.750	20.517
6	0.872	1.237	1.635	2.204	10.645	12.592	14.449	16.812	18.548	22.457
7	1.239	1.690	2.167	2.833	12.017	14.067	16.013	18.475	20.278	24.322
8	1.646	2.180	2.733	3.490	13.362	15.507	17.535	20.090	21.955	26.125
9	2.088	2.700	3.325	4.168	14.684	16.919	19.023	21.666	23.589	27.877
10	2.558	3.247	3.940	4.865	15.987	18.307	20.483	23.209	25.188	29.588
11	3.053	3.816	4.575	5.578	17.275	19.675	21.920	24.725	26.757	31.264
12	3.571	4.404	5.226	6.304	18.549	21.026	23.337	26.217	28.300	32.909
13	4.107	5.009	5.892	7.042	19.812	22.362	24.736	27.688	29.819	34.528
14	4.660	5.629	6.571	7.790	21.064	23.685	26.119	29.141	31.319	36.123
15	5.229	6.262	7.261	8.547	22.307	24.996	27.488	30.578	32.801	37.697
16	5.812	6.908	7.962	9.312	23.542	26.296	28.845	32.000	34.267	39.252
17	6.408	7.564	8.672	10.085	24.769	27.587	30.191	33.409	35.718	40.790
18	7.015	8.231	9.390	10.865	25.989	28.869	31.526	34.805	37.156	42.312
19	7.633	8.907	10.117	11.651	27.204	30.144	32.852	36.191	38.582	43.820
20	8.260	9.591	10.851	12.443	28.412	31.410	34.170	37.566	39.997	45.315
21	8.897	10.283	11.591	13.240	29.615	32.671	35.479	38.932	41.401	46.797
22	9.542	10.982	12.338	14.041	30.813	33.924	36.781	40.289	42.796	48.268
23	10.196	11.689	13.091	14.848	32.007	35.172	38.076	41.638	44.181	49.728
24	10.856	12.401	13.848	15.659	33.196	36.415	39.364	42.980	45.559	51.179
25	11.524	13.120	14.611	16.473	34.382	37.652	40.646	44.314	46.928	52.620
26	12.198	13.844	15.379	17.292	35.563	38.885	41.923	45.642	48.290	54.052
27	12.879	14.573	16.151	18.114	36.741	40.113	43.195	46.963	49.645	55.476
28	13.565	15.308	16.928	18.939	37.916	41.337	44.461	48.278	50.993	56.893
29	14.256	16.047	17.708	19.768	39.087	42.557	45.722	49.588	52.336	58.302
30	14.953	16.791	18.493	20.599	40.256	43.773	46.979	50.892	53.672	59.703
40	22.164	24.433	26.509	29.051	51.805	55.758	59.342	63.691	66.766	73.402
50	29.707	32.357	34.764	37.689	63.167	67.505	71.420	76.154	79.490	86.661
60	37.485	40.482	43.188	46.459	74.397	79.082	83.298	88.379	91.952	99.607
70	45.442	48.758	51.739	55.329	85.527	90.531	95.023	100.425	104.215	112.317
80	53.540	57.153	60.391	64.278	96.578	101.879	106.629	112.329	116.321	124.839

For example, for $\nu = 4$ and $\alpha = 0.05$, the critical value = 9.488, which means there is a 0.05 probability that this value will be exceeded.

Table A.4 Critical values for Kolmogorov-Smirnov test

	Significance level, α (see example below)				
n	**0.20**	**0.15**	**0.10**	**0.05**	**0.01**
1	0.900	0.925	0.950	0.975	0.995
2	0.684	0.726	0.776	0.842	0.929
3	0.565	0.597	0.642	0.708	0.828
4	0.494	0.525	0.564	0.624	0.733
5	0.446	0.474	0.510	0.565	0.669
6	0.410	0.436	0.470	0.521	0.618
7	0.381	0.405	0.438	0.486	0.577
8	0.358	0.381	0.411	0.457	0.543
9	0.339	0.360	0.388	0.432	0.514
10	0.322	0.342	0.368	0.410	0.490
11	0.307	0.326	0.352	0.391	0.468
12	0.295	0.313	0.338	0.375	0.450
13	0.284	0.302	0.325	0.361	0.433
14	0.274	0.292	0.314	0.349	0.418
15	0.266	0.283	0.304	0.338	0.404
16	0.258	0.274	0.295	0.328	0.392
17	0.250	0.266	0.286	0.318	0.381
18	0.244	0.259	0.278	0.309	0.371
19	0.237	0.252	0.272	0.301	0.363
20	0.231	0.246	0.264	0.294	0.356
25	0.210	0.220	0.240	0.270	0.320
30	0.190	0.200	0.220	0.240	0.290
35	0.180	0.190	0.210	0.230	0.270
40	0.169	0.180	0.193	0.210	0.257
45	0.160	0.170	0.182	0.203	0.243
50	0.151	0.161	0.172	0.192	0.231
>50	$1.07/\sqrt{n}$	$1.14/\sqrt{n}$	$1.22/\sqrt{n}$	$1.36/\sqrt{n}$	$1.63/\sqrt{n}$

Adapted from: Hoel, P.G. (1962) *Introduction to Mathematical Statistics*, 3rd edition. J. Wiley and Sons, New York, NY.

Example: With n = 17, at α = 0.05, the critical value $D_\alpha \approx 0.32$.Thus, there is a 0.95 probability that D_{max} computed from the K-S test will be less than this critical value.

Table A.5 Critical values of *F*-distribution $f_{0.025}$ (ν_1, ν_2)

	ν_1								
ν_2	1	2	3	4	5	6	7	8	9
1	647.8	799.5	864.2	899.6	921.9	937.1	948.2	956.7	963.3
2	38.51	39.00	39.17	39.25	39.30	39.33	39.36	39.37	39.39
3	17.44	16.04	15.44	15.10	14.88	14.73	14.62	14.54	14.47
4	12.22	10.65	9.98	9.60	9.36	9.20	9.07	8.98	8.91
5	10.01	8.43	7.76	7.39	7.15	6.98	6.85	6.76	6.68
6	8.81	7.26	6.60	6.23	5.99	5.82	5.70	5.60	5.52
7	8.07	6.54	5.89	5.52	5.29	5.12	4.99	4.90	4.82
8	7.57	6.06	5.42	5.05	4.82	4.65	4.53	4.43	4.36
9	7.21	5.71	5.08	4.72	4.48	4.32	4.20	4.10	4.03
10	6.94	5.46	4.83	4.47	4.24	4.07	3.95	3.85	3.78
11	6.72	5.26	4.63	4.28	4.04	3.88	3.76	3.66	3.59
12	6.55	5.10	4.47	4.12	3.89	3.73	3.61	3.51	3.44
13	6.41	4.97	4.35	4.00	3.77	3.60	3.48	3.39	3.31
14	6.30	4.86	4.24	3.89	3.66	3.50	3.38	3.29	3.21
15	6.20	4.77	4.15	3.80	3.58	3.41	3.29	3.20	3.12
16	6.12	4.69	4.08	3.73	3.50	3.34	3.22	3.12	3.05
17	6.04	4.62	4.01	3.66	3.44	3.28	3.16	3.06	2.98
18	5.98	4.56	3.95	3.61	3.38	3.22	3.10	3.01	2.93
19	5.92	4.51	3.90	3.56	3.33	3.17	3.05	2.96	2.88
20	5.87	4.46	3.86	3.51	3.29	3.13	3.01	2.91	2.84
21	5.83	4.42	3.82	3.48	3.25	3.09	2.97	2.87	2.80
22	5.79	4.38	3.78	3.44	3.22	3.05	2.93	2.84	2.76
23	5.75	4.35	3.75	3.41	3.18	3.02	2.90	2.81	2.73
24	5.72	4.32	3.72	3.38	3.15	2.99	2.87	2.78	2.70
25	5.69	4.29	3.69	3.35	3.13	2.97	2.85	2.75	2.68
26	5.66	4.27	3.67	3.33	3.10	2.94	2.82	2.73	2.65
27	5.63	4.24	3.65	3.31	3.08	2.92	2.80	2.71	2.63
28	5.61	4.22	3.63	3.29	3.06	2.90	2.78	2.69	2.61
29	5.59	4.20	3.61	3.27	3.04	2.88	2.76	2.67	2.59
30	5.57	4.18	3.59	3.25	3.03	2.87	2.75	2.65	2.57
40	5.42	4.05	3.46	3.13	2.90	2.74	2.62	2.53	2.45
60	5.29	3.93	3.34	3.01	2.79	2.63	2.51	2.41	2.33
120	5.15	3.80	3.23	2.89	2.67	2.52	2.39	2.30	2.22
∞	5.02	3.69	3.12	2.79	2.57	2.41	2.29	2.19	2.11

Adapted from Pearson, E. S. and Hartley, H. O. (1966) *Biometrika Tables for Statisticians*, 3rd edition. University Press, Cambridge, UK.

Table A.5 Critical values of F-distribution $f_{0.025}(\nu_1, \nu_2)$, continued

	ν_1									
ν_2	10	12	15	20	24	30	40	60	120	∞
1	969	976.7	984.9	993.1	997.2	1001.4	1005.6	1009.8	1014.0	1018.3
2	39.4	39.41	39.43	39.45	39.46	39.47	39.47	39.48	39.49	39.50
3	14.4	14.34	14.25	14.7	14.12	14.08	14.04	13.99	13.95	13.90
4	8.84	8.75	8.66	8.56	8.51	8.46	8.41	8.36	8.31	8.26
5	6.62	6.52	6.43	6.33	6.28	6.23	6.18	6.12	6.07	6.02
6	5.46	5.37	5.27	5.17	5.12	5.07	5.01	4.96	4.90	4.85
7	4.76	4.67	4.57	4.47	4.42	4.36	4.31	4.25	4.20	4.14
8	4.30	4.20	4.10	4.00	3.95	3.89	3.84	3.78	3.73	3.67
9	3.96	3.87	3.77	3.67	3.61	3.56	3.51	3.45	3.39	3.33
10	3.72	3.62	3.52	3.42	3.37	3.31	3.26	3.20	3.14	3.08
11	3.53	3.43	3.33	3.23	3.17	3.12	3.06	3.00	2.94	2.88
12	3.37	3.28	3.18	3.07	3.02	2.96	2.91	2.85	2.79	2.73
13	3.25	3.15	3.05	2.95	2.89	2.84	2.78	2.72	2.66	2.60
14	3.15	3.05	2.95	2.84	2.79	2.73	2.67	2.61	2.55	2.49
15	3.06	2.96	2.86	2.76	2.70	2.64	2.59	2.52	2.46	2.40
16	2.99	2.89	2.79	2.68	2.63	2.57	2.51	2.45	2.38	2.32
17	2.92	2.83	2.72	2.62	2.56	2.50	2.44	2.38	2.32	2.25
18	2.87	2.77	2.67	2.56	2.50	2.45	2.38	2.32	2.26	2.19
19	2.82	2.72	2.62	2.51	2.45	2.39	2.33	2.27	2.20	2.13
20	2.77	2.68	2.57	2.47	2.41	2.35	2.29	2.22	2.16	2.09
21	2.74	2.64	2.53	2.43	2.37	2.31	2.25	2.18	2.11	2.04
22	2.70	2.60	2.50	2.39	2.33	2.27	2.21	2.15	2.08	2.00
23	2.67	2.57	2.47	2.36	2.30	2.24	2.18	2.11	2.04	1.97
24	2.64	2.54	2.44	2.33	2.27	2.21	2.15	2.08	2.01	1.94
25	2.61	2.52	2.41	2.30	2.24	2.18	2.12	2.05	1.98	1.91
26	2.59	2.49	2.39	2.28	2.22	2.16	2.09	2.03	1.95	1.88
27	2.57	2.47	2.36	2.25	2.20	2.13	2.07	2.00	1.93	1.85
28	2.55	2.45	2.34	2.23	2.17	2.11	2.05	1.98	1.91	1.83
29	2.53	2.43	2.33	2.21	2.15	2.09	2.03	1.96	1.89	1.81
30	2.51	2.41	2.31	2.20	2.14	2.07	2.01	1.94	1.87	1.79
40	2.39	2.29	2.18	2.07	2.01	1.94	1.88	1.80	1.72	1.64
60	2.27	2.17	2.06	1.95	1.88	1.82	1.74	1.67	1.58	1.48
120	2.16	2.06	1.95	1.83	1.76	1.69	1.61	1.53	1.43	1.31
∞	2.05	1.95	1.83	1.71	1.64	1.57	1.48	1.39	1.27	1.00

Example: At $\alpha = 0.05$, $\alpha/2 = 0.025$ and with $\nu_1 = 9$ and $\nu_2 = 12$, $f_{\alpha/2}(9,12) = 3.44$ and $f_{\alpha/2}(12,9) = 3.87$. Thus, $f_{1-\alpha/2}(9,12) = 1/3.87 = 0.26$.

Table A.5 Critical values of *F*-distribution $f_{0.05}$ (ν_1, ν_2), continued

	ν_1								
ν_2	**1**	**2**	**3**	**4**	**5**	**6**	**7**	**8**	**9**
1	161.5	199.5	215.7	224.6	230.2	234.1	236.8	238.9	240.5
2	18.51	19.00	19.16	19.25	19.30	19.33	19.35	19.37	19.39
3	10.13	9.55	9.28	9.12	9.01	8.94	8.89	8.85	8.81
4	7.71	6.94	6.59	6.39	6.26	6.16	6.09	6.04	6.00
5	6.61	5.79	5.41	5.19	5.05	4.95	4.88	4.82	4.77
6	5.99	5.14	4.76	4.53	4.39	4.28	4.21	4.15	4.10
7	5.59	4.74	4.35	4.12	3.97	3.87	3.79	3.73	3.68
8	5.32	4.46	4.07	3.84	3.69	3.58	3.50	3.44	3.39
9	5.12	4.26	3.86	3.63	3.48	3.37	3.29	3.23	3.18
10	4.96	4.10	3.71	3.48	3.33	3.22	3.14	3.07	3.02
11	4.84	3.98	3.59	3.36	3.20	3.09	3.01	2.95	2.90
12	4.75	3.89	3.49	3.26	3.11	3.00	2.91	2.85	2.80
13	4.67	3.81	3.41	3.18	3.03	2.92	2.83	2.77	2.71
14	4.60	3.74	3.34	3.11	2.96	2.85	2.76	2.70	2.65
15	4.54	3.68	3.29	3.06	2.90	2.79	2.71	2.64	2.59
16	4.49	3.63	3.24	3.01	2.85	2.74	2.66	2.59	2.54
17	4.45	3.59	3.20	2.96	2.81	2.70	2.61	2.55	2.49
18	4.41	3.55	3.16	2.93	2.77	2.66	2.58	2.51	2.46
19	4.38	3.52	3.13	2.90	2.74	2.63	2.54	2.48	2.42
20	4.35	3.49	3.10	2.87	2.71	2.60	2.51	2.45	2.39
21	4.32	3.47	3.07	2.84	2.68	2.57	2.49	2.42	2.37
22	4.30	3.44	3.05	2.82	2.66	2.55	2.46	2.40	2.34
23	4.28	3.42	3.03	2.80	2.64	2.53	2.44	2.37	2.32
24	4.26	3.40	3.01	2.78	2.62	2.51	2.42	2.36	2.30
25	4.24	3.39	2.99	2.76	2.60	2.49	2.40	2.34	2.28
26	4.23	3.37	2.98	2.74	2.59	2.47	2.39	2.32	2.27
27	4.21	3.35	2.96	2.73	2.57	2.46	2.37	2.31	2.25
28	4.20	3.34	2.95	2.71	2.56	2.45	2.36	2.29	2.24
29	4.18	3.33	2.93	2.70	2.55	2.43	2.35	2.28	2.22
30	4.17	3.32	2.92	2.69	2.53	2.42	2.33	2.27	2.21
40	4.08	3.23	2.84	2.61	2.45	2.34	2.25	2.18	2.12
60	4.00	3.15	2.76	2.53	2.37	2.25	2.17	2.10	2.04
120	3.92	3.07	2.68	2.45	2.29	2.18	2.09	2.02	1.96
∞	3.84	3.00	2.60	2.37	2.21	2.10	2.01	1.94	1.88

Table A.5 Critical values of F-distribution $f_{0.05}(\nu_1, \nu_2)$, continued

	ν_1									
ν_2	**10**	**12**	**15**	**20**	**24**	**30**	**40**	**60**	**120**	∞
1	242	244	246	248	249	250	251	252	253	254
2	19.4	19.4	19.4	19.5	19.5	19.5	19.5	19.5	19.5	19.5
3	8.79	8.74	8.70	8.66	8.64	8.62	8.59	8.57	8.55	8.53
4	5.96	5.91	5.86	5.80	5.77	5.75	5.72	5.69	5.66	5.63
5	4.74	4.68	4.62	4.56	4.53	4.50	4.46	4.43	4.40	4.37
6	4.06	4.00	3.94	3.87	3.84	3.81	3.77	3.74	3.70	3.67
7	3.64	3.57	3.51	3.44	3.41	3.38	3.34	3.30	3.27	3.23
8	3.35	3.28	3.22	3.15	3.12	3.08	3.04	3.01	2.97	2.93
9	3.14	3.07	3.01	2.94	2.90	2.86	2.83	2.79	2.75	2.71
10	2.98	2.91	2.85	2.77	2.74	2.70	2.66	2.62	2.58	2.54
11	2.85	2.79	2.72	2.65	2.61	2.57	2.53	2.49	2.45	2.40
12	2.75	2.69	2.62	2.54	2.51	2.47	2.43	2.38	2.34	2.30
13	2.67	2.60	2.53	2.46	2.42	2.38	2.34	2.30	2.25	2.21
14	2.60	2.53	2.46	2.39	2.35	2.31	2.27	2.22	2.18	2.13
15	2.54	2.48	2.40	2.33	2.29	2.25	2.20	2.16	2.11	2.07
16	2.49	2.42	2.35	2.28	2.24	2.19	2.15	2.11	2.06	2.01
17	2.45	2.38	2.31	2.23	2.19	2.15	2.10	2.06	2.01	1.96
18	2.41	2.34	2.27	2.19	2.15	2.11	2.06	2.02	1.97	1.92
19	2.38	2.31	2.23	2.16	2.11	2.07	2.03	1.98	1.93	1.88
20	2.35	2.28	2.20	2.12	2.08	2.04	1.99	1.95	1.90	1.84
21	2.32	2.25	2.18	2.10	2.05	2.01	1.96	1.92	1.87	1.81
22	2.30	2.23	2.15	2.07	2.03	1.98	1.94	1.89	1.84	1.78
23	2.27	2.20	2.13	2.05	2.01	1.96	1.91	1.86	1.81	1.76
24	2.25	2.18	2.11	2.03	1.98	1.94	1.89	1.84	1.79	1.73
25	2.24	2.16	2.09	2.01	1.96	1.92	1.87	1.82	1.77	1.71
26	2.22	2.15	2.07	1.99	1.95	1.90	1.85	1.80	1.75	1.69
27	2.20	2.13	2.06	1.97	1.93	1.88	1.84	1.79	1.73	1.67
28	2.19	2.12	2.04	1.96	1.91	1.87	1.82	1.77	1.71	1.65
29	2.18	2.10	2.03	1.94	1.90	1.85	1.81	1.75	1.70	1.64
30	2.16	2.09	2.01	1.93	1.89	1.84	1.79	1.74	1.68	1.62
40	2.08	2.00	1.92	1.84	1.79	1.74	1.69	1.64	1.58	1.51
60	1.99	1.92	1.84	1.75	1.70	1.65	1.59	1.53	1.47	1.39
120	1.91	1.83	1.75	1.66	1.61	1.55	1.50	1.43	1.35	1.25
∞	1.83	1.75	1.67	1.57	1.52	1.46	1.39	1.32	1.22	1.00

Table A.5 Critical values of F-distribution $f_{0.01}$ (ν_1, ν_2), continued

	ν_1								
ν_2	**1**	**2**	**3**	**4**	**5**	**6**	**7**	**8**	**9**
1	4052.0	4999.5	5403.4	5624.6	5763.7	5859.0	5928.4	5981.1	6022.5
2	98.50	99.00	99.17	99.25	99.30	99.33	99.36	99.37	99.39
3	34.12	30.82	29.46	28.71	28.24	27.91	27.67	27.49	27.35
4	21.20	18.00	16.69	15.98	15.52	15.21	14.98	14.80	14.66
5	16.26	13.27	12.06	11.39	10.97	10.67	10.46	10.29	10.16
6	13.75	10.93	9.78	9.15	8.75	8.47	8.26	8.10	7.98
7	12.25	9.55	8.45	7.85	7.46	7.19	6.99	6.84	6.72
8	11.26	8.65	7.59	7.01	6.63	6.37	6.18	6.03	5.91
9	10.56	8.02	6.99	6.42	6.06	5.80	5.61	5.47	5.35
10	10.04	7.56	6.55	5.99	5.64	5.39	5.20	5.06	4.94
11	9.65	7.21	6.22	5.67	5.32	5.07	4.89	4.74	4.63
12	9.33	6.93	5.95	5.41	5.06	4.82	4.64	4.50	4.39
13	9.07	6.70	5.74	5.21	4.86	4.62	4.44	4.30	4.19
14	8.86	6.52	5.56	5.04	4.70	4.46	4.28	4.14	4.03
15	8.68	6.36	5.42	4.89	4.56	4.32	4.14	4.00	3.90
16	8.53	6.23	5.29	4.77	4.44	4.20	4.03	3.89	3.78
17	8.40	6.11	5.19	4.67	4.34	4.10	3.93	3.79	3.68
18	8.29	6.01	5.09	4.58	4.25	4.02	3.84	3.71	3.60
19	8.19	5.93	5.01	4.50	4.17	3.94	3.77	3.63	3.52
20	8.10	5.85	4.94	4.43	4.10	3.87	3.70	3.56	3.46
21	8.02	5.78	4.87	4.37	4.04	3.81	3.64	3.51	3.40
22	7.95	5.72	4.82	4.31	3.99	3.76	3.59	3.45	3.35
23	7.88	5.66	4.77	4.26	3.94	3.71	3.54	3.41	3.30
24	7.82	5.61	4.72	4.22	3.90	3.67	3.50	3.36	3.26
25	7.77	5.57	4.68	4.18	3.86	3.63	3.46	3.32	3.22
26	7.72	5.53	4.64	4.14	3.82	3.59	3.42	3.29	3.18
27	7.68	5.49	4.60	4.11	3.79	3.56	3.39	3.26	3.15
28	7.64	5.45	4.57	4.07	3.75	3.53	3.36	3.23	3.12
29	7.60	5.42	4.54	4.05	3.73	3.50	3.33	3.20	3.09
30	7.56	5.39	4.51	4.02	3.70	3.47	3.30	3.17	3.07
40	7.31	5.18	4.31	3.83	3.51	3.29	3.12	2.99	2.89
60	7.08	4.98	4.13	3.65	3.34	3.12	2.95	2.82	2.72
120	6.85	4.79	3.95	3.48	3.17	2.96	2.79	2.66	2.56
∞	6.64	4.61	3.78	3.32	3.02	2.80	2.64	2.51	2.41

Table A.5 Critical values of *F*-distribution $f_{0.01}$ (ν_1, ν_2), continued

	ν_1									
ν_2	**10**	**12**	**15**	**20**	**24**	**30**	**40**	**60**	**120**	∞
1	6055	6106	6157	6208	6234	6260	6286	6313	6339	6366
2	99.4	99.42	99.43	99.45	99.46	99.47	99.47	99.48	99.49	9950
3	27.2	27.05	26.87	26.69	26.60	26.51	26.41	26.32	26.22	26.13
4	14.6	14.37	14.20	14.02	13.93	13.84	13.75	13.65	13.56	13.46
5	10.1	9.89	9.72	9.55	9.47	9.38	9.29	9.20	9.11	9.02
6	7.87	7.72	7.56	7.40	7.31	7.23	7.14	7.06	6.97	6.88
7	6.62	6.47	6.31	6.16	6.07	5.99	5.91	5.82	5.74	5.65
8	5.81	5.67	5.52	5.36	5.28	5.20	5.12	5.03	4.95	4.86
9	5.26	5.11	4.96	4.81	4.73	4.65	4.57	4.48	4.40	4.31
10	4.85	4.71	4.56	4.41	4.33	4.25	4.17	4.08	4.00	3.91
11	4.54	4.40	4.25	4.10	4.02	3.94	3.86	3.78	3.69	3.60
12	4.30	4.16	4.01	3.86	3.78	3.70	3.62	3.54	3.45	3.36
13	4.10	3.96	3.82	3.67	3.59	3.51	3.43	3.34	3.26	3.17
14	3.94	3.80	3.66	3.51	3.43	3.35	3.27	3.18	3.09	3.00
15	3.81	3.67	3.52	3.37	3.29	3.21	3.13	3.05	2.96	2.87
16	3.69	3.55	3.41	3.26	3.18	3.10	3.02	2.93	2.85	2.75
17	3.59	3.46	3.31	3.16	3.08	3.00	2.92	2.84	2.75	2.65
18	3.51	3.37	3.23	3.08	3.00	2.92	2.84	2.75	2.66	2.57
19	3.43	3.30	3.15	3.00	2.93	2.84	2.76	2.67	2.58	2.49
20	3.37	3.23	3.09	2.94	2.86	2.78	2.70	2.61	2.52	2.42
21	3.31	3.17	3.03	2.88	2.80	2.72	2.64	2.55	2.46	2.36
22	3.26	3.12	2.98	2.83	2.75	2.67	2.58	2.50	2.40	2.31
23	3.21	3.07	2.93	2.78	2.70	2.62	2.54	2.45	2.35	2.26
24	3.17	3.03	2.89	2.74	2.66	2.58	2.49	2.40	2.31	2.21
25	3.13	2.99	2.85	2.70	2.62	2.54	2.45	2.36	2.27	2.17
26	3.09	2.96	2.82	2.66	2.59	2.50	2.42	2.33	2.23	2.13
27	3.06	2.93	2.78	2.63	2.55	2.47	2.38	2.29	2.20	2.10
28	3.03	2.90	2.75	2.60	2.52	2.44	2.35	2.26	2.17	2.06
29	3.01	2.87	2.73	2.57	2.50	2.41	2.33	2.23	2.14	2.03
30	2.98	2.84	2.70	2.55	2.47	2.39	2.30	2.21	2.11	2.01
40	2.80	2.67	2.52	2.37	2.29	2.20	2.11	2.02	1.92	1.81
60	2.63	2.50	2.35	2.20	2.12	2.03	1.94	1.84	1.73	1.60
120	2.47	2.34	2.19	2.04	1.95	1.86	1.76	1.66	1.53	1.38
∞	2.32	2.19	2.04	1.88	1.79	1.70	1.59	1.47	1.33	1.00

Table A.6 Critical values of multiple range test (r_p)

	α = 0.05					α = 0.01				
ν	r_2	r_3	r_4	r_5	r_6	r_2	r_3	r_4	r_5	r_6
1	17.97	17.97	17.97	17.97	17.97	90.03	90.03	90.03	90.03	90.03
2	6.085	6.085	6.085	6.085	6.085	14.04	14.04	14.04	14.04	14.04
3	4.501	4.516	4.516	4.516	4.516	8.261	8.321	8.321	8.321	8.321
4	3.927	4.013	4.033	4.033	4.033	6.512	6.677	6.740	6.756	6.756
5	3.635	3.749	3.797	3.814	3.814	5.702	5.893	5.898	6.040	6.065
6	3.461	3.587	3.649	3.680	3.604	5.243	5.439	5.549	5.614	5.655
7	3.344	3.477	3.548	3.588	3.611	4.949	5.145	5.260	5.334	5.383
8	3.261	3.399	3.475	3.521	3.549	4.746	4.939	5.057	5.135	5.189
9	3.199	3.339	3.420	3.470	3.502	4.596	4.787	4.906	4.986	5.043
10	3.151	3.293	3.376	3.430	3.465	4.482	4.671	4.790	4.871	4.931
11	3.113	3.256	3.342	3.397	3.435	4.392	4.579	4.697	4.780	4.841
12	3.082	3.225	3.313	3.370	3.410	4.320	4.504	4.622	4.706	4.767
13	3.055	3.200	3.289	3.348	3.389	4.260	4.442	4.560	4.644	4.706
14	3.033	3.178	3.268	3.329	3.372	4.210	4.391	4.508	4.591	4.654
15	3.014	3.160	3.250	3.312	3.356	4.168	4.347	4.463	4.547	4.610
16	2.998	3.144	3.235	3.298	3.343	4.131	4.309	4.425	4.509	4.572
17	2.984	3.130	3.222	3.285	3.331	4.099	4.275	4.391	4.475	4.539
18	2.971	3.118	3.210	3.274	3.321	4.071	4.246	4.362	4.445	4.509
19	2.960	3.107	3.199	3.264	3.311	4.046	4.220	4.335	4.419	4.483
20	2.050	3.097	3.190	3.255	3.303	4.024	4.197	4.312	4.395	4.459
21	2.941	3.088	3.181	3.247	3.295	4.004	4.177	4.291	4.374	4.438
22	2.933	3.080	3.173	3.239	3.288	3.986	4.158	4.272	4.355	4.419
23	2.926	3.072	3.166	23.233	3.282	3.970	4.141	4.254	4.337	4.402
24	2.919	3.066	3.160	3.226	3.276	3.955	4.126	4.239	4.322	4.386
25	2.913	3.059	3.154	3.221	3.271	3.942	4.112	4.224	4.307	4.371
30	2.888	3.035	3.131	3.199	3.250	3.889	4.056	4.168	4.250	4.314
40	2.858	3.005	3.102	3.171	3.224	3.825	3.988	4.098	4.180	4.243
60	2.829	2.976	3.073	3.143	3.198	3.762	3.922	4.030	4.111	4.174
120	2.800	2.947	3.045	3.116	3.172	3.702	3.858	3.964	4.044	4.107
∞	2.772	2.918	3.017	3.089	3.146	3.643	3.796	3.900	3.978	4.040

Adapted from: Harter, H.L. (1960) "Critical Values for Duncan's New Multiple Range Test," *Biometrics*, vol. 16, no. 4.
Note: The degree of freedom $\nu = \sum(n_i - 1)$ where the summation covers the group numbers and n_i is the sample size of group *i*. If, for example, there are 3 groups and each group has 8 data points, $\nu = 3 \times (8 - 1) = 21$. Thus, at $\alpha = 0.05$, $r_2 = 2.941$, $r_3 = 3.088$, etc.

Table A.7 Critical values for Wilcoxon single-rank test

	One-sided	α = 0.05	α = 0.025	α = 0.010	α = 0.005
n	Two-sided	α = 0.10	α = 0.050	α = 0.020	α = 0.010
5		1			
6		2	1		
7		4	2	0	
8		6	4	2	0
9		8	6	3	2
10		11	8	5	3
11		14	11	7	5
12		17	14	10	7
13		21	17	13	10
14		26	21	16	13
15		30	25	20	16
16		36	30	24	19
17		41	35	28	23
18		47	40	33	28
19		54	45	38	32
20		60	52	43	37
21		68	59	49	43
22		75	66	58	49
23		83	73	62	55
24		92	81	69	61
25		101	90	77	68
26		110	96	85	76
27		120	107	93	84
28		130	117	102	92
29		141	127	111	100
30		152	137	120	100
31		163	148	130	118
32		175	230	141	128
33		188	171	151	138
34		201	183	162	149
35		214	195	174	160
36		228	208	186	171
37		242	222	198	183
38		256	235	211	195
39		271	250	224	208
40		287	264	238	221
45		371	344	313	292
50		466	434	398	373

Adapted from: Beyer, W.H., ed. (1968) *CRC Handbook of Tables for Probability and Statistics*, 2nd ed. CRC Press, Boca Raton, FL.

Table A.8 Wilcoxon rank-sum test

	α = 0.05 for one-sided and α = 0.10 for two-sided test									
n	*m* = 3	*m* = 4	*m* = 5	*m* = 6	*m* = 7	*m* = 8	*m* = 9	*m* = 10	*m* = 11	*m* = 12
m	6,15	12,24	19,36	28,50	39,66	52,84	66,105	83,127	101,152	121,179
m + 1	7,17	13,27	20,40	30,54	41,71	54,90	69,111	86,134	105,159	125,187
m + 2	7,20	14,30	22,43	32,58	43,76	57,95	72,117	89,141	109,166	129,195
m + 3	8,22	15,33	24,46	33,63	46,80	60,100	75,123	93,147	112,174	134,202
m + 4	9,24	16,36	25,50	35,67	48,85	62,106	78,129	96,154	116,181	138,210
m + 5	9,27	17,39	26,54	37,71	50,90	65,111	81,135	100,160	120,188	142,218
m + 6	10,29	18,42	27,58	39,75	52,95	67,117	84,141	103,167	124,195	147,225
m + 7	11,31	19,45	29,61	41,79	54,100	70,122	87,147	107,173	128,202	151,233
m + 8	11,34	20,48	30,65	42,84	57,104	73,127	90,153	110,180	132,209	155,241
m + 9	12,36	21,51	32,68	44,88	59,109	75,133	93,159	114,186	136,216	159,249
m + 10	13,38	22,54	33,72	46,92	61,114	78,138	96,165	117,193	139,224	164,256
m + 11	13,41	23,57	34,76	48,96	63,119	80,144	100,170	120,200	143,231	168,264
m + 12	14,43	24,60	36,79	50,100	65,124	83,149	103,176	124,206	147,238	172,272
m + 13	15,45	25,63	37,83	52,104	68,128	86,154	106,182	127,213	151,245	177,279
m + 14	15,48	26,66	39,86	53,109	70,133	88,160	109,188	132,219	155,252	181,287
m + 15	16,50	27,69	40,90	55,113	72,138	91,165	112,194	134,226	159,259	185,295
m + 16	17,52	28,72	42,93	57,117	74,143	94,170	115,200	138,232	138,266	190,302
m + 17	17,55	29,75	43,97	59,121	77,147	96,176	118,206	141,239	167,273	194,310
m + 18	18,57	30,78	44,101	61,125	79,152	99,181	121,212	145,245	171,280	198,318
m + 19	19,59	31,81	46,104	62,130	81,157	102,186	124,218	148,252	175,287	203,325
m + 20	19,62	32,84	47,108	64,134	83,162	104,192	127,224	152,258	178,295	207,333

Adapted from: Beyer, W. H., ed. (1968) *CRC Handbook of Tables for Probability and Statistics*, 2nd ed. CRC Press, Boca Raton, FL.

Table A.8 Wilcoxon rank-sum test (continued)

	α = 0.05 for one-sided and α = 0.10 for two-sided test									
n	*m* = 13	*m* = 14	*m* = 15	*m* = 16	*m* = 17	*m* = 18	*m* = 19	*m* = 20	*m* = 21	*m* = 22
m	143,208	167,239	192,273	220,308	249,346	280,386	314,427	349,471	386,517	424,566
m + 1	148,216	172,248	198,282	226,318	256,356	287,397	321,439	356,484	394,530	433,579
m + 2	152,225	177,257	203,292	232,328	262,367	294,408	328,451	364,496	402,543	442,592
m + 3	157,233	182,266	209,301	238,338	268,378	301,419	336,462	372,508	410,556	450,606
m + 4	162,241	187,275	215,310	244,348	275,388	308,430	343,474	380,520	418,569	459,619
m + 5	166,250	192,284	220,320	250,358	281,399	315,341	550,486	387,533	427,581	486,632
m + 6	171,258	197,293	226,329	256,368	288,409	322,452	358,497	395,545	435,594	476,646
m + 7	176,256	203,301	231,339	262,378	294,420	329,463	365,509	403,557	443,607	485,659
m + 8	181,274	208,310	237,348	268,388	301,430	336,474	372,521	411,509	451,620	494,672
m + 9	185,283	213,319	242,358	274,398	307,441	342,486	380,532	419,581	459,633	502,686
m + 10	190,291	218,328	248,367	280,408	314,451	349,497	387,544	426,594	468,645	511,699
m + 11	195,299	223,337	254,376	286,418	320,462	356,508	394,556	434,606	476,658	520,712
m + 12	199,308	228,346	259,386	292,428	327,472	363,519	402,567	442,618	484,671	528,726
m + 13	204,316	234,354	265,395	298,438	333,483	370,530	409,579	450,630	492,684	537,739
m + 14	209,324	239,363	270,405	304,448	340,493	377,541	416,591	458,642	501,696	546,752
m + 15	214,332	244,372	276,414	310,458	346,504	384,552	424,602	465,655	509,709	554,766
m + 16	218,341	249,381	282,423	316,468	353,514	391,563	431,614	473,667	517,722	563,779
m + 17	223,349	254,390	287,433	322,478	359,525	398,574	438,626	481,679	526,734	572,792
m + 18	228,357	260,398	293,442	328,488	366,535	405,585	446,637	489,691	534,747	581,805
m + 19	233,365	265,407	299,451	334,498	372,546	412,596	453,649	497,703	542,760	589,819
m + 20	237,374	270,416	304,461	340,508	379,556	419,607	461,660	505,715	550,773	598,832

Table A.8 Wilcoxon rank-sum test (continued)

	α = 0.025 for one-sided and α = 0.05 for two-sided test									
n	$m = 3$	$m = 4$	$m = 5$	$m = 6$	$m = 7$	$m = 8$	$m = 9$	$m = 10$	$m = 11$	$m = 12$
m	5,16	11,25	18,37	26,52	37,68	49,87	63,103	79,131	96,157	116,184
$m + 1$	6,18	12,28	19,41	28,56	39,73	51,93	66,114	82,138	100,164	120,192
$m + 2$	6,21	12,32	20,45	29,61	41,78	54,98	68,121	85,145	103,172	124,200
$m + 3$	7,23	13,35	21,49	31,65	43,83	56,104	71,127	88,152	107,179	128,208
$m + 4$	7,26	14,38	22,53	32,70	45,88	58,110	74,133	91,159	110,187	131,217
$m + 5$	8,28	15,41	24,56	34,74	46,94	61,115	77,139	94,166	114,194	135,225
$m + 6$	8,31	16,44	25,60	38,78	48,99	63,121	79,146	97,173	118,201	139,233
$m + 7$	9,33	17,47	26,64	37,83	50,104	65,127	82,152	101,179	121,209	143,241
$m + 8$	10,35	17,51	27,68	39,87	52,109	68,132	85,158	104,186	125,216	147,249
$m + 9$	10,38	18,54	29,71	41,91	54,114	70,138	88,164	107,193	128,224	151,257
$m + 10$	11.40	19,57	30,75	42,96	56,119	72,144	90,171	110,200	132,231	155,265
$m + 11$	11,43	20,60	31,79	44,100	58,124	75,149	93,177	113,207	135,239	159,273
$m + 12$	12,45	21,63	32,83	45,105	60,129	77,155	96,183	117,213	139,246	163,281
$m + 13$	12,48	22,66	33,87	47,109	62,134	80,160	99,189	120,220	143,253	167,289
$m + 14$	13,50	23,69	35,90	49,113	64,139	82,166	101,196	123,227	146,261	171,297
$m + 15$	13,53	24,72	36,94	50,118	66,144	84,172	104,202	126,234	150,268	175,305
$m + 16$	14,55	24,76	37,98	52,122	68,149	87,177	107,208	129,241	153,276	179,313
$m + 17$	14,58	25,79	38,102	53,127	70,154	89,183	110,214	132,248	157,283	183,321
$m + 18$	15,60	26,82	40,105	55,131	72,159	92,188	113,220	136,254	161,290	187,329
$m + 19$	15,63	27,85	41,109	57,135	74,164	94,194	115,227	139,261	164,298	191,337
$m + 20$	16,65	28,88	42,113	58,140	76,169	96,200	118,233	142,268	168,305	195,345

Table A.8 Wilcoxon rank-sum test (continued)

n	α = 0.025 for one-sided and α = 0.05 for two-sided test									
	m = 13	*m* = 14	*m* = 15	*m* = 16	*m* = 17	*m* = 18	*m* = 19	*m* = 20	*m* = 21	*m* = 22
m	137,214	160,246	185,280	212,316	240,355	271,395	303,438	337,483	373,530	411,579
m + 1	141,223	165,255	190,290	217,327	246,366	277,407	310,450	345,495	381,543	419,593
m + 2	146,231	170,264	195,300	223,337	252,377	284,418	317,462	352,508	389,556	428,606
m + 3	150,240	174,274	201,309	229,347	258,388	290,430	324,474	359,521	397,589	436,620
m + 4	154,249	179,283	206,319	234,358	264,399	297,441	331,486	367,533	404,583	444,634
m + 5	159,257	184,292	211,329	240,368	271,409	303,453	338,498	374,546	412,596	452,648
m + 6	163,266	189,301	216,339	245,379	277,420	310,464	345,510	381,559	420,609	450,662
m + 7	168,274	194,310	221,349	251,389	283,431	316,476	351,523	389,571	428,622	469,675
m + 8	172,283	198,320	227,358	257,399	289,442	323,487	358,535	396,584	436,635	477,689
m + 9	176,292	203,329	232,363	262,410	295,453	329,499	365,547	403,597	443,649	485,703
m + 10	181,300	208,338	237,378	268,420	301,464	336,510	372,559	411,609	451,662	493,717
m + 11	185,309	213,347	242,388	274,430	307,475	342,522	379,571	418,622	459,675	502,730
m + 12	190,317	218,356	248,397	279,441	313,486	349,533	386,583	426,634	467,688	510,744
m + 13	194,326	222,365	253,407	285,451	319,497	355,545	393,595	433,647	475,701	518,758
m + 14	198,335	227,375	258,417	291,461	325,508	362,556	400,607	440,660	482,715	526,772
m + 15	203,343	232,384	263,427	296,472	331,519	368,568	407,619	448,672	490,728	535,785
m + 16	207,352	237,393	269,436	302,482	338,529	375,579	414,631	455,685	498,741	543,799
m + 17	212,360	242,402	274,446	308,492	344,540	381,591	421,643	463,697	506,754	551,813
m + 18	216,389	247,411	279,456	314,502	350,551	388,602	428,655	470,710	514,767	560,826
m + 19	221,377	252,420	284,466	319,513	356,562	395,613	435,667	477,723	522,780	568,840
m + 20	225,386	256,430	290,475	325,523	362,573	401,625	442,679	485,735	530,793	576,854

Table A.9 Random Numbers

39 65 76 45 45	19 90 69 64 61	20 26 36 31 62	58 24 97 14 97	95 06 70 99 00
73 71 23 70 90	65 97 60 12 11	31 56 34 19 19	47 83 75 51 33	30 62 38 20 46
72 18 47 33 84	51 67 47 97 19	98 40 07 17 66	23 05 09 51 80	59 78 11 52 49
75 12 25 69 17	17 95 21 78 58	24 33 45 77 48	69 81 84 09 29	93 22 70 45 80
37 17 79 88 74	63 52 06 34 30	01 31 60 10 27	35 07 79 71 53	28 99 52 01 41
02 48 08 16 94	85 53 83 29 95	56 27 09 24 43	21 78 55 09 82	71 61 88 73 61
87 89 15 70 07	37 79 49 12 38	48 13 93 55 96	41 92 45 71 51	09 18 25 58 94
98 18 71 70 15	89 09 39 59 24	00 06 41 41 20	14 36 59 25 47	54 45 17 24 89
10 83 58 07 04	76 62 16 48 68	58 76 17 14 86	59 53 11 52 21	66 04 18 72 87
47 08 56 37 31	71 82 13 50 41	27 55 10 24 92	28 04 67 53 44	95 23 00 84 47
93 90 31 03 07	34 18 04 52 35	74 13 39 35 22	68 95 23 92 35	36 63 70 35 33
21 05 11 47 99	11 20 99 45 18	76 51 94 84 86	13 79 93 37 55	98 16 04 41 67
95 89 94 06 97	27 37 83 28 71	79 57 95 13 91	09 61 87 25 21	56 20 11 32 44
97 18 31 55 73	10 65 81 92 59	77 31 61 95 46	20 44 90 32 64	26 99 76 75 63
69 08 88 86 13	59 71 74 17 32	48 38 75 93 29	73 37 32 04 05	60 82 29 20 25
41 26 10 25 03	87 63 93 95 17	81 83 83 04 49	77 45 85 50 51	79 88 01 97 30
91 47 41 63 62	08 61 74 51 69	92 79 43 89 79	29 18 94 51 23	14 85 11 47 23
80 94 54 18 47	08 52 85 08 40	48 40 35 94 22	72 65 71 08 86	50 03 42 99 36
67 06 77 63 99	89 85 84 46 06	64 71 06 21 66	89 37 20 70 01	61 65 70 22 12
59 72 24 13 75	42 29 72 23 19	06 94 76 10 08	81 30 15 39 14	81 33 17 16 33
63 62 06 34 41	79 53 36 02 95	94 61 09 43 62	20 21 14 68 86	84 95 48 46 45
78 47 23 53 90	79 93 96 38 63	34 85 52 05 09	85 43 01 72 73	14 93 87 81 40
87 68 62 15 43	97 48 72 66 48	53 16 71 13 81	59 97 50 99 52	24 62 20 42 31
47 60 92 10 77	26 97 05 73 51	88 46 38 03 58	72 68 49 29 31	75 70 16 08 24
56 88 87 59 41	06 87 37 78 48	65 88 69 58 39	88 02 84 27 83	85 81 56 39 38
22 17 68 65 84	87 02 22 57 51	68 69 80 95 44	11 29 01 95 80	49 34 35 36 47
19 36 27 59 46	39 77 32 77 09	79 57 92 36 59	89 74 39 82 15	08 58 94 34 74
16 77 23 02 77	28 06 24 25 93	22 45 44 84 11	87 80 61 65 31	09 71 91 74 25
78 43 76 71 61	97 67 63 99 61	30 45 67 93 82	59 73 19 85 23	53 33 65 97 21
03 28 28 26 08	69 30 16 09 05	53 58 47 70 93	66 56 45 65 79	45 56 20 19 47
04 31 17 21 56	33 73 99 19 87	26 72 39 27 67	53 77 57 68 93	60 61 97 22 61
61 06 98 03 91	87 14 77 43 96	43 00 65 98 50	45 60 33 01 07	98 99 46 50 47
23 68 35 26 00	99 53 93 61 28	52 70 05 48 34	56 65 05 61 86	90 92 10 70 80
15 39 25 70 99	93 86 52 77 65	15 33 59 05 28	22 87 26 07 47	86 96 98 29 06
58 71 96 30 24	18 46 23 34 27	85 13 99 24 44	49 18 09 79 49	74 16 32 23 02
93 22 53 64 39	07 10 63 76 35	87 03 04 79 88	08 13 13 85 51	55 34 57 72 69
78 76 58 54 74	92 38 70 96 92	52 06 79 79 45	82 63 18 27 44	69 66 92 19 09
61 81 31 96 82	00 57 25 60 59	46 72 60 18 77	55 66 12 62 11	08 99 55 64 57
42 88 07 10 05	24 98 65 63 21	46 21 61 88 32	27 80 30 21 60	10 92 35 36 12
77 94 30 05 39	28 10 99 00 27	12 73 73 99 12	49 99 57 94 82	96 88 57

Index